Weak Convergence of Measures

Mathematical
Surveys
and
Monographs

Volume 234

Weak Convergence of Measures

Vladimir I. Bogachev

AMERICAN
MATHEMATICAL
SOCIETY

Providence, Rhode Island

2010 *Mathematics Subject Classification.* Primary 60B10, 28C15, 46G12, 60B05, 60B11, 60B12, 60B15, 60E05, 60F05, 54A20.

For additional information and updates on this book, visit
www.ams.org/bookpages/surv-234

Library of Congress Cataloging-in-Publication Data

Names: Bogachev, V. I. (Vladimir Igorevich), 1961- author.
Title: Weak convergence of measures / Vladimir I. Bogachev.
Description: Providence, Rhode Island : American Mathematical Society, [2018] | Series: Mathematical surveys and monographs ; volume 234 | Includes bibliographical references and index.
Identifiers: LCCN 2018024621 | ISBN 9781470447380 (alk. paper)| Softcover ISBN 9781470477981
Subjects: LCSH: Probabilities. | Measure theory. | Convergence.
Classification: LCC QA273.43 .B64 2018 | DDC 519.2/3–dc23
LC record available at https://lccn.loc.gov/2018024621

Printed in the United States of America.
Reprinted by the American Mathematical Society, 2024.

∞ The paper used in this book is acid-free and falls within the guidelines
established to ensure permanence and durability.
Visit the AMS home page at `https://www.ams.org/`

11 10 9 8 7 6 5 4 3 2 29 28 27 26 25 24

Dedicated to the memory of
Yurii Vasil′evich Prohorov
and Anatolii Vladimirovich Skorohod

Contents

Preface

Many problems in measure theory, probability theory, and diverse applications are connected with various types of convergence of measures. The most frequently encountered is weak convergence, but often one has to deal with other modes of convergence, for example, in variation or setwise. In the form of convergence of distribution functions, weak convergence of measures appeared actually at the dawn of probability theory, and now it has become one of the most important tools in applied and theoretic statistics. Many of the key results in probability theory and mathematical statistics can be regarded as statements about weak convergence of probability distributions. The foundations of the theory of weak convergence of measures were laid by J. Radon, E. Helly, P. Lévy, S. Banach, A.N. Kolmogorov, V.I. Glivenko, N.N. Bogoliubov, N.M. Krylov, and other classics from the 1910s through the 1930s. The formation of this theory as a separate field at the junction of measure theory, probability theory, functional analysis, and general topology is connected with fundamental works of A.D. Alexandroff at the end of the 1930s and the beginning of the 1940s, and this theory gained its modern form after the appearance of the outstanding paper of Yu.V. Prohorov in 1956. An extremely important role was also played by the book by B.V. Gnedenko and A.N. Kolmogorov on limit theorems of probability theory and the works of L.V. Kantorovich on optimal transportation. More details are given in the comments.

Convergence of measures is the subject of a vast literature (see the comments), in particular, weak convergence of measures is discussed in detail in the author's two-volume book *Measure theory* (see [**81**]). However, already at the time of working on that book, it was clear that convergence of measures deserved a separate exposition, which was impossible in a book of broad thematic coverage such as [**81**]. In spite of the fact that all principal results related to convergence of measures are fully presented in Chapter 8 of [**81**], such a presentation cannot be qualified as exhaustive and sufficient for a broad readership. First of all, the presentation in [**81**] is oriented towards experienced readers and, by the necessity of keeping the size of the book within reasonable limits, is rather condensed. Secondly, due to the same constraint on book size, justifications of many interesting results and examples there were delegated to exercises, and although they contained hints, they were even more condensed. Finally, the discussion of applications in [**81**] is reduced to a minimum. The goal of this new book is a more accessible and paced presentation of the theory of weak convergence of measures and some other important types of convergence, oriented towards a broad circle of readers with different backgrounds. Certainly, the subject itself unavoidably presupposes certain minimum of prerequisites (presented in the first chapter), but the material is organized in a form which attempts to postpone for as long as possible the employment of any specialized knowledge. In this respect I followed the example of Billingsley, the author of a beautiful introductory book [**67**] on weak convergence of probability measures (I began my acquaintance with this subject using this book many years ago); though, unlike his text, this book includes considerably less elementary material for advanced readers. Thus, here we offer two levels of presentation: rather elementary material in the main sections of Chapters 1–3, and some more specialized information presented in the complements to all chapters and also in Chapters 4 and 5. Such a structure leads to the effect that some concepts and results appear first in relation to measures on the real line or on $\mathbb{R}^d$, next when considering measures on metric spaces, and finally in the general case of topological spaces. In this way the book combines features of a textbook and an advanced survey. Certainly, the material of the aforementioned Chapter 8 of [**81**] is completely covered by this book, but most of the proofs from that chapter have been reworked: following suggestions and corrections received from my readers, more details have been added and many gaps and inaccuracies have been corrected. In addition, a number of interesting results given in [**81**] only with formulations are now supplied with complete justifications. Although a number of classical principal results are given with exactly the same formulations as in [**81**] and slightly revised proofs (such examples can be found, e.g., in Sections 2.3, 2.6, 3.1, 4.2, 4.3, and 5.1), in many other cases the formulations have been altered as well. This is not because the old formulations were not satisfactory, but rather because the whole structure of the text has been changed significantly. In particular, the case of metric spaces is now studied first and does not come as a special case of the general situation as in [**81**]. Many relatively old and some very recent results included in the book have also contributed to its size being nearly three times more than that of Chapter 8 of [**81**]. Certainly, the bibliography has been considerably updated: More than 100 works in the references have been published over the last decade, and this is a small portion of the available literature. In particular, many authors presented in this bibliography have much longer lists of related publications so that I had to be very selective when preparing the bibliography.

In Chapter 1, after presenting some necessary facts from integration theory and functional analysis in the first three sections, we discuss the simplest notions and facts related to convergence of measures on the interval and the real line and also on $\mathbb{R}^d$. However, even these basic concepts are useful for a very broad circle of problems that ever touch on anything related to convergence in distribution and weak convergence of measures. Phenomena discussed here illuminate well the general situation. Specific for the one-dimensional case is analysis of convergence of distribution functions. In this chapter we also study the Fourier transform (the characteristic functionals).

In Chapter 2, still at a rather elementary level, the discussion moves to metric spaces, but here some topological concepts already show up. The central results of this chapter are connected with the theorem of Yu.V. Prohorov on weak compactness, the theorem of A.D. Alexandroff on convergence of probability measures, and the parametrization of weakly converging measures due to A.V. Skorohod. Separate sections or subsections are devoted to weak convergence of measures on various special spaces such as Hilbert, Banach or some concrete functional spaces. In this chapter weak convergence is considered not only for countable sequences, but also for more general uncountable nets. Mostly, this does not lead to any complications, but is useful from the point of view of general ideas (especially with a view towards the continuation of our discussion for topological spaces). However, in all places in Chapter 2 (but not in Chapters 4 and 5) where nets are mentioned, it is quite possible to assume that these are usual sequences.

In Chapter 3 we consider metrics on spaces of measures (in particular, we discuss the Prohorov, Kantorovich, Kantorovich–Rubinshtein, and Fortet–Mourier metrics), and we also give a brief introduction to the theory of Gromov metric triples. Separate subsections are devoted to Zolotarev metrics and certain special questions connected with various estimates.

In the past two decades this area has been intensively developing in close connection with another very popular modern direction—optimal transportation. However, this very important aspect is not touched on in the present book because any sufficiently detailed discussion would considerably increase the size of the text.

A more advanced exposition requiring some knowledge of basics of general topology and some experience of working with topological spaces starts in Chapter 4 and ends with a discussion of topological properties of spaces of measures in Chapter 5. Chapter 4 begins with a brief exposition of fundamentals of measure theory in general topological spaces, then the weak topology on spaces of measures on general spaces is discussed including A.D. Alexandroff's results in this general setting. Among other things, compactness in the weak topology is thoroughly studied. We return to Prohorov's theorem in this framework, which leads to an interesting class of topological spaces, the so-called Prohorov spaces. Fourier transforms of measures on locally convex spaces are introduced and considered in relation to weak convergence. These themes are continued in Chapter 5, where the main emphasis is on topological properties of spaces of measures equipped with the weak topology. Here we also return to Skorohod representations. Separate sections are devoted to setwise convergence topology and the ws-topology, which is a mixture of the weak and setwise convergence topologies. Both have interesting connections with our main subject. Uniformly distributed sequences in topological spaces is another related topic discussed in this chapter.

Each chapter ends with a collection of exercises including easy exercises and more subtle facts (with hints or references to the literature; some of such advanced exercises are in fact very difficult and, in principle, could be placed as theorems in the text with references to their sources, but their inclusion in the form of exercises may be regarded as an invitation to seek simpler solutions).

The book ends with brief historic and bibliographic comments, a list of references (with indications of all pages where they are cited), and the subject index (which begins with a list of notations).

For reading this book it is useful, although not necessary at all, to be acquainted with basics of probability theory, the problems, ideas, and methods of which are of great importance for the area we discuss. In addition to the known fundamental treatises, including Ash [**24**], Bauer [**44**], Billingsley [**66**], Borovkov [**108**], Chow, Teicher [**138**], Cramér [**146**], Dudley [**193**], Feller [**221**], Fristedt, Gray [**246**], Gänssler, Stute [**254**], Gnedenko [**281**], Hennequin, Tortrat [**318**], Hoffmann-Jørgensen [**328**], Kallenberg [**343**], Loève [**437**], Neveu [**482**], Rotar [**556**], Shiryaev [**581**], and Tortrat [**617**], I would note an elegant introduction by Lamperti [**408**].

A considerable part of the material in this book was presented by the author in lectures at the Department of Mechanics and Mathematics of Moscow State University, at the Independent Moscow University, at the Faculty of Mathematics of the Higher School of Economics in Moscow, and also in lectures and talks at other universities and mathematical institutes all over the world, including the Steklov Mathematical Institute of the Russian Academy of Science in Moscow and its St. Petersburg Department, Kiev, Berlin, Bonn, Bielefeld, Paris, Strasbourg, London, Cambridge, Warwick, Rome, Pisa, Copenhagen, Stockholm, Delft, Vienna, Barcelona, Lisbon, Athens, Berkeley, Boston, Minneapolis, Vancouver, Montreal, Edmonton, Haifa, Tokyo, Kyoto, Beijing, Sydney, and Santiago.

During many years of working on this book, I received considerable help from many persons in the form of remarks and corrections, additional references, and historic comments. I am particularly grateful to L. Ambrosio, T.O. Banakh, N.H. Bingham, D.B. Bukin, G.P. Chistyakov, G. Da Prato, A.N. Doledenok, R.M. Dudley, D. Elworthy, B.V. Gnedenko, I.A. Ibragimov, A.V. Kolesnikov, V.V. Kozlov, N.V. Krylov, P. Malliavin, I. Marshall, P.-A. Meyer, S.A. Molchanov, F.V. Petrov, S.N. Popova, Yu.V. Prohorov, M. Röckner, V.V. Sazonov, A.V. Shaposhnikov, S.V. Shaposhnikov, A.N. Shiryaev, A.V. Skorohod, O.G. Smolyanov, V.N. Sudakov, F. Topsøe, A. M.Vershik, A. D.Wentzel, and A.Yu. Zaitsev.

The book also includes results obtained in research supported by the Russian Science Foundation (Grant 17-11-01058 at Lomonosov Moscow State University).

Moscow, Russia
Spring 2018

CHAPTER 1

Weak convergence of measures on $\mathbb{R}^d$

In this chapter we consider the basic concepts and facts related to weak convergence of measures on the real line and on $\mathbb{R}^d$. In the first two sections we recall some facts from measure theory, Lebesgue integration and functional analysis.

1.1. Measures and integrals

In this section we present some basic concepts and facts from measure theory necessary below. We assume some knowledge of the basic notions related to metric spaces (such as completeness, separability, and compactness), occasionally some acquaintance with topological spaces is needed; one of the purposes to recall this information is just to agree about the terminology. In this chapter we deal with measures on $\mathbb{R}^d$. However, all general concepts introduced below do not become simpler even in the case of measures on an interval, since arbitrary Borel measures are involved, not just the classical Lebesgue measure. A thorough exposition of the fundamentals of measure theory can be found in the book Bogachev [**81**].

The inner product and norm in $\mathbb{R}^d$ are denoted by $\langle x, y\rangle$ and $|x|$. For an open set U in $\mathbb{R}^d$ the symbol $C^k(U)$, where $k \in \mathbb{N}$ or $k = \infty$, denotes the class of functions with continuous derivatives up to order k, $C_b^k(U)$ is its subclass consisting of functions whose derivatives up to order k are bounded; $C_0^\infty(U)$ is the subclass in $C^\infty(U)$ consisting of functions with compact support in U (i.e., vanishing outside a compact set in U).

A *topological space* is a set X with a distinguished family of subsets τ, called *open* by definition, if τ contains the empty set and the whole set X and also contains any finite intersections and arbitrary unions of its elements. The collection τ is called a topology. The *closed sets* are defined as the complements of open sets. A space is called *Hausdorff* (or separated) if for every pair of points $x \neq y$ there are sets $U, V \in \tau$ with $x \in U$, $y \in V$, $U \cap V = \varnothing$. Topologies will be needed in Chapters 4 and 5 and partly in Chapter 3.

An important subclass of topological spaces is formed by metric spaces. We recall that a *metric space* is a set X equipped with a *metric*, which is a function

$$d\colon\ X \times X \to [0, +\infty)$$

such that $d(x, y) = d(y, x)$, $d(x, y) = 0$ precisely when $x = y$ and the following triangle inequality is fulfilled: $d(x, z) \leqslant d(x, y) + d(y, z)$ for all $x, y, z \in X$. If all distances between different points equal 1, then such a metric is called *discrete*. The space $\mathbb{R}^d$ is equipped with the so-called standard metric

$$d(x, y) = \Bigl(\sum\nolimits_{i=1}^d |x_i - y_i|^2\Bigr)^{1/2}, \quad x = (x_1, \dots, x_d),\ y = (y_1, \dots, y_d).$$

A sequence of points x_n converges to a point x if $d(x_n, x) \to 0$; such a sequence is *fundamental* (is a *Cauchy sequence*), i.e., $d(x_n, x_m) \to 0$ as $n, m \to \infty$. A space is called *complete* if all fundamental sequences converge in it.

The *open ball* with center a and radius $r > 0$ in a metric space X is the set

$$U(a, r) := \{x \in X\colon\ d(x, a) < r\}.$$

The *closed ball* is determined by the non-strict inequality. The *diameter* of a set A is defined by the formula $\operatorname{diam} A := \sup\{d(a, b)\colon\ a, b \in A\}$.

Open sets in a metric space are by definition the empty set and all sets that can be represented as unions of open balls (in an arbitrary number). It is readily seen that the class of such sets satisfies the aforementioned conditions on a topology. In a metric space a set is closed precisely when it contains the limits of convergent sequences of its points.

A mapping f between metric spaces (X, d_X) and (Y, d_Y) is called continuous if $f^{-1}(U)$ is open in X for every open set $U \subset Y$. This is equivalent to the continuity of f at every point x, i.e., $f(x_n) \to f(x)$ as $x_n \to x$. If $d_Y\big(f(u), f(v)\big) \leqslant L d_X(u, v)$, then f is called *Lipschitz* with the Lipschitz constant L (or *L-Lipschitz*).

A one-to-one mapping continuous in both directions is called a *homeomorphism*.

A one-to-one mapping of metric spaces that preserves distances is called an *isometry*.

A set S in a metric (or topological) space is *everywhere dense* if it has a nonempty intersection with every nonempty open set. A set N is *nowhere dense* if each nonempty open set contains an nonempty open subset not intersecting N. A space is called *separable* if it contains a finite or countable everywhere dense set.

In a separable metric space the union of an arbitrary collection of open sets coincides with the union of some at most countable subcollection. Hence every open set is a countable union of open balls. These balls can be taken with rational radii and centers at points of any fixed countable everywhere dense set.

A set K in a Hausdorff space is called *compact* if every cover of this set by open sets contains a finite collection also covering K.

For a metric space (but not in the general case!) this is equivalent to the property that each infinite sequence in K contains a subsequence convergent to a point of K. Yet another equivalent characterization belongs to Hausdorff: K is complete (as a metric space) and is *totally bounded*, which means that for every $\varepsilon > 0$ its has a finite ε-net (a collection of points with the property that the open balls of radius ε centered at these points cover K).

In the space $\mathbb{R}^d$ the compact sets are precisely closed bounded sets. In the general case this is not the case at all. For example, the space of natural numbers with the discrete metric is not compact.

We recall that a *σ-algebra* in a given space X is a class of subsets of X containing X and closed with respect to taking complements and countable unions and intersections. The smallest class with this property is X and the empty set, the largest one is the collection 2^X of all subsets of X. We need σ-algebras as domains of definition of measures.

For every class $\mathcal{S}$ of subsets of a set X there is the smallest σ-algebra containing $\mathcal{S}$. It is denoted by the symbol $\sigma(\mathcal{S})$ and called the σ-algebra generated by the class $\mathcal{S}$. Formally its construction is very simple: we just take the intersection of all σ-algebras in X containing X. However, one can hardly ever constructively describe the elements of $\sigma(\mathcal{S})$. Here is a simple example where this can be done:

the σ-algebra generated by all finite sets in X consists of at most countable sets and their complements. Throughout we shall deal with σ-algebras generated by open sets (or balls in $\mathbb{R}^d$).

A function $\mu\colon \mathcal{A} \to \mathbb{R}$ on a σ-algebra $\mathcal{A}$ in a space X is called a *measure* if it is countably additive in the following sense: for every finite or countable collection of pairwise disjoint sets $A_n \in \mathcal{A}$, one has

$$\mu\Bigl(\bigcup_{n=1}^{\infty} A_n\Bigr) = \sum_{n=1}^{\infty} \mu(A_n).$$

Such a measure is uniquely represented in the form of the so-called *Hahn–Jordan decomposition*

$$\mu = \mu^+ - \mu^-,$$

where μ^+ and μ^- are nonnegative measures on $\mathcal{A}$ concentrated on disjoint sets X^+ and X^- (giving the *Hahn decomposition* of the space X) and called the positive and negative parts of the measure μ; the measure

$$|\mu| = \mu^+ + \mu^-$$

is called the *total variation* of the measure μ, and the quantity

$$\|\mu\| := \|\mu\|_{\mathrm{TV}} := |\mu|(X)$$

is called the *variation* or *variation norm* of μ.

If $\mu \geqslant 0$ and $\mu(X) = 1$, then μ is called a *probability measure.*

For any measure $\mu \geqslant 0$ on a σ-algebra $\mathcal{A}$ the outer measure μ^* is defined by the formula $\mu^*(E) = \inf\{\mu(A)\colon E \subset A,\ A \in \mathcal{A}\}$ for all E.

Given a nonnegative measure μ, we shall say that some property of points is fulfilled μ-almost everywhere (or μ-a.e.) if it holds for all points excepting points of a set of measure zero with respect to μ; in the case of a signed measure μ-a.e. will mean $|\mu|$-a.e. For example, one can say that $f(x) \geqslant 0$ almost everywhere. The complements of sets of measure zero are called sets of *full measure.*

Let now X be a topological space (in this chapter it will be only $\mathbb{R}^d$, in the next two chapters a metric space).

The symbol $\mathcal{B}(X)$ denotes the so-called σ-algebra of *Borel sets* in X defined as the smallest σ-algebra containing all open sets. Note that the definition of this smallest σ-algebra as the intersection of all σ-algebras in X containing open sets only seems to be simple. In all nontrivial cases the Borel sets have no constructive description. For example, a natural expectation to obtain all Borel sets on the real line with the aid of the inductive construction of increasing classes of sets $\mathcal{F}_n$, where $\mathcal{F}_0$ is the class of open sets, $\mathcal{F}_{n+1}$ is the class of complements, countable intersections and countable unions of sets in $\mathcal{F}_n$, is wrong (see Bogachev [**81**, Exercise 1.12.95]).

In this chapter the following definition will be applied to sets in $X \subset \mathbb{R}^d$, but actually it is universal and in the same form will be repeated for metric spaces in Chapter 2 and for topological spaces in Chapter 4.

1.1.1. DEFINITION. Any measure on $\mathcal{B}(X)$ is called a Borel measure. A Borel measure μ is called Radon if, for every $B \in \mathcal{B}(X)$ and every $\varepsilon > 0$, there exists a compact set $K \subset B$ such that $|\mu|(B\backslash K) < \varepsilon$.

It will be shown in Chapter 2 that all Borel measures on complete separable metric spaces are Radon and in this chapter it will be useful just to know that this is true for Borel measures on $\mathbb{R}^d$.

The simplest Radon probability measure is *Dirac's measure* δ_a at a point $a \in X$ assigning the value 1 to the point a and 0 to its complement. One can also define constructively a measure concentrated at countably many points a_n by setting

$$\mu(B) = \sum_{n=1}^{\infty} c_n \delta_{a_n}(B) = \sum_{n:\ a_n \in B} c_n,$$

where c_n are the elements of an absolutely convergent series. For other Radon measures there is no explicit definition on all Borel sets merely because of the absence of any explicit description of Borel sets. However, some formulas defining measures can be called explicit with stretching. For example, the classical Lebesgue measure on Borel sets in $[0,1]$ is defined by the formula

$$\lambda(B) = \inf\Bigl\{\sum_{i=1}^{\infty} \lambda(J_i)\Bigr\},$$

where inf is taken over all finite or countable collections of intervals J_i such that $B \subset \bigcup_{i=1}^{\infty} J_i$. Similarly, Lebesgue measure is defined on Borel sets in the cube $[0,1]^d$. However, the verification of the countable additivity of Lebesgue measure on Borel sets turns out to be a surprisingly nontrivial exercise: it cannot be done directly on the basis of the foregoing defining formula, and one has to make a long detour. Namely, first the above formula is used to define the outer Lebesgue measure $\lambda^*(B)$ on all sets $B \subset [0,1]$, not necessarily Borel. Next, one introduces the class $\mathcal{L}$ of Lebesgue measurable sets $B \subset [0,1]$ such that $\lambda^*(B) + \lambda^*([0,1]\backslash B) = 1$ (at this stage it is still not obvious that the Borel sets are measurable). Finally, as a result of rather thorough work, one shows that the class $\mathcal{L}$ is a σ-algebra and the outer measure is countably additive on it. Only after this, taking into account that the intervals belong to $\mathcal{L}$ (this also requires some verification, but it is easy), one also obtains the countable additivity on the Borel sets.

A set $A_0 \in \mathcal{A}$ is called an *atom* of a nonnegative measure μ on a σ-algebra $\mathcal{A}$ if $\mu(A_0) > 0$ and every set $A \in \mathcal{A}$ contained in A_0 has measure either 0 or $\mu(A_0)$. For a Radon measure, an atom is a point of positive measure. Measures without atoms are called *atomless*.

The smallest closed set S with $|\mu|(S) = |\mu|(X)$ is called the *topological support* of the measure μ and denoted by $\operatorname{supp}(\mu)$. Such a support exists for every Radon measure on a topological space and also for every Borel measure on a separable metric space. This is clear from the fact that in both cases the union of an arbitrary collection of open sets of measure zero has measure zero (in the first case due to the property that every compact set in this union is contained in the union of a finite collection of these sets, and in the second case because one can find a countable subfamily of these sets with the same union, see Exercise 1.7.6).

For a nonnegative countably additive measure μ on a σ-algebra $\mathcal{A}$ it is often useful to enlarge the domain of definition in order to obtain the following property of *completeness*: every subset of any set of measure zero belongs to the domain of definition. Originally this is not always the case. For example, this is false for Lebesgue measure on the Borel sets. Say, the classic *Cantor set* C has Lebesgue measure zero (this set consists of the points in the interval $[0,1]$ that can be written in the form $x = \sum_{n=1}^{\infty} x_n 3^{-n}$ with $x_n \in \{0,2\}$), but among its subsets there are non-Borel sets (this can be seen from the cardinality arguments, but also in other ways). The *completion* is achieved by a very simple procedure: just by adding to the sets in $\mathcal{A}$ all subsets of all sets of measure zero. It is readily seen that the new class of sets $\mathcal{A}_\mu$ is a σ-algebra and that the natural extension of the original measure μ to this class by the formula $\mu(A \cup Z) = \mu(A)$ if $A \in \mathcal{A}$ and Z is a subset of a set of measure zero is well-defined and countably additive. This class $\mathcal{A}_\mu$ appears at once as the class of μ-measurable sets in the described standard procedure of extending a countably additive measure from an algebra by means of the outer measure. However, it is not always at all necessary to pass to the completion of a measure: in many problems it is important to consider measures on a certain common domain of definition, which can be incompatible with completeness.

Under the assumption of the axiom of choice, not all sets on the real line are Lebesgue measurable. Moreover, with the aid of this axiom one can establish the existence of a set $E \subset [0,1]$ such that $\lambda^*(E) = 1$ and $\lambda^*([0,1]\backslash E) = 1$. In particular, all compact sets in E have zero Lebesgue measure. This set E can be turned into a separate separable metric space with a Borel probability measure that is not Radon. This can be done with the aid of the following general construction.

1.1.2. Example. (i) Let $(\Omega, \mathcal{B}, \mu)$ be a probability space and let $E \subset \Omega$ be such that $\mu^*(E) = 1$ (the set E can be nonmeasurable). Let us equip E with the induced σ-algebra $\mathcal{B}_E$ of sets of the form $B \cap E$ with $B \in \mathcal{B}$. On this σ-algebra we introduce the measure $\mu_{_E}$ by the equality

$$\mu_{_E}(B \cap E) := \mu(B), \quad B \in \mathcal{B}.$$

The definition is meaningful: if $B_1 \cap E = B_2 \cap E$, then $\mu(B_1) = \mu(B_2)$, since the symmetric difference $B_1 \triangle B_2$ belongs to $\Omega\backslash E$ and hence has measure zero (it is important here to have the condition $\mu^*(E) = 1$). It is easy to verify the countable additivity of the measure $\mu_{_E}$. It is clear that $\mu_{_E}(E) = 1$. This measure is called the restriction of the measure μ to E, because for measurable E this will be indeed the restriction to the class of measurable subsets in E, but for a nonmeasurable set E the sets $B \cap E$ need not be measurable with respect to μ.

(ii) In the aforementioned case of the set $E \subset [0,1]$ with

$$\lambda^*(E) = \lambda^*([0,1]\backslash E) = 1,$$

we obtain the Borel probability measure $\lambda_{_E}$ on the separable metric space E (with the original metric) that has no compact sets of positive measure, since any compact set in E will be compact in $[0,1]$ as well. Thus, the measure $\lambda_{_E}$ is not Radon. The fact that the Borel sets of the space E are precisely the intersections of E with Borel sets in the interval, is Exercise 1.7.10.

Let us recall the general concept of the Lebesgue integral with respect to a measure μ on a space X with a σ-algebra $\mathcal{A}$.

The *indicator* function I_A of a set A is the function that equals 1 on A and 0 outside A. A function f on a measurable space $(X, \mathcal{A})$ is called *measurable* with respect to the σ-algebra $\mathcal{A}$ if

$$\{x\colon\ f(x) < c\} \in \mathcal{A} \quad \text{for all } c \in \mathbb{R}.$$

Such a function is called simple if it assumes finitely many values. Given a nonnegative measure μ on $\mathcal{A}$, the Lebesgue integral of a simple function f with different values $c_1, \ldots, c_n$ on disjoint sets $A_1, \ldots, A_n$, i.e., $f = c_1 I_{A_1} + \cdots + c_n I_{A_n}$, is defined by the formula

$$\int_X f\, d\mu := c_1\mu(A_1) + \cdots + c_n\mu(A_n).$$

The integral of a measurable function $f \geqslant 0$ is defined by the formula

$$\int_X f\,d\mu := \sup \int_X \varphi\,d\mu,$$

where sup is taken over all simple functions $\varphi \geqslant 0$ for which $\varphi(x) \leqslant f(x)$ almost everywhere. We call f integrable only in the case where this quantity is finite. Finally, a general measurable function f is called integrable if so are its positive and negative parts

$$f^+ = \max(f, 0) \quad \text{and} \quad f^- = -\min(f, 0),$$

and we set

$$\int_X f\,d\mu := \int_X f^+\,d\mu - \int_X f^-\,d\mu.$$

Let us emphasize that this construction can be applied to $\mathcal{A}$ as well as to $\mathcal{A}_\mu$. Moreover, in the latter case it is useful to call a function f *μ-measurable* when it is defined not on all of X, but only outside a set of measure zero, and on its actual domain of definition is $\mathcal{A}_\mu$-measurable in the previous case. This does not affect the definition of the integral, since it is invariant with respect to redefinitions of a function on sets of measure zero. The integral with respect to a signed measure μ is defined as the difference of the integrals with respect to its positive and negative parts μ^+ and μ^- (their existence is required separately).

The following very useful *Chebyshev inequality* is seen from the definition:

$$\mu\bigl(x\colon\ |f(x)| \geqslant R\bigr) \leqslant \frac{1}{R}\int_X |f|\,d\mu, \quad R > 0. \tag{1.1.1}$$

The most important theorem on passage to the limit under the integral sign is the *Lebesgue dominated convergence theorem*: if integrable functions f_n converge almost everywhere to f and there is an integrable function Φ (a common majorant) for which $|f_n(x)| \leqslant \Phi(x)$ almost everywhere for all n, then f is integrable and

$$\int_X f\,d\mu = \lim_{n\to\infty} \int_X f_n\,d\mu.$$

The *Beppo Levi monotone convergence theorem* gives the same conclusion under a different condition: $\{f_n(x)\}$ monotonically increases to $f(x)$ and the integrals of f_n are uniformly bounded. Finally, *Fatou's theorem* gives a weaker assertion that if $f_n \geqslant 0$ and $f_n(x) \to f(x)$ a.e., then the inequality holds

$$\int_X f\,d\mu \leqslant \liminf_{n\to\infty} \int_X f_n\,d\mu.$$

But if the integrals of f_n converge to the integral of f, then $\|f - f_n\|_{L^1(\mu)} \to 0$ (*the Vitali–Scheffé theorem*, see Bogachev [**81**, Theorem 2.8.9]).

For a finite measure $\mu \geqslant 0$, we say that measurable functions f_n converge to a function f in measure if

$$\lim_{n\to\infty} \mu\bigl(x\colon\ |f(x) - f_n(x)| \geqslant \varepsilon\bigr) = 0 \quad \forall\, \varepsilon > 0.$$

For such convergence no integrability is needed and it follows from convergence almost everywhere as well as from convergence in the norm of $L^1(\mu)$, i.e., from the relation $\|f - f_n\|_{L^1(\mu)} \to 0$. On the other hand, one can construct a sequence converging in measure that converges at no point. However, as shown by F. Riesz, every sequence convergent in measure always contains a subsequence that converges almost everywhere (see Bogachev [**81**, p. 112]).

If we are given a measure μ on $\mathcal{A}$ (possibly, signed), then every function ϱ integrable with respect to it generates a new measure by the formula

$$\nu(A) = \int_A \varrho \, d\mu, \quad A \in \mathcal{A}.$$

In this case we say that the function ϱ is the *density* (or the *Radon–Nikodym density*) of the measure ν with respect to μ; it is denoted by $d\nu/d\mu$. We write $\nu = \varrho \cdot \mu$.

It is clear that $\nu(A) = 0$ for every A with $|\mu|(A) = 0$ (this property is called the *absolute continuity* of ν with respect to μ and denoted as follows: $\nu \ll \mu$). By the Radon–Nikodym theorem also the converse is true: if $\nu \ll \mu$, then $\nu = \varrho \cdot \mu$, where $\varrho \in L^1(\mu)$. The opposite relationship between measures is their *mutual singularity*, which is the property that these measures are concentrated on disjoint sets, i.e., there exists a set $X_0 \in \mathcal{A}$ such that $|\mu|(X \backslash X_0) = 0$, $|\nu|(X_0) = 0$. This symmetric relationship is denoted by $\mu \perp \nu$. The general case is a mixture of these two extreme cases: any measure ν on $\mathcal{A}$ can be written in the form $\nu = \nu_0 + \nu_1$, where the measure ν_0 is absolutely continuous with respect to μ and the measure ν_1 is mutually singular with μ. If $\nu \ll \mu$ and $\mu \ll \nu$, then the measures μ and ν are called equivalent, which is denoted by $\mu \sim \nu$. The equivalence is exactly the property that the Radon–Nikodym density $d\nu/d\mu$ is nonzero $|\mu|$-a.e. A measure on $\mathbb{R}^d$ concentrated on a set of Lebesgue measure zero is called *singular*.

The very special Lebesgue measure is of fundamental importance for the whole of measure theory. Most other widely employed measures are constructed with its aid. Firstly, one defines measures on $\mathbb{R}^d$ by densities with respect to Lebesgue measure (they are called *absolutely continuous*). Secondly, new measures are obtained from Lebesgue measure by measurable transformations. If we are given a measure μ on a measurable space X and a mapping F from X to a measurable space $(Y, \mathcal{B})$ that is *measurable*, i.e.,

$$F^{-1}(B) \in \mathcal{A} \quad \text{for all } B \in \mathcal{B},$$

then on $\mathcal{B}$ we obtain the measure $\mu{\circ}F^{-1}$ defined by the equality

$$\mu{\circ}F^{-1}(B) = \mu\bigl(F^{-1}(B)\bigr), \quad B \in \mathcal{B}$$

and called the *image measure* of μ under the mapping F and also the *measure induced* by F. It is readily seen that the function $\mu \circ F^{-1}$ is indeed countably additive. The integral of a bounded $\mathcal{B}$-measurable function f with respect to the measure $\mu{\circ}F^{-1}$ is evaluated by the formula

$$\int_Y f(y) \, \mu{\circ}F^{-1}(dy) = \int_X f\bigl(F(x)\bigr) \, \mu(dx), \tag{1.1.2}$$

called the *change-of-variables formula*. For the indicator functions of sets in $\mathcal{B}$ this formula is the definition of the image measure, so the general case follows by using uniform approximations of bounded measurable functions by simple functions.

A mapping $F\colon X \to Y$ of topological spaces is called *Borel* or *Borel measurable* if the set $F^{-1}(B)$ is Borel in X for every Borel set B in Y.

The images of Lebesgue measure under Borel mappings exhaust all Borel probability measures on complete separable metric spaces (in particular, on $\mathbb{R}^d$).

1.1.3. THEOREM. *For every Borel probability measure μ on a complete separable metric space X there exists a Borel mapping $F\colon [0,1] \to X$ such that μ coincides with the image of Lebesgue measure with respect to F. If the measure μ has no points of positive measure, then F can be chosen one-to-one.*

Yet another very important measure on $\mathbb{R}^d$ is the *standard Gaussian measure* γ_d given by the *standard Gaussian density*

$$(2\pi)^{-d/2}\exp(-|x|^2/2)$$

with respect to Lebesgue measure. A general Gaussian measure on $\mathbb{R}^d$ is defined as the image of the standard Gaussian measure under an affine mapping $x \mapsto a + \sigma x$, where a is a vector in $\mathbb{R}^d$, σ is a linear operator . Such a measure is given by a density precisely when $\det\sigma \neq 0$. Gaussian measures are also called Gaussian (or normal) distributions.

If we are given two measures μ and ν on measurable spaces $(X, \mathcal{A})$ and $(Y, \mathcal{B})$, then on the product $X\times Y$ we obtain the measure $\mu\otimes\nu$, called the product of the measures μ and ν and defined on the σ-algebra $\mathcal{A}\otimes\mathcal{B}$ generated by the products of sets $A\times B$, where $A \in \mathcal{A}$, $B \in \mathcal{B}$. On such products we set $\mu\otimes\nu(A\times B) := \mu(A)\nu(B)$. On general sets $E \in \mathcal{A}\otimes\mathcal{B}$ this measure is evaluated by the formula

$$\mu\otimes\nu(E) = \int_Y \mu(E_y)\,\nu(dy), \quad E_y := \{x \in X\colon\ (x, y) \in E\}.$$

It is verified that $E_y \in \mathcal{A}$ and the function $y \mapsto \mu(E_y)$ is measurable with respect to $\mathcal{B}$, so the formula above is meaningful. According to Fubini's theorem, the integral of a $\mu\otimes\nu$-integrable function f is calculated by the formula

$$\int_{X\times Y} f\,d(\mu\otimes\nu) = \int_X\int_Y f(x, y)\,\nu(dy)\,\mu(dx) = \int_Y\int_X f(x, y)\,\mu(dx)\,\nu(dy),$$

where the existence of the inner integrals (for almost all fixed free variables) follows automatically from the integrability of f.

By induction one defines the product of finitely many spaces with measures. Probability spaces can be multiplied in an infinite number of any cardinality. For a countable collection of probability spaces $(\Omega_n, \mathcal{B}_n, P_n)$ we obtain the probability measure $P = \bigotimes_{n=1}^{\infty} P_n$ on the countable product $\Omega = \prod_{n=1}^{\infty}\Omega_n$ that consists of all sequences $\omega = (\omega_1, \omega_2, \ldots)$, where $\omega_n \in \Omega_n$. The domain of definition of the measure P is the σ-algebra generated by the products $\prod_{n=1}^{\infty} B_n$, where $B_n \in \Omega_n$. On such product the value of P equals the products of $P_n(B_n)$. Certainly, the existence of such a measure requires justification.

An important example of a countable product is the countable power of the standard Gaussian measure on the real line, which is defined on the space $\mathbb{R}^\infty$ of all real sequences, i.e., on the countable power of the real line.

Let us introduce the operation of convolution of Borel measures on $\mathbb{R}^d$.

1.1.4. DEFINITION. The convolution $\mu*\nu$ of two Borel measures μ and ν on $\mathbb{R}^d$ is the image of their product $\mu\otimes\nu$ under the mapping $\mathbb{R}^d\times\mathbb{R}^d \to \mathbb{R}^d$, $(x, y) \mapsto x+y$.

It follows from the definition that the integral of a bounded Borel function f with respect to the measure $\mu * \nu$ is given by the formula

$$\int_{\mathbb{R}^d} f\,d(\mu * \nu) = \int_{\mathbb{R}^d}\int_{\mathbb{R}^d} f(x + y)\,\mu(dx)\nu(dy).$$

The values on Borel sets are evaluated by the formula

$$\mu * \nu(B) = \int_{\mathbb{R}^d} \mu(B - x)\,\nu(dx),$$

where $B - x = \{y - x\colon\ y \in B\}$. It is clear from these formulas that $\mu * \nu = \nu * \mu$.

Note that if at least one of the two measures μ and ν is absolutely continuous with respect to Lebesgue measure, then so is their convolution. For example, if the measure ν possesses a density ϱ_ν with respect to Lebesgue measure, then (using the Borel version of ϱ_ν), we find that the measure $\mu * \nu$ is given by the density

$$\varrho_{\mu*\nu}(x) = \int_{\mathbb{R}^d} \varrho_\nu(x-y)\,\mu(dy). \tag{1.1.3}$$

Convolutions with smooth densities are often used in the study of weak convergence. For example, if $\varrho_\nu \in C_0^\infty(\mathbb{R}^d)$, then one has the inclusion $\varrho_{\mu*\nu} \in C^\infty(\mathbb{R}^d)$ for every measure μ, which enables us to construct approximations of measures by measures with smooth densities.

A primary object of probability theory in Kolmogorov's axiomatics is a probability space $(\Omega, \mathcal{B}, P)$, where Ω is a nonempty set with a σ-algebra $\mathcal{B}$ of subsets of Ω and a probability measure P on $\mathcal{B}$, i.e., a nonnegative measure with $P(\Omega) = 1$. Functions measurable with respect to $\mathcal{B}$ are called random variables and their integrals with respect to the measure P in case of existence are called expectations (notation: $\mathbb{E}\,\xi$). For every random variable ξ the induced measure $P\circ\xi^{-1}$ on the real line is called the *distribution* of the random variable ξ. Formula (1.1.2) here takes the form

$$\mathbb{E}\,f(\xi) = \int_{\mathbb{R}} f(t)\,P\circ\xi^{-1}(dt)$$

for Borel functions f (under the condition of integrability of $f(\xi)$). Below we discuss connections between such integrals with distribution functions of random variables (see Definition 1.4.2) and with the Stieljes integral.

A important concept is independence of random variables $\xi_1, \ldots, \xi_n$ on a general probability space $(\Omega, \mathcal{B}, P)$ understood as the equality

$$P(\xi_1 \in B_1, \ldots, \xi_n \in B_n) = P(\xi_1 \in B_1) \cdots P(\xi_n \in B_n)$$

for all Borel sets B_i on the real line. Independence of an infinite collection of random variables is understood as independence of each finite subcollection in it. It should be noted that independence does not reduce to pairwise independence. An equivalent description of independence is that the image of the measure P under the mapping $(\xi_1, \ldots, \xi_n)$ to $\mathbb{R}^n$ is the product of the measures $P\circ\xi_1^{-1}, \ldots, P\circ\xi_n^{-1}$. This gives a simple way of constructing sequences of independent random variables with given distributions: it suffices to take coordinate functions on the countable power of the real line with the measure that equals the product of the given distributions.

Recall also that for a sequence of points of c_n (not necessarily different) the quantity $\limsup_n c_n$ is defined as the infimum of numbers c with the property that $c_n \leqslant c$ for all n starting from some number. If there are no such c, then $\limsup_n c_n = +\infty$ (this means that $c_{n_k} \to +\infty$ for some subsequence of indices). It can happen that $\limsup_n c_n = -\infty$ (this means that $c_n \to -\infty$).

Similarly, $\liminf_n c_n$ is the supremum of numbers c for which $c_n \geqslant c$ for all n starting from some number. If there are no such numbers, then $\liminf_n c_n = -\infty$. Note that

$$\inf_n c_n \leqslant \liminf_n c_n \leqslant \limsup_n c_n \leqslant \sup_n c_n,$$

but all these inequalities can be strict. The sequence $\{c_n\}$ has a limit precisely when $\liminf_n c_n = \limsup_n c_n$ (this is also true for infinite limits). Throughout convergence of a sequence will always mean existence of a finite limit.

1.2. Functions of bounded variation

Let us recall some facts related to functions of bounded variation on the real line.

The variation of a function F on an interval $I \subset \mathbb{R}$ (closed, open or semiopen) is defined by the formula

$$V(F, I) := \sup \sum_{i=1}^{n} |F(b_i) - F(a_i)|,$$

where sup is taken over all intervals $[a_i, b_i] \subset I$ with pairwise disjoint interiors. If $V(F, I) < \infty$, then F is called a *function of bounded variation* on I.

Every monotone function F on $[a, b]$ has bounded variation $|F(b) - F(a)|$. On the other hand, every function of bounded variation on $[a, b]$ is the difference of two increasing functions. This yields that every function of bounded variation on an interval has at most countably many points of discontinuity. In addition, such a function is almost everywhere differentiable. Indeed, both properties are fulfilled for increasing functions (see Bogachev [**81**, Chapter 5]). Note that there is a certain canonical decomposition of a function F of bounded variation into the difference of increasing functions F_1 and F_2 such that

$$V(F_1, [a, b]) \leqslant V(F, [a, b]), \quad V(F_2, [a, b]) \leqslant V(F, [a, b]).$$

To this end in case $F(b) \geqslant F(a)$ (otherwise we consider $-F$) we take the function $F_1(t) = V(F, [a, t])$, which is obviously increasing. Next we observe that the function $F_2 := F_1 - F$ is also increasing, since whenever $x < y$ we have

$$V(F, [a, x]) - F(x) \leqslant V(F, [a, y]) - F(y),$$

because $|F(y) - F(x)| + V(F, [a, x]) \leqslant V(F, [a, y])$ by the obvious inequality

$$|F(y) - F(x)| + \sum_{i=1}^{n-1} |F(t_{i+1}) - F(t_i)| \leqslant V(F, [a, y])$$

for every finite partition $a = t_1 < t_2 < \cdots < t_n = x$. Moreover, we have the indicated inequalities for the variations of F_1 and F_2, since $V(F_1, [a, b]) = F_1(b)$ equals $V(F, [a, b])$, and also

$$V(F_2, [a, b]) = F_2(b) - F_2(a) = V(F, [a, b]) - F(b) + F(a) \leqslant V(F, [a, b]).$$

We shall need the following fact about monotone functions.

1.2.1. LEMMA. *Suppose that a sequence of increasing functions f_n converges at every point of a countable everywhere dense set S in the compact interval $[a, b]$. Let us set $f(s) = \lim_{n\to\infty} f_n(s)$ for all $s \in S$ and $f(t) = \sup\{f(s)\colon\ s \in S, s \leqslant t\}$ for all $t \in (a, b]$, $f(a) = \inf\{f(s)\colon\ s \in S, s \geqslant a\}$. Then f is an increasing function and $f(t) = \lim_{n\to\infty} f_n(t)$ at all points of continuity of f (i.e., excepting at most countably many points).*

PROOF. It is clear that $f(t_1) \leqslant f(t_2)$ whenever $t_1 \leqslant t_2$ as all f_n are increasing. Let t be a point of continuity of f and $\varepsilon > 0$. There are points $s_1 < t$ and $s_2 > t$ in S for which

$$f(s_1) > f(t) - \varepsilon, \quad f(s_2) < f(t) + \varepsilon.$$

Therefore, there exists a number N such that

$$f_n(s_1) > f(t) - \varepsilon, \quad f_n(s_2) < f(t) + \varepsilon \quad \forall n \geqslant N.$$

The monotonicity of f_n yields the estimate $f(t)-\varepsilon < f_n(t) < f(t)+\varepsilon$ for all $n \geqslant N$. Thus, $f(t) = \lim_{n\to\infty} f_n(t)$. □

In our discussion functions of bounded variation will appear as distribution functions of measures.

1.2.2. Definition. For a bounded Borel measure μ on the real line, the function

$$F_\mu(t) = \mu\bigl((-\infty, t)\bigr)$$

is called the distribution function of the measure μ.

The function F_μ is obviously left-continuous: if points t_n increase to t, then we have $F_\mu(t_n) \to F_\mu(t)$. There might be no right-continuity: if $\mu \geqslant 0$ and a point τ has positive μ-measure, then

$$F_\mu(\tau + \varepsilon) \geqslant F_\mu(\tau) + \mu(\{\tau\}).$$

It is easy to verify (Exercise 1.7.11) that the continuity of F_μ is equivalent to the absence of points of nonzero μ-measure.

If the measure μ is nonnegative, then the function F_μ increases (the converse is also true, see Exercise 1.7.12). In the general case, decomposing μ into the difference of μ^+ and μ^-, we obtain that $F_\mu = F_{\mu^+} - F_{\mu^-}$ is a function of bounded variation. In addition,

$$V(F_\mu, \mathbb{R}) = \|\mu\|.$$

Indeed, $F_\mu(b) - F_\mu(a) = \mu\bigl([a, b)\bigr)$. Since the sum of absolute values of the measure on disjoint sets is not greater than its variation, we have $V(F_\mu, \mathbb{R}) \leqslant \|\mu\|$. On the other hand, $\|\mu\| = \mu(X^+) - \mu(X^-)$, where $\mathbb{R} = X^+ \cup X^-$ is the Hahn decomposition for the measure μ. For every $\varepsilon > 0$ there are compact sets $K \subset X^+$ and $S \subset X^-$ with $\mu(K) > \mu(X^+) - \varepsilon$ and $\mu(S) < \mu(X^-) + \varepsilon$. These compact sets are disjoint, hence one can find pairwise disjoint intervals $(a_1, b_1], \ldots, (a_n, b_n]$ and $(c_1, d_1], \ldots, (c_m, d_m]$ for which

$$K \subset A = \bigcup_{i=1}^{n} (a_i, b_i], \ S \subset C = \bigcup_{i=1}^{m} (c_i, d_i], \ |\mu(K) - \mu(A)| < \varepsilon, \ |\mu(S) - \mu(C)| < \varepsilon.$$

Then we obtain $\|\mu\| < \mu(A) - \mu(C) + 4\varepsilon \leqslant V(\mu, \mathbb{R}) + 4\varepsilon$.

The distribution functions of measures are not only left continuous, but also have zero limits at $-\infty$. These two properties uniquely characterize functions of bounded variation that are the distribution functions of Borel measures. Indeed, such a function can be written as the difference of two bounded increasing functions that are left continuous and have zero limits at $-\infty$ (Exercise 1.7.13). Now, if the function F is in addition increasing, then the measure μ for which it serves as the distribution function can be reconstructed as follows: on any interval $[\alpha, \beta)$ we set

$$\mu\bigl([\alpha, \beta)\bigr) := F(\beta) - F(\alpha),$$

and also $F(-\infty) = 0$, $F(+\infty) = \sup_t F(t)$. Next, we verify that the class of finite unions of such intervals is an algebra and the measure μ on it is countably additive (it is here that we need the left continuity of the function F). Finally, by the extension theorem for measures we obtain the desired Borel extension of μ.

For a measure on the interval $[a, b)$ the distribution function can be considered only on $[a, b)$, but it is often more convenient to extend this measure by zero on the complement and consider the distribution function on the whole real line.

A similar connection exists between left continuous functions of bounded variations on $[a, b]$ vanishing at the point a and Borel measures on the interval $[a, b)$ (it is important to exclude the right end b, because the value of the measure at this point cannot be defined by the values of the distribution function on $[a, b]$; say, for Dirac's measure at the point b the distribution function is identically zero on $[a, b]$).

The Riemann–Stieltjes integral of a continuous function f on $[a, b]$ with respect to a left continuous function of bounded variation F is defined similarly to the usual Riemann integral as the limit of the Riemann–Stieltjes sums

$$\sum_{i=1}^{n-1} f(t_i)[F(t_{i+1}) - F(t_i)]$$

with points $a = t_1 < t_2 < \cdots < t_n = b$ when the maximal length of the intervals $[t_i, t_{i+1}]$ tends to zero. This integral (more details can be found in Glivenko [**280**], Gohman [**283**], and Kamke [**347**]) is denoted by the symbol

$$\int_a^b f(t)\, dF(t).$$

One can assume that $F(a) = 0$ (subtracting $F(a)$ from F). Now we can take the measure μ on $[a, b)$ with the distribution function F (extended by zero to the left from a and by the value $F(b)$ to the right from b). Then we obtain the equality

$$\int_a^b f(t)\, dF(t) = \int_{[a,b)} f(t)\, \mu(dt).$$

If we extend the Riemann–Stieltjes integral to the whole real line, then for a bounded continuous function f we arrive at the equality

$$\int_{-\infty}^{+\infty} f(t)\, dF_\mu(t) = \int_{\mathbb{R}} f(t)\, \mu(dt). \tag{1.2.1}$$

This equality also remains valid for unbounded continuous functions that are integrable with respect to the measure μ. Moreover, by means of this equality one can define the Lebesgue–Stieltjes integral for Borel functions that are integrable with respect to the measure μ.

Note that in the definition of the Riemann–Stieltjes integral of a continuous function f over an interval the left continuity of F at the points of the interval (a, b) is not important (but it is important at b); this condition is accepted just for convenience of compatibility with distribution functions. By means of the limit of the indicated sums one can define the integral of f with respect to an arbitrary function of bounded variation F on $[a, b]$ that is left continuous at b and then verify that the value of the integral does not change if at the inner points of discontinuity of the function F we redefine it and make left continuous.

For every continuously differentiable function f the following integration by parts formula holds:

$$\int_{[a,b)} f(t)\, \mu(dt) = f(b)F_\mu(b) - \int_a^b f'(t)F_\mu(t)\, dt. \tag{1.2.2}$$

If we include the point b in the domain of integration in the left-hand side, then we obtain the formula

$$\int_{[a,b]} f(t)\, \mu(dt) = f(b)F_\mu(b+) - \int_a^b f'(t)F_\mu(t)\, dt, \tag{1.2.3}$$

where $F_\mu(b+) := F_\mu(b) + \mu(\{b\})$. The same formulae hold for a measure μ on the interval (a,b), since the difference appears only in the case of an atom of the measure μ at the point a and for Dirac's measure at a these formulae are true. We give a justification in the case of an open interval, where we also admit infinite values of a,b. First we observe that one has an even more symmetric formula.

Let μ and ν be bounded Borel measures (possibly, signed) on the interval (a,b) (bounded or unbounded) with the distribution functions F_μ and F_ν. These functions are bounded and are Borel. Hence they are integrable with respect to all Borel measures.

1.2.3. THEOREM. *One has the equality*

$$\int_{(a,b)} F_\mu(t)\,\nu(dt) = F_\mu(b)F_\nu(b) - \int_{(a,b)} F_\nu(t+)\,\mu(dt). \tag{1.2.4}$$

If at least one of the functions F_μ and F_ν is continuous (or they have common points of discontinuity), then in place of $F_\nu(t+)$ in this equality one can write $F_\nu(t)$.

PROOF. By Fubini's theorem for the product of the measures μ and ν, on account of the equality $I_{(a,t)}(s) = I_{(s,b)}(t)$ we have

$$\begin{aligned}\int_{(a,b)} F_\mu(t)\,\nu(dt) &= \int_{(a,b)}\int_{(a,b)} I_{(a,t)}(s)\,\mu(ds)\,\nu(dt)\\ &= \int_{(a,b)} \nu\bigl((s,b)\bigr)\,\mu(ds) = \int_{(a,b)} [F_\nu(b) - F_\nu(s+)]\,\mu(ds),\end{aligned}$$

as desired. The last assertion is obvious. □

In order to obtain from this equality formula (1.2.2) with (a,b) in place of $[a,b)$, we have to take for ν the measure with density f' with respect to Lebesgue measure on (a,b), which gives $f - f(a)$ for F_ν. Then formula (1.2.2) remains valid for f as well, since for the constant it is obviously true.

1.3. Facts from functional analysis

Here we recall some basic concepts and facts from functional analysis, of which in the first place we shall need the simplest facts related to compactness and also a number of examples of classic spaces of functions and sequences. Other facts will be actually needed only in Chapters 3–5.

A *normed space* is a linear space (over the field $\mathbb{R}$ or $\mathbb{C}$) equipped with a norm, which is a function $x \mapsto \|x\|$ from X to $[0,+\infty)$ with the following properties:

$$\|\lambda x\| = |\lambda|\cdot\|x\|, \quad \|x+y\| \leqslant \|x\| + \|y\|$$

for all $x, y \in X$ and all scalars λ and $\|x\| > 0$ if $x \neq 0$.

A normed space is a metric space with the metric

$$d(x,y) = \|x - y\|.$$

A complete normed space with respect to this metric is called a *Banach space*. A *Hilbert* space is a space with an inner product (a Euclidean space) that is complete with the norm $\|x\| = \sqrt{(x,x)}$.

The most important for us examples of Banach spaces are the space $B(\Omega)$ of bounded real functions on a nonempty set Ω with the norm

$$\|f\| = \sup_{\omega\in\Omega} |f(\omega)|,$$

and, in the case where Ω is a metric or topological space, the closed subspace $C_b(\Omega)$ in $B(\Omega)$ consisting of bounded continuous functions.

Of particular importance are the spaces $C(K)$ of continuous functions on compact metric spaces. Such a space with the indicated norm is separable, i.e., it has a countable everywhere dense set (Exercise 1.7.8). The most classic space is $C[a,b]$.

If the set Ω is infinite, the space $B(\Omega)$ is nonseparable. For example, in the case $\Omega = \mathbb{N}$ we obtain the nonseparable space l^∞ of all bounded real sequences with norm $\|x\|_\infty = \sup_n |x_n|$.

Another important Banach space is the set $\mathcal{M}(\mathcal{A})$ of all bounded measures on a given σ-algebra $\mathcal{A}$ equipped with the variation norm $\mu \mapsto \|\mu\|$. Convergence in this norm, i.e., $\|\mu_n - \mu\| \to 0$, is called *convergence in variation.* It is also called *convergence in total variation norm.*

The spaces of the form $B(\Omega)$ are universal in the following sense.

1.3.1. PROPOSITION. *Every nonempty metric space (Ω, d) is isometric to a subset of the space $B(\Omega)$, moreover, an isometric embedding can be defined by the formula*

$$\omega \mapsto f_\omega, \quad f_\omega(u) = d(u,\omega) - d(u,\omega_0),$$

where ω_0 is a fixed point in Ω.

PROOF. The function f_ω is bounded, since $|d(u,\omega) - d(u,\omega_0)| \leqslant d(\omega,\omega_0)$ by the triangle inequality. For bounded Ω we need not subtract $d(u,\omega_0)$, and in the general case subtraction does not influence the value of the distance between f_ω and $f_{\omega'}$ in the metric $B(\Omega)$, because

$$\sup_{u\in\Omega} |f_\omega(u) - f_{\omega'}(u)| = \sup_{u\in\Omega} |d(u,\omega) - d(u,\omega')| = d(\omega,\omega'),$$

since $|d(u,\omega) - d(u,\omega')| \leqslant d(\omega,\omega')$ for all u, and for $u = \omega$ one has the equality. □

If Ω is separable, then, certainly, one can pass to a separable closed linear subspace in $B(\Omega)$ containing the image of Ω under the indicated embedding. However, as discovered by Banach, for separable spaces there is a more convenient universal space (for a proof see, e.g., Bogachev, Smolyanov [**96**]). See also p. 123 about the Urysohn space.

1.3.2. THEOREM. *Every separable metric space is isometric to a subset in the space $C[0,1]$.*

In addition, every separable normed space is linearly isometric to a linear subspace in $C[0,1]$.

Under the indicated isometric embedding into $B(\Omega)$ the image of a complete space will be closed and the image of an incomplete space will be non-closed. However, incomplete metric spaces can be isometrically embedded as closed subsets in incomplete normed spaces (see Exercise 2.7.27).

For a given measure $\mu \geqslant 0$ the space $L^p(\mu)$, where $1 \leqslant p < \infty$, consists of equivalence classes of μ-measurable functions (equivalent functions are equal almost everywhere) for which

$$\|f\|_p := \left(\int |f|^p \, d\mu\right)^{1/p} < \infty.$$

Clearly, this quantity does not depend on our choice of representatives in equivalence classes.

A particular case of the spaces L^p is the space l^p of sequences $x = (x_n)$ (real or complex) for which

$$\|x\|_p = \Bigl(\sum_{n=1}^{\infty} |x_n|^p\Bigr)^{1/p} < \infty.$$

The dual (or topological dual) to a normed space X is the space X^* of continuous linear functionals on X (linear functions with values in the field of scalars). The norm of a functional f is given by the equality

$$\|f\| = \sup_{\|x\|\leqslant 1} |f(x)|.$$

For $X = l^p$ with $1<p<\infty$ the dual space is isomorphic to l^q with the conjugate exponent $q = p/(p-1)$. A general form of a continuous functional is this:

$$f(x) = \sum_{n=1}^{\infty} y_n x_n, \quad y = (y_n) \in l^q, \quad \|f\| = \|y\|_q.$$

For a compact metric space K, the dual space to $C(K)$ has an explicit description given by the following classic Riesz theorem (who proved it for a compact interval).

1.3.3. THEOREM. *A general form of a continuous linear functional on $C(K)$ for a compact metric space K is given by the formula*

$$\varphi \mapsto \int_K \varphi\, d\mu,$$

where μ is a Borel measure on K; the norm of this functional equals $\|\mu\|$. The latter is also true for every bounded Borel measure on a noncompact space X and the functional on $C_b(X)$ defined by this measure.

Theorems of this sort will be mentioned in this book without proof (which can be found in [**81**]), but it is instructive to see why the latter assertion about the equality of norms is true at least for measures on the real line. Certainly, if the measure is nonnegative, then everything is obvious, since the norm of the functional is attained at the function $\varphi = 1$. For signed measures usually there is no function at which the norm is attained. However, given ε, the sets X^+ and X^- from the Hahn decomposition of the real line into disjoint parts on which the measure μ assumes only nonnegative and nonpositive values, respectively, contain compact sets K and S with $\mu^+(X^+\backslash K) < \varepsilon$, $\mu^-(X^-\backslash S) < \varepsilon$. Then the remaining part of the real line is a set of measure less than 2ε with respect to $|\mu|$. Now we can take a continuous function φ such that $\varphi|_K = 1$, $\varphi|_S = -1$, $|\varphi| \leqslant 1$. The integral of this function over the complement of $K \cup S$ does not exceed 2ε, the integral over $K \cup S$ gives $\mu^+(K) + \mu^-(S)$, hence its integral over the whole real line differs from $\|\mu\|$ by less than 4ε. The reasoning is completely similar for $\mathbb{R}^d$ and for more general spaces (see Theorem 4.1.9).

For a compact metric space (K, d) the Ascoli–Arzelà theorem gives a simple description of compact sets in $C(K)$ with the usual sup-norm: a set S in $C(K)$ is compact precisely when it is bounded, closed and equicontinuous, i.e., for every $\varepsilon > 0$ there is $\delta > 0$ such that $|f(t) - f(s)| \leqslant \varepsilon$ for all $f \in S$ whenever $d(t, s) \leqslant \delta$.

One more useful fact related to $C(K)$ (for all compact spaces, not only metric) is expressed by the Stone–Weierstrass theorem: if a sub-algebra $A \subset C(K)$ (a linear subspace closed with respect to products) contains 1 and separates points in K, i.e., for every pair $x \neq y$ there exists $f \in A$ with $f(x) \neq f(y)$, then A is everywhere dense in $C(K)$. The proofs of these facts can be found in many books, e.g., see Bogachev, Smolyanov [**96**] and Dunford, Schwartz [**197**].

A linear operator between two normed spaces X and Y is just a linear mapping $A\colon X \to Y$, i.e., $A(\alpha x + \beta y) = \alpha Ax + \beta Ay$. However, in case of an infinite-dimensional space X not every linear mapping is continuous. The continuity of a linear operator is equivalent to having a finite norm

$$\|A\| := \sup_{\|x\|\leqslant 1} \|Ax\|.$$

The set $\mathcal{L}(X,Y)$ of all continuous (also called *bounded*) linear operators is a normed space with the indicated norm (it is complete if Y is complete).

Let us mention the important Banach–Steinhaus theorem (or the uniform boundedness principle).

1.3.4. THEOREM. *Suppose that a set $\mathcal{A} \subset \mathcal{L}(X,Y)$ possesses the property that*

$$\sup_{A\in\mathcal{A}} \|Ax\| < \infty \quad \textit{for every } x \in X.$$

If X is a Banach space, then $\mathcal{A}$ is norm bounded, i.e., $\sup_{A\in\mathcal{A}} \|A\| < \infty$.

One more necessary classic result is the Hahn–Banach theorem.

1.3.5. THEOREM. *Let X be a real linear space, $p\colon X \to [0,+\infty)$ a function such that $p(x+y) \leqslant p(x)+p(y)$ and $p(tx) = tp(x)$ for all $x, y \in X$, $t \geqslant 0$. Suppose that on a linear subspace $X_0 \subset X$ we are given a linear function l_0 for which $l_0(x) \leqslant p(x)$ for all $x \in X_0$. Then it can be extended to a linear function l on all of X such that $l(x) \leqslant p(x)$ for all $x \in X$.*

It follows from this theorem that every continuous linear functional defined on a linear subspace of a normed space extends to a linear functional on the whole space without increasing its norm.

Every normed space X possesses not only the topology generated by its metric (obtained from its norm), but also the so-called weak topology, which is denoted by the symbol $\sigma(X, X^*)$. The weak topology consists of the empty set and all possible unions of the basic cylindrical sets of the form

$$U(x_0, f_1, \ldots, f_n, \varepsilon) = \{x \in X\colon\ |f_i(x - x_0)| < \varepsilon, i = 1, \ldots, n\}, \tag{1.3.1}$$

where $x_0 \in X$, $f_1, \ldots, f_n \in X^*$, $\varepsilon > 0$.

Similarly, the dual space X^* (which is also normed) can be equipped with the weak-$*$ topology denoted by the symbol $\sigma(X^*, X)$. Nonvoid open sets in this topology are all possible unions of the basic cylindrical sets of the form

$$U(f_0, x_1, \ldots, x_n, \varepsilon) = \{f \in X^*\colon\ |f(x_i) - f_0(x_i)| < \varepsilon, i = 1, \ldots, n\},$$

where $f_0 \in X^*$, $x_1, \ldots, x_n \in X$, $\varepsilon > 0$. Convergence of a sequence $\{f_n\}$ to a functional f in the weak-$*$ topology is simply convergence on every element in X (without uniformity on any sets).

Both these topologies are special cases of the so-called *topology of duality* $\sigma(E, F)$ on a linear space E generated by some linear space F of linear functions on E with the property that the elements of F separate points in E, i.e., for every pair of distinct points x and y, there is a functional $f \in F$ with $f(x) \neq f(y)$. Basic neighborhoods of a point $x_0 \in E$ are defined by the same formula (1.3.1), where now the functionals f_i are taken in F. Unlike the previous situation, there might be no original topology on E.

When the space E is equipped with the topology $\sigma(E, F)$, the set of all continuous in this topology linear functions on E turns out to be exactly F.

In our book this construction will be applied (starting from Chapter 3) to the pair of spaces consisting of some space of measures (this will be E) on a metric or topological space X and the space of bounded continuous functions on X (this will be F), which act as functionals on measures by means of the integrals with respect to these measures.

The space E with the topology $\sigma(E, F)$ is in turn a very special case of a *locally convex space*, i.e., a real or complex linear space E equipped with a family $\mathcal{P}$ of *seminorms* (functions $p\colon E \to [0, +\infty)$ for which $p(x + y) \leqslant p(x) + p(y)$ and $p(\lambda x) = |\lambda| p(x)$). The topology $\sigma(E, F)$ corresponds to the family of seminorms $x \mapsto |f(x)|$, where $f \in F$. Although the topology we discuss has a very special form, it also leads to interesting problems from the point of view of the general theory of locally convex spaces, as we shall see below.

A set V in a linear space is called *convex* if $tv + (1-t)u \in V$ for all $u, v \in V$ and $t \in [0, 1]$. If in addition $\lambda v \in V$ for all $v \in V$ and $|\lambda| \leqslant 1$, then V is called *absolutely convex*. The *convex hull* of a set is the intersection of all convex sets containing it. The *absolutely convex hull* of a set is the intersection of all absolutely convex sets containing it.

It should be noted that the weak and weak-$*$ topologies of a Banach space and of its dual are not metrizable unless X is finite-dimensional. For example, if the topology $\sigma(X, X^*)$ is metrizable, then every ball of radius $1/n$ in this metric centered at the origin contains a basic neighborhood determined by finitely many functionals, whose union gives a countable collection $\{f_i\}$ with the following property: every basic neighborhood of zero contains a neighborhood determined by a finite subfamily in $\{f_i\}$. In particular, for every $f \in X^*$ the set $f^{-1}(-1, 1)$ must contain some neighborhood $U(0, f_1, \ldots, f_n, \varepsilon)$. This is only possible if f vanishes on the intersection of the kernels of the functionals $f_1, \ldots, f_n$, which is equivalent to f being a linear combination of $f_1, \ldots, f_n$ (this is easily verified by induction, see Bogachev, Smolyanov [**96**, p. 279]). Thus, the whole space X^* is the linear span of $\{f_i\}$, i.e., the union of a sequence of finite-dimensional subspaces. In the infinite-dimensional case this is impossible by Baire's theorem, since X^* is a Banach space.

More generally, the reasoning above shows that the topology of duality $\sigma(E, F)$ is not metrizable if F is not the union of a sequence of finite-dimensional space. The latter, certainly, is also possible for infinite-dimensional spaces. For example, the space $\mathbb{R}^\infty$ of all infinite real sequences can be equipped with the topology of coordinate-wise convergence, which is exactly the topology of the form $\sigma(\mathbb{R}^\infty, F)$, where F is the linear span of the coordinate functions. This topology is metrizable and one can take the metric

$$d(x, y) = \sum_{n=1}^{\infty} 2^{-n} \min(|x_n - y_n|, 1), \quad x = (x_n),\ y = (y_n).$$

However, some bounded sets are metrizable in the weak or weak-$*$ topology, which is often rather useful. In particular, the following Banach–Alaoglu theorem holds.

1.3.6. THEOREM. *Let X be a Banach space. Then the closed balls in X^* are compact in the weak-$*$ topology, and if X is separable, they are metrizable in this topology.*

It follows from this theorem (but can be easily verified directly without any theorems) that a norm bounded sequence of functionals on a separable Banach space always contains a pointwise convergent subsequence. For a straightforward verification we take a countable set $\{x_i\}$ dense in the unit ball in X, extract in the given sequence of functionals f_n a subsequence converging on x_1, extract in it a further subsequence converging on x_2, and so on. Using this sequence of embedded subsequences, one can easily construct a subsequence that converges on every elements x_i, and then by the density of $\{x_i\}$ in the ball and the norm boundedness of $\{f_n\}$ we obtain the pointwise convergence on the ball (hence on the whole space).

However, for nonseparable spaces this is false. For example, the sequence of coordinate functionals $x \mapsto x_n$ on l^∞ does not contain a pointwise convergent subsequence (for every its subsequence one can find an element of l^∞ on which it will assume infinitely many values 0 and 1).

For a metric defining the weak-$*$ topology on the ball of the dual space to a separable Banach space one can take the function

$$d(f,g) = \sum_{n=1}^{\infty} 2^{-n}|f(x_n) - g(x_n)|,$$

where $\{x_n\}$ is a sequence everywhere dense in the unit ball in X.

Thus, in the case of the space $C(K)$, where K is compact metric, we obtain on the account of the Riesz theorem that the set of Borel measures on K bounded in variation norm by a number r is metrizable compact in the weak-$*$ topology. This fact plays an important role in the study of weak convergence of measures.

Although the weak topology of an infinite-dimensional Banach space is always strictly weaker than the norm topology, it can happen that the collections of convergent sequences in these different topologies coincide. For us the most important case where this happens is the space l^1 of absolutely convergent series, i.e., sequences $x = (x_n)$ with finite norm $\|x\|_1 = \sum_{n=1}^{\infty} |x_n|$. The dual space to l^1 can be identified with l^∞: the general form of a continuous functional on l^1 is given by the formula

$$f(x) = \sum_{n=1}^{\infty} y_n x_n, \quad y = (y_n) \in l^\infty,$$

and $\|f\| = \|y\|_\infty$. It is known (see Exercise 1.7.15) that every weakly convergent sequence in l^1 also converges in norm, moreover, any weakly compact set is norm compact. A bit more is even true.

1.3.7. LEMMA. *If a sequence of vectors $a_n \in l^1$ is such that for every functional $f \in l^\infty$ the sequence of numbers $f(a_n)$ converges, then $\{a_n\}$ converges in norm.*

The situation is different for $L^1(\mu)$, the dual space to which is isomorphic to the space $L^\infty(\mu)$ of equivalence classes of bounded measurable functions with norm $\|f\|_\infty := \inf_g \sup_x |g(x)|$, where inf is taken over all functions g equivalent to f. For example, the sequence of functions $\sin(2\pi nt)$ converges weakly to zero (which follows from the Riemann–Lebesgue theorem or Bessel's inequality), but does not converge in norm.

It follows from the Banach–Steinhaus theorem that for any Banach space X every sequence in X^* that is fundamental in the topology $\sigma(X^*, X)$, i.e., is pointwise convergent, converges in X^* in this topology. This is false for the weak topology $\sigma(X, X^*)$: the fact that on vectors x_n the values of every functional f in X^* converge does not imply the existence of a vector $x \in X$ for which these values converge to $f(x)$. As an example one can take the space c_0 of sequences tending to zero with norm $\|x\| = \sup_n |x_n|$ and the dual space l^1 (for vectors x_n one can take the vectors with 1 on the first n positions and 0 on the rest). However, some important Banach spaces are sequentially complete in the weak topology. This is true for all $L^p(\mu)$ with $1 \leqslant p < \infty$, including l^1 (which is not completely obvious).

Among nontrivial facts related to the weak topology, the most important for us is the following *Eberlein–Shmulian theorem* (see Bogachev, Smolyanov [**96**, Theorem 6.10.13], [**97**, Theorem 3.4.8], Dunford, Schwartz [**197**]).

1.3.8. THEOREM. *A set A in a Banach space has compact closure in the weak topology precisely when every infinite sequence in it contains a weakly convergent subsequence.*

Since measures are often related to subsets of L^1, it is useful to have a criterion of weak compactness in $L^1(\mu)$, $\mu \geqslant 0$.

1.3.9. DEFINITION. A set of functions $\mathcal{F} \subset L^1(\mu)$ is called uniformly integrable if we have

$$\lim_{C\to+\infty} \sup_{f\in\mathcal{F}} \int_{\{|f|>C\}} |f|\, d\mu = 0. \tag{1.3.2}$$

1.3.10. THEOREM. *Let μ be a finite measure and let $\mathcal{F}$ be a set of μ-integrable functions. The set $\mathcal{F}$ is uniformly integrable precisely when it has compact closure in the weak topology of $L^1(\mu)$.*

In case of a separable space $L^1(\mu)$, this set is metrizable in the weak topology (see Theorem 5.6.8).

Let us mention the following useful criterion of uniform integrability that is due to Ch.-J. de la Vallée-Poussin.

1.3.11. THEOREM. *Let μ be a finite nonnegative measure. A family $\mathcal{F}$ of μ-integrable functions is uniformly integrable if and only if there exists a nonnegative increasing function G on $[0, +\infty)$ such that*

$$\lim_{t\to+\infty} \frac{G(t)}{t} = \infty \quad \textit{and} \quad \sup_{f\in\mathcal{F}} \int G\big(|f(x)|\big)\, \mu(dx) < \infty. \tag{1.3.3}$$

In this case the function G can be taken convex.

In more detail these questions are discussed in Bogachev [**81**, Chapter 4].

1.3.12. THEOREM. *The closed absolutely convex hull of a weakly compact set in a Banach space is weakly compact (and metrizable in a separable space).*

A proof of a more general fact can be found in Bogachev, Smolyanov [**97**, Corollary 5.6.12]; on a separable space there is a sequence of continuous functionals separating points (see also Proposition 4.5.12).

1.4. Weak convergence of measures on the real line and on $\mathbb{R}^d$

Here we discuss in detail weak convergence of measures on the real line and on $\mathbb{R}^d$. Throughout by measures on the real line or on $\mathbb{R}^d$on we understand *bounded Borel measures.* The following definition is fundamental.

1.4.1. DEFINITION. A sequence $\{\mu_n\}$ of Borel measures on $\mathbb{R}^d$ is called weakly convergent to a Borel measure μ if, for every bounded continuous real function f, one has the equality

$$\lim_{n\to\infty} \int_{\mathbb{R}^d} f(x)\,\mu_n(dx) = \int_{\mathbb{R}^d} f(x)\,\mu(dx). \tag{1.4.1}$$

Notation: $\mu_n \Rightarrow \mu$.

1.4.2. EXAMPLE. If a sequence of measures μ_n converges in variation to a measure μ, i.e., $\|\mu_n - \mu\| \to 0$, then it also converges to μ weakly. More generally, if $\lim\limits_{n\to\infty} \mu_n(B) = \mu(B)$ for every set $B \in \mathcal{B}(\mathbb{R}^d)$ or, provided that $\{\mu_n\}$ is bounded in variation, at least for every set B of the form $B = \{f < c\}$, where $f \in C_b(\mathbb{R}^d)$ and $|\mu|(\{f = c\}) = 0$, then $\mu_n \Rightarrow \mu$.

PROOF. The first assertion is obvious from the estimate

$$\biggl|\int_{\mathbb{R}^d} f(x)\,\mu_n(dx) - \int_{\mathbb{R}^d} f(x)\,\mu(dx)\biggr| \leqslant \|\mu_n - \mu\| \sup_x |f(x)|.$$

Let us prove the more general last assertion (clearly, convergence in variation yields convergence on sets and the latter yields boundedness in variation by Nikodym's theorem (see, e.g., Corollary 5.6.11 below). By its assumption there exists $C > 0$ such that $\|\mu_n\| \leqslant C$ and $\|\mu\| \leqslant C$ for all n. Let $f \in C_b(\mathbb{R}^d)$ and $\varepsilon > 0$. One can assume that $|f(x)| < 1$. There are numbers $c_i \in [-1, 1]$, $i = 1, \ldots, k$, such that $0 < c_{i+1} - c_i < \varepsilon$, $c_1 = -1$, $c_k = 1$ and $|\mu|(\{f = c_i\}) = 0$. This follows from the fact that any interval contains points of zero $|\mu|$-measure (otherwise this measure would be infinite). Let $g(x) = c_i$ if $c_i \leqslant f(x) < c_{i+1}$. Then $|f(x) - g(x)| < \varepsilon$, hence

$$\biggl|\int_{\mathbb{R}^d} [f(x) - g(x)]\,\mu_n(dx)\biggr| \leqslant \int_{\mathbb{R}^d} |f(x) - g(x)|\,|\mu_n|(dx) \leqslant C\varepsilon$$

and similarly for μ in place of μ_n. There is an index n_0 such that for all $n \geqslant n_0$ we have

$$\biggl|\int_{\mathbb{R}^d} g(x)\,\mu_n(dx) - \int_{\mathbb{R}^d} g(x)\,\mu(dx)\biggr| \leqslant \varepsilon,$$

since $\lim\limits_{n\to\infty} \mu_n(\{c_i \leqslant f < c_{i+1}\}) = \mu(\{c_i \leqslant f < c_{i+1}\})$ by our condition and the equality $\{c_i \leqslant f < c_{i+1}\} = \{f < c_{i+1}\} \backslash \{f < c_i\}$. Hence, whenever $n \geqslant n_0$, the absolute value of the difference between the integrals of f against the measures μ and μ_n does not exceed $(2C + 1)\varepsilon$. See also Example 4.4.2. □

However, weak convergence does not imply convergence on all open sets. The next simple example is very typical.

1.4.3. EXAMPLE. (i) Let p be a probability density on the real line and let ν_n be the probability measures with densities

$$p_n(t) = np(nt).$$

Then the measures ν_n converge weakly to Dirac's measure δ at zero, although their values on the open set $\mathbb{R}\backslash\{0\}$ are equal to 1 and do not converge to $\delta(\mathbb{R}\backslash\{0\}) = 0$. Indeed, if $f \in C_b(\mathbb{R})$, then

$$\lim_{n\to\infty} \int_{-\infty}^{+\infty} f(t)p_n(t)\,dt = \lim_{n\to\infty} \int_{-\infty}^{+\infty} f(s/n)p(s)\,ds = f(0).$$

(ii) The sequence of measures ν_n in (i) is called a delta sequence. On $\mathbb{R}^d$ a delta sequence of measures is constructed on the basis of a fixed probability density p by the formula

$$p_n(t) = \varepsilon_n^{-d} p(t/\varepsilon_n), \quad \varepsilon_n \to 0.$$

In the same manner we verify that $\nu_n \Rightarrow \delta_0$. Convolutions with a delta sequence give approximations $\mu * \nu_n$ for every bounded Borel measure μ on $\mathbb{R}^d$, since for $f \in C_b(\mathbb{R}^d)$ we have

$$\begin{aligned}\int_{\mathbb{R}^d} f(x)\,\mu * \nu_n(dx) &= \int_{\mathbb{R}^d}\int_{\mathbb{R}^d} f(x+y)\,\mu(dx)\,\nu_n(dy)\\ &= \int_{\mathbb{R}^d}\int_{\mathbb{R}^d} f(x+\varepsilon_n y)\,\mu(dx)\,p(y)\,dy,\end{aligned}$$

which tends to the integral of f with respect to the measure μ by the Lebesgue dominated convergence theorem. In different situations these approximations are frequently used with an appropriate choice of the density p and numbers $\varepsilon_n \to 0$.

Weak convergence of probability measures has been used for a long time in the form of the so-called *convergence in distribution* of random variables. We recall that a random variable is just a measurable function on a probability space $(\Omega, \mathcal{B}, P)$. As any measurable function, a random variable ξ induces a Borel measure on the real line by the formula

$$P_\xi(B) = P\bigl(\xi^{-1}(B)\bigr), \quad B \in \mathcal{B}(\mathbb{R}).$$

Even before creation of a general concept of measure it was customary to consider probabilities of hitting intervals of the form (a, b) and $(-\infty, c)$ by a random variable ξ. Actually, before appearance of the modern concept of a random variable (in Kolmogorov's axiomatics) it was informally generally accepted to regard as random variables quantities for which the words "to assume values in a given interval with a certain probability" were meaningful (see Lyapunov [**448**, p. 127]). For this reason, technically the most important characteristic of a random variable was its distribution function

$$F_\xi(t) = P(\xi < t), \tag{1.4.2}$$

i.e., in our terminology the distribution function of the induced measure. By convergence in distribution the usual convergence of distribution functions was meant (pointwise for continuous distributions and outside discontinuity points for distributions with jumps). Importance of this type of convergence was already understood at the dawn of probability theory; two great laws of probability theory were discovered that are still cornerstones of all applications in statistics: the law of large numbers and the central limit theorem.

Let us recall that the *law of large numbers* asserts that if we are given a sequence of independent random variables ξ_n with identical distributions and zero means (expectations), then the arithmetic means

$$\Sigma_n = \frac{\xi_1 + \cdots + \xi_n}{n}$$

converge to zero almost surely. Though in a such form the law of large numbers was established only by Kolmogorov (and its proof is rather difficult), much earlier this fundamental law had been known in more special cases. For example, if these random variables are square integrable, then it is very easy to prove convergence of the arithmetic means in probability. To this end, it suffices to calculate the variances: $\mathbb{E}\,|\Sigma_n|^2 = n^{-1}\,\mathbb{E}\,|\xi_1|^2$, where independence is used to show that the expectations of $\xi_i\xi_j$ with $i \neq j$ vanish. Now it remains to apply Chebyshev's inequality (1.1.1).

The *central limit theorem* deals with a completely different convergence. One considers the sums

$$S_n = \frac{\xi_1 + \cdots + \xi_n}{\sqrt{n}}$$

with another normalization. Typically such sums do not converge almost everywhere or in probability. For example, if ξ_n has the standard normal distribution, then the set of points of convergence of S_n has zero probability. However, if $\mathbb{E}\,\xi_n^2 = 1$, then the distribution functions of S_n converge pointwise to the distribution function of the standard normal random variable, i.e., one has convergence in distribution to the standard Gaussian distribution. Thus, we obtain that the limit distribution does not depend at all on the distribution of ξ_n (only the variance of ξ_n matters). A very large part of concepts, methods and facts of the theory of weak convergence grew from analysis of this very special case of convergence in distribution that has become a separate chapter in probability theory. Even in the simplest case of random variables with only two values -1 and 1 (covered by the Moivre–Laplace theorem, historically the first version of the central limit theorem), the proof of convergence in distribution is not completely trivial and requires some tools (say, characteristic functions and Taylor's expansion).

Different versions of these two laws with relaxed conditions of independence (which, certainly, cannot be removed without any compensation) are the subject of extensive literature, see Araujo, Giné [**19**], Fischer [**237**], Gnedenko, Kolmogorov [**282**], Hennequin, Tortrat [**318**], Ibragimov, Linnik [**331**], Loève [**437**], Petrov [**510**], Senatov [**576**], and Stroock [**598**]. In these investigations special features of the considered situations become so important that it is not reasonable to touch upon this subject in this book; however, examples and motives related to the central limit theorem will appear in our discussion.

Note that if for equally distributed independent random variables ξ_n there exist numbers a_n such that the distributions of the sums $(\xi_1+\cdots+\xi_n)/\sqrt{n}-a_n$ converge weakly, then necessarily $\mathbb{E}\,\xi_n^2 < \infty$ (see [**331**, Theorem 2.1.1, Theorem 2.6.6] and also Lemma 4.8.16 below). This fact is not obvious.

1.4.4. PROPOSITION. *Any weakly convergent sequence of Borel measures on the space $\mathbb{R}^d$ is bounded in the variation norm.*

PROOF. This follows from the Banach–Steinhaus theorem (see § 1.3), since the variation of a Borel measure μ on $\mathbb{R}^d$ equals the norm of the generated functional on the space $C_b(\mathbb{R}^d)$ (see Theorem 1.3.3). □

Certainly, for nonnegative measures this assertion is completely obvious.

Now we establish a number of simple criteria for weak convergence of general (possibly, signed) measures and then obtain even simpler conditions for weak convergence of nonnegative measures. First we give a straightforward corollary of the result mentioned in § 1.3 on the weak-$*$ compactness of balls in the space dual to a separable Banach space.

1.4.5. THEOREM. *Every sequence of measures on the compact interval that is bounded in variation contains a weakly convergent subsequence.*

The same assertion is true for measures on every compact set in $\mathbb{R}^d$.

From this theorem one can also obtain the following assertion called the *Helly theorem* or *Helly–Bray theorem*, which we prove, however, by a straightforward elementary reasoning with the aid of the known "Cantor's diagonal principle", frequently used for selecting subsequences.

1.4.6. THEOREM. *Every uniformly bounded sequence of increasing functions on a compact interval contains a pointwise convergent subsequence.*

The same is true for every uniformly bounded sequence of functions on a compact interval possessing uniformly bounded variations.

PROOF. Suppose we are given increasing functions f_n on the interval $[a, b]$. The set of their points of discontinuity is at most countable. Hence its complement contains a countable set of points t_n that is everywhere dense in $[a, b]$. The sequence $\{f_n(t_1)\}$ is bounded, hence it possesses a convergent subsequence. In the obtained sequence of indices we pick a further subsequence for which the values of our functions converge at t_2. Continuing this process inductively we obtain embedded subsequences of indices with the property that the subsequence with the number k gives converging values of our functions at the points $t_1, \ldots, t_k$. Finally, we take the "diagonal" subsequence: the first index in the first subsequence, the second one in the second subsequence and so on. We obtain a subsequence of the original sequence that converges at all points t_k. We can assume that so is the whole original sequence $\{f_n\}$. Let us take the increasing function f defined in Lemma 1.2.1 to which the functions f_n converge at all points of its continuity. Since the set of discontinuity points of f is finite or countable, the reasoning at the beginning of our proof gives a subsequence in $\{f_n\}$ that converges on this set as well.

The case of functions of bounded variation reduces to the considered one, since one has the decomposition $f_n = g_n - h_n$, where g_n and h_n are increasing and uniformly bounded (see § 1.2; for this it suffices to have the estimate $\sup_n V(f_n, [a, b]) < \infty$, $\sup_n |f_n(a)| < \infty$). Now one can pick a pointwise convergent subsequence in $\{g_n\}$ and then take a further subsequence giving convergence for $\{h_n\}$ as well. □

In the next theorem convergence of functions of bounded variation is connected with weak convergence of signed measures.

1.4.7. THEOREM. *A sequence of signed measures μ_n on the interval $[a, b]$ converges weakly to a measure μ precisely when* $\sup_n \|\mu_n\| < \infty$ *and every subsequence in the sequence of distribution functions F_{μ_n} of the measures μ_n contains a further subsequence converging to F_μ at all points excepting points of an at most countable set. In case of nonnegative measures the whole sequence F_{μ_n} converges to the function F_μ at the continuity points of the latter.*

An equivalent condition: $\sup_n \|\mu_n\| < \infty$ *and, for every compact interval* $[c,d] \subset [a,b]$ *and every* $\varepsilon > 0$*, there exists a number* N *such that*

$$\inf_{t\in[c,d]} |F_\mu(t) - F_{\mu_n}(t)| < \varepsilon \quad \textit{for all } n \geqslant N.$$

In the case of measures on the whole real line $\mathbb{R}$ *the listed conditions must be complemented by the following condition of uniform tightness: for every* $\varepsilon > 0$*, there is a compact interval* $[a,b]$ *such that* $|\mu_n|(\mathbb{R}\backslash[a,b]) < \varepsilon$ *for all* n*.*

PROOF. Suppose that the measures μ_n are uniformly bounded in variation and satisfy the indicated condition with subsequences, but do not converge weakly to μ. Since each continuous function f on $[a,b]$ can be uniformly approximated by smooth functions, taking into account the uniform boundedness of $\|\mu_n\|$ we obtain that there exists a smooth function f such that the integrals of f with respect to the measures μ_n do not converge to the integral of f with respect to measure μ. Passing to a subsequence one can assume that the difference between the indicated integrals remains greater than some $\delta > 0$. Yet another passage to a subsequence enables us to assume that $\lim\limits_{n\to\infty} F_{\mu_n}(t) = F_\mu(t)$ everywhere excepting a finite or countable set. By the integration by parts formula (1.2.2) for the smooth function f we obtain convergence of the right sides of the equality

$$\int_a^b f(t)\,\mu_n(dt) = f(b)F_{\mu_n}(b+) - \int_a^b f'(t)F_{\mu_n}(t)\,dt \tag{1.4.3}$$

to $f(b)F_\mu(b+) - \int_a^b f'(t)F_\mu(t)\,dt = \int_a^b f(t)\,\mu(dt)$, since F_μ and F_{μ_n} are constant on $(b,+\infty)$. This leads to a contradiction. In case of nonnegative measures the functions F_{μ_n} are nondecreasing. By Theorem 1.4.6 every subsequence in $\{F_{\mu_n}\}$ contains a subsequence converging to F_μ at the points of continuity of F_μ, whence convergence of the whole sequence at such points is obvious.

Conversely, suppose that the measures μ_n converge weakly to μ. Then, as shown above, $\sup\limits_n \|\mu_n\| < \infty$. This also gives the uniform boundedness of the variations of the functions F_{μ_n}. By Theorem 1.4.6 a given subsequence in $\{F_{\mu_n}\}$ contains a further subsequence that converges at every point. Thus, one can assume that $\{F_{\mu_n}\}$ converges pointwise to some function G. From (1.4.3) and the equality

$$\lim_{n\to\infty} F_{\mu_n}(b+) = \lim_{n\to\infty} \mu_n([a,b]) = \mu([a,b]) = F_\mu(b+)$$

on account of weak convergence and the Lebesgue theorem we obtain

$$\int_a^b f(t)\,\mu(dt) = f(b)F_\mu(b+) - \int_a^b f'(t)G(t)\,dt,$$

whence

$$\int_a^b f'(t)G(t)\,dt = \int_a^b f'(t)F_\mu(t)\,dt$$

for every polynomial f. Hence $G(t) = F_\mu(t)$ a.e. Therefore, the functions G and F_μ coincide at all points where both are continuous, i.e., on the complement of an at most countable set (depending on G, in particular, on the original subsequence).

Let us turn to the second condition. If it is violated, then either our measures are not uniformly bounded and then there is no weak convergence, or one can find an interval $[c,d]$, $\varepsilon > 0$ and a subsequence $\{n_k\}$ with $|F_\mu(t) - F_{\mu_{n_k}}(t)| > \varepsilon$ for all points $t \in [c, d]$, which contradicts the condition with subsequences. Suppose that the second condition is fulfilled. According to Theorem 1.4.5, every bounded sequence of measures contains a weakly convergent subsequence. Hence if there is no weak convergence of μ_n to μ, then there is a subsequence in $\{\mu_n\}$ that is weakly convergent to some measure ν on $[a, b]$ different from μ. According to what has already been proved, this subsequence contains a further subsequence with indices $\{n_k\}$ for which the functions $F_{\mu_{n_k}}$ converge to F_ν on the complement to some at most countable set. Passing to a subsequence we can assume that $\{\mu_{n_k}^+\}$ and $\{\mu_{n_k}^-\}$ have weak limits ν_1 and ν_2 and that $n_k = k$.

The set of convergence of F_{μ_k} contains a point τ_1 at which all functions F_μ, F_ν, F_{ν_1} and F_{ν_2} are continuous and such that

$$|F_\mu(\tau_1) - F_\nu(\tau_1)| = \varepsilon > 0,$$

since otherwise $\mu = \nu$. There is a point $\tau_2 \in [a, b]$ and a number $N \in \mathbb{N}$ such that $|F_\nu(t) - F_\mu(t)| > \varepsilon/2$ and $|F_{\nu_i}(t) - F_{\nu_i}(s)| \leqslant \varepsilon/16$ for all t, s in $I = [\tau_1, \tau_2]$, $|F_{\mu_k^+}(\tau_i) - F_{\nu_1}(\tau_i)| \leqslant \varepsilon/16$ and $|F_{\mu_k^-}(\tau_i) - F_{\nu_2}(\tau_i)| \leqslant \varepsilon/16$ whenever $k \geqslant N$. Then

$$\sup_{t\in I} |F_{\mu_k^+}(t) - F_{\nu_1}(t)| \leqslant 3\varepsilon/16, \ \sup_{t\in I} |F_{\mu_k^-}(t) - F_{\nu_2}(t)| \leqslant 3\varepsilon/16,$$

whence $\sup_{t\in I} |F_{\mu_k}(t) - F_\nu(t)| \leqslant 3\varepsilon/8$, i.e.,

$$\inf_{t\in I} |F_\mu(t) - F_{\mu_k}(t)| \geqslant \varepsilon/8 \quad \text{for all } k \geqslant N.$$

The case of the real line is similar in the part of sufficiency of the indicated condition. In the proof of necessity we have to verify the aforementioned condition of the uniform tightness of measures. It is completely trivial in case of nonnegative measures, but becomes much less obvious for signed measures. For probability measures it suffices to find an interval $[-N, N]$ such that $\mu(\mathbb{R}\backslash[-N, N]) < \varepsilon$, then take a continuous function f with $0 \leqslant f \leqslant 1$ equal to 1 on $[-N, N]$ and 0 outside $[-N - 1, N + 1]$ and find a number n_1 such that the integral of f with respect to the measure μ_n is greater than $1 - \varepsilon$ for all $n \geqslant n_1$. This is possible by weak convergence and the fact that the integral of f with respect to the measure μ is greater than $1 - \varepsilon$. Then $\mu_n([-N-1, N+1]) \geqslant 1 - \varepsilon$ whenever $n \geqslant n_1$. Increasing the interval, it is easy to obtain the desired estimate for all n.

However, for signed measures this reasoning does not work. Let us consider the general case. Suppose that the measures μ_n converge weakly to the measure $\mu = 0$ (to this case we pass considering the measures $\mu_n - \mu$, when we need not take care of nonnegativity), but are not uniformly tight. This means that for some $\varepsilon > 0$ we can find an increasing sequence of numbers n_k and a sequence of disjoint intervals $I_k = [a_k, b_k]$ with increasing $a_k \to +\infty$ (or $b_k \to -\infty$) such that $|\mu_{n_k}|(I_k) \geqslant \varepsilon$ and $|\mu_{n_i}|(\mathbb{R}\backslash[-b_k, b_k]) < \varepsilon/8$ if $i = 1, \ldots, k-1$. These numbers and intervals are constructed by induction. Now we can assume that $n_k = k$. However, we pick yet another subsequence. Let us take a continuous function f_1 with $|f_1| \leqslant 1$ that vanishes outside a small neighborhood of I_1 disjoint with I_2 and satisfies the estimate

$$\int_{\mathbb{R}} f_1\, d\mu_1 \geqslant 3\varepsilon/4.$$

This is possible due to our assumption that $|\mu_1|(I_1) \geqslant \varepsilon$. Since the integrals of f_1 with respect to the measures μ_n tend to zero, one can find an odd number n_2 starting from which they become less than $\varepsilon/8$ in absolute value. Then for μ_{n_2} we find a continuous function f_2 with $|f_2| \leqslant 1$ that vanishes outside a small neighborhood of the interval I_{n_2} disjoint with all other intervals I_n and satisfies the estimate

$$\int_{\mathbb{R}} f_2 \, d\mu_{n_2} \geqslant 3\varepsilon/4.$$

By construction $|\mu_1|\bigl([b_{n_2}, +\infty)\bigr) < \varepsilon/8$. Next we find an odd number $n_3 > n_2$ starting from which the integrals of $f_1 + f_2$ with respect to the measures μ_n become less than $\varepsilon/64$ in absolute value. Continuing this process, we obtain continuous functions f_k with $|f_k| \leqslant 1$ and disjoint supports and possessing the property that the function f on the whole real line defined by zero outside their supports and equal to these functions on their supports is continuous and bounded. In addition, we obtain measures μ_{n_k} for which the integrals of f_k with respect to μ_{n_k} are greater than $3\varepsilon/4$ and the integrals of $f_1 + \cdots + f_{k-1}$ with respect to the measures μ_m with $m \geqslant n_k$ are less than $\varepsilon/8^k$ in absolute value. The integral of f with respect to the measure μ_{n_k} remains not less than $3\varepsilon/4 - \varepsilon/8^k - \varepsilon/8$ (here we also take into account that $|\mu_{n_k}|\bigl([b_{k+1}, +\infty)\bigr) < \varepsilon/8$), which gives a contradiction. □

Another possible proof and a number of useful related assertions can be found in Theorem 1.7.1 and Theorem 1.7.2.

It should be noted that in case of signed measures weak convergence does not imply the pointwise convergence of distribution functions on an everywhere dense set. Moreover, it is easy to give an example of a weakly convergent sequence of signed measures μ_n on $[0, 1]$ for which the distribution functions converge at no point in the interval $(0, 1)$.

1.4.8. EXAMPLE. Let us consider the measures $\mu_n := \delta_{x_n} - \delta_{y_n}$, where the sequence of intervals $[x_n, y_n]$ is obtained in the following way: for every n we take consecutive intervals of length 2^{-n} with the ends at points of the form $k2^{-n}$ and number all these intervals consecutively. The measures μ_n converge weakly to the zero measure, but the functions F_{μ_n} do not converge at points of the interval $(0, 1)$. Indeed, the integral of a continuous function f on $[0, 1]$ with respect to the measure μ_n equals $f(x_n) - f(y_n)$, which tends to zero as n increases by the uniform continuity of f. The function F_{μ_n} is the indicator function of $[x_n, y_n)$. Hence for every $t \in (0, 1)$ the sequence $\{F_{\mu_n}(t)\}$ contains infinitely many zeros and units.

Let us give a simple sufficient condition for weak convergence (as shown in the next section, in case of nonnegative measures this condition is also necessary).

1.4.9. PROPOSITION. *Suppose that a sequence of signed Borel measures μ_n on the interval $[a, b]$ is bounded in variation. Then a sufficient (but not necessary!) condition for weak convergence of μ_n to a measure μ is convergence of $F_{\mu_n}(t)$ to $F_\mu(t)$ at the points of an everywhere dense set.*

PROOF. Let f be a continuous function on $[a, b]$ and let $\varepsilon > 0$. Let us consider the function

$$f_m(t) = \sum_{k=1}^{m} f(a_{k,m}) I_{[a_{k,m}, a_{k+1,m})}(t),$$

where the points $a_{k,m}$ belong to the set of points of convergence of the functions F_{μ_n} to F_μ, $a_{1,m} = a$, $a_{k,m} < a_{k+1,m}$, $a_{m,m} = b + m^{-1}$ and

$$\sup_k |a_{k,m} - a_{k+1,m}| \to 0 \quad \text{as } m \to \infty,$$

and we set $f(t) := f(b)$ for $t > b$. The functions f_m converge uniformly to f by the uniform continuity of the function f. In addition, for every m we have

$$\lim_{n\to\infty} \int f_m\, d\mu_n = \int f_m\, d\mu,$$

since

$$\begin{aligned}\lim_{n\to\infty} \mu_n\bigl([a_{k,m}, a_{k+1,m})\bigr) &= F_{\mu_n}(a_{k+1,m}) - F_{\mu_n}(a_{k,m}) \\ &= F_{\mu}(a_{k+1,m}) - F_{\mu}(a_{k,m}) = \mu\bigl([a_{k,m}, a_{k+1,m})\bigr)\end{aligned}$$

by our choice of the points $a_{k,m}$. Hence we have convergence of the integrals of the function f. □

The case of an interval considered in some results above is actually specific only in the respect that we have employed distribution functions. It will be clear below that the case of general metric compact spaces is completely similar and that it plays a key role when combined with the property of uniform tightness of families of measures (this property has already been encountered above) and having an exceptional importance in all aspects of weak convergence.

1.4.10. DEFINITION. A family of Borel measures $\mathcal{M}$ on a metric (or topological) space X is called tight (or uniformly tight) if, for every $\varepsilon > 0$, there exists a compact set $K_\varepsilon \subset X$ such that

$$|\mu|(X\backslash K_\varepsilon) \leqslant \varepsilon \quad \forall\, \mu \in \mathcal{M}.$$

From the proof of the last assertion of Theorem 1.4.7 related to the whole real line one can derive with the same justification the following very important fact (in the next chapter it will be extended to general metric spaces in the Prohorov theorem).

1.4.11. THEOREM. *If Borel measures μ_n on $\mathbb{R}^d$ (possibly, signed) converge weakly to a Borel measure μ, then the sequence $\{\mu_n\}$ is uniformly tight.*

If a sequence of Borel measures on $\mathbb{R}^d$ (possibly, signed) is bounded in variation and uniformly tight, then it contains a weakly convergent subsequence.

PROOF. We explain only the second assertion. By using Theorem 1.4.5, one can pick in the given sequence a subsequence of measures μ_n weakly convergent to some measure ν_k on every cube of the form $Q_k = [-k, k]^d$. It is readily seen that the measure ν_{k+1} coincides with ν_k on the cube Q_k. Hence one can define a bounded Borel measure ν on all of $\mathbb{R}^d$. By using the uniform tightness of the sequence $\{\mu_n\}$ and weak convergence on the cubes Q_k it is now easy to verify that we have $\mu_n \Rightarrow \nu$. □

A typical example of using this property is the following assertion.

1.4.12. PROPOSITION. *Suppose we are given a sequence of Borel measures μ_n (possibly, signed) and a Borel measure μ on $\mathbb{R}^d$. Suppose that*

$$\lim_{n\to\infty} \int_{\mathbb{R}^d} \varphi\, d\mu_n = \int_{\mathbb{R}^d} \varphi\, d\mu \quad \forall\, \varphi \in C_0^\infty(\mathbb{R}^d).$$

Then for weak convergence of the measures μ_n to the measure μ it is necessary and sufficient to have the uniform boundedness and the uniform tightness of the sequence $\{\mu_n\}$.

PROOF. The necessity of the indicated conditions is already known. Their sufficiency is clear from the fact that by the previous theorem one can pick weakly convergent subsequences, but their limits will be the same, since two measures with equal integrals of the functions of class $C_0^\infty(\mathbb{R}^d)$ coincide. Note also that sufficiency can be easily verified directly. Let $f \in C_b(\mathbb{R}^d)$ and $\varepsilon > 0$. One can assume that $|f| \leqslant 1$ and $\|\mu_n\| \leqslant 1$. By our condition there exists a compact set K with $|\mu|(\mathbb{R}^d\backslash K) \leqslant \varepsilon$ and $|\mu_n|(\mathbb{R}^d\backslash K) \leqslant \varepsilon$ for all n. One can also find a function φ in $C_0^\infty(\mathbb{R}^d)$ with $|\varphi| \leqslant 1$, $\sup_{x\in K} |f(x) - \varphi(x)| \leqslant \varepsilon$. Let us take n_1 such that

$$\left|\int_{\mathbb{R}^d} \varphi\, d\mu_n - \int_{\mathbb{R}^d} \varphi\, d\mu\right| \leqslant \varepsilon \quad \forall n \geqslant n_1.$$

Then, whenever $n \geqslant n_1$, we have

$$\begin{aligned}\left|\int_{\mathbb{R}^d} f\, d\mu_n - \int_{\mathbb{R}^d} f\, d\mu\right| &\leqslant \left|\int_K f\, d\mu_n - \int_K f\, d\mu\right| + 2\varepsilon \\ &\leqslant \left|\int_K \varphi\, d\mu_n - \int_K \varphi\, d\mu\right| + 4\varepsilon \leqslant 7\varepsilon,\end{aligned}$$

since

$$\left|\int_K \varphi\, d\mu_n - \int_K \varphi\, d\mu\right| \leqslant \left|\int_{\mathbb{R}^d} \varphi\, d\mu_n - \int_{\mathbb{R}^d} \varphi\, d\mu\right| + 2\varepsilon$$

due to our choice of K and the estimate $|\varphi| \leqslant 1$. □

Let us emphasize that the uniform boundedness and tightness do not follow automatically from the first condition of this proposition (but see Theorem 1.5.3).

1.5. Weak convergence of nonnegative measures

The case of nonnegative measures has significant specific features. They are so important that in the study of weak convergence of measures it is quite common to consider only this case. Since a number of basic properties of weak convergence have already been discussed in the general case, here we concentrate on those properties for which positivity is essential. The next theorem follows from Theorem 1.4.7, but we include a simple direct proof.

1.5.1. THEOREM. *If nonnegative measures μ_n on $\mathbb{R}$ converge weakly to a measure μ, then at all points of continuity of the function F_μ one has convergence of the distribution functions $F_{\mu_n}(t) \to F_\mu(t)$.*

PROOF. Clearly, the measure μ is also nonnegative, since our condition yields that the integrals of nonnegative bounded continuous functions with respect to this measure are nonnegative. Let t be a point of continuity of the function F_μ and let $\varepsilon > 0$. Take $\delta > 0$ such that $\mu([t-\delta, t+\delta]) < \varepsilon$. Next, let us take a continuous function f for which $0 \leqslant f \leqslant 1$, $f(s) = 1$ if $s \leqslant t - \delta$ and $f(s) = 0$ if $s \geqslant t$. Then

$$F_{\mu_n}(t) \geqslant \int f\, d\mu_n \geqslant \int f\, d\mu - \varepsilon > F_\mu(t) - 2\varepsilon$$

for all sufficiently large n. Let us take a continuous function g such that $0 \leqslant g \leqslant 1$, $g(s) = 1$ if $s \leqslant t$ and $g(s) = 0$ if $s \geqslant t + \delta$. Then

$$F_{\mu_n}(t) \leqslant \int g\, d\mu_n \leqslant \int g\, d\mu + \varepsilon \leqslant F_\mu(t) + 2\varepsilon$$

for all sufficiently large n. □

The next two theorems are presented for probability measures just for simplification of their formulations, but they can be easily modified to apply to nonnegative measures. The following criterion is due to A.D. Alexandroff and holds in general spaces. We also consider the proof, although in the next chapter it will be given in the case of general metric spaces and in Chapter 4 extended to topological spaces.

1.5.2. THEOREM. *Suppose that we are given a sequence of Borel probability measures* $\{\mu_n\}$ *and a Borel probability measure* μ *on the real line. Then the following conditions are equivalent:*

(i) *the sequence* $\{\mu_n\}$ *converges weakly to* μ*;*

(ii) *for every closed set* F *one has the inequality*

$$\limsup_n \mu_n(F) \leqslant \mu(F); \tag{1.5.1}$$

(iii) *for every open set* U *one has the inequality*

$$\liminf_n \mu_n(U) \geqslant \mu(U). \tag{1.5.2}$$

PROOF. We first observe that conditions (ii) and (iii) are equivalent, since $\mathbb{R}\backslash F$ is open, $\mathbb{R}\backslash U$ is closed, and

$$\mu_n(\mathbb{R}\backslash F) = 1 - \mu_n(F)$$

and similarly for μ, and the same is true for the set U. If (ii) and (iii) hold, then for the open rays $U = (-\infty, t)$, where t is not an atom of the measure μ, we obtain convergence $\mu_n(U) \to \mu(U)$, since $\mu_n(U) \leqslant \mu_n(\overline{U})$, $\mu(U) = \mu(\overline{U})$. Hence one has convergence of the distribution functions at the points of continuity of F_μ, which, as we know, implies weak convergence.

Conversely, suppose that the measures μ_n converge weakly to μ, F is closed and $\varepsilon > 0$. There is a continuous function f with $0 \leqslant f \leqslant 1$, equal to 1 on F and vanishing outside an open set $W \supset F$ such that $\mu(W\backslash F) < \varepsilon$. Then

$$\mu_n(F) \leqslant \int_{\mathbb{R}} f\,d\mu_n \leqslant \int_{\mathbb{R}} f\,d\mu + \varepsilon \leqslant \mu(F) + 2\varepsilon$$

for all sufficiently large n, which completes the proof, because ε was arbitrary. □

1.5.3. THEOREM. *Suppose that we are given a sequence of Borel probability measures* $\{\mu_n\}$ *and a Borel probability measure* μ *on* $\mathbb{R}^d$*. Then weak convergence of the measures* μ_n *to the measure* μ *is equivalent to convergence of the integrals of functions of class* $C_0^\infty(\mathbb{R}^d)$ *with respect to the measures* μ_n *to their integrals with respect to* μ*. In addition, it implies convergence* $\|\mu_n * \sigma - \mu * \sigma\| \to 0$ *for every absolutely continuous probability measure* σ *on* $\mathbb{R}^d$*.*

PROOF. The necessity is clear and for the proof of sufficiency we establish the uniform tightness of $\{\mu_n\}$ (see Proposition 1.4.12). Let $\varepsilon > 0$. Let us take $N \in \mathbb{N}$ such that $\mu(U_N) > 1 - \varepsilon$, where U_N is the ball of radius N centered t zero. Let us find a function $f \in C_0^\infty(\mathbb{R}^d)$ such that $0 \leqslant f \leqslant 1$, $f(x) = 1$ if $|x| \leqslant N$, $f(x) = 0$ if $|x| \geqslant N+1$. Then the integral of f with respect to μ is greater than $1-\varepsilon$, hence there is a number n_1 such that the integral of f with respect to the measure μ_n will be greater than $1 - \varepsilon$ for all $n \geqslant n_1$. Hence $\mu_n(U_{N+1}) > 1 - \varepsilon$ if $n \geqslant n_1$, whence the uniform tightness of the whole sequence follows.

If $\mu_n \Rightarrow \mu$, then $\|\mu_n * \sigma - \mu * \sigma\| \to 0$, provided that σ belongs to $C_0^\infty(\mathbb{R}^d)$: indeed, the densities of the measures $\mu_n * \sigma$ converge pointwise to the density of $\mu * \sigma$

(see (1.1.3)), so we can apply the Vitali–Scheffé theorem (see p. 6). In the general case there are measures σ_j with densities of the indicated type that converge to σ in variation, moreover, $\|\mu_n * \sigma - \mu_n * \sigma_j\| \leqslant \|\sigma - \sigma_j\|$. $\square$

It is easy to see that these theorems do not extend to signed measures. For example, the measures $\delta_{n+1/n} - \delta_n$ do not converge weakly.

One more important characterization of weak convergence of nonnegative measures is given below in terms of Fourier transforms.

1.6. Connections with Fourier transforms

We recall that the *Fourier transform* (or the *characteristic functional*) of a Borel measure μ on $\mathbb{R}^d$ (with the inner product $\langle x, y\rangle$) is the complex function

$$\widetilde{\mu}(y) = \int_{\mathbb{R}^d} \exp\bigl(i\langle y, x\rangle\bigr)\, \mu(dx).$$

This function is bounded and continuous (by the Lebesgue theorem). We emphasize that the measure can be signed.

For measures on the real line the formula takes the form

$$\widetilde{\mu}(y) = \int_{\mathbb{R}} \exp(iyx)\, \mu(dx).$$

Although an explicit calculation of the Fourier transform is possible only in some exceptional cases, it plays a very important role in the study of measures on $\mathbb{R}^d$.

For the standard Gaussian measure γ_d on $\mathbb{R}^d$ with density

$$(2\pi)^{-d/2} \exp(-|x|^2/2)$$

the Fourier transformation equals $\exp(-|y|^2/2)$, and for Dirac's measure at the point a we have $\widetilde{\delta_a}(y) = \exp\bigl(i\langle y, a\rangle\bigr)$. The image of the standard Gaussian measure under the linear mapping T has the Fourier transform $\exp\bigl(-\langle T^*y, T^*y\rangle/2\bigr)$, which is clear from the change of variables formula.

It is known (see Bogachev [**81**, §3.8]) that the equality $\widetilde{\mu} = \widetilde{\nu}$ of the Fourier transforms yields the equality of measures $\mu = \nu$.

Note that the Fourier transform $\widetilde{\mu}$ is real precisely when the measure μ is *symmetric*, i.e., $\mu(B) = \mu(-B)$ for all sets $B \in \mathcal{B}(\mathbb{R}^d)$. This follows from the fact that the image of the measure μ under the mapping $x \mapsto -x$ has the Fourier transform equal the complex conjugate of $\widetilde{\mu}$.

This definition is quite traditional, but it does not coincide with the definition of the Fourier transform for integrable functions that is given by the formula

$$\widehat{f}(x) = (2\pi)^{-d/2} \int_{\mathbb{R}^d} \exp\bigl(-i\langle x, y\rangle\bigr)\, f(y)\, dy,$$

i.e., differs by a factor and the sign of the argument. The mapping $f \mapsto \widehat{f}$ preserves the norm in $L^2(\mathbb{R}^d)$, so it extends to a unitary operator on $L^2(\mathbb{R}^d)$.

If the measure μ is given by a density ϱ with respect to Lebesgue measure and, in addition, the function $\widetilde{\mu}$ is integrable over $\mathbb{R}^d$ (which, certainly, is not always the case, e.g., for $I_{[0,1]}$), then almost everywhere there holds the inversion formula

$$\varrho(x) = \frac{1}{(2\pi)^d} \int_{\mathbb{R}^d} \exp(-i\langle x, y\rangle)\, \widetilde{\mu}(y)\, dy. \tag{1.6.1}$$

Moreover, ϱ possesses a continuous version for which this equality is true everywhere.

If the function $\widetilde{\mu}$ is not integrable, then one can use the formula

$$\varrho(x) = \lim_{\sigma\to 0+} \frac{1}{(2\pi)^d} \int_{\mathbb{R}^d} \exp(-i\langle x, y\rangle)\, \widetilde{\mu}(y) e^{-\sigma|y|^2}\, dy. \tag{1.6.2}$$

The inverse Fourier transform on $L^1(\mathbb{R}^d)$ is defined as $(2\pi)^{-d/2}\widetilde{f}(x)$, i.e., as $\widehat{f}(-x)$. The class $\mathcal{S}(\mathbb{R}^d)$ of complex smooth functions f for which $(1+|x|^m)\partial_{x_1}^{k_1}\cdots\partial_{x_n}^{k_n} f(x)$ is bounded for all m and k_j is mapped by the Fourier transform onto itself.

For the convolution of measures Fubini's theorem yields the formula

$$\widetilde{\mu * \nu} = \widetilde{\mu}\widetilde{\nu}. \tag{1.6.3}$$

The equality of measures with equal Fourier transforms yields that two measures on $\mathbb{R}^d$ with equal one-dimensional images under all functionals $l_u\colon\ x \mapsto \langle u, x\rangle$ are equal. Indeed, the equality

$$\widetilde{\mu}(u) = \widetilde{\mu\circ l_u^{-1}}(1)$$

shows that their Fourier transforms coincide.

There is no any constructive description of the class of Fourier transforms of all measures, but such a description is available for the Fourier transforms of probability measures. It is given by the *Bochner theorem* and is this: a complex function φ on $\mathbb{R}^d$ coincides with the Fourier transform of a probability measure precisely when $\varphi \in C_b(\mathbb{R}^d)$, $\varphi(0) = 1$ and φ is *positive definite* in the following sense: for all vectors $y_i \in \mathbb{R}^n$ and numbers $c_i \in \mathbb{C}$, where $i = 1, \ldots, k$, one has the inequality $\sum_{i,j=1}^k c_i\overline{c_j}\varphi(y_i - y_j) \geqslant 0$.

A positive definite function need not be continuous, but if it is Lebesgue measurable, then it coincides almost everywhere with a continuous positive definite function.

If for a probability measure μ on the real line the function $|x|^k$ is μ-integrable, then the function $\widehat{\mu}$ has k derivatives and

$$\widetilde{\mu}^{(k)}(0) = \int_{\mathbb{R}} (ix)^k\, \mu(dx).$$

This is verified with the aid of the Lebesgue theorem on differentiation of integrals with respect to a parameter. A similar assertion is true in $\mathbb{R}^d$:

$$\partial_{y_{i_1}}\cdots\partial_{y_{i_k}}\widetilde{\mu}(0) = \int_{\mathbb{R}^d} (ix_{i_1})\cdots(ix_{i_k})\, \mu(dx).$$

If we are given two Borel measures μ and ν on $\mathbb{R}^d$, then the following Parseval equality holds:

$$\int_{\mathbb{R}^d} \widetilde{\nu}(x)\, \mu(dx) = \int_{\mathbb{R}^d} \widetilde{\mu}(x)\, \nu(dx). \tag{1.6.4}$$

For the proof it suffices to integrate the bounded complex function $\exp(i\langle x, y\rangle)$ with respect to the measure $\mu\otimes\nu$ and apply Fubini's theorem on interchanging the limits of integration.

With the aid of this equality one can reconstruct the values of the measure μ on some sets by means of its Fourier transform. For example, taking for ν all measures with Fourier transforms from $C_0^\infty(\mathbb{R}^d)$, we find the integrals of such functions with respect to the measure μ. Approximating indicator functions of sets by functions of this class, in principle we can also find the values of μ on certain sets. We cannot obtain indicator functions directly as Fourier transforms (excepting for constant zero and one) due to the discontinuity of indicator functions and continuity of the Fourier transforms of measures. For example, for finding the measure of an interval of the real line one can apply the following formula.

1.6.1. PROPOSITION. *Let μ be a bounded Borel measure on the real line without atoms at the points a, b. Then*

$$\mu([a,b]) = \lim_{\varepsilon\to 0} \frac{1}{2\pi} \int_{-\infty}^{+\infty} \widetilde{\mu}(t) \frac{e^{-itb} - e^{-ita}}{-it} e^{-\varepsilon^2 t^2/2}\, dt.$$

The Parseval equality can be applied for deriving estimates enabling one to obtain the uniform tightness of a family of probability measures with Fourier transforms equicontinuous at the origin. Let us recall that a family of functions $\mathcal{F}$ is called equicontinuous at a point x_0 if $\lim\limits_{x\to x_0} \sup_{f\in\mathcal{F}} |f(x) - f(x_0)| = 0$.

For Fourier transforms of probability measures the equicontinuity at the origin implies the equicontinuity at any other point, since one has the inequality

$$|\widetilde{\mu}(x) - \widetilde{\mu}(y)|^2 \leqslant 2\bigl(1 - \operatorname{Re}\widetilde{\mu}(x-y)\bigr), \tag{1.6.5}$$

which follows immediately from the Cauchy–Bunyakovskii inequality and the equality $|\exp(i\langle x,z\rangle) - \exp(i\langle y,z\rangle)|^2 = 2 - 2\cos(\langle x-y, z\rangle)$.

1.6.2. PROPOSITION. *Let μ and ν be Borel probability measures on $\mathbb{R}^d$. If the function $\widetilde{\nu}$ is real, then*

$$\mu\bigl(x\colon \widetilde{\nu}(x) \leqslant t\bigr) \leqslant \frac{1}{1-t} \int_{\mathbb{R}^d} [1 - \widetilde{\mu}(y)]\, \nu(dy) \quad \forall\, t \in (0,1), \tag{1.6.6}$$

where the right-hand side is automatically real.

PROOF. The left-hand side equals $\mu\bigl(x\colon 1-\widetilde{\nu}(x) \geqslant 1-t\bigr)$, which by the Chebyshev inequality does not exceed

$$\frac{1}{1-t} \int_{\mathbb{R}^d} [1 - \widetilde{\nu}(x)]\, \mu(dx).$$

It remains to apply (1.6.4), which also shows that the right-hand side of (1.6.6) is real. □

We emphasize that the function $\widetilde{\mu}$ itself need not be real; only its integral with respect to the measure ν is real.

1.6.3. COROLLARY. *For every Borel probability measure μ on $\mathbb{R}^d$ one has*

$$\mu\bigl(x\colon |x| \geqslant t\bigr) \leqslant \frac{\sqrt{e}}{\sqrt{e}-1} \int_{\mathbb{R}^d} [1 - \widetilde{\mu}(y/t)]\, \gamma(dy) \quad \forall\, t > 0, \tag{1.6.7}$$

where γ is the standard Gaussian measure on $\mathbb{R}^d$.

PROOF. We know that $\widetilde{\gamma}(x) = e^{-|x|^2/2}$. Let γ_t be the image of γ under the mapping $x \mapsto x/t$. Then

$$\widetilde{\gamma_t}(x) = \exp\bigl(-t^{-2}|x|^2/2\bigr).$$

Therefore, by (1.6.6) we obtain

$$\mu\bigl(x\colon |x| \geqslant t\bigr) = \mu\bigl(x\colon \widetilde{\gamma_t}(x) \leqslant e^{-1/2}\bigr) \leqslant \frac{1}{1-e^{-1/2}} \int_{\mathbb{R}^d} [1 - \widetilde{\mu}(y)]\, \gamma_t(dy).$$

The right-hand side of this inequality equals the right-hand side of (1.6.7) by the definition of γ_t. □

1.6.4. COROLLARY. *Let μ a Borel probability measure on $\mathbb{R}^d$ and $r > 0$. Then*

$$\mu\bigl(x\colon\ |x| \geqslant r^{-2}\bigr) \leqslant 6dr^2 + 3\sup_{|z|\leqslant r} |1 - \widetilde{\mu}(z)|. \tag{1.6.8}$$

Therefore, any family of probability measures whose Fourier transforms are equicontinuous at the origin is uniformly tight.

PROOF. The left-hand side of estimate (1.6.8) does not exceed the integral of $3|1 - \widetilde{\mu}(r^2y)|$ with respect to the measure γ, since $\sqrt{e}(\sqrt{e}-1)^{-1} < 3$. The integral over the ball of radius r^{-1} does not exceed $3\sup_{|z|\leqslant r} |1 - \widetilde{\mu}(z)|$, since $|r^2y| \leqslant r$ whenever $|y| \leqslant r^{-1}$. By the Chebyshev inequality

$$\gamma\bigl(y\colon\ |y| > r^{-1}\bigr) \leqslant r^2 \int_{\mathbb{R}^d} |y|^2\,\gamma(dy) = dr^2.$$

It remains to observe that $|1 - \widetilde{\mu}| \leqslant 2$. □

It is useful (in particular, for the infinite-dimensional case) to have the following modifications of these estimates. Let $\operatorname{tr} A := \sum_{i=1}^d a^{ii}$ be the *trace of an operator* $A = (a^{ij})_{i,j\leqslant d}$.

1.6.5. PROPOSITION. *Let A and B be nonnegative definite operators on $\mathbb{R}^d$ and let μ be a Borel probability measure on $\mathbb{R}^d$. Set*

$$\varepsilon := \sup\bigl\{1 - \operatorname{Re}\widetilde{\mu}(y)\colon\ \langle Ay, y\rangle \leqslant 1\bigr\}.$$

Then

$$\mu\bigl(x\colon\ \langle Bx, x\rangle > t\bigr) \leqslant \frac{\sqrt{e}}{\sqrt{e}-1}(\varepsilon + 2t^{-1}\operatorname{tr} AB) \quad \forall\, t > 0. \tag{1.6.9}$$

PROOF. One can assume that $t = 1$, replacing the operator B by $t^{-1}B$. Let us take the centered Gaussian measure ν with the Fourier transform $\exp\bigl(-\langle By, y\rangle/2\bigr)$, which is the image of the standard Gaussian measure under the operator $\sqrt{B}$. The bound $\langle Bx, x\rangle > 1$ is equivalent to the inequality $\exp\bigl(-\langle By, y\rangle/2\bigr) < \exp(-1/2)$. Hence by virtue of (1.6.6) and taking into account that the integral of $\widetilde{\mu}$ with respect to the measure ν is real we see that the left-hand side of (1.6.9) does not exceed

$$\frac{\sqrt{e}}{\sqrt{e}-1} \int_{\mathbb{R}^d} \bigl[1 - \operatorname{Re}\widetilde{\mu}(y)\bigr]\,\nu(dy).$$

Here the integral equals the sum of the integrals over the set $\{y\colon\ \langle Ay, y\rangle \leqslant 1\}$, on which the integrated function does not exceed ε, and also over its complement. The integral over the complement is estimated by the integral of $\langle Ay, y\rangle$ over the whole space $\mathbb{R}^d$, which equals $\operatorname{tr} AB$, which is verified directly by passing to the eigenbasis of the operator A. □

Weak convergence of nonnegative measures admits a convenient description by means of Fourier transforms. This is the content of the following theorem due to Paul Lévy (called sometimes the "continuity theorem").

1.6.6. THEOREM. *A sequence $\{\mu_n\}$ of probability measures on $\mathbb{R}^d$ converges weakly to a probability measure μ precisely when $\widetilde{\mu}(y) = \lim_{n\to\infty} \widetilde{\mu}_n(y)$ for each $y \in \mathbb{R}^d$.*

PROOF. Weak convergence of measures yields the pointwise convergence of their characteristic functionals. Let us prove the converse. We give two proofs. First we show that the sequence of measures μ_n is uniformly tight. This follows from (1.6.7), since, for any given $\varepsilon > 0$, by the Lebesgue dominated convergence theorem there exists $t_1 > 0$ such that the right-hand side of (1.6.7) is less than ε for $t = t_1$. Applying the Lebesgue theorem once again, we conclude that this also holds true for the measures μ_n in place of μ for all sufficiently large n. This gives weak convergence of $\{\mu_n\}$ to μ. Indeed, every weakly convergent subsequence in $\{\mu_n\}$ can converge only to μ by convergence of the Fourier transforms. The uniform tightness yields that every subsequence in $\{\mu_n\}$ contains a further subsequence that is weakly convergent. Hence the whole original sequence converges weakly to μ. The second proof is even shorter, but uses (1.6.4) and the inversion formula (1.6.1). For $f \in C_0^\infty(\mathbb{R}^d)$ we take the measure σ with density f and let $g(x) = (2\pi)^{-d}\widetilde{\sigma}(-x)$ and $\nu = g\,dx$. Then $f(x) = \widetilde{\nu}(x)$ by (1.6.1), hence by (1.6.4) the integral of f against μ_n is the integral of $\widetilde{\mu}_n$ against ν and similarly for μ, but the integrals of $\widetilde{\mu}_n$ against ν converge to the integral of $\widetilde{\mu}$ against ν by the Lebesgue dominated convergence theorem, so Theorem 1.5.3 applies. □

1.6.7. COROLLARY. *A sequence $\{\mu_n\}$ of probability measures on $\mathbb{R}^d$ converges weakly precisely when the sequence of their characteristic functionals $\widetilde{\mu}_n$ converges at every point and the function $\varphi(y) := \lim_{n\to\infty} \widetilde{\mu}_n(y)$ is continuous at the origin. In this case φ is the characteristic functional of the probability measure μ to which the measures μ_n converge weakly.*

PROOF. As explained above, under the stated assumptions the function φ is the characteristic functional of some Borel probability measure. □

1.6.8. REMARK. (i) In this corollary one cannot omit the condition of continuity of φ. Indeed, for every n, the function $(\cos x)^{2n}$ is the characteristic functional of the $2n$-fold convolution of the probability measure ν assigning the value $1/2$ to the points -1 and 1. These functions converge pointwise to the function φ equal to 1 at the points πk and 0 at all other points. Clearly, φ is not the characteristic functional of a measure due to its discontinuity. Note that any function φ obtained as the pointwise limit of a sequence of Fourier transforms of probability measures has always a continuous modification that is also the characteristic functional of some nonnegative measure μ, but this measure need not be a probability measure (in the considered example $\mu = 0$). Hence in place of continuity of φ one can require that φ be almost everywhere equal to the characteristic functional of a probability measure. Another equivalent condition: uniform convergence of $\widetilde{\mu}_n$ on compact sets, which will be seen from the corollary below.

(ii) The main implication of the Lévy theorem is not valid for signed measures. For example, one can find a sequence of integrable functions on the real line that is not norm bounded in $L^1(\mathbb{R})$, but has Fourier transforms uniformly convergent to zero (Exercise 1.7.21). It is clear that the sequence of measures with such densities is not weakly convergent.

In § 3.5(iii) we consider some quantitative bounds for weak convergence in terms of Fourier transforms.

Lévy's theorem can be further reinforced in terms of the property of equicontinuity.

1.6.9. COROLLARY. (i) *If a sequence of Borel measures on* $\mathbb{R}^d$ *converges weakly, then their Fourier transforms are equicontinuous at all points and converge uniformly on compact sets.*

(ii) *A family* $\mathcal{M}$ *of probability measures on* $\mathbb{R}^d$ *is uniformly tight precisely when the set of their Fourier transforms is equicontinuous at the origin (and then at all points).*

PROOF. (i) Let $\mu_n \Rightarrow \mu$ and $x_n \to x_0$ in $\mathbb{R}^d$. We show that $\widetilde{\mu_n}(x_n) \to \widetilde{\mu}(x_0)$. This implies the equicontinuity of the functions $\widetilde{\mu_n}$ at x_0 and uniform convergence on all compact sets. One can assume that $\|\mu_n\| \leqslant 1$, $\|\mu\| \leqslant 1$. Let $\varepsilon > 0$. By Theorem 1.4.11 there exists a cube K such that $|\mu_n|(\mathbb{R}^d\backslash K) < \varepsilon$ for all n. The complex functions $f_n(x) = \exp(i\langle x, x_n\rangle)$ converge to $f(x) = \exp(i\langle x, x_0\rangle)$ uniformly on K. Hence there exists n_1 such that $|f_n(x) - f(x)| < \varepsilon$ for all $x \in K$ and $n \geqslant n_1$. Increasing n_1, we obtain that the integrals of f with respect to the measures μ and μ_n with $n \geqslant n_1$ differ in absolute value by less than ε. At the same time the integrals of f_n and f with respect to the measures μ_n with $n \geqslant n_1$ differ in absolute value by less than 3ε, since $|f_n - f| < \varepsilon$ on K and the measure of the complement is less than ε. Hence $|\widetilde{\mu_n}(x_n) - \widetilde{\mu}(x_0)| < 4\varepsilon$.

(ii) Since we now deal with probability measures, by estimate (1.6.5) it suffices to prove the equicontinuity at the origin. If the family of functions $\widetilde{\mu}$ for $\mu \in \mathcal{M}$ is equicontinuous at the origin, then estimate (1.6.8) yields the uniform tightness of measures in $\mathcal{M}$.

Conversely, let a family of probability measures $\mathcal{M}$ be uniformly tight. If the set of their Fourier transforms is not equicontinuous at the origin, then one can find a sequence of measures $\mu_n \in \mathcal{M}$ and a sequence of vectors $x_n \to 0$ for which the numbers $\widetilde{\mu_n}(x_n)$ are separated from 1. Picking in $\{\mu_n\}$ a weakly convergent subsequence by Theorem 1.4.11, we arrive at a contradiction by assertion (i). □

Note that in (ii) the equicontinuity of the Fourier transforms of a tight family of measures also holds for signed measures (which follows from the proven assertion), but the converse is false. For example, the measures $\delta_{n+1/n} - \delta_n$ on the real line are not uniformly tight, but their Fourier transforms are equicontinuous at each point. One can also construct an example of a sequence of signed measures that is not uniformly tight, but possesses uniformly convergent to zero Fourier transforms (Exercise 1.7.21).

Let us prove one more assertion that enables one to verify the uniform tightness of measures by means of their Fourier transforms.

1.6.10. COROLLARY. *Let* $\{\mu_n\}$ *and* $\{\nu_n\}$ *be two sequences of probability measures on* $\mathbb{R}^d$ *such that the sequence of their convolutions* $\mu_n * \nu_n$ *is uniformly tight. If the measures* ν_n *have real Fourier transforms (which is equivalent to their symmetry), then both sequences* $\{\mu_n\}$ *and* $\{\nu_n\}$ *are uniformly tight.*

PROOF. Since $\widetilde{\nu_n}$ is real, the estimate $|\widetilde{\mu_n}\widetilde{\nu_n}(x) - 1| < \varepsilon$ yields the estimate $|\widetilde{\mu_n}(x) - 1| < \varepsilon$, which ensures the equicontinuity at the origin of the functions $\widetilde{\mu_n}$. This also yields the equicontinuity at the origin of the functions $\widetilde{\nu_n}$. Hence the sequences $\{\mu_n\}$ and $\{\nu_n\}$ are uniformly tight. The condition of symmetry of the measures ν_n is important: for example, the convolution of the Dirac measures $\mu_n = \delta_n$ and $\nu_n = \delta_{-n}$ is Dirac's measure at the origin. □

In § 4.8(iv) we return to this phenomenon in a more general situation.

Suppose that on $\mathbb{R}^d$ we are given a Borel probability measure μ with respect to which the square of the norm is integrable (then μ is called a measure with *finite second moment*). The norm is also integrable in this case, hence the coordinate functions are integrable as well, which enables us to introduce the *mean of the measure* μ by the vector equality

$$a_\mu := \int_{\mathbb{R}^d} x\,\mu(dx),$$

which is understood coordinate-wise: the ith component of the vector a_μ is the integral of x_i with respect to the measure μ. If we move the origin to a_μ, then we obtain a measure with zero mean. Next, the *covariance operator* R_μ of the measure μ is defined by the equality

$$\langle R_\mu u, v\rangle := \int_{\mathbb{R}^d} \langle u, x - a_\mu\rangle\langle v, x - a_\mu\rangle\,\mu(dx).$$

For measures with zero mean the quadratic form $\langle R_\mu u, u\rangle$ coincides with the integral of $\langle u, x\rangle^2$ with respect to the measure μ and hence is nonnegative.

A Gaussian measure is uniquely determined by its mean and covariance operator. For other measures, these two characteristics are also very important, they are among the most frequently used characteristics of distributions in theoretical and applied statistics.

The mean and covariance operator of a random element with values in $\mathbb{R}^d$ are the respective mean and covariance operator of its distribution. If ξ and η are two independent random elements in $\mathbb{R}^d$ with zero mean and covariance operators R_ξ and R_η, then the covariance operator of the vector $\xi + \eta$ also exists and equals $R_{\xi+\eta} = R_\xi + R_\eta$. This follows from the equality $\mathbb{E}\langle \xi, u\rangle\langle \eta, v\rangle = 0$, which holds due to independence of ξ and η and vanishing of their means.

Note also the equality

$$\int_{\mathbb{R}^d} |x|^2\,\mu(dx) = \sum_{i=1}^d \langle R_\mu e_i, e_i\rangle = \operatorname{tr} R_\mu,$$

which along with the Chebyshev inequality yields the uniform tightness of the distributions μ_n of random vectors ξ_n with zero means and uniformly bounded traces of their covariance operators: $\mu_n(x\colon\ |x| \geqslant R) \leqslant R^{-2}\operatorname{tr} R_{\xi_n}$.

These facts play an important role in the study of sums of random vectors.

1.6.11. Example. (*The central limit theorem*) Suppose we are given a sequence $\{\xi_n\}$ of independent random vectors in $\mathbb{R}^d$ with identical distributions with zero mean and a finite second moment. Let us consider the random vectors

$$S_n = \alpha_{n,1}\xi_1 + \cdots + \alpha_{n,n}\xi_n,$$

where $\alpha_{n,k}$ are real numbers such that $\alpha_{n,1}^2 + \cdots + \alpha_{n,n}^2 \leqslant C$ for some number C. Then the sequence of distributions μ_n of the vectors S_n is uniformly tight and possesses a weakly convergent subsequence. In the special case

$$S_n = \frac{\xi_1 + \cdots + \xi_n}{\sqrt{n}}$$

the central limit theorem holds: the measures μ_n converge weakly to the Gaussian measure with zero mean and the covariance operator equal to R_{ξ_1}. Indeed, it follows from our discussion above that

$$R_{S_n} = \alpha_{n,1}^2 R_{\xi_1} + \cdots + \alpha_{n,n}^2 R_{\xi_n} = (\alpha_{n,1}^2 + \cdots + \alpha_{n,n}^2) R_{\xi_1} \leqslant C R_{\xi_1},$$

which yields that $\operatorname{tr} R_{\mu_n} \leqslant C \operatorname{tr} R_{\xi_1}$, i.e., the integrals of $|x|^2$ with respect to the measures μ_n are uniformly bounded. Thus, the measures μ_n are uniformly tight, hence $\{\mu_n\}$ contains a weakly convergent subsequence. In case of the central limit theorem all such subsequences have one and the same limit, which proves that the whole original sequence converges to it. In order to verify the coincidence of limits for all convergent subsequences it suffices to establish the central limit theorem in the one-dimensional case, since two measures on $\mathbb{R}^d$ with equal one-dimensional images under all functionals $l_u\colon x \mapsto \langle u, x\rangle$ coincide. Certainly, one can declare the central limit theorem in the one-dimensional case a known fact, but actually it is not difficult to derive it with the aid of the Fourier transform. One can assume that ξ_1 has variance 1. The distribution of $(\xi_1 + \cdots + \xi_n)/\sqrt{n}$ equals the n-fold convolution of the distribution of $\xi_1/\sqrt{n}$, hence its Fourier transform coincides with the nth power of the Fourier transform of the distribution of $\xi_1/\sqrt{n}$, i.e.,

$$\widetilde{\mu}_n(t) = \varphi(t/\sqrt{n})^n, \quad \varphi(t) = \mathbb{E}\, e^{it\xi_1}.$$

Our goal is to show that the limit of $\widetilde{\mu}_n(t)$ equals $e^{-t^2/2}$. To this end it suffices to verify that $\varphi(t/\sqrt{n}) = 1 - t^2/(2n) + o(1/n)$ for every fixed t. It remains to use Taylor's expansion for the twice differentiable function φ at zero and take into account that its first derivative at zero vanishes, because the expectation is zero.

There are much more refined versions of the central limit theorem where the distributions of ξ_n can be different, in addition, condition of independence can be relaxed. Certainly, some additional conditions are needed. For example, in case of independent random variables ξ_n with different distributions and variances $\sigma_n^2 = \mathbb{E}\,\xi_n^2$, it is usual to consider the normalized sums $S_n = (\xi_1 + \cdots + \xi_n)/B_n$, where $B_n^2 = \sum_{i=1}^n \sigma_i^2$. Such normalizations fit the condition above that guarantees the uniform tightness of distributions and existence of weakly convergent subsequences. However, for weak convergence of the whole sequence $\{S_n\}$ to the standard normal distribution it is also necessary to impose the Lindeberg condition

$$\lim_{n\to\infty} \frac{1}{B_n^2} \sum_{i=1}^n \int_{|x|\geqslant \varepsilon B_n} x^2\, P_{\xi_i}(dx) = 0.$$

Under the condition of uniform smallness $\max_{i\leqslant n} \sigma_i/B_n \to 0$, the Lindeberg condition is necessary and sufficient for weak convergence to the standard normal distribution. Hence there exist examples where the uniform tightness of distributions of S_n holds, but there is no their weak convergence. If ξ_1 also has a third moment, then the Berry–Esseen theorem gives a rate of convergence: there is C such that for identically distributed independent ξ_n one has the inequality

$$\sup_t |F_{S_n}(t) - \Phi(t)| \leqslant C \frac{\mathbb{E}\,|\xi_1|^3}{\sigma_1^3 \sqrt{n}},$$

where Φ is the standard normal distribution function, see Bhattacharia, Ranga Rao [**64**]. The least possible constant C is unknown at present, but it is established that it lies between 0,4097 and 0,4784 (see Korolev, Shevtsova [**382**], [**383**], and Tyurin [**621**] for some recent advances).

If ξ_1 has a bounded distribution density, then S_n also has a bounded distribution density p_n and

$$\lim_{n\to\infty} \sup_t |p_n(t) - \varrho(t)| = 0,$$

where ϱ is the standard normal density.

However, the existence of the second moment is not necessary for weak convergence of the normalized sums with different normalizations. For example, the so-called stable distributions arise as weak limits of the sums $(\xi_1 + \cdots + \xi_n)n^{-1/\alpha}$ with $\alpha \in (0, 2]$, see § 1.7(ii).

1.7. Complements and exercises

(i) Convergence of distribution functions (38). (ii) Stable and infinitely divisible distributions (39). (iii) Convex measures (40). Exercises (42).

1.7(i). Convergence of distribution functions

Here we discuss connections of weak convergence with certain other modes of convergence of distribution functions different from pointwise convergence. The next result was obtained by A.N. Kolmogorov (see Glivenko [**280**, p. 157]).

1.7.1. THEOREM. *A sequence of Borel measures μ_n (possibly, signed) on the interval $[a, b]$ converges weakly to a Borel measure μ precisely when*

(i) *the variations of the measures μ_n are uniformly bounded,*

(ii) $\mu([a, b]) = \lim_{n\to\infty} \mu_n([a, b])$,

(iii) *for the corresponding distribution functions we have*

$$\lim_{n\to\infty} \int_a^b |F_{\mu_n}(t) - F_\mu(t)|\, dt = 0.$$

Furthermore, (iii) *can be replaced by the condition*

$$\|F_{\mu_n} - F_\mu\|_{L^p[a,b]} \to 0$$

with some (and then with every) $p \in [1, +\infty)$.

Similar assertions are true for the cube $[a, b]^d$ in $\mathbb{R}^d$ when the distribution functions are defined by

$$F_{\mu_n}(t_1, \ldots, t_d) := \mu_n\bigl([a, t_1)\times \cdots \times [a, t_d)\bigr), \quad (t_1, \ldots, t_d) \in [a, b]^d,$$

and similarly for μ.

PROOF. Weak convergence yields conditions (i) and (ii); condition (iii) follows from the uniform boundedness of the functions F_{μ_n} and Theorem 1.4.7. If condition (i) is fulfilled, then for weak convergence it suffices to verify convergence of the integrals of smooth functions, since by such functions one can uniformly approximate all continuous functions on a compact interval. By the integration by parts formula and the equality $F_\mu(b+) = \lim_{n\to\infty} F_{\mu_n}(b+)$ it remains to observe that the integral of $f'(F_\mu - F_{\mu_n})$ over $[a, b]$ tends to zero for every smooth function f. The equivalence of convergence in L^1 to convergence in L^p (and also to convergence in measure) for a uniformly bounded sequence is obvious.

Let us give another justification, which easily extends to the multidimensional case. Let $[a, b] = [0, 2\pi]$, $\varphi_k(t) = \exp(ikt)$, where $k \in \mathbb{Z}$. Set

$$f_{n,k} := (\varphi_k, F_{\mu_n})_{L^2[0,2\pi]}, \quad f_k := (\varphi_k, F_\mu)_{L^2[0,2\pi]}.$$

If we have $\sup_n \|\mu_n\| \leqslant C < \infty$, then the absolute value of the integral of φ_k with respect to the measure μ_n does not exceed C. By the integration by parts formula

$$ik \int_0^{2\pi} \varphi_k(t) F_{\mu_n}(t)\, dt = F_{\mu_n}(2\pi+) - \int_0^{2\pi} \varphi_k(t)\, dF_{\mu_n}(t)\, dt.$$

Hence $|kf_{n,k}| \leqslant 2C$. Thus, the sequence $\{F_{\mu_n}\}$ is precompact in $L^2[0, 2\pi]$, hence is contained in a compact set. This follows from the known and simply verified criterion of precompactness of a bounded set S in a Hilbert space with an orthonormal basis $\{e_k\}$: it is necessary and sufficient to have the equality

$$\lim_{N\to\infty} \sup_{s\in S} \sum_{k=N}^{\infty} |(s, e_k)|^2 = 0$$

(see Exercise 1.7.26). In our case this condition is fulfilled by convergence of the series of k^{-2}.

If the measures μ_n converge weakly to μ, then $f_{n,k} \to f_k$ for each k, which is clear from the integration by parts formula indicated above (for $k = 0$ we use the equality $t' = 1 = \varphi_0(t)$). This gives convergence of F_{μ_n} to F_μ in $L^2[0, 2\pi]$. Since $|F_{\mu_n}(t)| \leqslant C$, convergence in $L^2[0, 2\pi]$ is equivalent to convergence in every space $L^p[0, 2\pi]$ with $p < \infty$, and also to convergence in measure. In the case of a cube we take the basis formed by the products

$$\varphi_{k_1,\ldots,k_d}(t_1, \ldots, t_d) := \varphi_{k_1}(t_1) \cdots \varphi_{k_d}(t_d)$$

and estimate $(F_{\mu_n}, \varphi_{k_1,\ldots,k_d})_{L^2([0,1]^d)}$ by $const \cdot k_1^{-1} \cdots k_d^{-1}$. □

The next result, close to the previous theorem, was obtained for intervals by G.M. Fichtenholz, see Glivenko [**280**, p. 154]. For a measure μ on $\mathbb{R}^d$ the distribution function is defined by the formula

$$F_\mu(t_1, \ldots, t_d) := \mu_n\bigl((-\infty, t_1) \times \cdots \times (-\infty, t_d)\bigr).$$

1.7.2. THEOREM. *A sequence of bounded Borel measures μ_n on $\mathbb{R}^d$ converges weakly to a bounded Borel measure μ precisely when*

(i) *the sequence $\{\mu_n\}$ is uniformly bounded in variation and uniformly tight,*

(ii) *the sequence $\{F_{\mu_n}\}$ converges to F_μ in measure with respect to Lebesgue measure on every cube.*

PROOF. If $\mu_n \Rightarrow \mu$, then the sequence $\{\mu_n\}$ is uniformly tight, and the second proof of the previous theorem yields convergence of distribution functions in L^2 on cubes, which also gives convergence in measure.

Conversely, if the stated conditions are fulfilled, then every subsequence in $\{\mu_n\}$ contains a further subsequence that is weakly convergent. By condition (ii) the distribution functions of the limit measures coincide almost everywhere, which means that only one limit measure can exist, and the whole original sequence must converge to this measure. □

Since convergence in measure for equivalent bounded measures is the same, condition (ii) is equivalent to convergence in measure with respect to an arbitrary probability measure equivalent to Lebesgue measure.

1.7(ii). Infinitely divisible and stable distributions

In the central limit theorem the summation of independent random variables employs the normalization with the square root of the number of variables. However, other normalizations are possible that lead to interesting nondegenerate limit distributions. Such limit distributions arise in the so-called triangular arrays (schemes of series), where in place of a single sequence of independent random variables we have a sequence of series of independent (in each series with the number n) random variables $\xi_{n,1}, \xi_{n,2}, \ldots$, but in total they are not independent. Suppose that we have some numbers k_n and A_n and form the sums $Z_n = \xi_{n,1} + \cdots + \xi_{n,k_n} - A_n$. The class of limit distributions (called *infinitely divisible*) for such sums Z_n admits a rather constructive description in terms of Fourier transforms. An important special case arises if we still deal with a single original sequence of independent random variables ξ_n, but Z_n is given by the formula $Z_n = (\xi_1 + \cdots + \xi_n)/B_n - A_n$ with some numbers $B_n > 0$ and A_n. This case leads to a more narrow class of stable distributions. They are also described in terms of Fourier transforms, but admit a simple description in terms of convolutions.

A distribution μ on $\mathbb{R}$ is called *stable of order* $\alpha \in (0,2]$ if, for each $n \in \mathbb{N}$, there exists a number a_n such that the distribution of the random variable

$$n^{-1/\alpha}(\xi_1 + \cdots + \xi_n) - a_n,$$

where $\xi_1, \ldots, \xi_n$ are independent and have the same distribution μ, also coincides with μ. In terms of convolutions this can be written as

$$\mu(B) = \mu_n * \cdots * \mu_n(B + a_n),$$

where we have the n-fold convolution of the measure μ_n obtained by scaling μ: $\mu_n(B) = \mu(n^{1/\alpha}B)$. If $a_n = 0$, then this distribution is called strictly stable. It is interesting that for α only this interval of values is possible, and the case $\alpha = 2$ corresponds precisely to Gaussian distributions. The class of symmetric stable distributions is more narrow: these are measures with the Fourier transform of the form $\exp(-c|y|^\alpha)$, $c \geqslant 0$, $\alpha \in (0,2]$. About the theory of infinitely divisible and stable distributions see Feller [**221**], Gnedenko [**281**], Gnedenko, Kolmogorov [**282**], Ibragimov, Linnik [**331**], Kruglov [**393**], Kruglov, Korolev [**394**], Linde [**432**], Linnik, Ostrovskii [**434**], Lukacs [**446**], [**447**], Ramachandran [**538**], Steutel, van Harn [**595**], and Zolotarev [**672**], [**674**]. See also § 4.8(iii).

1.7(iii). Convex measures

A Borel probability measure on $\mathbb{R}^d$ is called convex (or logarithmically concave) if it is given by a density of the form e^{-V} with some convex function V (taking values in $(-\infty, +\infty]$) on all of $\mathbb{R}^d$ or on an affine subspace in $\mathbb{R}^d$. The density vanishes outside the effective domain $\{V < \infty\}$, which is a convex set (and V is locally Lipschitz in the interior of this set). Redefining V on the boundary of $\{V < \infty\}$ (which has measure zero) one can always assume that the sets $\{V \leqslant R\}$ are closed, i.e., V is lower semicontinuous (see Rockaffellar [**553**, Section 7]). For example, Lebesgue measure on $[0,1]$ regarded as a measure on the real line is convex as well as a Gaussian measure. For an equivalent characterization of convex measures, see Exercise 2.7.52.

The next theorem was proved in Medvedev [**454**], in a special case a similar result was proved in Meckes, Meckes [**453**].

1.7.3. THEOREM. *Suppose that convex measures μ_j on $\mathbb{R}^d$ converge weakly to an absolutely continuous convex measure μ. Then $\|\mu_j - \mu\| \to 0$ and the measures μ_j are absolutely continuous for all sufficiently large j.*

PROOF. The last assertion is clear from the fact that a singular convex measure is concentrated on a hyperplane. Let $\mu_j = e^{-V_j}\,dx$ and $\mu = e^{-V}\,dx$, where V_j and V are lower semicontinuous convex functions. It suffices to show that every sequence in $\{\mu_j\}$ contains a subsequence that converges in variation. It is known (see Rockaffellar [**553**, Theorem 10.9]) that if a sequence of convex functions on an open convex set C is bounded at every point of some dense set, then it has a subsequence convergent to some convex function uniformly on compact sets in C. Hence it suffices to show that every closed ball B in the interior U of the set $\{V < \infty\}$ contains a point at which some subsequence of $\{V_j\}$ is bounded: this will imply that every subsequence in $\{e^{-V_j}\}$ contains a further subsequence that converges uniformly on compact subsets in U, whence convergence in variation of the whole original sequence follows. Thus, we have to show that $\{V_j\}$ cannot tend on B to $+\infty$ or to $-\infty$. The latter is obvious, since by Fatou's theorem the integral of $\liminf_j e^{-V_j}$ does not exceed 1, so $\liminf_j e^{-V_j}$ cannot be infinite on a positive measure set. Hence it remains to exclude the case where $V_j(x) \to +\infty$ a.e. on B.

Let us consider the convex compact sets $W_j = B \cap \{W_j \leqslant \ln(1/\alpha)\}$, where $\alpha = \mu(B)/(2\lambda(B))$. For all j sufficiently large these sets are not empty, since otherwise by weak convergence we would have $\mu(B) \leqslant \alpha\lambda(B) = \mu(B)/2$. By the Blaschke selection theorem (see, e.g., Polovinkin, Balashov [**519**, Theorem 1.3.3]) the sequence of sets W_j contains a subsequence that converges in the sense of Hausdorff (see p. 123) to a convex compact set W. This result also follows from less elementary facts on compactness in the class BV of functions of bounded variation (see Evans, Gariepy [**209**, Chapter 5]). Thus, for each $\varepsilon > 0$ there is a number k_0 such that for all $k \geqslant k_0$ the sets W_{j_k} and W belong to the ε-neighborhoods of each other. Since $V_j(x) \to +\infty$ a.e. on B, we have $\lambda(W_j) \to 0$, hence $\lambda(W) = 0$. This means that W is contained in a hyperplane. Hence there is a ball B' in B disjoint with W_{j_k} for all k sufficiently large. On B' the functions $e^{-V_{j_k}}$ are uniformly bounded, which yields convergence of their integrals over B' to zero. So $\mu(B') = 0$, which is a contradiction because $B' \subset U$. □

Note that a weak limit of convex measures is always convex (see Exercise 2.7.52).

If K is a convex compact set in $\mathbb{R}^d$ of positive Lebesgue measure λ_d, then the normalized measure $\lambda_d|_K/\lambda_d(K)$ is convex. It turns out that any convex measure on $\mathbb{R}^d$ is a weak limit of projections to $\mathbb{R}^d$ of such uniform measures on spaces of large dimensions.

1.7.4. THEOREM. *Every convex measure μ on $\mathbb{R}^d$ is the limit of a weakly convergent sequence of projections on $\mathbb{R}^d$ of the uniform distributions of the indicated form on some convex compact sets K_n in $\mathbb{R}^n$, where $n \to \infty$.*

PROOF. It suffices to consider the case of a measure $\mu = e^{-V}\,dx$ with a finite convex function V on a convex compact set K. Then V is continuous on K, hence for $n > \max_K V$ the set

$$K_n := \big\{(x, y_1, \ldots, y_n)\colon\ x \in K,\ 0 \leqslant y_i \leqslant 1 - V(x)/n,\ i = 1, \ldots, n\big\} \subset \mathbb{R}^{d+n}$$

is compact and convex (the functions $y_i + V(x)/n$ are convex). Let $P_d\colon \mathbb{R}^{d+n} \to \mathbb{R}^d$ be the natural projection. Then $P_d(K_n) = K$ and the preimage of every point $x \in K$ is the cube $x \times [0, 1 - V(x)/n]^n$ of volume $\big(1 - V(x)/n\big)^n$. Hence the projection of the normalized uniform distribution on K_n is the measure with the density $\big(1 - V(x)/n\big)^n$ on K, and these measures converge in variation to the measure μ due to the uniform convergence of the indicated functions to $e^{-V(x)}$. □

Concerning weak convergence of convex measures, see also p. 129 and Exercise 2.7.52.

Exercises

1.7.5. Prove that in any separable metric space every subset is a separable space.

1.7.6. Prove that in any separable metric space every collection of open sets contains an at most countable subcollection with the same union.

HINT: take a countable everywhere dense set S in this union U and for every point $s \in S$ and every open ball $U(s,r) \subset U$ with a rational radius take one of the given open sets containing this ball if such exists.

1.7.7. Prove that in any metric space every closed set is a countable intersection of open sets, and every open set is a countable union of closed sets.

HINT: for a closed set F in a given space consider the sets $U_n = \{x\colon \operatorname{dist}(x, F) < 1/n\}$ (see p. 45), prove that they are open and that their intersection is F.

1.7.8. Let K be a compact metric space. Prove that the space $C(K)$ with its sup-norm is separable.

HINT: prove that $C(K)$ is separable as part of $B(K)$, considering the countable family of functions with finitely many values of the following type: for every $n \in \mathbb{N}$, the space K is partitioned into finitely many parts of diameter at most $1/n$; next, take the countable set F_n of functions on K each of which is constant on every of these parts and assumes there a rational value; verify that each function in $C(K)$ can be approximated in $B(K)$ by functions from the union of all F_n.

1.7.9. Let X be a metric space. Prove that the space $C_b(X)$ is separable precisely when X is compact.

HINT: consider the case where there is a sequence of points with mutual distances separated from zero and take bounded continuous functions whose values at these points give arbitrary sequences of 0 and 1; consider also the case of a Cauchy sequence that has no limit (and construct similar functions).

1.7.10. Let X be a metric space and let Y be its subset with the induced metric. Prove that the Borel subsets of Y (as a separate space) have the form $B \cap Y$, where $B \in \mathcal{B}(X)$.

HINT: verify that the class of such sets is a σ-algebra and contains all sets open in Y.

1.7.11. Prove that the continuity of the distribution function F_μ of a signed Borel measure μ on the real line is equivalent to the absence of points of nonzero μ-measure.

1.7.12. Prove that a Borel measure μ on the real line is nonnegative precisely when its distribution function F_μ is increasing.

HINT: if $\mu([a,b)) \geqslant 0$ for every semi-interval, then $\mu(K) \geqslant 0$ for every compact set K.

1.7.13. Prove that every left-continuous function of bounded variation on the real line having zero limit at $-\infty$ can be written as the difference of two bounded increasing left-continuous functions having zero limits at $-\infty$.

1.7.14. Let μ and ν be two probability measures on a σ-algebra $\mathcal{A}$ such that for some $\alpha \in (0,1)$ one has $\|\alpha\mu - (1-\alpha)\nu\| = 1$. Prove that $\mu \perp \nu$.

HINT: let $\mu = f \cdot \sigma$, $\nu = g \cdot \sigma$, where $\sigma = (\mu+\nu)/2$; then the integral of $|\alpha f - (1-\alpha)g|$ with respect to the measure σ equals 1, which is only possible if $fg = 0$ σ-a.e., since the integral of $\alpha f + (1-\alpha)g$ is also 1.

1.7.15. (i) Prove that every weakly convergent sequence in l^1 also converges in norm. (ii) Prove Lemma 1.3.7. (iii) Prove that all weakly compact sets in l^1 are compact in norm.

HINT: (i) consider a sequence that converges weakly to zero; arguing from the opposite, construct a functional on l^1 (an element of l^∞) that does not tend to zero on this sequence; (ii) use the weak sequential completeness of l^1 (see § 1.3) and (i); (iii) apply Theorem 1.3.8 and (i).

1.7.16. Construct an example of a sequence of probability measures μ_n on the interval $[0, 1]$ given by smooth uniformly bounded densities ϱ_n with respect to Lebesgue measure and converging weakly to a measure μ with a smooth density ϱ such that the functions ϱ_n do not converge in measure.

HINT: consider $\varrho_n(x) = 1 + \sin(2\pi n x)$ and $\varrho(x) = 1$.

1.7.17. (G. Pólya) Let μ be a probability measure and let f_n and f be measurable functions such that the measures $\mu \circ f_n^{-1}$ converge weakly to the measure $\mu \circ f^{-1}$ that has no atoms. Prove that the corresponding distribution functions converge uniformly.

1.7.18. (A. Khinchine) (i) Let F be a continuous distribution function on the real line such that $F(t) \equiv F(at + b)$ for some numbers a and b. Show that $a = 1$ and $b = 0$. Observe that for Dirac's measure at the origin one can take $b = 0$ and any $a > 0$.

(ii) Let F_n be distribution functions such that there exist continuous distribution functions G and H and numbers a_n, b_n, c_n, d_n for which

$$F_n(a_n t + b_n) \to G(t), \quad F_n(c_n t + d_n) \to H(t) \quad \forall t.$$

Prove that there exist finite limits $a = \lim_{n\to\infty} c_n/a_n$ and $b = \lim_{n\to\infty} (d_n - b_n)/a_n$ and there holds the identity $H(t) = G(at + b)$.

HINT: (i) consider a point $t \neq at + b$ such that F is not constant on any $(t, t + 1/k)$ and $(t - 1/k, t)$; see Bovier [**118**, p. 7].

1.7.19. Let μ_n be Borel measures on $\mathbb{R}^d$ with $\sup_n \|\mu_n\| < \infty$ such that there exists a bounded Borel measure μ such that the characteristic functionals $\widetilde{\mu}_n$ of the measures μ_n converge pointwise to the characteristic functional $\widetilde{\mu}$ of the measure μ. Prove that

$$\lim_{n\to\infty} \int_{\mathbb{R}^d} \varphi \, d\mu_n = \int_{\mathbb{R}^d} \varphi \, d\mu$$

for every continuous function φ with compact support. Show that in this situation weak convergence can fail.

HINT: it suffices to prove this equality for $\varphi \in C_0^\infty(\mathbb{R}^d)$, then it remains to observe that the Fourier transform $\widehat{\varphi}$ of the function φ is integrable and by the Lebesgue dominated convergence theorem

$$\int_{\mathbb{R}^d} \varphi \, d\mu_n = (2\pi)^{d/2} \int_{\mathbb{R}^d} \widetilde{\mu_n}(y)\widehat{\varphi}(y) \, dy \to (2\pi)^{d/2} \int_{\mathbb{R}^d} \widetilde{\mu}(y)\widehat{\varphi}(y) \, dy.$$

1.7.20. Let μ be a probability measure and let f_n and f be μ-integrable functions such that the measures $\mu \circ f_n^{-1}$ converge weakly to the measure $\mu \circ f^{-1}$. Show that if the sequence $\{f_n\}$ is uniformly μ-integrable, then $\int f_n \, d\mu_n \to \int f \, d\mu$.

HINT: for $\varepsilon > 0$ find $C > 0$ with

$$\int_{\{|f_n| \geqslant C\}} |f_n| \, d\mu < \varepsilon/3, \qquad \int_{\{|f| \geqslant C\}} |f| \, d\mu < \varepsilon/3,$$

set $\varphi(t) = \operatorname{sign}(t) \min(|t|, C)$, take N such that for all $n \geqslant N$ the integrals of $\varphi \circ f_n$ and $\varphi \circ f$ differ by at most $\varepsilon/3$, and observe that

$$\int f_n \, d\mu_n - \int f \, d\mu = \int t \, \mu \circ f_n^{-1}(dt) - \int t \, \mu \circ f^{-1}(dt),$$

$$\int_{|t| \geqslant C} |t| \, \mu \circ f_n^{-1}(dt) < \varepsilon/3, \qquad \int_{|t| \geqslant C} |t| \, \mu \circ f^{-1}(dt) < \varepsilon/3.$$

1.7.21. Construct a sequence of integrable functions on the real line that is not norm bounded in $L^1(\mathbb{R})$, but has Fourier transforms uniformly convergent to zero. Construct also an example of a bounded sequence of signed measures that are not uniformly tight, but have Fourier transforms uniformly convergent to zero.

HINT: observe that there is a uniformly bounded sequence of smooth functions f_j with support in $[-1, 1]$ pointwise convergent to $f = I_{[-1,1]}$; the sequence of their Fourier transforms $\widehat{f_j}$ cannot be bounded in L^1, since $\widehat{f}$ is not integrable, so the desired sequence can be found in the form $g_j = c_j\widehat{f_j}$ with c_j tending to zero sufficiently slowly. For the second example use the same method, but pick numbers c_j to obtain functions g_j with unit norm in L^1 and then consider $g_j(x - a_j)$ with a_j increasing to infinity.

1.7.22. (Visintin [**650**]) If μ is a probability measure and functions f_n converge to f weakly in $L^1(\mu)$ and on the range of all these functions there is a strictly convex function Φ (i.e., its graph contains no intervals) such that the integrals of $\Phi(f_n)$ converge to the integral of $\Phi(f)$, then $f_n \to f$ in norm in $L^1(\mu)$. Useful special cases: $\Phi(t) = |t|^p$ with $p > 1$ and $\Phi(t) = t\ln t$, $t \geqslant 0$. For other results of this type, see Reshetnyak [**548**], Giaquinta, Modica, Souček [**265**].

1.7.23. Construct a probability measure ν on $\mathbb{R}^d$ with $\widetilde{\nu} \in C_0^\infty(\mathbb{R}^d)$ and two probability measures $\mu_1 \neq \mu_2$ such that $\mu_1 * \nu = \mu_2 * \nu$.

HINT: see Sasvári [**562**, Chapter 3], Ushakov [**624**, p. 266].

1.7.24. Suppose that we are given two sequences of Borel probability measures $\{\mu_n\}$ and $\{\nu_n\}$ on $\mathbb{R}^d$ such that $\widetilde{\mu_n}(x) - \widetilde{\nu_n}(x) \to 0$ for each $x \in \mathbb{R}^d$ and one of the two sequences is uniformly tight. Prove that both sequences are uniformly tight and $\mu_n - \nu_n \Rightarrow 0$.

HINT: it suffices to show that $\mu_n - \nu_n \Rightarrow 0$, for which it suffices to verify that every subsequence in $\mu_n - \nu_n$ has a further subsequence with this property; if $\{\mu_n\}$ is tight, then there exists a subsequence $\{\mu_{n_k}\}$ that is weakly convergent to a probability measure μ, which gives convergence of $\widetilde{\mu_{n_k}}$ to $\widetilde{\mu}$, hence also weak convergence of ν_{n_k} to μ.

1.7.25. Suppose that we are given a sequence of measurable spaces $(X_n, \mathcal{A}_n)$ and for every n on $\mathcal{A}_n$ we are given two probability measures P_n and Q_n. The sequences $\{Q_n\}$ and $\{P_n\}$ are called contiguous (or mutually contiguous) if for all $A_n \in \mathcal{A}_n$ the condition $P_n(A_n) \to 0$ is equivalent to the condition $Q_n(A_n) \to 0$. Set

$$\lambda_n = (P_n + Q_n)/2, \quad f_n = dP_n/d\lambda_n, \quad g_n = dQ_n/d\lambda_n,$$

and also $\Lambda_n = \ln(g_n/f_n)$ if $f_n g_n > 0$ and $\Lambda_n = 0$ otherwise. Prove that the following conditions are equivalent:

(i) $\{Q_n\}$ and $\{P_n\}$ are contiguous,

(ii) the sequence $\{P_n \circ \Lambda_n^{-1}\}$ is uniformly tight on the real line,

(iii) the sequence $\{Q_n \circ \Lambda_n^{-1}\}$ is uniformly tight on the real line.

HINT: see Roussas [**557**, Chapter 1].

1.7.26. (i) Prove that a bounded set S in a Hilbert space with an orthonormal basis $\{e_k\}$ is contained in a compact set precisely when $\lim\limits_{N\to\infty} \sup_{s\in S} \sum_{k=N}^\infty |(s, e_k)|^2 = 0$.

(ii) Prove that a bounded set S in a Hilbert space with an orthonormal basis $\{e_k\}$ is contained in a compact set precisely when it is contained in a compact ellipsoid of the form

$$W = \bigl\{x\colon\ \textstyle\sum_{k=1}^\infty C_k|(x, e_k)|^2 \leqslant 1\bigr\}, \quad \text{where } C_k > 0,\ \lim\limits_{k\to\infty} C_k = \infty.$$

HINT: observe that the condition in (i) implies that S is totally bounded. Conversely, it holds for finite sets, hence for totally bounded sets. The ellipsoid in (ii) satisfies the condition in (i) if $\lim\limits_{k\to\infty} C_k = \infty$. Conversely, if S is totally bounded, then by (i) there are increasing numbers N_k such that $\sum_{n=N_k}^\infty |(s, e_n)|^2 \leqslant 4^{-k}$ for all k. Then we can take $C_n = 2^k$ if $N_k \leqslant n < N_{k+1}$.

CHAPTER 2

Convergence of measures on metric spaces

In this chapter the basic facts related to weak convergence of measures that we discussed in Chapter 1 are extended to the case of general metric spaces. The most important results of this chapter are Alexandroff's theorem about conditions for weak convergence of probability measures and Prohorov's theorem on conditions for weak compactness. In § 2.1 we present some basic facts about Borel measures on metric spaces.

2.1. Measures on metric spaces

Here we present some basic information about measures on metric spaces. Although this chapter deals mainly with complete separable spaces, the most fundamental concepts and facts are discussed in the general case.

Let (X, d) be a metric space (see Chapter 1). For every set $B \subset X$ its open ε-neighborhood is defined by the formula

$$B^{\varepsilon} := \{x\colon \operatorname{dist}(x, B) < \varepsilon\},$$

where $\operatorname{dist}(x, B)$ is the distance from the point to the set defined by the formula

$$\operatorname{dist}(x, B) := \inf\{d(x, b)\colon\ b \in B\}.$$

Complete separable metric space are usually called *Polish* spaces. Their topological structure is given by the following theorem (see Engelking [**203**, Theorem 4.2.10, Theorem 4.3.24, Corollary 4.3.25]). Recall that a countable intersection of open sets is called a G_δ-set.

2.1.1. THEOREM. *Each Polish space is homeomorphic to a G_δ-set in $[0,1]^\infty$ and also to a closed subset in $\mathbb{R}^\infty$.*

Recall that for a space X the symbol X^∞ denotes the countable power of X, which is a special case of the product $\prod_{n=1}^\infty X_n$ of metric spaces (X_n, d_n), i.e., the space of infinite sequences $x = (x_n)$ with $x_n \in X_n$ equipped with the metric $\varrho(x,y) = \sum_{n=1}^\infty 2^{-n} \min\bigl(d_n(x_n, y_n), 1\bigr)$. If all X_n are complete and separable, then so is their product; for compact X_n the product is compact as well.

We shall also need the following classical results (see, for example, Bogachev, Smolyanov [**96**, § 1.9], Engelking [**203**]).

2.1.2. THEOREM. (i) (*The Urysohn theorem*) *For every nonempty closed set F in a metric space X and every open neighborhood W of F, there exists a function $f \in C_b(X)$ such that $f = 1$ on F, $f = 0$ outside W and $0 \leqslant f \leqslant 1$.*

(ii) (*The Tietze–Urysohn theorem*) *Every bounded continuous function on a closed set in a metric space extends to a continuous function on the whole space with the same sup-norm.*

For the study of weak convergence of measures and weak topology (especially for general topological spaces) it is useful to employ the concept of a net, generalizing that of a countable sequence.

A nonempty set T is called *directed* if it is partially ordered (some pairs are in relation $t \leqslant s$ such that if $s \leqslant t$ and $t \leqslant u$, then $s \leqslant u$, in addition, $t \leqslant t$ for all t), and for every pair $s, t \in T$ there exists $\tau \in T$ with $s \leqslant \tau$ and $t \leqslant \tau$. We write $t \geqslant s$ to express that $s \leqslant t$. We write $s < t$ if $s \leqslant t$ and $s \neq t$. For example, the set of all open neighborhoods of a point in a metric space is partially ordered by the inverse inclusion and is directed (although not linearly ordered, i.e., not all elements are comparable).

A *net* $\{x_t\}_{t\in T}$ in a set is a mapping from a directed set T to this set. A particular case is a usual sequence.

In a somewhat peculiar way (as compared to sequences) one introduces the notion of a subnet $\{y_s\}_{s\in S}$ of a net $\{x_t\}_{t\in T}$: it is required that there is a mapping $F\colon S \to T$ such that $y_s = x_{F(s)}$ for all $s \in S$ and, for each $t \in T$, there is $s_t \in S$ such that $F(s) > t$ whenever $s > s_t$ (certainly, a subsequence in a sequence satisfies this condition).

For example, in the countable net $\mathbb{Z}$ of integer numbers (indexed by the same set $\mathbb{Z}$ with its usual order) the subset of negative numbers is not a subnet, but the subset of natural numbers is a subnet. A subnet of a countable sequence can be uncountable.

Not every countable net is isomorphic to a sequence indexed by natural numbers with its usual order (say, it can occur in a countable net that for every index there are infinitely many smaller indices).

Convergence of a net $\{x_t\}_{t\in T}$ in a metric space to a point x is defined as follows: for every neighborhood U of the point x, there exists an index τ such that $x_t \in U$ for all $t \geqslant \tau$.

Unlike convergence of a sequence this does not mean that outside U there are only finitely many points of the net. For example, if $\mathbb{Z}$ is the set of integer numbers with its usual order, then the net defined by $x_t = 2$ if $t < 0$, $x_t = 0$ if $t \geqslant 0$ converges to zero, although outside the 1-neighborhood of zero there are infinitely many its elements.

For a net of numbers c_α the definition of superior and inferior limits $\limsup_\alpha c_\alpha$ and $\liminf_\alpha c_\alpha$ is somewhat different from the case of sequences (although the formal definition is the same). Namely: $\limsup_\alpha c_\alpha$ is the exact upper bound of numbers c such that, for every $\alpha_0 \in \Lambda$, there exists $\alpha > \alpha_0$ with $c_\alpha \geqslant c$. Similarly one defines the inferior limit of the net $\{c_\alpha\}$ by $\liminf_\alpha c_\alpha$ (one can also set $\liminf_\alpha c_\alpha = -\limsup_\alpha(-c_\alpha)$).

The difference (even for countable nets) as compared to sequences is that, for example, it can happen that there are infinitely many numbers c_α for which $c_\alpha > \limsup_\alpha c_\alpha + 1$ (since there can be infinitely many indices α smaller than α_0).

In this chapter employment of nets is not crucial, so everywhere where they are mentioned one can assume that we talk about usual sequences. On the other hand, no additional complications arise in the proofs of those facts where nets are involved, which seems to be a sufficient justification for their consideration already here and not only in Chapters 4 and 5, where involvement of nets becomes crucial when dealing with convergence in topologies.

Exactly as in the case of $\mathbb{R}^d$ and also in the case of a general topological space, the *Borel σ-algebra* $\mathcal{B}(X)$ of a space X is the smallest σ-algebra containing all open sets, i.e., the intersection of all such σ-algebras. It is clear that this is the smallest σ-algebra containing all closed sets.

2.1.3. LEMMA. *In every metric space the Borel σ-algebra coincides with the smallest σ-algebra making measurable all continuous functions.*

PROOF. The latter σ-algebra is part of the Borel one, since the set $f^{-1}(-\infty, c)$ is open for $f \in C(X)$. On the other hand, it contains every nonempty closed set Z, since $Z = f^{-1}(0)$, where $f(x) = \operatorname{dist}(x, Z)$. □

It is clear from our discussion in § 1.1 that in case of a separable space X the Borel σ-algebra is generated by any countable collection of open balls with rational radii and centers at points of an arbitrary countable everywhere dense set. There are nonseparable spaces in which the balls generate the Borel σ-algebra (Exercise 2.7.26), but in the general case this is false. For example, if on the real line we define all distances between different points to be equal to 1, then the balls will be the single points or the whole real line. Hence the σ-algebra generated by them consists of at most countable sets and their complements. It does not contain intervals (as any subset of the real line with the indicated metric, any interval is both open and closed).

Yet another σ-algebra competing with the Borel one is generated by compact sets. This σ-algebra $\mathcal{K}(X)$ is often much smaller than the Borel one. We recall that a set is called nowhere dense if every open ball contains an open ball not intersecting this set. *First category* sets are countable unions of nowhere dense sets.

2.1.4. EXAMPLE. Let X be an infinite-dimensional Banach space. Then $\mathcal{K}(X) \neq \mathcal{B}(X)$. Indeed, since the balls in X are noncompact, every compact set is nowhere dense, hence is a first category set. Hence $\mathcal{K}(X)$ is contained in the σ-algebra generated by first category sets, which has a simple explicit description (Exercise 2.7.28): it consists of first category sets and their complements. By Baire's theorem a ball of positive radius in X is not a first category set and the same is true for its complement.

2.1.5. REMARK. Note that all compact sets belong to the σ-algebra $\mathcal{B}_S(X)$ generated by balls. Indeed, any compact set K can be covered by finitely many open (or closed) balls of radius $1/n$. The intersection of the obtained sequence of sets belongs to $\mathcal{B}_S(X)$ and is precisely K.

By using countable covers by balls we prove similarly that $\mathcal{B}_S(X)$ contains all separable closed sets. Moreover, all separable Borel sets belong to $\mathcal{B}_S(X)$. Indeed, for every set A in this class, its closure Z is also separable, hence belongs to $\mathcal{B}_S(X)$ as shown above. It remains to observe that $A \in \mathcal{B}(Z)$ and that in the separable space Z the balls (that are the intersections of Z with balls in X with centers in Z) generate the Borel σ-algebra.

For the reader's convenience we repeat the universal definitions of Borel and Radon measures.

2.1.6. DEFINITION. A bounded real measure (possibly, signed) on the Borel σ-algebra is called a Borel measure.

A Borel measure μ is called Radon if, for every Borel set B and every $\varepsilon > 0$, there exists a compact set $K_\varepsilon \subset B$ such that $|\mu|(B \backslash K_\varepsilon) < \varepsilon$.

Unlike the case of a general topological space discussed in Chapter 4, for Borel measures on metric spaces the Radon property follows from the following property of tightness (already encountered above).

2.1.7. DEFINITION. A Borel measure μ is called tight if, for every $\varepsilon > 0$, there exists a compact set K_ε such that $|\mu|(X\backslash K_\varepsilon) < \varepsilon$.

2.1.8. PROPOSITION. *Every Borel measure μ on a metric space (X, d) is regular in the following sense: for every $B \in \mathcal{B}(X)$ and every $\varepsilon > 0$, there exists a closed set $Z \subset B$ and an open set $U \supset B$ with $|\mu|(U\backslash Z) < \varepsilon$. Therefore, a tight Borel measure on X is Radon.*

PROOF. One can assume that $\mu \geqslant 0$. Let $\mathcal{E}$ be the class of all sets $B \in \mathcal{B}(X)$ for which for each $\varepsilon > 0$ there exists a closed set Z and an open set U with $Z \subset B \subset U$ and $\mu(U\backslash Z) < \varepsilon$. The closed sets belong to $\mathcal{E}$, since for any closed B one can take $Z = B$ and the open set

$$U_n := \{x\colon \operatorname{dist}(x, B) < 1/n\}$$

for some n, because the sets U_n decrease to B. The class $\mathcal{E}$ is obviously closed with respect to complementation. In addition, $B = \bigcup_{n=1}^\infty B_n \in \mathcal{E}$ if $B_n \in \mathcal{E}$. Indeed, for any given $\varepsilon > 0$ we can find an open set $U_n \supset B_n$ with $\mu(U_n\backslash B_n) < \varepsilon 4^{-n}$ and take $U = \bigcup_{n=1}^\infty U_n$. Then we find N such that $\mu\bigl(B\backslash \bigcup_{n=1}^N B_n\bigr) < \varepsilon/4$ and take closed sets $Z_n \subset B_n$ with $\mu(B_n\backslash Z_n) < \varepsilon 4^{-n}$. For the closed set $Z = \bigcup_{n=1}^N Z_n \subset B$ we have $\mu(U\backslash Z) < \varepsilon$. Hence $\mathcal{E}$ is a σ-algebra containing all closed sets, which means that $\mathcal{E} = \mathcal{B}(X)$. □

However, the regularity is not equivalent to the Radon property, as is seen from Example 1.1.2, where for X a Lebesgue nonmeasurable set in $[0, 1]$ is taken with zero inner measure and outer measure 1, and a Borel measure μ on X is defined by the formula $\mu(B \cap X) = \lambda(B)$.

Such an unpleasant phenomenon cannot happen in a complete separable space, which is asserted in the following theorem of Ulam that plays an important role in measure theory.

2.1.9. THEOREM. *On a complete separable metric space all Borel measures are Radon.*

PROOF. Let μ be a nonnegative Borel measure on a complete separable metric space X. We already know that this measure is regular. By using the separability and completeness of X, we establish tightness of μ. Let $\varepsilon > 0$ and let $\{x_n\}$ be a countable everywhere dense set in X. For every $k \in \mathbb{N}$, the union of the open balls $U(x_n, 1/k)$ of radius $1/k$ is the whole space X. Hence there exists a number N_k such that $\mu\bigl(X\backslash \bigcup_{n=1}^{N_k} U(x_n, 1/k)\bigr) < \varepsilon 2^{-k}$. Set $B := \bigcap_{k=1}^\infty \bigcup_{n=1}^{N_k} U(x_n, 1/k)$. Then we have $\mu(X\backslash B) \leqslant \varepsilon$. The set B is totally bounded, since for every k it is covered by finitely many balls of radius $1/k$. Hence B has compact closure (by completeness of X) with the desired property. □

The question of whether one can omit the separability of the space in this theorem cannot be answered without involving additional set-theoretic axioms (see Bogachev [**81**, § 1.12(x), § 7.2]).

One more regularity property of measures is τ-additivity.

2.1.10. DEFINITION. A Borel measure $\mu \geqslant 0$ is called τ-additive if every collection of open sets U_α contains a countable part $\{U_{\alpha_n}\}$ with union of the same measure as the whole union $\bigcup_\alpha U_\alpha$. A signed measure μ is τ-additive if so is its total variation $|\mu|$.

Since in a separable metric space every collection of open sets contains a countable part with the same union, here all Borel measures are τ-additive. It is also obvious that for the τ-additivity it suffices that the measure be concentrated on a separable part of the space. Actually, for metric spaces, the τ-additivity is equivalent to the separability of the topological support of the measure: for every n, one can cover the topological support by balls of radius $1/n$ and pick a countable collection of full measure, then intersect the obtained countable unions, which gives a separable set of full measure.

Let us give a simple consequence of the regularity of Borel measures.

2.1.11. LEMMA. *Let μ be a Borel measure on a metric space X. Then, for every open set $U \subset X$, one has*

$$|\mu|(U) = \sup\biggl\{\int_X f\,d\mu\colon\ f \in C_b(X), |f| \leqslant 1,\ f|_{X\backslash U} = 0\biggr\}.$$

PROOF. Let $\varepsilon > 0$. Let $X = X^+ \cup X^-$ be the Hahn decomposition for the measure μ. By the regularity of the measures μ^+ and μ^- we have closed sets $Z_1 \subset U \cap X^+$ and $Z_2 \subset U \cap X^-$ such that

$$\mu^+\bigl((U \cap X^+)\backslash Z_1\bigr) < \varepsilon, \quad \mu^-\bigl((U \cap X^-)\backslash Z_2\bigr) < \varepsilon.$$

Since $Z_1 \cap Z_2 = \varnothing$, according to Theorem 2.1.2 there is a continuous function $f\colon X \to [-1, 1]$ that equals 1 on Z_1, -1 on Z_2 and 0 outside U. It is readily seen that the integral of this function is not less than $|\mu|(U) - 2\varepsilon$. □

For μ-measurable mappings from a space with a probability measure μ to a separable metric space E, i.e., mappings for which the preimages of Borel sets are μ-measurable, an analog of the classical *Egorov theorem* holds: if such mappings f_n converge almost everywhere to a mapping f, then, for every $\varepsilon > 0$, there exists a set Ω_ε of measure greater than $1 - \varepsilon$ on which convergence is uniform.

A set $A \subset X$ is called *universally measurable* if it is measurable with respect to every Borel measure μ on X, i.e., belongs to $\mathcal{B}(X)_\mu$.

Sometimes one has to extend a measure to the Borel σ-algebra from a more narrow σ-algebra. Below a sufficient condition for the existence of extensions is given, but an extension does not always exist.

2.1.12. EXAMPLE. Let us consider the σ-algebra $\mathcal{E}$ consisting of Borel subsets of the real line that are either first category sets (see p. 47) or their complements. On the sets of the first type the measure μ is defined by 0, and on the sets of the second type by 1. By Baire's theorem the whole real line is not a first category set, hence its measure equals 1. The obtained probability measure is obviously countably additive, but has no Borel extensions, since otherwise the extension would be concentrated on a countable union of compact sets without inner points (Exercise 2.7.29), but such a union is a first category set, so that the original measure equals zero on it.

Similarly we conclude that a probability measure μ defined on the σ-algebra $\mathcal{K}(X)$ generated by compact sets may fail to have Borel extensions.

2.1.13. EXAMPLE. On the Hilbert space $X = l^2$ consider the measure μ that is the restriction to $\mathcal{K}(X)$ of the measure with the σ-algebra $\mathcal{E}$ defined similarly to the previous example. For the same reason it has no Borel extensions.

It is not always possible to find a Borel extension of a measure on the σ-algebra $\mathcal{B}_S(X)$. For example, if the cardinality of X equals the least uncountable cardinal (under the continuum hypothesis this is the cardinality of the real line), then, by a theorem due to Ulam, on X with the discrete metric (all nonzero distances equal 1) there is no nonzero measure vanishing on all points; see Bogachev [**81**, Theorem 1.12.40]. As noted above, here $\mathcal{B}(X)$ is the class of all sets (all sets are open), and $\mathcal{B}_S(X)$ consists of at most countable sets and their complements. Hence the measure equal to 0 on the sets of the first type to 1 on the sets of the second type has no Borel extensions.

In many questions of measure theory (including extensions of measures) it is very useful to employ the concept of a *Souslin space*, which is a Hausdorff space that is the image of a complete separable metric space under a continuous mapping. In particular, all complete separable metric spaces are Souslin, but the latter class is much broader. For example, any Borel set in a complete separable metric space is a Souslin space, see Bogachev [**81**, § 6.6]. It is known that the image of a Borel set in a complete separable metric space under a continuous injective mapping to a metric space is Borel, but without the injectivity condition this is false (although the image is Souslin). In § 4.1 we mention some basic facts about Souslin spaces in relation to measures.

2.1.14. THEOREM. *Suppose that X is an arbitrary metric space and $\mathcal{A}$ is a sub-σ-algebra in $\mathcal{B}(X)$. Let μ be a probability measure on $\mathcal{A}$. For the existence of a Borel extension of μ either of the following conditions is sufficient:*

(i) *for every set $A \in \mathcal{A}$ and every $\varepsilon > 0$, there exists a compact set $K_\varepsilon \subset A$ such that $\mu^*(A \backslash K_\varepsilon) < \varepsilon$,*

(ii) *the space X is complete and separable or is Souslin and $\mathcal{A}$ is generated by a countable family of sets.*

For a proof, see Bogachev [**81**, § 9.8].

The concept of a conditional measure is also useful. Let X and Y be two Souslin spaces (say, complete separable metric spaces) and let μ be a Borel probability measure on their product $X \times Y$ such that μ_Y is its projection on Y. For every set $B \in \mathcal{B}(X \times Y)$, let

$$B^y := \{x \in X\colon\ (x,y) \in B\}, \quad y \in Y.$$

Then one can find for μ the so-called conditional measures μ^y on X (denoted also by $\mu(dx|y)$) with the following property: for every set $B \in \mathcal{B}(X \times Y)$, the function $y \mapsto \mu^y(B^y)$, measuring the μ^y-measure of the section of B, is measurable with respect to the measure μ_Y, and

$$\mu(B) = \int_Y \mu^y(B^y)\, \mu_Y(dy). \tag{2.1.1}$$

For bounded Borel functions on $X \times Y$ this gives the equality

$$\int_{X\times Y} f\, d\mu = \int_Y \int_X f(x,y)\, \mu^y(dx)\, \mu_Y(dy). \tag{2.1.2}$$

Conditional measures are uniquely defined μ_Y-almost everywhere.

In § 1.3 we mentioned the Riesz theorem that identifies Borel measures on a compact metric space with continuous linear functionals on the space of continuous functions. This theorem with the same formulation is false for noncompact spaces: for example, the zero functional on the space of continuous functions with compact support on the real line can be extended to a nonzero continuous functional on $C_b(\mathbb{R})$, which is not given by a measure on the real line (extending first to the subspace of functions having a limit at infinity by the value of this limit). However, the Riesz theorem has useful modifications that hold for arbitrary spaces. Let us give three such results (proofs can be found in Bogachev [**81**, Chapter 7]).

2.1.15. THEOREM. *Let X be a metric space.*

(i) *A continuous linear functional L on the Banach space $C_b(X)$ has the form*

$$L(f) = \int_X f\,d\mu$$

with a Borel measure μ precisely when $L(f_n)\to 0$ for every sequence of functions $f_n \in C_b(X)$ pointwise decreasing to zero (note that such a sequence is automatically uniformly bounded).

(ii) *A continuous linear functional L on the Banach space $C_b(X)$ has the indicated form with a Radon measure μ precisely when, for every $\varepsilon > 0$, there exists a compact set $K_\varepsilon \subset X$ such that $|L(f)| \leqslant \varepsilon \sup_{x\in X} |f(x)|$ for every function $f \in C_b(X)$ vanishing on K_ε.*

(iii) *A continuous linear functional L on the Banach space $C_b(X)$ has the indicated form with a τ-additive measure μ precisely when $L(f_\alpha) \to 0$ for every uniformly bounded net of functions $f_\alpha \in C_b(X)$ pointwise decreasing to zero.*

In all these cases $\|L\| = \|\mu\|$.

The latter equality in all these cases is verified by analogy with the case of the real line discussed in Chapter 1 (though, for non-Radon measures it is not enough to deal with compact sets, instead one considers closed sets, which requires to employ the Tietze–Urysohn theorem).

2.2. Definition and properties of weak convergence

In this section X is a metric space with a metric ϱ. In our discussion below there will be some difference between the case where X is complete and separable and the general case (certainly, the latter is more involved).

Throughout $\mathcal{M}(X)$ denotes the set of all Borel measures on X, $\mathcal{M}^+(X)$ the subset of nonnegative measures, $\mathcal{P}(X)$ the subset of probability measures. Let also $\mathcal{M}_r(X)$, $\mathcal{P}_r(X)$ and $\mathcal{M}_\tau(X)$, $\mathcal{P}_\tau(X)$ denote the corresponding sets of Radon measures and τ-additive measures; let $\mathcal{M}_r^+(X)$ and $\mathcal{M}_\tau^+(X)$ be the subsets of nonnegative measures in $\mathcal{M}_r(X)$ and $\mathcal{M}_\tau(X)$. For a complete separable space

$$\mathcal{M}(X) = \mathcal{M}_r(X) = \mathcal{M}_\tau(X),$$

for a merely separable space $\mathcal{M}(X) = \mathcal{M}_\tau(X)$.

2.2.1. DEFINITION. A net $\{\mu_\alpha\} \subset \mathcal{M}(X)$ is called weakly convergent to a measure $\mu \in \mathcal{M}(X)$ if, for every bounded continuous real function f on X, we have

$$\lim_\alpha \int_X f(x)\,\mu_\alpha(dx) = \int_X f(x)\,\mu(dx). \tag{2.2.1}$$

Notation: $\mu_\alpha \Rightarrow \mu$.

In particular, weak convergence of a sequence $\mu_n \Rightarrow \mu$ means, as in the previous chapter, that

$$\lim_{n\to\infty} \int_X f(x)\,\mu_n(dx) = \int_X f(x)\,\mu(dx) \quad \forall\, f \in C_b(X).$$

2.2.2. DEFINITION. We shall say that a sequence of Borel measures μ_n on a space X is weakly fundamental if, for every bounded continuous function f on X, the sequence of integrals $\displaystyle\int_X f\,d\mu_n$ is fundamental (hence converges).

We shall see below that every weakly fundamental sequence of Borel measures converges weakly to some Borel measure (see Theorem 2.3.9 and Theorem 5.1.10). However, it is false that a sequence of Radon measures converges weakly to a Radon measure. This is seen from Example 2.2.12 and also from the fact that on a separable metric space every Borel measure (possibly, non-Radon) is the limit of a sequence of measures with finite supports (Example 2.2.7). However, a weak limit μ of a sequence of τ-additive measures μ_n on a metric space is a τ-additive measure, since the closure S of the union of separable topological supports of the measures μ_n is separable and the measure μ is concentrated on it, i.e., the $|\mu|$-measure of the complement of S is zero. The latter follows from Lemma 2.1.11, since the integrals with respect to the measures μ_n of every bounded continuous function vanishing on S equal zero, hence the integral of such a function with respect to the measure μ equals zero as well.

2.2.3. EXAMPLE. (i) A net $\{x_\alpha\}$ of elements of a metric space X converges to an element $x \in X$ precisely when the Dirac measures δ_{x_α} converge weakly to δ_x (recall that $\delta_x(A) = 1$ if $x \in A$, $\delta_x(A) = 0$ if $x \notin A$).

Indeed, if $x_\alpha \to x$, then, for every function $f \in C_b(X)$ and every $\varepsilon > 0$, one can take a neighborhood U of the point x such that $|f(x) - f(y)| < \varepsilon$ for all $y \in U$. Next, one can take an index τ such that $x_\alpha \in U$ whenever $\tau \leqslant \alpha$, which yields the inequality $|f(x_\alpha) - f(x)| < \varepsilon$, but this is the desired estimate for the integrals with respect to the Dirac measures.

Conversely, if $\delta_{x_\alpha} \Rightarrow \delta_x$, then one has convergence of the integrals of the function $f(y) = \min\bigl(1, \varrho(y, x)\bigr)$, which gives convergence $\varrho(x_\alpha, x) \to 0$.

(ii) Let μ be a Borel measure on a separable Hilbert space X and let us set $P_n(x) = \sum_{i=1}^n (x, e_i)e_i$, where $\{e_n\}$ is an orthonormal basis. Then $\mu\circ P_n^{-1} \Rightarrow \mu$. This is obvious from the Lebesgue dominated convergence theorem, because we have $f\bigl(P_n(x)\bigr) \to f(x)$ for all continuous functions f.

(iii) A sequence of measures μ_n on the space $\mathbb{N}$ with its usual metric converges weakly to a measure μ precisely when we have $\|\mu - \mu_n\| \to 0$. This is seen from Corollary 1.3.7, since $C_b(\mathbb{N}) = l^\infty$.

As in the case of measures on $\mathbb{R}^d$, the following fact is true.

2.2.4. PROPOSITION. *Let $\mathcal{M} \subset \mathcal{M}(X)$ be a family of measures such that*

$$\sup_{\mu\in\mathcal{M}} \int_X f\,d\mu < \infty \quad \forall\, f \in C_b(X).$$

Then $\sup_{\mu\in\mathcal{M}} \|\mu\| < \infty$. In particular, every weakly convergent sequence of measures is bounded in variation.

PROOF. We apply the Banach–Steinhaus theorem and the fact that the variation of a Borel measure μ equals the norm of the functional it generates on $C_b(X)$ (see Theorem 2.1.15). □

Recall that a convergent net need not be bounded (unlike a sequence).

Let us give a version of Alexandroff's theorem (see Theorem 1.5.2) for metric spaces (its general version for topological spaces is given in Chapter 4 in Theorem 4.3.2). The reader may wish to consider sequences in place of nets.

2.2.5. THEOREM. *Suppose we are given a net (e.g., a sequence) of Borel probability measures $\{\mu_\alpha\}$ and a Borel probability measure μ on a metric space. Then following conditions are equivalent:*

(i) *the net $\{\mu_\alpha\}$ converges weakly to μ;*

(ii) *for every closed set F one has*

$$\limsup_{\alpha} \mu_\alpha(F) \leqslant \mu(F); \tag{2.2.2}$$

(iii) *for every open set U one has*

$$\liminf_{\alpha} \mu_\alpha(U) \geqslant \mu(U). \tag{2.2.3}$$

The same equivalence remains in force in case of nonnegative measures under the additional condition $\lim\limits_{\alpha} \mu_\alpha(X) = \mu(X)$.

The proof is the same as for measures on the real line, one should only use the fact that for every nonempty closed set F in the space X and its arbitrary open neighborhood W there exists a function $f \in C_b(X)$ such that $f = 1$ on F, $f = 0$ outside W and $0 \leqslant f \leqslant 1$. The case of nonnegative measures reduces to the case of probability measures by normalization (the case $\mu = 0$ is trivial).

A function f on a space X is called *lower semicontinuous* if all sets $\{f > c\}$ are open. If the sets $\{f < c\}$ are open, then f is called *upper semicontinuous*. For example, the indicator function of an open set is lower semicontinuous and the indicator function of a closed set is upper semicontinuous.

2.2.6. COROLLARY. *Suppose that a net of Borel probability measures μ_α on a metric space X converges weakly to a Borel probability measure μ. If f is a bounded upper semicontinuous function, then*

$$\limsup_{\alpha} \int_X f\, d\mu_\alpha \leqslant \int_X f\, d\mu.$$

If f is a bounded lower semicontinuous function, then

$$\liminf_{\alpha} \int_X f\, d\mu_\alpha \geqslant \int_X f\, d\mu.$$

On the other hand, for weak convergence of a net of measures $\mu_\alpha \in \mathcal{P}(X)$ to a measure $\mu \in \mathcal{P}(X)$ it suffices to have the identity

$$\int_X f\, d\mu = \lim_{\alpha} \int_X f\, d\mu_\alpha \quad \forall\, f \in \mathrm{Lip}_1(X) \cap C_b(X). \tag{2.2.4}$$

Finally, if (X, d) is separable (or we consider τ-additive measures), then for weak convergence of a sequence of measures $\mu_n \in \mathcal{P}(X)$ to a measure $\mu \in \mathcal{P}(X)$ it suffices to have this equality for functions of the form

$$f(x) = \max\bigl(c_0, c_1 - d(x, y_1), \dots, c_n - d(x, y_n)\bigr), \tag{2.2.5}$$

where $c_0, \ldots, c_n \in \mathbb{R}$ *(or* $c_i \in \mathbb{Q}$*),* $y_1, \ldots, y_n$ *belong to an everywhere dense set in* X*, which gives a countable family in case of separable* X*. If* X *is bounded, then it suffices to verify this equality for all polynomials in* $d(x, y_1), \ldots, d(x, y_n)$*, where* $y_1, \ldots, y_n \in X$*.*

PROOF. We can assume that $0 < f < 1$. For a fixed number n, let us set $U_k := \{x\colon f(x) > k/n\}$, $k = 1, \ldots, n$. For a lower semicontinuous function f the sets U_k are open. Hence for $f_n := n^{-1}\sum_{k=1}^n I_{U_k}$ we have

$$\liminf_{\alpha} \int_X f_n\, d\mu_\alpha \geqslant \int_X f_n\, d\mu.$$

It remains to observe that $|f(x) - f_n(x)| \leqslant n^{-1}$ for all $x \in X$. Indeed, if we have $m/n < f(x) \leqslant (m+1)/n$, where $m \geqslant 1$, then $I_{U_k}(x) = 1$ if $k \leqslant m$ and $I_{U_k}(x) = 0$ if $k > m$, whence we obtain $f_n(x) = m/n$. If $m = 0$, then $f_n(x) = 0$.

For the proof of weak convergence of probability measures under the condition of convergence of the integrals of bounded Lipschitz functions (following from convergence of the integrals of bounded functions of class $\mathrm{Lip}_1(X)$) it suffices to observe that, for every closed set F and every $\varepsilon > 0$, one can find a Lipschitz function f such that $0 \leqslant f \leqslant 1$, $f|_F = 1$ and the integral of f with respect to the measure μ is less than $\mu(F) + \varepsilon$. For such a function one can take

$$1 - \delta^{-1} \min\bigl(\mathrm{dist}(x, F), \delta\bigr),$$

where $\delta > 0$ is such that $\mu(F^\delta) < \mu(F) + \varepsilon$. Then, for some α_0, for all $\alpha \geqslant \alpha_0$ the integral of f with respect to the measure μ_α is also less than $\mu(F) + \varepsilon$, which gives the estimate $\mu_\alpha(F) \leqslant \mu(F) + \varepsilon$. Hence $\limsup_\alpha \mu_\alpha(F) \leqslant \mu(F)$, so assertion (ii) of the previous theorem applies. The second assertion is proven.

For the proof of the last assertion we take in X a countable everywhere dense set $\{y_i\}_{i=1}^\infty$; in the case of τ-additive measures in place of X one can consider the union of their topological supports. It is readily seen that for every bounded function $f \in \mathrm{Lip}_1(X)$ with $f \geqslant c$ we have the equality

$$f(x) = \sup_{j \geqslant 1} \max\bigl(c, f(y_j) - d(x, y_j)\bigr).$$

In case of a bounded space X one can take the sup of the functions $f(y_j) - d(x, y_j)$. Set

$$f_k(x) = \max_{j \leqslant k} \max\bigl(c, f(y_j) - d(x, y_j)\bigr).$$

By our assumption, for every $k \in \mathbb{N}$ we have

$$\lim_{n\to\infty} \int_X f_k(x)\, \mu_n(dx) = \int_X f_k(x)\, \mu(dx).$$

Since, as $k \to \infty$, the right-hand side tends to the integral of f with respect to the measure μ by the Lebesgue dominated convergence theorem and

$$\int_X f_k(x)\, \mu_n(dx) \leqslant \int_X f(x)\, \mu_n(dx),$$

we obtain the inequality

$$\int_X f(x)\, \mu(dx) \leqslant \liminf_{n\to\infty} \int_X f(x)\, \mu_n(dx).$$

Replacing the function f by $-f$, we obtain the opposite inequality, which completes the proof. In the case of bounded X every continuous function of variables

$d(x,y_1),\dots,d(x,y_n)$ can be uniformly approximated by polynomials in these variables, so it suffices to verify convergence of integrals of such polynomials. □

Exercise 2.7.37 extends the last assertion to the case of nets.

Note that weak convergence on the whole ball in $\mathcal{M}(\mathbb{N}) = l^1$ (unlike $\mathcal{P}(X)$ with separable X) cannot be determined by a countable family of functions f_j: otherwise for $\|f_j\|_\infty = 1$ the norm $p(\mu) = \sum_{j=1}^\infty 2^{-j}|\langle f_j,\mu\rangle|$ on l^1 would be equivalent to the standard norm by Exercise 1.7.15, which is impossible, as one can easily see (e.g., from the fact that p is approximated uniformly on the ball by finite sums of the series).

2.2.7. EXAMPLE. Every Borel measure μ on a separable metric space X is the limit of a weakly convergent sequence of finite linear combinations of the Dirac measures at points of a given countable everywhere dense set.

PROOF. Let $\{x_n\}$ be an everywhere dense sequence in X. It suffices to prove our assertion for a probability measure μ. For every $k\in\mathbb{N}$ one can cover X by a countable family of open balls $U_{k,j}$ of diameter less than $1/k$. In each ball $U_{k,j}$ we pick a point $x_{k,j}$ from the given sequence. For every fixed k consider the disjoint partition of X in countably many Borel parts $B_{k,1} = U_{k,1}$, $B_{k,2} = U_{k,2}\backslash U_{k,1}$, $B_{k,m} = U_{k,m}\backslash(U_{k,1}\cup\dots\cup U_{k,m-1})$ and so on of diameter less than $1/k$. Let us take the measures

$$\mu_k = \sum_{m=1}^\infty \mu(B_{k,m})\delta_{x_{k,m}},$$

which are concentrated on a countable set. We observe that these are probability measures and $\mu_k \Rightarrow \mu$. Since the measures μ_k and μ are nonnegative, due to the previous corollary it suffices to verify convergence of the integrals (2.2.4) for every bounded 1-Lipschitz function f. For such a function, its integral over the set $B_{k,m}$ with respect to the measure μ differs from $f(x_{k,m})\mu(B_{k,m})$ by at most $\mu(B_{k,m})/k$, because $|f(x) - f(x_{k,m})| \leqslant d(x,x_{k,m}) < 1/k$ for all $x\in B_{k,m}$ by the inclusions $B_{k,m}\subset U_{k,m}$ and $x_{k,m}\in U_{k,m}$. Hence the integrals of f with respect to the measures μ and μ_k differ by at most $1/k$. It remains to replace the measure μ_k by a finite sum ν_k of the series satisfying the estimate $\|\mu_k-\nu_k\| < 1/k$. □

The next theorem due to Ranga Rao [**541**] gives a practical sufficient condition for uniform convergence of the integrals with respect to weakly convergent measures (see also Exercise 2.7.33 and Exercise 2.7.34).

2.2.8. THEOREM. *Suppose that a net $\{\mu_\alpha\}$ (e.g., a sequence) of Borel probability measures on a separable metric space X converges weakly to a Borel probability measure μ. If $\Gamma\subset C_b(X)$ is a uniformly bounded and pointwise equicontinuous family of functions, i.e., for every x and $\varepsilon>0$ there exists a neighborhood U of x with $|f(x)-f(y)|<\varepsilon$ for all $y\in U$ and $f\in\Gamma$, then*

$$\lim_\alpha \sup_{f\in\Gamma}\biggl|\int_X f\,d\mu_\alpha - \int_X f\,d\mu\biggr| = 0. \tag{2.2.6}$$

In particular, for every weakly convergent sequence of Borel probability measures on a separable space (or Radon measures on an arbitrary metric space) we obtain uniform convergence of the integrals of functions satisfying the Lipschitz condition with a common constant and bounded in absolute value by a common constant.

PROOF. We can assume that $|f| \leqslant 1$ for all $f \in \Gamma$. Let $\varepsilon > 0$. Each point x has an open neighborhood U_x (one can take a ball) such that $\mu(\partial U_x) = 0$ and $|f(x) - f(y)| < \varepsilon$ for all $y \in U_x$ and $f \in \Gamma$. Since X is separable, some countable collection of sets U_{x_n} covers X. Let $V_n = U_{x_n} \backslash \bigcup_{i=1}^{n-1} V_i$, $V_1 = U_{x_1}$. It is readily seen that the pairwise disjoint sets V_n cover X and $\mu(\partial V_n) = 0$. Let $\nu = \sum_{n=1}^{\infty} \mu(V_n)\delta_{x_n}$, $\nu_\alpha = \sum_{n=1}^{\infty} \mu_\alpha(V_n)\delta_{x_n}$. We observe that

$$\limsup_{\alpha} \sup_{f\in\Gamma} \left| \int_X f\, d\nu_\alpha - \int_X f\, d\nu \right| \leqslant \lim_{\alpha} \sum_{n=1}^{\infty} |\mu_\alpha(V_n) - \mu(V_n)| = 0. \tag{2.2.7}$$

The last equality in (2.2.7) follows from the equality

$$\lim_{\alpha} \mu_\alpha(V_n) = \mu(V_n)$$

for every fixed index n (which holds by Theorem 2.2.5) and the obvious equality $\sum_{n=1}^{\infty} \mu_\alpha(V_n) = \sum_{n=1}^{\infty} \mu(V_n) = 1$. We observe that

$$\begin{aligned}
&\left| \int_X f\, d\mu_\alpha - \int_X f\, d\mu \right| \\
&\leqslant \left| \int_X f\, d\mu_\alpha - \int_X f\, d\nu_\alpha \right| + \left| \int_X f\, d\nu_\alpha - \int_X f\, d\nu \right| + \left| \int_X f\, d\nu - \int_X f\, d\mu \right| \\
&\leqslant \sum_{n=1}^{\infty} \int_{V_n} |f(x) - f(x_n)|\, (\mu_\alpha + \mu)(dx) + \left| \int_X f\, d\nu_\alpha - \int_X f\, d\nu \right| \\
&\leqslant 2\varepsilon + \left| \int_X f\, d\nu_\alpha - \int_X f\, d\nu \right|,
\end{aligned}$$

because $|f(x) - f(x_n)| \leqslant \varepsilon$ for all $x \in V_n$, since $V_n \subset U_{x_n}$. As ε was arbitrary, equality (2.2.6) follows from (2.2.7). □

Concerning signed measures, see Exercises 2.7.31 and 2.7.32, which contain negative and positive results.

Let us consider the behaviour of weak convergence of measures under mappings. By the very definition, weak convergence of measures μ_α to a measure μ is weak convergence of all images $\mu_\alpha \circ f^{-1}$ to $\mu \circ f^{-1}$ for functions $f \in C_b(X)$. The next assertion gives a bit finer results.

2.2.9. THEOREM. *Suppose that a net of Borel measures μ_α on a metric space X converges weakly to a measure μ. Then the following assertions are true.*

(i) *For every continuous mapping F from X to a metric space Y, the net of measures $\mu_\alpha \circ F^{-1}$ converges weakly to a measure $\mu \circ F^{-1}$.*

(ii) *Let $\mu_\alpha \geqslant 0$, $\mu \geqslant 0$, and let F be a Borel mapping from X to a metric space Y such that F is continuous μ-almost everywhere. Then $\mu_\alpha \circ F^{-1} \Rightarrow \mu \circ F^{-1}$.*

(iii) *If X is separable, $\mu_\alpha \geqslant 0$ and $\{F_\alpha\}$ is a pointwise equicontinuous family of mappings from X to a metric space Y such that there is a Borel mapping $F\colon X \to Y$ for which the measures $\mu \circ F_\alpha^{-1}$ converge weakly to the measure $\mu \circ F^{-1}$, then the measures $\mu_\alpha \circ F_\alpha^{-1}$ also converge weakly to $\mu \circ F^{-1}$.*

PROOF. Assertion (i) is obvious. Let us verify assertion (ii). Let Z be a closed set in Y and let D_F be the set of discontinuity points of F. We observe that $\overline{F^{-1}(Z)} \subset F^{-1}(Z) \cup D_F$, where $\overline{A}$ is the closure of A. Then by Theorem 2.2.5 we have the inequalities

$$\limsup_{\alpha} \mu_\alpha \circ F^{-1}(Z) \leqslant \limsup_{\alpha} \mu_\alpha\Bigl(\overline{F^{-1}(Z)}\Bigr) \leqslant \mu\Bigl(\overline{F^{-1}(Z)}\Bigr) = \mu\Bigl(F^{-1}(Z)\Bigr),$$

which gives $\mu_\alpha \circ F^{-1} \Rightarrow \mu \circ F^{-1}$.

For the proof of assertion (iii) we fix a uniformly continuous bounded function φ on Y. The functions $\varphi \circ F_\alpha$ on X are uniformly bounded and pointwise equicontinuous. By Theorem 2.2.8, for every $\varepsilon > 0$, there exists an index α_0 such that for all $\alpha \geqslant \alpha_0$ one has

$$\left|\int_X \varphi \circ F_\alpha \, d\mu_\alpha - \int_X \varphi \circ F_\alpha \, d\mu\right| < \varepsilon/2.$$

It follows from our condition that there exists an index $\alpha_1 \geqslant \alpha_0$ such that for all $\alpha \geqslant \alpha_1$ we have

$$\left|\int_X \varphi \circ F_\alpha \, d\mu - \int_X \varphi \circ F \, d\mu\right| < \varepsilon/2.$$

The two obtained estimates yield our claim. □

2.2.10. COROLLARY. *Let $\{\mu_\alpha\}$ be a net of Borel probability measures on a metric space X and let μ be a Borel probability measure. Then $\{\mu_\alpha\}$ converges weakly to μ precisely when the equality*

$$\lim_{\alpha} \int_X f \, d\mu_\alpha = \int_X f \, d\mu$$

is valid for every bounded Borel function f that is μ-almost everywhere continuous.

PROOF. The sufficiency of the indicated condition is clear. Its necessity follows from assertion (ii) of the previous theorem by the equality

$$\int_X f \, d\mu_\alpha = \int_{\mathbb{R}} h \, d(\mu_\alpha \circ f^{-1}),$$

where $h \in C_b(\mathbb{R})$ and $h(t) = t$ if $|t| \leqslant \sup|f|$, and by a similar equality for μ. □

It is readily seen that in general weak convergence is not preserved by elements of the Jordan–Hahn decomposition and is not interchangeable with taking the total variation. For this reason a study of signed measures does not reduce to the case of nonnegative measures. Let us consider some examples.

2.2.11. EXAMPLE. (i) Let μ_n be the measure on the interval $[0, 2\pi]$ defined as follows: $\mu_n = 0$ for odd n and $\mu_n = \sin(nx)\,dx$ for even n. It is readily seen that the measures μ_n converge weakly to zero and that the measures $|\mu_n|$ have no weak limit.

(ii) The measures $\delta_0 - \delta_{1/n}$ on the real line converge weakly to zero, but their total variations $|\delta_0 - \delta_{1/n}| = \delta_0 + \delta_{1/n}$ converge weakly to $2\delta_0$.

The next example due to Le Cam [**417**] exhibits another interesting aspect of a similar phenomenon.

2.2.12. EXAMPLE. Let X be a subset of $[0, 1]$ containing all binary rational numbers and having outer measure 1 and inner measure zero (i.e., $\lambda^*([0,1]\backslash X) = 1$). Let us equip X with the induced topology and the measure μ that is the restriction of Lebesgue measure λ to X. This means (see Example 1.1.2) that $\mu(B \cap X) = \lambda(B)$

for every Borel set B in $[0,1]$ (their intersections with X give the Borel σ-algebra of X). Set

$$\nu_n(k2^{-n}) = 2^{-n} \quad \text{for } k = 1, \ldots, 2^n, \quad \mu_n = \nu_{n+1} - \nu_n.$$

The sequence $\{\mu_n\}$ of Radon measures on X converges weakly to zero, since $\nu_n \Rightarrow \mu$, while the measures $|\mu_n| = \nu_{n+1}$ converge weakly to the measure μ on X, which is τ-additive, but not Radon.

Our discussion of convergence of total variations of weakly convergent measures continues in § 2.7(ii).

2.3. The Prohorov theorem and weak compactness

In Chapter 1 we have seen a useful property of probability measures on a compact interval: every sequence of such measures contains a weakly convergent subsequence. For the whole real line this is obviously false, but it turns out that also for general spaces there is an efficient method of verifying this property.

This method is based on the concept of uniform tightness of a family of measures introduced in Definition 1.4.10. Let us recall it: a family of Borel measures $\mathcal{M}$ on a metric space X is called *tight* (or *uniformly tight*) if, for every $\varepsilon > 0$, there exists a compact set $K_\varepsilon \subset X$ such that

$$|\mu|(X \backslash K_\varepsilon) \leqslant \varepsilon \quad \forall\, \mu \in \mathcal{M}.$$

Certainly, every measure in such a family is Radon.

2.3.1. REMARK. For a complete space X the uniform tightness is equivalent to the property that for every $\varepsilon > 0$ and every $r > 0$ there exists a finite number of open balls of radius r such that for their union $S_{\varepsilon,r}$ one has the estimate

$$|\mu|(X \backslash S_{\varepsilon,r}) \leqslant \varepsilon \quad \forall\, \mu \in \mathcal{M}.$$

Indeed, taking the sets $S_{\varepsilon 2^{-n}, 2^{-n}}$, we obtain the set $S_\varepsilon = \bigcap_{n=1}^\infty S_{\varepsilon 2^{-n}, 2^{-n}}$. It is totally bounded and $|\mu|(X \backslash S_\varepsilon) \leqslant \varepsilon$ for all $\mu \in \mathcal{M}$. Its closure is compact.

This yields the following simple fact.

2.3.2. EXAMPLE. A family $\mathcal{M}$ of Borel probability measures on a Banach space X is tight precisely when for every functional $f \in X^*$ the set of measures $\mu \circ f^{-1}$ with $\mu \in \mathcal{M}$ is uniformly tight and for every $\varepsilon > 0$ there exists a finite-dimensional subspace $L_\varepsilon \subset X$ such that $\mu(L_\varepsilon^\varepsilon) > 1 - \varepsilon$ for all $\mu \in \mathcal{M}$, where $L_\varepsilon^\varepsilon$ is the ε-neighborhood of L_ε.

We give first a sufficient condition in order that a weakly fundamental sequence of Radon measures be weakly convergent to a Radon measure.

2.3.3. LEMMA. *If a sequence of Radon measures μ_n on a metric space is weakly fundamental and tight, then it converges weakly to some Radon measure μ.*

PROOF. On the space $C_b(X)$ the formula

$$L(f) = \lim_{n\to\infty} \int_X f\, d\mu_n$$

defines a functional that is continuous by the Banach–Steinhaus theorem. By using Theorem 2.1.15(ii) we verify that this functional is generated by a Radon measure. Without loss of generality we can assume that $\|\mu_n\| \leqslant 1$. Let $\varepsilon > 0$. By assumption, there exists a compact set K such that $|\mu_n|(X\backslash K) < \varepsilon$ for all n. Suppose that $f \in C_b(X)$, $|f| \leqslant 1$ and $f|_K = 0$. Then the integral of f with respect to the measure μ_n does not exceed ε in absolute value for all n, whence $|L(f)| \leqslant \varepsilon$, so that the hypothesis of the cited theorem is fulfilled. □

The most important for applications is the following fundamental result due to Yu.V. Prohorov [**525**] (who considered probability measures).

2.3.4. THEOREM. *Let X be a complete separable metric space and let $\mathcal{M}$ be a family of Borel measures on X. Then the following conditions are equivalent:*

(i) *every sequence in $\{\mu_n\} \subset \mathcal{M}$ contains a weakly convergent subsequence;*

(ii) *the family $\mathcal{M}$ is uniformly tight and uniformly bounded in variation.*

The indicated conditions are equivalent for an arbitrary complete metric space X if $\mathcal{M} \subset \mathcal{M}_r(X)$.

PROOF. Let (i) be fulfilled. The boundedness of $\mathcal{M}$ in variation is already known from the Banach–Steinhaus theorem. Suppose that the family $\mathcal{M}$ is not uniformly tight. We show that there exists $\varepsilon > 0$ with the following property: for every compact set $K \subset X$, one can find a measure $\mu^K \in \mathcal{M}$ such that

$$|\mu^K|(X\backslash K^\varepsilon) > \varepsilon, \tag{2.3.1}$$

where K^ε is the closed ε-neighborhood of K. Indeed, otherwise for every $\varepsilon > 0$ there exists a compact set $K(\varepsilon) \subset X$ such that

$$|\mu|\bigl(X\backslash K(\varepsilon)^\varepsilon\bigr) \leqslant \varepsilon \quad \forall\, \mu \in \mathcal{M}.$$

For a fixed number $\delta > 0$, we take the set $K_n = K(\delta 2^{-n})^{\delta 2^{-n}}$ and consider the set $K = \bigcap_{n=1}^\infty K_n$, which is compact (being closed and totally bounded, which is easily seen from the compactness of $K(\delta 2^{-n})$) and

$$|\mu|(X\backslash K) \leqslant \sum_{n=1}^\infty |\mu|(X\backslash K_n) \leqslant \delta \quad \forall\, \mu \in \mathcal{M},$$

which is a contradiction. Now with the aid of (2.3.1) by induction we find a sequence of pairwise disjoint compact sets K_j and a sequence of measures $\mu_j \in \mathcal{M}$ with the following two properties:

1) $|\mu_j|(K_j) > \varepsilon$, 2) $K_{j+1} \subset X\backslash \bigcup_{i=1}^j K_i^\varepsilon$.

For μ_1 we take an arbitrary measure in $\mathcal{M}$ with $\|\mu_1\| > \varepsilon$ (its existence follows from (2.3.1)), for K_1 we take a compact set with $|\mu_1|(K_1) > \varepsilon$, next by means of K_1 we find μ_2 with the aid of (2.3.1). Then we take a compact set $K_2 \subset X\backslash K_1^\varepsilon$ with $|\mu_2|(K_2) > \varepsilon$, consider the set $Q_2 = K_1 \cup K_2$ and find a measure μ_3 with $|\mu_3|(X\backslash Q_2^\varepsilon) > \varepsilon$, and so on. Property 2) yields that the sets $U_j := K_j^{\varepsilon/4}$ are pairwise disjoint. There exist continuous functions f_j such that $f_j = 0$ outside U_j, $|f_j| \leqslant 1$ and

$$\int_{U_j} f_j \, d\mu_j > \varepsilon.$$

By our condition the sequence $\{\mu_j\}$ contains a weakly convergent subsequence. For notational simplicity we shall assume that the whole sequence $\{\mu_j\}$ is weakly convergent. Having in mind to apply Lemma 1.3.7 on convergence in l^1, we let

$$a_n^i = \int_X f_i(x)\, \mu_n(dx).$$

Then $a_n := (a_n^1, a_n^2, \ldots)$ belongs to l^1, because $\sum_{i=1}^\infty |f_i| \leqslant 1$. For every element $\lambda = (\lambda_i) \in l^\infty$ the function $f^\lambda = \sum_{i=1}^\infty \lambda_i f_i$ is continuous on X and $|f^\lambda| \leqslant \sup_i |\lambda_i|$. Since the sequence of numbers

$$\langle \lambda, a_n \rangle = \int_X f^\lambda \, d\mu_n$$

converges, we obtain that the sequence $\{a_n\}$ is fundamental in the topology $\sigma(l^1, l^\infty)$. According to Lemma 1.3.7 the sequence $\{a_n\}$ converges in norm of l^1. Hence $\lim_{n\to\infty} a_n^n = 0$, which contradicts our choice of f_n. Thus, $\mathcal{M}$ is uniformly tight.

Suppose now that condition (ii) is fulfilled and $\{\mu_n\} \subset \mathcal{M}$. Let $C = \sup_{\mu\in\mathcal{M}} \|\mu\|$. We first recall that every uniformly bounded sequence of measures on a metrizable compact space K contains a weakly convergent subsequence. Let us now take an increasing sequence of compact sets K_j such that

$$|\mu_n|(X \backslash K_j) < 2^{-j} \quad \text{for all } n.$$

It is clear from what has been said above that by means of the diagonal process one can find a sequence of measures μ_{n_i} whose restrictions to every compact K_j are weakly convergent. Let $f \in C_b(X)$. Let us show that the sequence of integrals of f with respect to the measures μ_{n_i} is fundamental. Let $\varepsilon > 0$. One can assume that $|f| \leqslant 1$. We find j with $2^{-j} < \varepsilon$. Then

$$\left| \int_X f \, d\mu_{n_i} - \int_X f \, d\mu_{n_m} \right| \leqslant \left| \int_{K_j} f \, d\mu_{n_i} - \int_{K_j} f \, d\mu_{n_m} \right| + 2\varepsilon,$$

whence our claim follows on account of Lemma 2.3.3. Finally, in case of a non-separable space every sequence of Radon measures is concentrated on a separable subspace. □

From the proof of the Prohorov theorem it is easy to derive the following assertion.

2.3.5. COROLLARY. *Every weakly fundamental sequence of Radon measures μ_n on a complete metric space X is uniformly tight and hence converges weakly to some Radon measure.*

Moreover, if the measures μ_n are nonnegative, then their uniform tightness (hence also weak convergence) follows from the condition that for every bounded Lipschitz function f the sequence of integrals $\int_X f \, d\mu_n$ is Cauchy.

PROOF. The first assertion has actually been proved (taking into account Lemma 2.3.3). Let us explain necessary changes in our reasoning that enable us to cover the second assertion too. To this end it suffices to take functions f_j in such a way that they are Lipschitz with a common constant, $0 \leqslant f_j \leqslant 1$, $f_j = 1$ on K_j and $f_j = 0$ outside U_j. This is possible, since $U_j = K_j^{\varepsilon/4}$. The functions f^λ in this case are Lipschitz. Since μ_n and f_n are nonnegative, the integrals of f_n with respect to the measures μ_n are greater than $\mu_n(K_n) > \varepsilon$. □

For nonnegative measures the Prohorov theorem is proved in a shorter way. Moreover, as discovered by Le Cam, in case of nonnegative measures completeness of X is not required if the limit measure is also tight. The nonnegativity of measures is important here: recall that in Example 2.2.12 a sequence of signed tight measures μ_n on a separable metric space X (a subset of an interval) is constructed that converges weakly to zero, but the measures $|\mu_n|$ converge weakly to a measure that is not tight. Clearly, such a sequence $\{\mu_n\}$ cannot be uniformly tight. The next theorem due to Le Cam applies to arbitrary metric spaces.

2.3.6. THEOREM. *If a sequence of nonnegative Radon measures μ_n on a metric space X converges weakly to a Radon measure μ, then this sequence is uniformly tight.*

PROOF. Let $\varepsilon > 0$. We can find a compact set K such that $\mu(X\backslash K) < \varepsilon/4$. Set $G_k = \{x\colon \operatorname{dist}(x, K) < 1/k\}$. By Theorem 2.2.5 there exists an increasing sequence of indices n_k such that

$$\mu_n(X\backslash G_k) < \mu(X\backslash G_k) + \varepsilon/4 < \varepsilon/2 \quad \forall n \geqslant n_k. \tag{2.3.2}$$

For every n with $n_k \leqslant n \leqslant n_{k+1}$ we take a compact set $K_n \subset G_k$ such that

$$\mu_n(G_k\backslash K_n) < \varepsilon/4.$$

Set $Q_k = K \cup \bigl(\bigcup_{n=n_k}^{n_{k+1}} K_n\bigr)$ and $K_\varepsilon = \bigcup_{k=1}^{\infty} Q_k$. We observe that the sets Q_k are compact, $K \subset Q_k \subset G_k$ and $\mu_n(G_k\backslash Q_k) < \varepsilon/4$ if $n_k \leqslant n \leqslant n_{k+1}$. Inequality (2.3.2) yields that $\mu_n(X\backslash Q_k) < \varepsilon$ if $n_k \leqslant n \leqslant n_{k+1}$, whence we have $\mu_n(X\backslash K_\varepsilon) < \varepsilon$ for all n. It remains to verify that K_ε is compact.

Indeed, let $\{x_j\} \subset K_\varepsilon$. If some of the sets Q_k contains an infinite part of $\{x_j\}$, then in Q_k, hence in K_ε, there is a limit point of this sequence. If we have no such Q_k, then there exist two infinite sequences of indices j_m and i_m such that $x_{j_m} \in Q_{i_m}$. Since $Q_{i_m} \subset G_{i_m}$, there exist points $z_m \in K$ such that the distance between x_{j_m} and z_m does not exceed i_m^{-1}. The sequence $\{z_m\}$ has some limit point $z \in K$, which is obviously a limit point for $\{x_{j_m}\}$ as well. □

The Prohorov theorem gives a condition for the weak sequential compactness of a set of measures on a complete separable metric space. In the next chapter the space of measures will be equipped with a topology and a metric, hence it will be appropriate to discuss topological forms of compactness of measures (not always reducing to sequences). For general topological spaces, our discussion will be continued in Chapter 4. Let us consider an example that helps to verify the uniform tightness of measures.

2.3.7. EXAMPLE. A family $\mathcal{M}$ of Borel probability measures on a complete separable metric space X is uniformly tight is and only if there exists a Borel function $V\colon X \to [0, +\infty]$ such that the sets $\{V \leqslant c\}$ with $c < +\infty$ are compact (V is called *compact*), $\mu(V = +\infty) = 0$ for all $\mu \in \mathcal{M}$, and

$$\sup_{\mu\in\mathcal{M}} \int_X V(x)\,\mu(dx) < \infty.$$

PROOF. The sufficiency of this condition follows by the Chebyshev inequality, according to which $C\mu(V > C)$ does not exceed the integral of V with respect to μ, so the measures of $\{V > C\}$ are uniformly small for large C. To prove the necessity of the stated condition, we take an increasing sequence of compact sets K_n with $\mu(K_n) > 1 - 2^{-n}$ for all $\mu \in \mathcal{M}$ and let $V = +\infty$ on the complement to the union of K_n, $V = 1$ on K_1, $V = n$ on $K_{n+1}\backslash K_n$ for $n \geqslant 1$. Then for all $\mu \in \mathcal{M}$ we have

$$\int_X V\,d\mu = \mu(K_1) + \sum_{n=1}^{\infty} n\mu(K_{n+1}\backslash K_n) \leqslant 1 + \sum_{n=1}^{\infty} n2^{-n},$$

whence our claim follows. □

2.3.8. EXAMPLE. A subset K of a metric space X has compact closure precisely when every sequence in the family of Dirac measures $\{\delta_x, x \in K\}$ has a weakly convergent subsequence.

With the aid of the Prohorov theorem we shall prove in Chapter 5 an important result of A.D. Alexandroff about the sequential completeness of the space of measures with the weak topology for any completely regular spaces (Theorem 5.1.10). Here we formulate its version in the case of metric spaces (the proof in this case does not change and does not become simpler).

2.3.9. THEOREM. *Let a sequence of Borel measures on a metric space X be weakly fundamental. Then it converges weakly to some Borel measure on X.*

However, in the case of a complete separable space this theorem can be easily derived from Corollary 2.3.5 (Exercise 2.7.35).

Note that for convergent nets the conclusion of the Prohorov theorem can fail. However, in the next chapter the Prohorov theorem will be established for sets of measures compact in the weak topology. Nevertheless, the very idea of uniform tightness is also useful for convergence of nets, as the next theorem shows.

2.3.10. THEOREM. *Let $\{\mu_\alpha\}$ be a net of Radon measures on a metric space X that is uniformly bounded in variation and uniformly tight and let μ be a Radon measure on X such that*

$$\int_X f\,d\mu = \lim_\alpha \int_X f\,d\mu_\alpha$$

for all functions f from a subalgebra of functions in $C_b(X)$ containing 1 and separating points. Then $\mu_\alpha \Rightarrow \mu$.

PROOF. Let $f \in C_b(X)$ and $\varepsilon > 0$. We can assume that $\|\mu_\alpha\| \leqslant 1$, $\|\mu\| \leqslant 1$ and $|f| \leqslant 1$. By assumption, there is a compact set K with $(|\mu_\alpha| + |\mu|)(X\backslash K) < \varepsilon$ for all α. It is easy to derive from the Stone–Weierstrass theorem (Exercise 4.8.24) that there exists a function g in the given algebra for which $\sup_{x\in K} |f(x) - g(x)| \leqslant \varepsilon$ and $\sup_{x\in X} |g(x)| \leqslant 1$. By assumption, there exists an index α_1 such that the integrals of g with respect to the measures μ and μ_α for all $\alpha \geqslant \alpha_1$ differ by less than ε. By the estimate of the measures of the complement of K and the inequalities $|f| \leqslant 1$, $|g| \leqslant 1$ the integrals of the functions f and g with respect to the measures μ_α differ by less than 3ε. The same is also true for μ. Hence the integrals of f with respect to the measures μ_α and μ for all $\alpha \geqslant \alpha_1$ differ by less than 7ε. □

2.4. Connections with convergence on sets

One can see from Theorem 2.2.5 that weak convergence ensures convergence on some "sufficiently regular" sets. Let us discuss this circumstance in more detail.

Let μ be a nonnegative Borel measure on a metric space X. Denote by Γ_μ the class of Borel sets $E \subset X$ having boundaries of μ-measure zero. The boundary ∂E of a set E is defined as the closure of E without the interior of E, hence is a Borel set for any E. Sets from Γ_μ are called *continuity sets of the measure μ*.

2.4.1. THEOREM. *A net $\{\mu_\alpha\}$ of Borel probability measures on a metric space X converges weakly to a Borel probability measure μ precisely when the equality*

$$\lim_\alpha \mu_\alpha(E) = \mu(E) \tag{2.4.1}$$

is fulfilled for every set $E \in \Gamma_\mu$.

In addition, if $P \in \mathcal{P}(X)$ is an arbitrary measure, then for weak convergence of $\{\mu_\alpha\}$ to μ it suffices to have this equality for all sets $E \in \Gamma_P$.

PROOF. A Borel set E belongs to Γ_μ precisely when one cab find an open set W and a closed set F in X such that $W \subset E \subset F$ and $\mu(F\backslash W) = 0$. In case of weak convergence we have

$$\limsup_\alpha \mu_\alpha(E) \leqslant \limsup_\alpha \mu_\alpha(F) \leqslant \mu(F) = \mu(E).$$

Similarly, $\liminf_\alpha \mu_\alpha(E) \geqslant \mu(E)$, whence (2.4.1) follows at once. Suppose now that (2.4.1) is true. Let $U = \{f > 0\}$, where $f \in C(X)$, and let $\varepsilon > 0$. It is readily seen that there exists $c > 0$ such that $\mu(U) < \mu(\{f > c\}) + \varepsilon$ and $\mu(\{f > c\}) = \mu(\{f \geqslant c\})$. Then for $E = \{f > c\}$ we have (2.4.1), since we can take the sets $W = E$ and $F = \{f \geqslant c\}$, the first of which is open and second is closed. Thus, $\liminf_\alpha \mu_\alpha(U) \geqslant \mu(U) - \varepsilon$, which yields (2.2.3), because ε was arbitrary. In this implication in place of Γ_μ also Γ_P works for an arbitrary measure $P \in \mathcal{P}(X)$ (see also Corollary 4.4.2 for signed measures and Corollary 2.4.5). □

2.4.2. PROPOSITION. *The class Γ_μ is a subalgebra in $\mathcal{B}(X)$ and contains a topology base.*

PROOF. The first assertion follows from the fact that E and $X\backslash E$ have the same boundary and the boundary of the union of two sets is contained in the union of their boundaries. In order to prove the second assertion, for every bounded continuous function f on X we let $U(f,c) = \{x\colon\ f(x) > c\}$ and observe that the set

$$M_f = \bigl\{c \in \mathbb{R}\colon\ \mu\bigl(\partial U(f,c)\bigr) > 0\bigr\}$$

is at most countable, since $\partial U(f,c) \subset f^{-1}(c)$ and the measure $\mu\circ f^{-1}$ has an at most countable set of atoms. The sets $U(f,c)$ for all $c \in \mathbb{R}\backslash M_f$ belong to Γ_μ and constitute a topology base. Indeed, for every point x and every open set U containing the point x, there exists a continuous function $f\colon X \to [0,1]$ with $f(x) = 1$ that equals 0 outside U. Thus, U contains the set $U(f,c)$ for some number $c \in \mathbb{R}\backslash M_f$. □

2.4.3. EXAMPLE. Suppose that Borel probability measures μ_n on $\mathbb{R}^d$ converge weakly to a Borel probability measure μ that is absolutely continuous. Then $\lim\limits_{n\to\infty} \mu_n(E) = \mu(E)$ for every Jordan measurable Borel set E. In particular, if μ_n and μ are Borel probability measures on $[0,1]$ such that μ is absolutely continuous and $\lim\limits_{n\to\infty} \mu_n([a,b]) = \mu([a,b])$ for every interval $[a,b] \subset [0,1]$, then convergence holds on all Jordan measurable Borel sets.

The last assertion in the case of absolutely continuous measures μ_n was obtained by G.M. Fichtenholz [**231**], who also constructed an example showing that in this case convergence on some Borel sets can fail. Such an example can be easily constructed by using basic properties of weak convergence. Namely, let $E \subset [0,1]$ be a nowhere dense compact set of positive Lebesgue measure. It is clear from Example 2.2.7 that one can find probability measures ν_n on $[0,1]$ weakly converging to Lebesgue measure λ on $[0,1]$ and concentrated on finite sets of points from the complement of E. Hence there exist probability measures μ_n weakly converging to λ and given by smooth probability densities vanishing on E. It suffices to take such measures μ_n in balls of radius $1/n$ centered at ν_n with respect to any metric on the space of probability measures defining weak convergence and discussed in the next chapter. It would also be possible just to take measures $\sigma_n * \nu_n$, where the measure σ_n is given by a probability density $p_n(x) = np(x/n)$, where p is a smooth probability density with support in $[0,1]$. Then weak convergence of $\sigma_n * \nu_n$ to λ is seen from Corollary 2.2.6, since for every 1-Lipschitz function f one has the estimate

$$\left|\int_{\mathbb{R}} f\,d(\nu_n - \sigma_n * \nu_n)\right| \leqslant \int_0^1\int_0^1 |f(x) - f(x+y/n)|p(y)\,dy\,\nu_n(dx) \leqslant \frac{1}{n},$$

because $|f(x) - f(x+y/n)| \leqslant n^{-1}$ for all $x, y \in [0,1]$.

One more sufficient condition for weak convergence in terms of convergence on some sets is given in the following theorem from the classical paper Prohorov [**525**].

2.4.4. THEOREM. *Let $\mathcal{E}$ be a class of Borel sets in a metric space X such that $\mathcal{E}$ is closed with respect to finite intersections and every open set is representable in the form of a finite or countable union of sets from $\mathcal{E}$. Let $\{\mu_\alpha\}$ be a net (e.g., a sequence) of Borel probability measures on X and μ_n a Borel probability measure on X such that $\mu_\alpha(E) \to \mu(E)$ for all $E \in \mathcal{E}$. Then $\{\mu_\alpha\}$ converges weakly to μ.*

PROOF. We observe that

$$\lim_\alpha \mu_\alpha\Bigl(\bigcup_{j=1}^k E_j\Bigr) = \mu\Bigl(\bigcup_{j=1}^k E_j\Bigr) \quad \text{for all } E_1, \ldots, E_k \in \mathcal{E}.$$

Indeed, for $k = 2$ by assumption we have convergence on the sets E_1, E_2 and $E_1 \cap E_2$, which gives convergence on $E_1\backslash(E_1 \cap E_2)$ and $E_2\backslash(E_1 \cap E_2)$, hence also on the set $E_1 \cup E_2$ that equals the disjoint union of $E_1 \cap E_2$, $E_1\backslash(E_1 \cap E_2)$ and $E_2\backslash(E_1 \cap E_2)$. By induction on k we obtain our assertion. Indeed, if it is true for some k, then it remains true for $k+1$ too, since the set $(E_1 \cup \ldots \cup E_k) \cap E_{k+1}$ is the union of k sets $E_i \cap E_{k+1} \in \mathcal{E}$, which gives convergence on this set. If now we are given a set $U = \{f > 0\}$, where $f \in C_b(X)$, then it can be represented in the form of an at most countable union of sets $E_j \in \mathcal{E}$. Hence

$$\mu(U) = \lim_{k\to\infty} \mu\Bigl(\bigcup_{j=1}^k E_j\Bigr) = \lim_{k\to\infty} \lim_\alpha \mu_\alpha\Bigl(\bigcup_{j=1}^k E_j\Bigr) \leqslant \liminf_\alpha \mu_\alpha(U),$$

whence our claim follows. □

It is easy to conclude from the proof that this theorem remains in force if U is representable in the form of a union of sets from $\mathcal{E}$ up to a set of μ-measure zero.

2.4.5. COROLLARY. *Let μ be a Borel probability measure on a metric space X and let $\mathcal{E}$ be a class of Borel sets in X that is closed with respect to finite intersections. Suppose that for every open set U and every $\varepsilon > 0$, one can find sets $E_1, \ldots, E_k \in \mathcal{E}$ such that $\bigcup_{i=1}^k E_i \subset U$ and $\mu\bigl(U\backslash\bigcup_{i=1}^k E_i\bigr) < \varepsilon$. If a net (e.g., a sequence) of Borel probability measures μ_α is such that $\lim\limits_\alpha \mu_\alpha(E) = \mu(E)$ for all sets $E \in \mathcal{E}$, then the measures μ_α converge weakly to μ.*

PROOF. It suffices to observe that in the proof of the theorem above it was sufficient to represent U as the union of a sequence of sets of class $\mathcal{E}$ up to a set of μ-measure zero. Note that this also yields the second assertion of Theorem 2.4.1. □

Let us consider examples of classes $\mathcal{E}$ satisfying the hypotheses of the proven theorem.

2.4.6. COROLLARY. *Let $\mathcal{E}$ be a class of Borel sets in a separable metric space X such that $\mathcal{E}$ is closed with respect to finite intersections and, for every point $x \in X$ and every neighborhood U of x, one can find a set $E_x \in \mathcal{E}$ containing some neighborhood of x and contained in U. Then convergence of a net (e.g., a sequence) of Borel probability measures on all sets from $\mathcal{E}$ implies its weak convergence.*

PROOF. Let U be open. By assumption, for every point $x \in U$ there exists a set $E_x \in \mathcal{E}$ such that $x \in E_x \subset U$ and x possesses a neighborhood $V_x \subset E_x$. By the separability of X the cover of U by the sets V_x contains an at most countable subcover $\{V_{x_n}\}$, which means that $U = \bigcup_{n=1}^{\infty} E_{x_n}$. □

2.4.7. COROLLARY. *Let $\{\mu_\alpha\}$ be a net (e.g., a sequence) of Borel probability measures on a separable metric space X and let μ be a Borel probability measure on X such that $\mu_\alpha(E) \to \mu(E)$ for every continuity set E for μ (i.e., $E \in \Gamma_\mu$) and is representable as a finite intersection of open balls. Then $\mu_\alpha \Rightarrow \mu$.*

PROOF. The family of sets with the indicated properties satisfies the hypotheses of the previous corollary. Indeed, a finite intersection of such sets is also a continuity set. In addition, for every point x and every $\varepsilon > 0$, there exists $r \in (0, \varepsilon)$ such that the boundary of the ball of radius r centered at x has μ-measure zero, since for different r these boundaries do not intersect (the boundary of a ball of radius r is contained in the sphere of radius r with the same center). □

2.4.8. COROLLARY. *A net $\{\mu_\alpha\}$ (e.g., a sequence) of Borel probability measures on $\mathbb{R}^\infty$ converges weakly to a probability measure μ precisely when the finite-dimensional projections of the measures μ_α, i.e., the images of the measures μ_α under the projection operators $\pi_d\colon \mathbb{R}^\infty \to \mathbb{R}^d$, $(x_i) \mapsto (x_1, \ldots, x_d)$, converge weakly to the corresponding projections of μ.*

More generally, a net or sequence of Borel probability measures on the product $\prod_{i=1}^{\infty} X_i$ of a sequence of separable metric spaces X_i converges weakly to a Borel probability measure precisely when one has weak convergence of the projections on all finite products.

PROOF. The necessity of weak convergence of finite-dimensional projections is obvious. Its sufficiency follows by Corollary 2.4.6 applied to the class of open cylinders of the form

$$C = \{x\colon (x_1, \ldots, x_d) \in U\}, \quad \text{where } U \subset \mathbb{R}^d \text{ is open},$$

with μ-zero boundaries. Convergence $\mu_\alpha(C) \to \mu(C)$ is clear from the fact that $\mu \circ \pi_d^{-1}(\partial U) = \mu(\partial C) = 0$. The indicated class is obviously closed with respect to finite intersections. Let us verify that each ball centered at x contains a cylinder containing x and having boundary of μ-measure zero. In this ball we take a cylindrical neighborhood of x the base of which is an open ball V in $\mathbb{R}^d$ with some center y. Next we take a ball in $\mathbb{R}^d$ of a smaller radius the boundary of which has measure zero with respect to $\mu \circ \pi_d^{-1}$. Now we can take the cylinder generated by this smaller ball. The second assertion is proved similarly. □

A simple consequence of the proven result is that any Borel measure μ (possibly, signed) on $\mathbb{R}^\infty$ is the weak limit of its finite-dimensional projections under the mappings

$$P_n\colon x \mapsto (x_1, \ldots, x_n, 0, 0, \ldots).$$

However, the corollary itself does not extend to signed measures: weak convergence of finite-dimensional projections is not sufficient for weak convergence of signed measures. It is not difficult to construct an example in which for every fixed k the projections of the measures μ_n on $\mathbb{R}^k$ converge weakly to zero, but the measures are not even uniformly bounded in norm (see Exercise 2.7.51).

The last corollary enables one to derive the following interesting and rather unexpected fact discovered by Mehler [**455**], Borel [**105**] and Gateaux (Gâteaux) [**258**] in the form of convergence of integrals of cylindrical functions.

2.4.9. THEOREM. *Let μ_n be the probability measure on $\mathbb{R}^n$ obtained by the normalization of the standard surface measure on the sphere of radius $\sqrt{n}$ centered at zero. Then the measures μ_n, regarded as measures on $\mathbb{R}^\infty$ (after the natural embedding of $\mathbb{R}^n$ into $\mathbb{R}^\infty$), converge weakly to the Gaussian measure γ equal to the countable power of the standard Gaussian measure on $\mathbb{R}$.*

PROOF. Let us verify weak convergence of the projections to every space $\mathbb{R}^d$. The coordinate functions x_n on $(\mathbb{R}^\infty, \gamma)$ constitute a sequence of independent standard Gaussian random variables. Let us consider on $(\mathbb{R}^\infty, \gamma)$ the random variables

$$\zeta_n := \sqrt{x_1^2 + \cdots + x_n^2}.$$

By using the known fact (see Bogachev [**81**, Exercise 9.12.56]) that the surface measure on the sphere is a unique spherically invariant measure on the sphere up to a constant factor, we conclude that the image of the measure γ under the mapping

$$(x_i) \mapsto \sqrt{n}(x_1, \ldots, x_n)/\zeta_n$$

is the normalized surface measure on the sphere of radius $\sqrt{n}$ in $\mathbb{R}^n$. Hence the projection of this surface measure to $\mathbb{R}^d$ coincides with the image of the measure γ under the mapping $f_n = \sqrt{n}(x_1, \ldots, x_d)/\zeta_n$ from $\mathbb{R}^\infty$ to $\mathbb{R}^d$. As $n \to \infty$, by the law of large numbers we have $\zeta_n^2/n \to 1$ a.e., whence $\zeta_n/\sqrt{n} \to 1$ a.e. Hence the measures $\gamma \circ f_n^{-1}$ on $\mathbb{R}^d$ converge weakly to the projection of the measure γ on $\mathbb{R}^d$. See also Exercises 2.7.39 and 2.7.40. □

From Corollary 2.4.5, reinforcing Theorem 2.4.4, one can easily derive the following result due to Kolmogorov and Prohorov [**380**]. However, we give a simple straightforward argument.

2.4.10. THEOREM. *Let $\{\mu_\alpha\}$ be a net of Borel probability measures on a metric space X and let μ be a τ-additive probability measure on X. Suppose that the equality*

$$\lim_\alpha \mu_\alpha(U) = \mu(U)$$

is fulfilled for all elements U of some topology base $\mathcal{O}$ closed with respect to finite intersections. Then the net $\{\mu_\alpha\}$ converges weakly to μ.

PROOF. Denote by $\mathcal{U}$ the family of all finite unions of sets from $\mathcal{O}$. Convergence on $\mathcal{O}$ and closedness of the family $\mathcal{O}$ with respect to finite intersections yield that $\lim_\alpha \mu_\alpha(U) = \mu(U)$ for $U \in \mathcal{U}$. For every open set G and every set $U \in \mathcal{U}$ contained in it we have

$$\mu(U) = \lim_\alpha \mu_\alpha(U) \leqslant \liminf_\alpha \mu_\alpha(G),$$

whence by the τ-additivity of μ we obtain the estimate

$$\mu(G) = \sup\{\mu(U)\colon\ U \subset G,\, U \in \mathcal{U}\} \leqslant \liminf_{\alpha} \mu_\alpha(G),$$

since G is the union of the directed family of all sets $U \subset G$ from $\mathcal{U}$ (recall that $\mathcal{O}$ is a topology base). As we know, the obtained estimate is equivalent to weak convergence of μ_α to μ. □

Let us apply the established result to the one-dimensional case already considered in Chapter 1.

2.4.11. COROLLARY. *A net $\{\mu_\alpha\}$ of probability measures on the real line converges weakly to a probability measure μ precisely when the corresponding distribution functions F_{μ_α} converge to the distribution function F_μ of the measure μ at the points of continuity of F_μ, where $F_\mu(t) = \mu\bigl((-\infty, t)\bigr)$.*

PROOF. For a topology base closed with respect to finite intersections we take the family of open intervals with ends at the points of continuity of F_μ (i.e., at points of zero μ-measure). Our assumption gives convergence not on such intervals, but on the semi-closed intervals $[a, b)$. However, in fact we have convergence on (a, b) as well. Indeed, let $\varepsilon > 0$. There is $a_1 \in (a, b)$ for which $\mu\bigl([a_1, b)\bigr) > \mu\bigl([a, b)\bigr) - \varepsilon$. Hence $\mu_\alpha\bigl([a_1, b)\bigr) > \mu\bigl([a, b)\bigr) - 2\varepsilon$ for all $\alpha \geqslant \alpha_0$ for some α_0, whence we conclude that $\mu_\alpha\bigl((a, b)\bigr) > \mu\bigl((a, b)\bigr) - 2\varepsilon$. On the other hand,

$$\mu_\alpha\bigl((a, b)\bigr) \leqslant \mu_\alpha\bigl([a, b)\bigr) \leqslant \mu\bigl([a, b)\bigr) + \varepsilon = \mu\bigl((a, b)\bigr) + \varepsilon$$

for all $\alpha \geqslant \alpha_1$ for some α_1. □

For a *Fréchet space* (a locally convex space the topology of which is generated by a complete metric), the Prohorov theorem yields the following criterion.

2.4.12. EXAMPLE. A sequence of Radon measures μ_n on a Fréchet space X converges weakly to a Radon measure μ precisely when this sequence is uniformly tight and one has convergence $\mu_n \circ l^{-1} \Rightarrow \mu \circ l^{-1}$ for every continuous linear functional l on X.

PROOF. By the Prohorov theorem these conditions follow from weak convergence. Conversely, suppose that they are fulfilled. One can assume that X is separable, since all measures μ_n and μ are concentrated on a separable closed linear subspace, because they are Radon. By the Prohorov theorem every subsequence in $\{\mu_n\}$ has a further subsequence which is weakly convergent. Hence we only need to show that the measure μ can be a unique limit for all such weakly convergent subsequences. Let ν be another limit. Then the measures μ and ν have equal images under each finite-dimensional operator of the form $P\colon x \mapsto \bigl(l_1(x), \ldots, l_k(x)\bigr)$, where l_i are continuous linear functionals. This follows by the equality of the Fourier transforms of the measures $\mu\circ P^{-1}$ and $\nu\circ P^{-1}$ on $\mathbb{R}^k$ on every vector $y \in \mathbb{R}^k$, since by the change of variables formula

$$\begin{aligned}\widetilde{\mu\circ P^{-1}}(y) &= \int_X \exp\bigl(i(y_1 l_1 + \cdots + y_k l_k)\bigr)\, d\mu \\ &= \lim_{n\to\infty} \int_X \exp\bigl(i(y_1 l_1 + \cdots + y_k l_k)\bigr)\, d\mu_n,\end{aligned}$$

and the same equality is also true for $\nu\circ P^{-1}$. Hence every function of the form $\varphi(l_1,\dots,l_k)$, where $\varphi\in C_b(\mathbb{R}^k)$, has equal integrals with respect to the measures μ and ν. This implies the equality $\mu=\nu$ (Exercise 2.7.30). □

Note that the space $\mathbb{R}^\infty$ is a special case of a Fréchet space, but in this space, according to Example 2.4.8, convergence of the images of measures under continuous linear functionals yields by itself the uniform tightness, which is false in the general case. For example, on the Hilbert space l^2 the Dirac measures at the elements of the standard basis $\{e_n\}$ do not converge weakly, although their images under continuous linear functionals converge weakly to the Dirac measure at zero, since for every $x\in l^2$ we have $(x,e_n)\to 0$.

2.5. The case of a Hilbert space

Here we discuss weak convergence of measures on a Hilbert space. This case exhibits certain similarities as compared to the case of $\mathbb{R}^n$, as well as instructive distinctions. Let H be a separable real Hilbert space with the inner product $(\cdot,\cdot)$.

We start by recalling some concepts and facts related to linear operators on Hilbert spaces. More details can be found in Bogachev, Smolyanov [**96**]. The norm of a continuous linear operator A on H is defined by the formula

$$\|A\| = \sup_{\|h\|\leqslant 1} \|Ah\|.$$

The space $\mathcal{L}(H)$ of all continuous linear operators on H is Banach with this norm (but is not Hilbert).

The adjoint A^* of a continuous linear operator A on H is defined by the formula $(Ax,y) = (x,A^*y)$. If $A^*=A$, then A is called *selfadjoint*. Every selfadjoint operator A on H is isomorphic to some operator A_φ of multiplication by a bounded real Borel function φ on the space $L^2(\sigma)$ for some bounded nonnegative Borel measure σ on an interval, i.e., the operator of the form

$$A_\varphi x(t) = \varphi(t)x(t).$$

Here an isomorphism means a linear operator J mapping H onto $L^2(\sigma)$ with preservation of the inner product such that $A=J^{-1}A_\varphi J$.

If A is selfadjoint and $(Ax,x)\geqslant 0$, then A is called *nonnegative*, which is denoted by $A\geqslant 0$. In this case there exists a unique nonnegative operator $\sqrt{A}$ such that $A=\sqrt{A}\sqrt{A}$. We write $A\geqslant B$ in case $A-B\geqslant 0$. If A is written in the form of multiplication by φ, then the nonnegativity of A is equivalent to the nonnegativity of φ, in addition, $\sqrt{A}$ acts as multiplication by $\sqrt{\varphi}$.

An operator A on H is called *compact* if it takes the closed unit ball to a compact set. By the Hilbert–Schmidt theorem each compact selfadjoint operator A has an orthonormal basis $\{e_n\}$ consisting of eigenvectors (an eigenbasis): $Ae_n=\alpha_n e_n$, where $\alpha_n\to 0$. Conversely, any operator of such a form is compact and selfadjoint (for real H). The set $\mathcal{K}(H)$ of compact operators is a closed linear subspace in the space of all bounded operators $\mathcal{L}(H)$.

A bounded operator A is called a *Hilbert–Schmidt operator* if

$$\sum_{n=1}^\infty \|Ae_n\|^2 < \infty$$

for some orthonormal basis $\{e_n\}$; in this case the series converges for every orthonormal basis and the sum is the same. The set $\mathcal{H}(H)$ of Hilbert–Schmidt operators on H is a linear subspace in $\mathcal{K}(H)$. It is equipped with the Hilbert–Schmidt norm

$$\|A\|_{HS} := \Bigl(\sum_{n=1}^{\infty} \|Ae_n\|^2\Bigr)^{1/2},$$

with respect to which it becomes a separable Hilbert space. The corresponding inner product has the form

$$(A, B)_{HS} = \sum_{n=1}^{\infty} (Ae_n, Be_n).$$

It is readily verified that A belongs to $\mathcal{H}(H)$ precisely when A^* belongs to $\mathcal{H}(H)$ and that $\|A\|_{HS} = \|A^*\|_{HS}$.

A compact selfadjoint operator is called *nuclear* or a *trace class operator* if the series of its eigenvalues converges absolutely. The class $\mathcal{H}_1(H)$ of nuclear operators consists of bounded operators A such that the selfadjoint operator $\sqrt{A^*A}$ is nuclear. The space $\mathcal{H}_1(H)$ is Banach with the nuclear (or trace) norm

$$\|A\|_1 := \|(A^*A)^{1/4}\|_{HS}^2.$$

For a symmetric nuclear operator, the nuclear norm equals the sum of absolute values of its eigenvalues (counted with multiplicities). For any nuclear operator A the trace

$$\operatorname{tr} A = \sum_{n=1}^{\infty} (Ae_n, e_n)$$

is well-defined and does not depend on our choice of the basis. Note that the nuclearity of a bounded operator A is equivalent to convergence of the series of (Ae_n, e_n) for every orthonormal basis (but convergence just for some basis is not enough!). It is known that

$$\|A\|_1 = \sup \sum_{n=1}^{\infty} |(Ae_n, \varphi_n)|,$$

where sup is taken over all pairs of orthonormal bases $\{e_n\}$ and $\{\varphi_n\}$.

An invariant description of nonnegative nuclear operators is this: $A = R^*R$, where R is a Hilbert–Schmidt operator. All nuclear operators can be described as the products ST of Hilbert–Schmidt operators, moreover, the previous representation of the nuclear norm and the Cauchy–Bunyakovskii inequality yield the inequality

$$\|ST\|_1 \leqslant \|S\|_{HS}\|T\|_{HS}. \tag{2.5.1}$$

Note the following property of the operation of taking the square root: if we have $A \geqslant B \geqslant 0$, then $\sqrt{A} \geqslant \sqrt{B}$.

2.5.1. LEMMA. (i) *The mapping $A \mapsto \sqrt{A}$ is a homeomorphism of the set of nonnegative operators on $\mathcal{L}(H)$, in addition,*

$$\|\sqrt{A} - \sqrt{B}\| \leqslant \sqrt{\|A - B\|}.$$

(ii) *The mapping $A \mapsto \sqrt{A}$ is a homeomorphism between the set of nonnegative nuclear operators with the metric generated by the nuclear norm and the set of* *nonnegative Hilbert–Schmidt operators with the metric generated by the Hilbert–Schmidt norm.*

PROOF. (i) Let $A, B \geqslant 0$, $D = B - A$, $r = \|D\|$. Then $B \leqslant A + rI$, whence

$$\sqrt{B} \leqslant \sqrt{A + rI} \leqslant \sqrt{A} + \sqrt{r}I,$$

where the last inequality is verified directly in the case where A is represented as the multiplication operator by a function $\varphi \geqslant 0$, since in this case $\sqrt{A + rI}$ is the operator of multiplication by $\sqrt{\varphi + r}$ for all $r \geqslant 0$. Thus, $\sqrt{B} - \sqrt{A} \leqslant \sqrt{r}I$. Similarly, $\sqrt{B} - \sqrt{A} \geqslant -\sqrt{r}I$. The operator $\sqrt{B} - \sqrt{A}$ is also isomorphic to the operator of multiplication by a bounded function ψ, for which the obtained estimates mean the inequality $-\sqrt{r} \leqslant \psi \leqslant \sqrt{r}$, whence the desired estimate follows.

Thus, if $A_n \to A$, then $\sqrt{A_n} \to \sqrt{A}$ in $\mathcal{L}(H)$. The inverse mapping of taking the square is also continuous with respect to the operator norm, which is easily verified.

(ii) If T_n, T are nonnegative Hilbert–Schmidt operators and $\|T_n - T\|_{HS} \to 0$ as $n \to \infty$, then, letting $S_n = T_n - T$, we obtain $(T + S_n)^2 - T^2 = TS_n + S_nT + S_n^2$, which tends to zero in the nuclear norm by (2.5.1).

Conversely, if $\|A_n - A\|_1 \to 0$, where A_n, A are nonnegative nuclear operators, then by (i) we have $\|\sqrt{A_n} - \sqrt{A}\| \to 0$. In addition,

$$\|\sqrt{A_n}\|_{HS}^2 = \operatorname{tr} A_n \to \operatorname{tr} A = \|\sqrt{A}\|_{HS}^2.$$

In the Hilbert space $\mathcal{H}(H)$ this gives convergence of norms and weak convergence, whence convergence in norm follows. □

As in the case of $\mathbb{R}^d$, the Fourier transform (the characteristic functional) of a Borel measure μ on H is given by the formula

$$\widetilde{\mu}(y) = \int_H \exp\bigl(i(y, x)\bigr)\, \mu(dx), \quad y \in H.$$

In § 4.6 we discuss the Fourier transform of measures on a locally convex space that is defined on the dual space. Here the Fourier transform is defined on the original space, because the dual spaces is identified with the original space by means of the Riesz theorem. The *convolution of measures* $\mu * \nu$ is defined as on $\mathbb{R}^d$ and on locally convex spaces (see Definition 1.1.4 and Definition 4.6.2).

If a measure $\mu \geqslant 0$ has finite first moment, which means that the norm is μ-integrable, then one can define the Bochner vector integral

$$a_\mu = \int_H x\, \mu(dx).$$

The vector a_μ is called the *mean* of the measure μ. By shifting to a_μ one can pass to the measure with zero mean: $\mu_0(B) = \mu(B + a_\mu)$.

If the second moment is finite as well (the square of the norm is integrable), then we can define the *covariance operator* K_μ:

$$(K_\mu u, v) = \int_H (u, x - a_\mu)(v, x - a_\mu)\, \mu(dx).$$

The measure μ_0 with zero mean has the same covariance operator. If we do not subtract a_μ, then we obtain the *correlation operator* of the measure μ. In the vector form one can write

$$K_\mu u = \int_H (x - a_\mu)(u, x - a_\mu)\,\mu(dx).$$

Let $a_\mu = 0$. Taking an orthonormal basis $\{e_n\}$ in H, we obtain

$$\int_H (x, x)\,\mu(dx) = \sum_{n=1}^{\infty} \int_H (x, e_n)^2\,\mu(dx) = \sum_{n=1}^{\infty} (K_\mu e_n, e_n) = \operatorname{tr} K_\mu.$$

Covariance operators are useful for establishing the uniform tightness of families of measures. In a Hilbert space with a basis $\{e_n\}$ any compact set is contained in a compact ellipsoid of the form

$$S = \Bigl\{x\colon \sum_{n=1}^{\infty} C_n^2 (x, e_n)^2 \leqslant 1\Bigr\},$$

where $C_n > 0$, $C_n \uparrow +\infty$. Such an ellipsoid is the unit ball of the Hilbert space obtained as the image of H under the compact linear operator A with the eigenbasis $\{e_n\}$ and eigenvalues C_n.

2.5.2. PROPOSITION. (i) *If a family $\mathcal{P}$ of Borel probability measures with zero means on H is such that*

$$\sup_{\mu\in\mathcal{P}} \sum_{n=1}^{\infty} C_n^2 \int_H (x, e_n)^2\,\mu(dx) = M < \infty,$$

where $C_n > 0$, $C_n \uparrow +\infty$, then this family is uniformly tight.

This condition is equivalent to the following: the measures in $\mathcal{P}$ have uniformly bounded second moments and, for some numbers $C_n > 0$ increasing to infinity, the series $\sum_{n=1}^{\infty} C_n^2 (K_\mu e_n, e_n)$ converge uniformly in $\mu \in \mathcal{P}$, where K_μ is the covariance operator of the measure μ. This is also equivalent to the precompactness of the set of covariance operators of these measures in the nuclear norm.

For measures with arbitrary means, it is necessary to add to the indicated condition the precompactness of the means.

(ii) *If a sequence of probability measures μ_i converges weakly to a measure μ and their second moments are finite and converge to the finite second moment of the measure μ, then $\{\mu_i\}$ satisfies the condition in* (i), *moreover, the covariance operators of the measures μ_i converge to the covariance operator of the measure μ in the nuclear norm and the means of the measures μ_i converge in norm to the mean of the measure μ.*

PROOF. (i) It follows from the stated conditions that the quadratic function $q(x)^2 = \sum_{n=1}^{\infty} C_n^2 (x, e_n)^2$ has uniformly bounded integrals with respect to the measures from $\mathcal{P}$ (in case of zero means) and is finite almost everywhere with respect to these measures. Since the sets $\{q^2 \leqslant R\}$ are compact ellipsoids, the Chebyshev inequality yields the uniform tightness of $\mathcal{P}$. Since the numbers C_n increase, we also have the uniform boundedness of the series corresponding to $C_n = 1$. The series obtained by replacing C_n by $\sqrt{C_n}$ will converge uniformly. Conversely, the uniform convergence of the indicated series yields the uniform boundedness of their sums in n from some number N, and the uniform boundedness of the sums from 1 to N is ensured by the uniform boundedness of the second moments.

Let us explain the connection with the compactness of the set of covariance operators in the nuclear norm. Under the stated condition, one can approximate the set of covariance operators of measures in $\mathcal{P}$ up to a given $\varepsilon > 0$ by a bounded finite-dimensional set of operators, namely, operators of the form P_nKP_n, where P_n is the orthogonal projection onto the linear span of $e_1, \dots, e_n$. Indeed, the nuclear norm of $K - P_nKP_n$ is $\sum_{i=n+1}^{\infty}(Ke_i, e_i)$, which is not greater than MC_{n+1}^{-2}. On the other hand, if the set of these covariance operators K is precompact in the nuclear norm, then for every m one can find a finite 2^{-m}-net $\mathcal{K}_m$ in the nuclear norm. There is a number n_1 such that $\sum_{n>n_1}(Ke_n, e_n) \leqslant 2^{-1}$ for all $K \in \mathcal{K}_1$. Then $\sum_{n>n_1}(Ke_n, e_n) \leqslant 1$ for all our operators K. By induction we find increasing numbers n_m such that $\sum_{n>n_m}(Ke_n, e_n) \leqslant 2^{1-m}$ for all our operators K. Now we take $C_n = 1$ if $n < n_1$, $C_n = m$ if $n_m \leqslant n < n_{m+1}$, which ensures the indicated condition.

It is clear that the shifts of measures from a uniformly tight family by vectors from a precompact set form a uniformly tight family.

(ii) First we consider the case of zero means. Convergence of the integrals of (x, x) by Corollary 2.7.3 gives convergence of the integrals of $(x, e_n)^2$ for each fixed n. Let us consider the vectors $\lambda_i = (\lambda_{i,n}) \in l^1$ whose components equal the integrals of the functions $(x, e_n)^2$ with respect to the measures μ_i, and also a similar vector $\lambda = (\lambda_n)$ obtained from the measure μ. We have coordinate-wise convergence of the vectors λ_i to λ and convergence of norms in l^1 as well. This gives convergence in norm in l^1, i.e., $\sum_{n=1}^{\infty}|\lambda_n - \lambda_{i,n}| \to 0$ as $i \to \infty$. Now we can find numbers $C_n > 0$ increasing to infinite sufficiently slowly such that $\sum_{n=1}^{\infty} C_n^2|\lambda_n - \lambda_{i,n}| \leqslant 1$ and $\sum_{n=1}^{\infty} C_n^2\lambda_n < \infty$. These are the desired numbers.

In case of nonzero means a_n of the measures μ_n it is enough to observe that they converge in norm to the mean a of the measure μ. Indeed, the functions $x \mapsto (x, y)$ with $\|y\| \leqslant 1$ are uniformly Lipschitz and $|(y, x)| \leqslant \|x\|$. This gives the uniform convergence of their integrals with respect to the measures μ_i to the integral with respect to the measure μ according to Theorem 2.7.1. Thus, we have obtained the uniform convergence of the quantities (y, a_n) to (y, a), i.e., convergence in norm. □

It is possible to derive from the proof an even stronger assertion: convergence of the second moments of the measures μ_i holds for the stronger norm with the square $\sum_{n=1}^{\infty} C_n^2(x, e_n)^2$ for some sequence of numbers $C_n \uparrow +\infty$.

Unlike the finite-dimensional case, the uniform boundedness of means and second moments does not give the uniform tightness. For example, the sequence $\{\delta_{e_n}\}$, where $\{e_n\}$ is a basis, is not uniformly tight, although these measures have zero covariance operators and means of unit norm. The measures $(\delta_{e_n} + \delta_{-e_n})/2$ have zero means and the second moment 1, but they are not uniformly tight as well.

2.5.3. Example. From assertion (i) in the proposition one can obtain a sufficient condition for weak convergence of a sequence of measures μ_i to a measure μ by adding the pointwise convergence of the Fourier transforms: $\widetilde{\mu_n}(y) \to \widetilde{\mu}(y)$. This question in a more general situation is considered in § 4.6, but here it suffices to observe only that this follows from the existence of weakly convergent subsequences and the fact that they have a common limit μ due to the equality of measures with equal Fourier transforms.

The assumption of existence of second moments is rather restrictive. It can be weakened, though, by making it less convenient for applications.

2.5.4. COROLLARY. *A family $\mathcal{P}$ of Borel probability measures on H is uniformly tight precisely when for every $\varepsilon > 0$ there exists a ball B_ε such that $\mu(H\backslash B_\varepsilon) \leqslant \varepsilon$ for all $\mu \in \mathcal{P}$ and for the restrictions of the measures in $\mathcal{P}$ to this ball the means belong to a compact set and the family of covariance operators is precompact in the nuclear norm.*

PROOF. Sufficiency of this condition is clear from the proposition, since the set of restrictions to B_ε approximates $\mathcal{P}$ in variation up to ε. Necessity is obvious, since this condition is fulfilled for measures on a compact set. The latter is seen from the fact that a compact set is contained in a compact ellipsoid on which the function $\sum_{n=1}^\infty C_n^2(x, e_n)^2$ with some $C_n \uparrow +\infty$ is bounded, hence the integrals of this function with respect to all probability measures on this ellipsoid are uniformly bounded. □

The following simple criterion of the uniform tightness is due to Prohorov [**525**].

2.5.5. PROPOSITION. *Let $\{e_n\}$ be an orthonormal basis in H and let P_n be the orthogonal projection onto the linear span H_n of the vectors $e_1, \ldots, e_n$. A family $\mathcal{P}$ of Borel probability measures on H is uniformly tight precisely when for each n the family of projections of measures from $\mathcal{P}$ on H_n is uniformly tight and*

$$\lim_{n\to\infty} \sup_{\mu\in\mathcal{P}} J_n(\mu) = 0, \quad J_n(\mu) := \int_H \Bigl(1 - \exp(-\|x - P_n x\|^2/2)\Bigr)\,\mu(dx).$$

PROOF. The necessity of this condition is obvious from the Dini theorem, according to which the functions $\|x - P_n x\|^2$ converge to zero uniformly on every compact set. For the proof of sufficiency we use Remark 2.3.1 (or Example 2.3.2). Fix $\varepsilon, \delta \in (0,1)$. Let us take n such that $\sup_{\mu\in\mathcal{P}} J_n(\mu) \leqslant \varepsilon\delta^2/16$. Then the estimate $\delta^2/8 \leqslant 1 - \exp(-\delta^2/4)$ and the Chebyshev inequality give

$$8^{-1}\delta^2\mu\bigl(x\colon \|x - P_n x\|^2 \geqslant \delta^2/2\bigr) \leqslant J_n(\mu) \leqslant \varepsilon\delta^2/16$$

for all $\mu \in \mathcal{P}$, i.e., $\mu\bigl(x\colon \|x - P_n x\|^2 < \delta^2/2\bigr) > 1 - \varepsilon/2$. We now observe that if $\|x - P_n x\|^2 < \delta^2/2$, then $\operatorname{dist}(x, H_{n-1}) < \delta/\sqrt{2}$. Along with the uniform tightness of projections on every H_n this yields the condition indicated in Remark 2.3.1. One could also refer to Example 2.3.2, having verified the uniform tightness of the images of measures from $\mathcal{M}$ for every functional $x \mapsto (x, u)$; to this end it suffices to take n in such a way that the norm of $u - P_n u$ becomes small. □

With the aid of Proposition 1.6.5 one can obtain a sufficient condition for the uniform tightness of a family of probability measures on H in terms of their Fourier transforms.

2.5.6. THEOREM. *Let $\mathcal{P}$ be a family of Borel probability measures on H such that for every $\varepsilon > 0$ there exists a Hilbert–Schmidt operator $T_\varepsilon \geqslant 0$ such that the inequality $\|T_\varepsilon y\| \leqslant 1$ yields the inequality $|1 - \widetilde{\mu}(y)| \leqslant \varepsilon$. Then the family $\mathcal{P}$ is uniformly tight.*

In this case the operators T_ε can be taken in the form $\delta_\varepsilon^{-1}T$ with some numbers $\delta_\varepsilon > 0$ and some nonnegative Hilbert–Schmidt operator T.

PROOF. For T one can take $\sum_{n=1}^\infty c_n T_{1/n}$, where $c_n = 2^{-n}(\|T_{1/n}\|_1 + 1)^{-1}$. Let $\varepsilon > 0$ and let $\{e_n\}$ be an eigenbasis of the operator T, $Te_n = t_n e_n$. There are numbers $C_n \uparrow +\infty$ such that the series of $C_n^2 t_n^2$ converges. Let $\mu \in \mathcal{P}$. The projection of μ onto the linear span H_n of the vectors $e_1,\dots,e_n$ is denoted by μ_n. According to Proposition 1.6.5 we have

$$\mu_n\biggl(x\colon \sum_{i=1}^n C_i^2(x,e_i)^2>t\biggr)\leqslant 6\biggl(\varepsilon+2t^{-1}\sum_{i=1}^n C_i^2t_i^2\biggr),$$

since the Fourier transform of μ_n on H_n coincides with the restriction of $\widetilde{\mu}$. Hence the left-hand side does not exceed 7ε for $t=2\varepsilon^{-1}\sum_{i=1}^\infty C_i^2t_i^2$. Therefore, for the compact ellipsoid $S=\bigl\{x\colon \sum_{i=1}^\infty C_i^2(x,e_i)^2\leqslant t\bigr\}$ we have $\mu(H\backslash S)\leqslant 7\varepsilon$ by convergence $\mu_n\Rightarrow\mu$. Thus, $\mathcal{P}$ is uniformly tight. □

Note that the indicated condition is sufficient, but not necessary for the uniform tightness of $\mathcal{P}$. As shown in Prohorov, Sazonov [**527**], on an infinite-dimensional space H there is no topology such that the equicontinuity of the Fourier transforms of a family of probability measures in this topology is equivalent to the uniform tightness of this family. Let us give a justification borrowed from Mushtari [**475**, p. 205] (it is given here for the condition with nuclear operators, but can be easily modified for the general case). Let us write the element of the space l^2 with components $1/n$ as the sum of the series of elements x_k obtained by partitioning the set of natural numbers into consecutive disjoint finite intervals M_k, $k\in\mathbb{N}$, chosen such that the sum S_k of the numbers $1/n$ with $n\in M_k$ exceeds k^2. Clearly, the Dirac measures at the points a_n of the form $\sum_{i\in N_k}x_i$ obtained for all possible subsets $N_k\subset M_k$ and numbered consecutively considering all finite sets M_k converge weakly to the Dirac measure at zero. Suppose that there is an ellipsoid $V=\{y\colon (Ty,y)\leqslant 1\}$ with a nuclear operator $T\geqslant 0$ such that $|1-\widetilde{\delta_{a_n}}(y)|\leqslant 1/4$ for all $y\in V$. Since $\widetilde{\delta_{a_n}}(y)=\exp\bigl(i(a_n,y)\bigr)$, we have $|(a_n,y)|\leqslant 1/2$. Due to our possibility to take arbitrary sets in M_k we have $\sum_{i\in M_k}|(x_i,y)|\leqslant 1$ whenever $y\in V$.

Now for every k we consider the probability measure μ_k that is concentrated at the points $kx_i/\|x_i\|$ with $i\in M_k$ and assigns the value $\|x_i\|/S_k$ to such a point. For every $y\in V$ we have

$$\int_H|(y,x)|\,\mu_k(dx)=\sum_{i\in M_k}\frac{k|(y,x_i)|}{S_k}\leqslant k^{-1}.$$

Hence $|1-\widetilde{\mu}_k(y)|\leqslant 1/k$, but the sequence $\{\mu_k\}$ is not uniformly tight, since μ_k is concentrated on the sphere of radius k.

However, one can modify the formulation of Theorem 2.5.6 and obtain the following result due to Prohorov [**525**].

2.5.7. THEOREM. *A family $\mathcal{P}$ of Borel probability measures on H is uniformly tight precisely when, for every $\varepsilon>0$, for every measure $\mu\in\mathcal{P}$ there exists a nuclear operator $T_\mu\geqslant 0$ such that the set $\{T_\mu\}$ is precompact in the nuclear norm and*

$$|1-\widetilde{\mu}(y)|\leqslant\varepsilon+(T_\mu y,y)\quad \forall\, y\in H,\forall\,\mu\in\mathcal{P}.$$

PROOF. The necessity of this condition follows easily from Corollary 2.5.4. We verify its sufficiency with the aid of Proposition 2.5.5. Let $\varepsilon>0$. For a fixed number $n<m$, the integral of $1-\exp\bigl(-\sum_{i=n+1}^m(x,e_i)^2/2\bigr)$ with respect to the measure μ equals the integral of the same function against the projection $\mu_{n,m}$ of the measure μ on the linear span $H_{n,m}$ of the vectors $e_n,\dots,e_m$, which coincides with the integral of $1-\operatorname{Re}\widetilde{\mu}_{n,m}$ with respect to the standard Gaussian measure $\gamma_{n,m}$ on $H_{n,m}$ (see §1.6). Therefore, this integral does not exceed the integral

of $\varepsilon + (T_\mu y, y)$ with respect to $\gamma_{n,m}$, which equals $\varepsilon + \sum_{i=n}^m (T_\mu e_i, e_i)$. By our condition there exists n such that the latter quantity will be less than 2ε for all $m > n$ and all $\mu \in \mathcal{P}$. Letting $m \to +\infty$ we obtain for $J_n(\mu)$ the condition from Proposition 2.5.5. The uniform tightness of the projections on H_n follows from Corollary 1.6.4. $\square$

2.6. The Skorohod representation

Let us start with a simple observation: almost everywhere convergent mappings induce weakly convergent measures, i.e., if P is a probability measure on a measurable space $(\Omega, \mathcal{F})$ and ξ, ξ_n, where $n \in \mathbb{N}$, are measurable mappings from Ω to a metric space X equipped with its Borel σ-algebra $\mathcal{B}(X)$ such that $\xi(\omega) = \lim_{n\to\infty} \xi_n(\omega)$ for P-a.e. $\omega \in \Omega$, then the induced measures $\mu_n = P\circ\xi_n^{-1}$ converge weakly to the measure $\mu = P\circ\xi^{-1}$. Indeed, for all $\varphi \in C_b(X)$ by the Lebesgue theorem we have

$$\lim_{n\to\infty} \int_\Omega \varphi\bigl(\xi_n(\omega)\bigr)\, P(d\omega) = \int_\Omega \varphi\bigl(\xi(\omega)\bigr)\, P(d\omega).$$

Skorohod [**584**], [**585**] discovered that every weakly convergent sequence of probability measures on a complete separable metric space X admits the described representation, moreover, for P one can take Lebesgue measure on $[0,1]$ (for measures on $X = \mathbb{R}^d$ this had been shown in Hammersley [**308**]). It was later shown in Blackwell, Dubins [**70**] and Fernique [**226**] that one can simultaneously parametrize all Borel probability measures on X by mappings from $[0,1]$ such that to weakly convergent sequences of measures there will correspond almost everywhere convergent sequences of mappings. Here we give a simple derivation of these results with the aid of functional-topological considerations. It will be useful to employ the following concept introduced in Bogachev, Kolesnikov [**87**] (for topological spaces, which is discussed in § 5.4) that is of independent interest.

2.6.1. DEFINITION. A metric space X possesses the *strong Skorohod property* for Radon measures if to every Radon probability measure μ on X one can associate a Borel mapping $\xi_\mu\colon [0,1] \to X$ such that μ is the image of Lebesgue measure under the mapping ξ_μ and if measures μ_n converge weakly to μ, then $\xi_{\mu_n}(t) \to \xi_\mu(t)$ a.e.

If such a parametrization exists for the class of all Borel probability measures on X, then this property will be called the strong Skorohod property for Borel measures. For Souslin spaces, there is no difference between these two properties. Similarly one can define the strong Skorohod property for other classes of measures (for example, discrete).

2.6.2. LEMMA. *Let X be a metric space with the strong Skorohod property for Radon measures. Then*

(i) *every subset Y of X possesses this property;*

(ii) *if F is a continuous mapping from X onto a metric space Y and there is a mapping $\Psi\colon \mathcal{P}_r(Y) \to \mathcal{P}_r(X)$ such that $\Psi(\nu)\circ F^{-1} = \nu$ for all $\nu \in \mathcal{P}_r(Y)$ and $\Psi(\nu_n) \Rightarrow \Psi(\nu)$ whenever $\nu_n \Rightarrow \nu$, then Y possesses the strong Skorohod property for Radon measures.*

PROOF. (i) Each Radon measure μ on Y extends uniquely to a Radon measure on X and Y is measurable with respect to this extension, since Y contains compact sets K_n (these sets will be compact in X as well) whose union has measure one.

Suppose that we have fixed a parametrization of $\mathcal{P}_r(X)$ and $\xi_\mu\colon [0,1] \to X$ is the Borel mapping corresponding to μ in the definition of the strong Skorohod property. As noted above, there exists a set $B \subset Y$ of μ-measure one that is σ-compact in X and Y. Set

$$\eta_\mu(t) = \xi_\mu(t) \text{ if } t \in \xi_\mu^{-1}(B), \ \eta_\mu(t) = z \text{ if } t \notin \xi_\mu^{-1}(B),$$

where z is an arbitrary point in Y. Then $\lambda\bigl(\xi_\mu^{-1}(B)\bigr) = 1$ and hence $\eta_\mu(t) = \xi_\mu(t)$ for almost all t in $[0,1]$, whence we obtain the equality $\lambda\circ\eta_\mu^{-1} = \lambda\circ\xi_\mu^{-1}$. If probability measures μ_n on Y converge weakly to a measure μ, then their extensions to X converge weakly to the extension of μ, whence we have $\lim\limits_{n\to\infty} \xi_{\mu_n}(t) = \xi_\mu(t)$ almost everywhere. Therefore, almost everywhere $\lim\limits_{n\to\infty} \eta_{\mu_n}(t) = \eta_\mu(t)$.

(ii) For any $\nu \in \mathcal{P}_r(Y)$ set $\eta_\nu(t) = F\bigl(\xi_{\Psi(\nu)}(t)\bigr)$, where ξ is a parametrization of measures in $\mathcal{P}_r(X)$ by Borel mappings of the interval $[0,1]$ to X. Then

$$\lambda\circ\eta_\nu^{-1} = \bigl(\lambda\circ\xi_{\Psi(\nu)}^{-1}\bigr)\circ F^{-1} = \Psi(\nu)\circ F^{-1} = \nu.$$

If measures ν_n converge weakly to the measure ν on Y, then the measures $\Psi(\nu_n)$ converge weakly to the measure $\Psi(\nu)$ on X, hence $\xi_{\Psi(\nu_n)}(t) \to \xi_{\Psi(\nu)}(t)$ for almost all t in the interval $[0,1]$, whence we obtain $\eta_{\nu_n}(t) \to \eta_\nu(t)$ for all such t by the continuity of F with respect to weak convergence of sequences. □

The mapping Ψ in assertion (ii) of the lemma is called a *continuous right inverse* (it is indeed continuous in the weak topology on the space of measures studied in Chapter 3) to the induced mapping

$$\widehat{F}\colon \mathcal{P}_r(X) \to \mathcal{P}_r(Y), \quad \mu \mapsto \mu\circ F^{-1}.$$

Let $F\colon X \to Y$ be a continuous surjection of compact spaces X and Y (not necessarily metrizable).

A linear operator $U\colon C(X) \to C(Y)$ is called a *regular averaging operator* for F if $U\psi \geqslant 0$ for all functions $\psi \geqslant 0$ and $U(\varphi\circ F) = \varphi$ for all functions $\varphi \in C(Y)$. Such an operator is automatically continuous and has unit norm.

It is readily seen that the operator

$$V = U^*\colon \mathcal{M}_r(Y) = C(Y)^* \to \mathcal{M}_r(X) = C(X)^*$$

takes $\mathcal{P}_r(Y)$ to $\mathcal{P}_r(X)$ and $\widehat{F}\circ V$ is the identity mapping on $\mathcal{M}_r(Y)$, i.e., V is a continuous right inverse for $\widehat{F}$. Indeed, for all $\nu \in \mathcal{M}_r(Y)$ and $\varphi \in C(Y)$ we have

$$\begin{aligned}\int_Y \varphi(y)\, \widehat{F}\bigl(V(\nu)\bigr)(dy) &= \int_X \varphi\bigl(F(x)\bigr)\, V(\nu)(dx)\\ &= \int_Y U(\varphi\circ F)(y)\, \nu(dy) = \int_Y \varphi(y)\, \nu(dy).\end{aligned}$$

A compact space S is called a *Milyutin space* if, for some cardinality τ, there exists a continuous surjection $F\colon \{0,1\}^\tau \to S$, where $\{0,1\}$ is the two-point space, possessing a regular averaging operator.

In this section we are interested only in metric spaces; the case of topological spaces is considered in Chapter 5.

According to the well-known Milyutin lemma (see Pełczyński [**508**, Theorem 5.6], Fedorchuk, Filippov [**217**, Chapter 8, §4]), any compact interval is a Milyutin space. In addition, it is known that the product of an arbitrary family of compact metric spaces is a Milyutin space. In particular, $S = [0,1]^\infty$ is a Milyutin space, moreover, for τ one can take $\mathbb{N}$. Since the countable power $\{0,1\}^\infty$ of the two-point set is homeomorphic to the classical Cantor set $C \subset [0,1]$, which consists of all numbers of the interval $[0,1]$ the triadic expansion of which does not contain 1 (see § 1.1 or Bogachev, Smolyanov [**96**, p. 62], Engelking [**203**, Example 3.1.28]), the next result follows from what has been said.

2.6.3. LEMMA. *Let S be a nonempty compact metric space and C the Cantor set. Then there exists a continuous surjection $F\colon C \to S$ for which the mapping $\widehat{F}$ possesses a linear continuous right inverse.*

Now everything is prepared for the proof of the following theorem on parametrization of probability measures.

2.6.4. THEOREM. *Let X be a universally measurable set in a complete separable metric space. Then to every Borel probability measure μ on X one can associate a Borel mapping $\xi_\mu\colon [0,1] \to X$ such that $\mu = \lambda\circ\xi_\mu^{-1}$, where λ is Lebesgue measure, and if measures μ_n converge weakly to a measure μ, then $\xi_{\mu_n}(t)\to\xi_\mu(t)$ for almost all $t\in[0,1]$. If X is an arbitrary set in a complete separable metric space, then an analogous assertion is true for the space of Radon probability measures.*

PROOF. Since any Polish space is homeomorphic to a G_δ-set in $[0,1]^\infty$ (see Theorem 2.1.1), by Lemma 2.6.2(i) one can assume that $X \subset [0,1]^\infty$. Therefore, part (ii) of this lemma and Lemma 2.6.3 reduce our assertion to the case of subsets of $[0,1]$, which further reduces everything to the case $X = [0,1]$. In the latter case the required mapping is given by the explicit formula

$$\xi_\mu(t) = \sup\Big\{x \in [0,1]\colon\ \mu\big([0,x)\big) \leqslant t\Big\}. \tag{2.6.1}$$

Indeed, it is easy to see that for every point c we have the equality

$$\lambda\circ\xi_\mu^{-1}\big([0,c)\big) = \mu\big([0,c)\big).$$

Hence $\lambda\circ\xi_\mu^{-1} = \mu$. If measures μ_n converge weakly to a measure μ, then their distribution functions $F_{\mu_n}(t) = \mu_n\big([0,t)\big)$ converge to the distribution function F_μ of the measure μ at all continuity points of F_μ. Let $t \in [0,1]$ and $\varepsilon > 0$. If $\limsup\limits_{n\to\infty} \xi_{\mu_n}(t) > \xi_\mu(t) + 2\varepsilon$, then there is a point $x_0 \in \big(\xi_\mu(t)+\varepsilon, \xi_\mu(t)+2\varepsilon\big)$ with $F_\mu(x_0) = \lim\limits_{n\to\infty} F_{\mu_n}(x_0)$. For an infinite sequence of indices n_k we have $\xi_{\mu_{n_k}}(t) > x_0$, i.e., $F_{\mu_{n_k}}(x_0) \leqslant t$, whence $F_\mu(x_0) \leqslant t$. Hence $\xi_\mu(t) \geqslant x_0$, which is a contradiction. Similarly we consider the case $\liminf\limits_{n\to\infty} \xi_{\mu_n}(t) \leqslant \xi_\mu(t) - 2\varepsilon$. Thus, $\lim\limits_{n\to\infty} \xi_{\mu_n}(t) = \xi_\mu(t)$. In case of Radon measures an analogous reasoning applies to arbitrary subsets of Polish spaces. □

Thus, any subspace of a Polish space possesses the strong Skorohod property for Radon measures (and universally measurable subspaces possess the strong Skorohod property for Borel measures). In the paper Bogachev, Kolesnikov [**87**] the following more difficult result is proved, to which along with some additional discussion we return in Chapter 5.

2.6.5. THEOREM. *All complete metric spaces possess the strong Skorohod property for Radon measures.*

In relation to the Skorohod representation we observe that for weak convergence of the distributions $P\circ\xi_n^{-1}$ of random elements ξ_n in a separable metric space (X,d)

to the distribution $P\circ\xi^{-1}$ of a random element ξ it suffices to have convergence of ξ_n to ξ not almost surely, but only in probability, which means that

$$P\bigl(d(\xi_n, \xi) \geqslant c\bigr) \to 0$$

for every $c > 0$. Indeed, for every 1-Lipschitz function f with $|f| \leqslant 1$ we have

$$|\,\mathbb{E}\, f(\xi_n) - \mathbb{E}\, f(\xi)| \leqslant 2\,\mathbb{E} \min\bigl(1, d(\xi_n, \xi)\bigr) \to 0$$

by the Lebesgue dominated convergence theorem. In the general case convergence in probability is stronger than convergence in distribution, however, it turns out, as shown in Padmanabhan [**501**], that if we use all measures equivalent to the given probability measure, then both modes of convergence are equivalent.

2.6.6. PROPOSITION. *Let $(\Omega, \mathcal{B}, P)$ be a probability space and let X be a complete separable metric space. A sequence of measurable mappings $\xi_n\colon \Omega \to X$ converges in probability to a mapping ξ precisely when, for every measure Q equivalent to P, the measures $Q\circ\xi_n^{-1}$ converge weakly to the measure $Q\circ\xi^{-1}$.*

PROOF. Let us consider the case $X = [0, 1]$. If $P\circ\xi_n^{-1} \Rightarrow P\circ\xi^{-1}$, then we have $\|\xi_n\|_2 \to \|\xi\|_2$, since $\mathbb{E}\,\xi_n^2 \to \mathbb{E}\,\xi^2$ by convergence of the integrals of the function x^2 on $[0, 1]$ with respect to the distributions of ξ_n. In addition, for every set $A \in \mathcal{B}$ we have $(\xi_n, I_A)_2 \to (\xi, I_A)_2$. Indeed, by our condition one can take the equivalent measure Q with density $\bigl(1 + P(A)\bigr)^{-1}(1 + I_A)$. Convergence of the expectations of the variables ξ_n and ξ with respect to Q and also with respect to the measure P gives the indicated relation. Hence $\|\xi_n - \xi\|_2 \to 0$ (see Exercise 2.7.38 or Bogachev [**81**, Corollary 4.7.16]).

The case $X = \mathbb{R}$ reduces to the considered one by replacing the metric with a bounded one. This also yields the case $X = \mathbb{R}^\infty$, because in $\mathbb{R}^\infty$ we deal with the coordinate-wise convergence. In the general case one can use that X is homeomorphic to a set in $\mathbb{R}^\infty$. □

2.7. Complements and exercises

2.7(i). Uniform integrability

Let us prove one more result on convergence of integrals of unbounded functions with respect to weakly convergent measures under the condition of the so-called uniform integrability.

2.7.1. THEOREM. *If a net $\{\mu_\alpha\}$ of Borel probability measures on a metric space X converges weakly to a Borel probability measure μ, then, for every continuous function f on X satisfying the condition*

$$\lim_{R\to\infty} \sup_{\alpha} \int_{|f|\geqslant R} |f|\, d\mu_\alpha = 0, \qquad (2.7.1)$$

one has $\displaystyle\lim_\alpha \int_X f\, d\mu_\alpha = \int_X f\, d\mu.$

This convergence is uniform on any family of functions f that is equicontinuous at every point (as in Theorem 2.2.8) and such that also (2.7.1) is fulfilled uniformly.

PROOF. Note that $f \in L^1(\mu)$. Indeed, set $f_n = \min(|f|, n)$. Then $f_n \leqslant |f|$ and hence by the condition of the lemma we obtain

$$M := \sup_{n,\alpha} \int_X f_n \, d\mu_\alpha < \infty.$$

Since $f_n \in C_b(X)$, one has

$$\int_X f_n \, d\mu \leqslant M \quad \text{for all } n,$$

whence it follows that $f \in L^1(\mu)$. Let now $\varepsilon > 0$. Choose $R > 0$ such that

$$\int_{|f| \geqslant R} |f| \, d\mu_\alpha + \int_{|f| \geqslant R} |f| \, d\mu < \varepsilon \quad \forall \alpha.$$

Set $g = \max\bigl(\min(f, R), -R\bigr)$. For all α such that the absolute value of the integral of g with respect to the measure $\mu_\alpha - \mu$ is less than ε we obtain

$$\left| \int_X f \, d\mu_\alpha - \int_X f \, d\mu \right| \leqslant 3\varepsilon,$$

since $|g(x)| \leqslant |f(x)|$ and $g(x) = f(x)$ whenever $|f(x)| \leqslant R$. The second assertion can be easily seen from Theorem 2.2.8, which should be applied to the cut-off functions $f_R = \max\bigl(\min(f, R), -R\bigr)$. □

2.7.2. REMARK. Note that in case $f \geqslant 0$ the established equality implies the stated condition of uniform integrability. Indeed, let $\varepsilon > 0$. Let us take $R > 0$ such that the integral of f with respect to the measure μ over the set $X_R = \{f \geqslant R\}$ becomes smaller than ε. Then $R\mu(X_R) < \varepsilon$, hence $R\mu_\alpha(X_R) \leqslant \varepsilon$ for all $\alpha \geqslant \alpha_1$ for some index α_1. Increasing α_1, one can assume that the integral of the function $f_R = \min(f, R)$ with respect to the measure μ_α will differ from its integral with respect to the measure μ by less than ε, and also the difference of the integrals of f will be less than ε. Then the integrals of $\min(f, R)$ and f with respect to the measure μ_α with $\alpha \geqslant \alpha_1$ will differ by less than 2ε. This yields that the integral of f over X_R with respect to the measure μ_α is estimated by $2\varepsilon + R\mu_\alpha(X_R) \leqslant 3\varepsilon$.

From this remark and the theorem we obtain the following assertion.

2.7.3. COROLLARY. *Suppose that a net $\{\mu_\alpha\}$ of Borel probability measures on a metric space X converges weakly to a Borel probability measure μ and a continuous function $f \geqslant 0$ on X is such that the integrals of f with respect to the measures μ_α converge to the integral of f with respect to the measure μ. Then this convergence of integrals holds also for every continuous function g such that $|g| \leqslant f$.*

2.7(ii). Weak convergence of restrictions and total variations

Here we discuss the behavior of weak convergence under restricting measures to subsets. Some of these results will be repeated with proofs in Chapter 4 for topological spaces, so here we include only the formulations. It is clear that in the general case there is no convergence of restrictions: in Example 1.4.3, convergent measures vanish on the set $\{0\}$, but the limiting Dirac measure is concentrated on this set.

Suppose that a net $\{\mu_\alpha\}$ of Borel probability measures on a metric space X converges weakly to a Borel probability measure μ. Let us equip the set $X_0 \subset X$ with the induced topology (it is not assumed to be Borel). If X_0 has outer measure 1 for all measures μ_α and μ, then the induced measures μ_α^0 on X_0 are probability measures and converge weakly to the measure μ^0 induced by μ (the measure μ_α^0 is defined by the formula $\mu_\alpha^0(B\cap X_0) := \mu_\alpha(B)$ for $B \in \mathcal{B}(X)$, where the sets of the form $B\cap X_0$ constitute exactly $\mathcal{B}(X_0)$ and the measure μ_α^0 is well-defined, see Example 1.1.2). This is seen, for example, from Theorem 2.2.5, since the open and closed sets in X_0 have the form of intersections of X_0 with sets of the respective type in X.

Another case when we have convergence of the measures μ_α^0 arises if the set X_0 is open or closed and $\mu_\alpha(X_0) \to \mu(X_0)$. Here the cited theorem applies as well.

The next result from Varadarajan [**635**, Part 2, Theorem 3] is useful for the study of weak convergence of signed measures. Its justification will be given in Theorem 4.8.1 for general topological spaces.

2.7.4. THEOREM. *Suppose that a net $\{\mu_\alpha\}$ of Borel measures on a metric space X converges weakly to a Borel measure μ. Then, for every open set U, we have*

$$\liminf_\alpha |\mu_\alpha|(U) \geqslant |\mu|(U).$$

In addition, the net of measures $|\mu_\alpha|$ converges weakly to $|\mu|$ precisely when $|\mu_\alpha|(X)$ converges to $|\mu|(X)$.

2.7.5. COROLLARY. *Suppose that a net $\{\mu_\alpha\}$ of Borel measures on a metric space X converges weakly to a Borel measure μ and*

$$\lim_\alpha |\mu_\alpha|(X) = |\mu|(X).$$

Let $\mu_\alpha = \mu_\alpha^+ - \mu_\alpha^-$ and $\mu = \mu^+ - \mu^-$. Then the nets $\{\mu_\alpha^+\}$ and $\{\mu_\alpha^-\}$ converge weakly to μ^+ and μ^-, respectively.

Let us consider the preservation of weak convergence under multiplication by a function. It follows from the very definition that if measures μ_α converge weakly to a measure μ, then for every bounded continuous function f the measures $f\cdot\mu_\alpha$ converge weakly to the measure $f\cdot\mu$. However, there are less obvious results of this sort. For example, Corollary 2.2.10 and the fact that for every function $f \in C_b(X)$ the measures $f\cdot\mu_\alpha$ converge weakly to the measure $f\cdot\mu$ yield the following assertion.

2.7.6. PROPOSITION. *Suppose that a net of Borel probability measures μ_α on a metric space X converges weakly to a Borel probability measure μ and a bounded Borel function g is continuous μ-a.e. Then the measures $g\cdot\mu_\alpha$ converge weakly to the measure $g\cdot\mu$.*

2.7(iii). Convergence of products

Let us consider weak convergence of products of measures. We recall that the product of two Borel measures μ and ν on metric spaces X and Y is defined as a measure on the Borel σ-algebra provided that both spaces are separable (otherwise the product $\mathcal{B}(X)\otimes\mathcal{B}(Y)$ is smaller than $\mathcal{B}(X\times Y)$). It also suffices that both measures have separable supports, since in this case their product is a Borel measure on the product of supports, after which it extends to a Borel measure on $X\times Y$

by zero outside the product of supports. Similarly the product of countably many probability τ-additive measures μ_n on spaces X_n extends to a τ-additive probability measure on the product of X_n.

2.7.7. PROPOSITION. *Let $\{\mu_\alpha\}$ and $\{\nu_\alpha\}$ be two nets of Borel probability measures on separable metric spaces X and Y converging weakly to measures μ and ν, respectively. Then the measures $\mu_\alpha\otimes\nu_\alpha$ on $X\times Y$ converge weakly to $\mu\otimes\nu$.*

If, for every n, a net of τ-additive measures μ_α^n on a metric space X_n converges weakly to a τ-additive measure μ^n, then the net of products $\bigotimes_{n=1}^\infty \mu_\alpha^n$ converges weakly to the product $\bigotimes_{n=1}^\infty \mu^n$.

PROOF. This assertion is also proved in Chapter 4 in the topological case, but here one can use the criterion of weak convergence by means of bounded Lipschitz functions. Indeed, if f is such a function on $X\times Y$, then the functions

$$g_\alpha\colon\ y\mapsto \int_X f(x,y)\,\mu_\alpha(dx)$$

converge uniformly on Y to an analogous function g for the measure μ according to Theorem 2.2.8, since the functions $x\mapsto f(x,y)$ are uniformly bounded and uniformly Lipschitz. Since the function g is continuous on Y (by the Lebesgue dominated convergence theorem), the integrals of g_α against μ_α converge to the integral of g against μ. Note that this proof works if at least one of the two spaces is separable (if both are nonseparable, then the problem already arises with a Borel extension of the measure $\mu\otimes\nu$, which is defined on the σ-algebra $\mathcal{B}(X)\otimes\mathcal{B}(Y)$).

The case of a countable product reduces to the case of finite products by Corollary 2.4.8. □

Let us obtain a similar result for signed measures in the case of sequences (for its version for nets, see Theorem 4.3.18).

2.7.8. PROPOSITION. *Let $\{\mu_n\}$ and $\{\nu_n\}$ be two sequences of Borel measures on complete separable metric spaces X and Y converging weakly to measures μ and ν, respectively. Then the measures $\mu_n\otimes\nu_n$ on $X\times Y$ converge weakly to the measure $\mu\otimes\nu$.*

PROOF. Both sequences are bounded in variation and tight by the Prohorov theorem. Hence the sequence of measures $\mu_n\otimes\nu_n$ is bounded in variation and tight as well. The integrals of the functions of the form $f_1(x)g_1(y)+\cdots+f_k(x)g_k(y)$, where $f\in C_b(X)$, $g\in C_b(Y)$, with respect to these measures converge (by Fubini's theorem) to the integral with respect to the measure $\mu\otimes\nu$. It is clear that the class of such functions is an algebra, contains 1 and separates points. Hence Theorem 2.3.10 applies. □

2.7(iv). Weak convergence of measures on Banach spaces

Here we show that a weakly convergent sequence of Radon measures on a Banach space (or even on a Fréchet space, i.e., a locally convex space metrizable by a complete metric) in fact converges weakly in a stronger sense, namely, on a compactly embedded separable reflexive Banach space. We recall that a linear subspace E in a Fréchet space X is called a compactly embedded Banach space if E possesses a norm making it a Banach space the unit ball of which (with respect to this norm) has compact closure in X. The proof employs the Grothendieck construction associating with a centrally symmetric convex compact set K in a locally convex space X a certain Banach space: this is the Banach space E_K that is the linear span of K equipped with the norm in which K is the closed unit ball (see Bogachev, Smolyanov [**97**, § 2.5]).

2.7.9. THEOREM. *Let $\mathcal{M}$ be a uniformly tight family of Radon measures on a Fréchet space X. Then there exists a reflexive separable Banach space $E \subset X$ such that the closed unit ball of E is compact in X, $|\mu|(X\backslash E) = 0$ for all $\mu \in \mathcal{M}$, and the family $\mathcal{M}$ is also uniformly tight with respect to the norm of E.*

PROOF. It is clear that it suffices to consider nonnegative measures. The topology of the space X is defined by a metric ϱ. For every n there is a compact set K_n with $\mu(X\backslash K_n) < 1/n$ for all $\mu \in \mathcal{M}$. Then $\mu\bigl(X\backslash \bigcup_{n=1}^{\infty} K_n\bigr) = 0$ for all $\mu \in \mathcal{M}$. Pick numbers $c_n > 0$ such that $c_n K_n$ is contained in the ball of radius $1/n$ centered at the origin. It is readily verified that the closure S of the set $\bigcup_{n=1}^{\infty} c_n K_n$ is compact (every sequence in this set has a limit point).

The closed absolutely convex hull K_0 of the set S is compact, but this set may be not suitable, since the corresponding Grothendieck space E_{K_0} need not be even separable. However, by Corollary 2.5.12 in [**97**] one can take a larger centrally symmetric convex compact set W such that E_W will be a separable reflexive space and K_0 will be also compact as a subset of E_W. The measures $\mu \in \mathcal{M}$ can be restricted to $E = E_W$, since the Borel sets in E_0 are Borel in X as well (see p. 50). It remains to observe that the family $\mathcal{M}$ is uniformly tight on E_W, since S is compact in E_W and $\mu(E_W\backslash c_n^{-1}S) < 1/n$ for all measures $\mu \in \mathcal{M}$. □

If the space X is Hilbert, then E can be chosen Hilbert too. To this end, for compact sets we take ellipsoids. However, for a non-Hilbert Banach space X it is not always possible to find a Hilbert space E even for a single measure.

A separable Banach space X is called a space with the approximation property if there exists a sequence of finite-dimensional continuous operators P_n on X such that $\|P_n x - x\| \to 0$ for each x. This property holds in every space with a Schauder basis (a sequence of vectors φ_n such that for each x there are unique numbers c_n with $x = \sum_{n=1}^{\infty} c_n \varphi_n$), but there exist spaces without this property.

2.7.10. THEOREM. *Let X be a Banach space with the approximation property. A set $\mathcal{M} \subset \mathcal{P}(X)$ is uniformly tight (equivalently, has compact closure in the weak topology, see Chapter 3) precisely when*

$$\lim_{r\to\infty} \sup_{\mu\in\mathcal{M}} \mu(x\colon \|x\| > r) = 0, \qquad \lim_{n\to\infty} \sup_{\mu\in\mathcal{M}} \mu(x\colon \|x - P_n x\| > \varepsilon) = 0$$

for all $\varepsilon > 0$, where $\{P_n\}$ are operators from the approximation property.

PROOF. The necessity of these conditions is clear from the Prohorov theorem, since $\|x - P_n x\| \to 0$ uniformly on compact sets (this can be easily obtained from the uniform boundedness of $\|P_n\|$ that holds by the Banach–Steinhaus theorem). The converse follows from Example 2.3.2, the first condition of which is ensured by the first of the indicated equalities and the second condition by the second one, since $P_n(X)$ is finite-dimensional. □

Let us consider yet another class of spaces. A Banach space X is called uniformly convex if, for every $\varepsilon > 0$, there exists $\delta > 0$ such that for all $x, y \in X$ with $\|x\| = \|y\| = 1$ we have $\|(x+y)/2\| \leqslant 1 - \delta$ whenever $\|x - y\| \geqslant \varepsilon$; for example, the space L^p with $p \in (1, +\infty)$ has this property. The uniform convexity implies the so-called Radon–Riesz property (or property H): weak convergence $x_n \to x$ along with convergence of norms $\|x_n\| \to \|x\|$ yields convergence in norm $\|x - x_n\| \to 0$. It is also known that all uniformly convex Banach spaces are reflexive (see Diestel [**168**, Chapter 2, § 4] or Bogachev, Smolyanov [**96**, § 6.10(iii)]).

For every probability measure P on a space Ω and every separable Banach space X, for any $p \geqslant 1$ we can define the Banach space $L^p(P, X)$ of equivalence classes of measurable mappings $f\colon \Omega \to X$ with finite norm $\|f\|_p$ for which $\|f\|_p^p$ is the integral of $\|f(x)\|^p$. If X is uniformly convex, then so is $L^p(P, X)$ whenever $1 < p < \infty$ (see, e.g., McShane [**452**]), but for property H this is false (see Smith, Turett [**588**]).

2.7.11. THEOREM. *Let X be a separable uniformly convex Banach space and let Radon probability measures μ_n and μ on X be such that $\mu_n \circ l^{-1} \Rightarrow \mu \circ l^{-1}$ for each $l \in X^*$. If for some $p \in (1, +\infty)$ the integrals of $\|x\|^p$ with respect to the measures μ_n converge to the integral with respect to the measure μ, then $\mu_n \Rightarrow \mu$.*

PROOF. The space X^* is separable due to reflexivity, hence its unit sphere contains a dense sequence $\{l_j\}$ defining a continuous injection $h\colon X \to \mathbb{R}^\infty$. By our condition the measures $\mu_n \circ h^{-1}$ converge weakly to $\mu_n \circ h^{-1}$ on $\mathbb{R}^\infty$. By the Skorohod theorem there exist Borel mappings $\xi_n, \xi\colon [0, 1] \to \mathbb{R}^\infty$ such that $\mu_n = \lambda \circ \xi_n^{-1}$ and $\mu = \lambda \circ \xi^{-1}$, where λ is Lebesgue measure on $[0, 1]$, and $\xi_n(s) \to \xi(s)$ a.e., which means convergence $l_j\bigl(\xi_n(s)\bigr) \to l_j\bigl(\xi(s)\bigr)$ a.e. We observe that these mappings a.e. take values in X (identified with $h(X)$), since the measures μ_n and μ are concentrated on $h(X)$. As noted above, the space $L^p(\lambda, X)$ is also uniformly convex. The dual space coincides with $L^q(\lambda, X^*)$, $q = p/(p-1)$, and the continuous functionals have the form of integrals of $g(s)\bigl(f(s)\bigr)$, where $g \in L^q(\lambda, X^*)$. By our condition the quantities $\|\xi_n\|_p^p$, equal to the integrals of $\|x\|^p$ with respect to the measures μ_n, converge to $\|\xi\|_p^p$ (the integral of $\|x\|^p$ with respect to the measure μ). In order to obtain convergence $\|\xi_n - \xi\|_p \to 0$, it suffices to show that one has convergence $g(\xi_n) \to g(\xi)$ for every $g \in L^q(\lambda, X^*)$. By convergence of norms it suffices to do this for elements of an everywhere dense set. For such a set we can take the linear span of mappings of the form $\psi \cdot l_j$, where ψ is a bounded Borel function. So we have to verify convergence of the integrals of $\psi l_j(\xi_n)$ to the integral of $\psi l_j(\xi)$ for each j. It holds by convergence almost everywhere and the uniform boundedness of the integrals of $|l_j(\xi_n)|^p$. $\square$

A more general fact is proved in Baushev [**47**] (see also [**46**]), in particular, in place of convergence of the integrals of $\|x\|^p$ it suffices to require convergence of the integrals of $\varphi(\|x\|)$, where φ is a continuous positive strictly increasing function on $[0, +\infty)$ such that the integrals of $\varphi(\|x\|)I_{\|x\| \geqslant R}$ with respect to the measures μ_n tend to 0 uniformly in n as $R \to +\infty$. In the same paper, the condition of uniform convexity is relaxed.

2.7(v). Weak convergence on C and L^p

In the theory of random processes one often uses weak convergence of finite-dimensional distributions of processes. From the point of view of distributions in path spaces this corresponds to weak convergence of measures on the path space equipped with the topology of pointwise convergence.

We recall that if $\{\xi_t\}_{t\in T}$ is a *random process* on a probability space $(\Omega, \mathcal{B}, P)$ with a nonempty parameter set T, i.e., just a family of $\mathcal{B}$-measurable random variables indexed by elements of T, then its finite-dimensional distributions, determined by collections of points $t_1, \dots, t_n \in T$, are defined by the following formula:

$$P_{t_1,\dots,t_n}(B) = P\bigl(\omega \in \Omega\colon\ (\xi_{t_1}(\omega), \dots, \xi_{t_n}(\omega)) \in B\bigr), \quad B \in \mathcal{B}(\mathbb{R}^n).$$

It is easy to verify that the set of points ω in the right-hand side belongs to $\mathcal{B}$, so that its probability is defined. Hence $P_{t_1,\dots,t_n}$ is a Borel probability measure on $\mathbb{R}^n$ (independently of the nature of the set Ω). Sometimes it is assumed that the points t_i are distinct, but this is not important. This assumption is convenient in the case where finite-dimensional distributions are determined not by ordered collections of points, but by finite subsets $S \subset T$. By the classical Kolmogorov theorem, on the space $\mathbb{R}^T$ of all real functions on T we obtain a probability measure P_ξ on the σ-algebra $\sigma(\mathbb{R}^T)$ generated by all cylindrical sets of the form

$$C_{t_1,\dots,t_n,B} := \bigl\{x\colon\ (x(t_1), \dots, x(t_n)) \in B\bigr\}, \quad B \in \mathcal{B}(\mathbb{R}^n),$$

for which $P_\xi(C_{t_1,\dots,t_n,B}) = P_{t_1,\dots,t_n}(B)$. The measure P_ξ is called the *distribution of the process in the path space.*

Finite-dimensional distributions possess the property that the measure corresponding to a set of indices is mapped to the measure corresponding to a subset of this set of indices under the naturally arising projection of the finite-dimensional space. The second compatibility property of finite-dimensional distributions is that for permutations of indices $t_1, \dots, t_n$ the measures $P_{t_1,\dots,t_n}$ change according to the linear transformations of $\mathbb{R}^n$ generated by these permutations.

By the Kolmogorov theorem every collection of Borel probability measures $P_{t_1,\dots,t_n}$ on the spaces $\mathbb{R}^n$ with these two compatibility properties is the collection of finite-dimensional distributions of some random process. This process can be constructed in a very simple way if we prove that $\mathbb{R}^T$ carries a probability measure the values of which on the cylinders $C_{t_1,\dots,t_n,B}$ are exactly the values $P_{t_1,\dots,t_n}(B)$ (the proof of this fact is not quite trivial); this measure on the space of all trajectories $\mathbb{R}^T$ can be taken for the measure P by setting $\xi_t(\omega) := \omega(t)$.

This simple and important construction is not always convenient, since usually the trajectories of the considered processes have certain additional properties like continuity or integrability, and it would be desirable to define the distribution of such a process not on all paths, but only on those possessing the required property.

It is often impossible to merely restrict the measure to the set of such paths, because this set turns out to be nonmeasurable with respect to P_ξ. For example, this is the case if $T = [0,1]$ and we are interested in the set of continuous paths. Even if the process ξ is identically zero, so that it seems that its distribution must be concentrated on the single zero path (i.e., be the Dirac measure at the zero function), an unpleasant thing occurs: the zero function is nonmeasurable with respect to the introduced measure P_ξ on $\sigma(\mathbb{R}^T)$! Indeed, the outer measure of the point 0 is 1, since every cylinder containing it has measure 1. However, its inner measure from the point of view of $\sigma(\mathbb{R}^T)$ equals zero, since the point contains no nonempty sets from $\sigma(\mathbb{R}^T)$ due to the fact that any set from $\sigma(\mathbb{R}^T)$ has the form $\bigl\{x\colon\ (x(t_i))_{i=1}^\infty \in B\bigr\}$ for some countable set of points t_i and some set $B \in \mathcal{B}(\mathbb{R}^\infty)$ (Exercise 2.7.46). For this reason extensions of distributions of processes to more special sets of paths (when this is possible) are done by other means. Below we discuss the case of spaces L^p, and here we note that in the case of the space of continuous paths $C[a,b]$ the distribution of a process ξ is constructed on $C[a,b]$ with the aid of the "restriction" from Example 1.1.2 provided that $P_\xi^*(C[a,b]) = 1$. There are sufficient conditions for the latter equality. The best known is the Kolmogorov condition:

$$\mathbb{E}\,|\xi_t - \xi_s|^\alpha \leqslant L|t-s|^{1+\beta}, \quad \text{where } \alpha, \beta, L > 0.$$

However, in many applications weak convergence of finite-dimensional distributions turns out to be too weak, so it becomes necessary to complement it by various conditions in order to ensure convergence of the distributions of processes in functional spaces with norms or metrics. For example, for continuous processes on $[0,1]$, a natural space of paths is $C[0,1]$; convergence of finite-dimensional distributions does not yield convergence of distributions in $C[0,1]$ when $C[0,1]$ is equipped with the usual sup-norm, so one has to require additionally the uniform tightness of distributions. These questions are thoroughly studied in Billingsley [**67**], Borovkov [**106**], [**107**], Gihman, Skorohod [**273**], Cremers, Kadelka [**149**], [**150**], Pollard [**518**], and van der Vaart, Wellner [**625**].

A great number of the results obtained is covered by the following scheme: if a path space X is equipped with a norm or metric (in [**149**], the whole class of metrics is considered convergence in which yields convergence in measure), then for weak convergence in X of the distributions P_n of processes ξ_n it suffices to have weak convergence of finite-dimensional distributions and the uniform tightness of the measures P_n. The latter condition usually cannot be omitted. However, as we shall see below, this condition can be omitted in the case where the space of continuous paths is equipped with the metric of convergence in measure (in place of its usual uniform metric). It is worth noting that this property cannot be true for a norm in place of the metric of convergence in measure. Indeed, for the non random processes $x_n \in C[0,1]$ with disjoint supports in the intervals $[2^{-n}, 2^{-n} + 8^{-n}]$ the finite-dimensional distributions defined by their values at points $t_1, \ldots, t_k$ converge weakly to the corresponding distributions of the zero process. The same is true for the processes $C_n x_n$ for any choice of constants C_n. If $C[0,1]$ is equipped with a norm p, then one can take C_n such that $p(C_n x_n) \to +\infty$, so that there will be no weak convergence of the Dirac measures at the functions $C_n x_n$. Hence in the large class of metrics on path spaces considered in [**149**] and ensuring convergence in measure (such as the uniform metric and the L^p-metric) only the metric of convergence in measure possesses the property that it is not necessary to require in addition the uniform tightness.

First we give sufficient conditions for weak convergence on the space of continuous functions $C[a,b]$ with its usual sup-norm. For $\alpha \in (0,1]$ let H^α denote the set of α-Hölder functions h on $[a,b]$ with its Hölder norm

$$\|h\|_\alpha := \sup_{t\neq s} |h(t)-h(s)|/|t-s|^\alpha + \sup_t |h(t)|.$$

The space H^α with this norm is Banach (but nonseparable). For $\beta > 0$ we have $H^\beta \subset H^\alpha$ and this embedding is compact. In addition, the natural embedding of H^α into $C[a,b]$ is compact (by the Ascoli–Arzelà theorem closed balls in H^α are compact in $C[a,b]$, see p. 15).

2.7.12. THEOREM. *Borel probability measures μ_n on $C[a,b]$ converge weakly to a probability measure μ precisely when one has weak convergence of finite-dimensional distributions and, for every $\varepsilon > 0$, there exists a number $C_\varepsilon > 0$ and a modulus* *of continuity* ω_ε *such that for the set of functions*

$$F_\varepsilon = \bigl\{x\colon\ |x(t)-x(s)| \leqslant \omega_\varepsilon(t-s),\ |x(0)| \leqslant L_\varepsilon\bigr\}$$

we have the estimate $\mu_n(F_\varepsilon) > 1-\varepsilon$ *for all* n.

A sufficient condition for that is given by the equality $\mu_n(H^\alpha)=1$ *for all* n *with some* $\alpha \in (0,1]$ *along with the estimate*

$$\sup_n \int V(\|h\|_\alpha)\,\mu_n(dh) < \infty$$

with some function $V \geqslant 0$ *on* $[0,+\infty)$ *increasing to* $+\infty$.

PROOF. The necessity of the indicated conditions is obvious from the Prohorov theorem. The sufficiency also follows the Prohorov theorem and Example 2.4.12. The last assertion is clear from Example 2.3.7 and compactness of the embedding of H^α to $C[a,b]$. □

2.7.13. COROLLARY. *For the uniform tightness of a sequence of Borel probability measures* μ_n *on* $C[0,1]$ *the following conditions are sufficient: for every* $\varepsilon > 0$, *one can find* $R > 0$, $\delta \in (0,1)$ *and* $N_\varepsilon \in \mathbb{N}$ *such that*

$$\sup_n \mu_n\bigl(x\colon\ |x(0)| > R\bigr) \leqslant \varepsilon,$$

$$\mu_n\bigl(x\colon \sup_{s\in t,t+\delta]} |x(t)-x(s)| \geqslant \varepsilon\bigr) \leqslant \varepsilon\delta \quad \forall\, n \geqslant N_\varepsilon, \forall\, t \in [0,1].$$

PROOF. The set B of points x such that $\omega(x,\delta) \geqslant 3\varepsilon$ belongs to the union of sets $\bigl\{x\colon\ \sup_{s\in[i\delta,(i+1)\delta]} |x(s)-x(i\delta)| \geqslant \varepsilon\bigr\}$ with $i < \delta^{-1}$. Hence for all $n \geqslant N_\varepsilon$ we have $\mu_n(B) \leqslant (1+\delta^{-1})\varepsilon\delta < 2\varepsilon$. □

The second condition in this corollary is not necessary even for a single measure: for example, if P is the distribution of the process $t\xi$, where ξ has distribution density $2^{-1}t^{-3/2}I_{[1,+\infty)}$, then the second condition is not fulfilled. However, for justification of the invariance principle it is easier to verify this stronger condition.

Let $(T,\mathcal{B},\lambda)$ be a measurable space with a finite nonnegative measure λ and let $(\Omega,\mathcal{F},P)$ be a probability space. Denote by L^0 the space of measurable real functions on T with the metric $d(\cdot,\cdot)$ of convergence in measure λ on T defined by the formula

$$d(x,y) = \int_T \min\{|x(t)-y(t)|,1\}\,\lambda(dt).$$

We recall that in nontrivial cases (say, for Lebesgue measure) convergence in this metric cannot be generated by a norm and the space of measurable functions with it is not locally convex, since it has no nonzero continuous linear functionals (see Bogachev [**81**, Exercise 4.7.49]). We observe that the metric d is bounded.

We shall assume that the measure λ is separable, i.e., (L^0,d) is separable; this is equivalent to the separability of $L^1(\lambda)$ or $L^2(\lambda)$.

Suppose that we have a certain space $\mathcal{F}$ of functions on T that are measurable with respect to the measure λ (not equivalence classes like L^1, but individual functions) such that the equality of two functions in $\mathcal{F}$ almost everywhere yields their pointwise equality. Examples of such spaces $\mathcal{F}$ are the class of continuous functions $C(T)$ for a topological space T with a Borel measure λ on T not vanishing on nonempty open sets and also the class of left continuous functions on $[a,b]$ with Lebesgue measure.

The space $\mathcal{F}$ will be equipped with the metric d of convergence in measure λ, i.e., the metric from L^0 (although L^0 consists of equivalence classes). This space will be denoted by $\mathcal{F}_d$.

Suppose additionally that, for all $t \in T$, the evaluation functions $x \mapsto x(t)$ on $\mathcal{F}$ are measurable with respect to all Borel measures on $\mathcal{F}_d$. For example, this assumption is fulfilled if λ is Lebesgue measure on the interval $[0,1]$ and $\mathcal{F} = C[0,1]$ (in this case the indicated functions are Borel measurable).

The finite-dimensional distribution of a measure $\mu \in \mathcal{P}(\mathcal{F}_d)$ generated by points $t_1, \ldots, t_k \in T$ is defined as the image of the measure μ under the mapping from $\mathcal{F}_d$ to $\mathbb{R}^k$ given by the equality $x \mapsto \bigl(x(t_1), \ldots, x(t_k)\bigr)$. The image of the measure under this mapping is well-defined due to our assumption about the measurability of the evaluation functions.

2.7.14. THEOREM. *Suppose that we are given a measure $\mu \in \mathcal{P}(\mathcal{F}_d)$ and a sequence of measures $\mu_n \in \mathcal{P}(\mathcal{F}_d)$. If the finite-dimensional distributions of the measures μ_n converge weakly to the corresponding finite-dimensional distributions of the measure μ, then the measures μ_n converge weakly to μ on the space $\mathcal{F}_d$.*

PROOF. By assumption the space $\mathcal{F}_d$ is separable. Let us take a countable everywhere dense set $\{y_i\}_{i=1}^\infty$ in it. By the last assertion of Corollary 2.2.6 it suffices to verify (2.2.4) for polynomials in bounded functions

$$d(x, y_i) = \int_T \min(1, |x(t) - y_i(t)|)\, \lambda(dt), \quad \text{where } y_1, \ldots, y_k \in E_d.$$

Weak convergence of finite-dimensional distributions of the given measures means that for all $t_1, \ldots, t_k \in T$ and $\varphi \in C_b(\mathbb{R}^k)$ we have

$$\lim_{n\to\infty} \int_X \varphi\bigl(x(t_1), \ldots, x(t_k)\bigr)\, \mu_n(dx) = \int_X \varphi\bigl(x(t_1), \ldots, x(t_k)\bigr)\, \mu(dx).$$

By the Lebesgue dominated convergence theorem, for every bounded Borel function ψ on $\mathbb{R}^k \times \mathbb{R}^k$ that is continuous in the first k variables we obtain the equality

$$\lim_{n\to\infty} \int_X \int_{T^k} \psi\bigl(x(t_1), \ldots, x(t_k), t_1, \ldots, t_k\bigr)\, \lambda(dt_1) \cdots \lambda(dt_k) \mu_n(dx)$$
$$= \int_X \psi\bigl(x(t_1), \ldots, x(t_k), t_1, \ldots, t_k\bigr)\, \lambda(dt_1) \cdots \lambda(dt_k) \mu(dx).$$

In particular, this equality is true for all functions of the form

$$\varphi(s_1, \ldots, s_k, t_1, \ldots, t_k) = \textstyle\prod_{i=1}^k \min\bigl(1, |s_i - y_i(t_i)|\bigr)$$

and their linear combinations, which gives all polynomials in $d(x, y_i)$. □

In terms of random processes we obtain the following assertions.

2.7.15. COROLLARY. *Let ξ and $\{\xi_n\}_{n=1}^\infty$ be random processes on the set T with paths in the space $\mathcal{F}_d$. Suppose that the finite-dimensional distributions of ξ_n converge weakly to the corresponding finite-dimensional distributions of ξ. Then the measures P_{ξ_n} converge weakly to the measure P_ξ on the space $\mathcal{F}_d$.*

It is worth noting that for a uniformly bounded sequence of functions convergence in measure λ is equivalent to convergence in $L^2(\lambda)$, and if the sequence is bounded in $L^p(\lambda)$, then its convergence in measure is equivalent to convergence in $L^r(\lambda)$ whenever $r < p$. Hence, although for the L^p-norm an analog of the proven theorem is false, as noted above, we obtain the following assertion.

2.7.16. COROLLARY. *Let ξ and $\{\xi_n\}_{n=1}^\infty$ be random processes on the set T with paths in $\mathcal{F}_d$ such that the finite-dimensional distributions of ξ_n converge weakly to the corresponding finite-dimensional distributions of ξ. If*

$$\sup_n \mathbb{E} \int_T |\xi_n(t)|^p \, \lambda(dt) < \infty$$

for some $p \in (1, +\infty)$, then the measures P_{ξ_n} converge weakly to the measure P_ξ on the space $L^r(\lambda)$ for every r in the interval $[1, p)$.

PROOF. The Skorohod theorem gives Borel mappings $\eta_n, \eta\colon [0,1] \to L^0(\lambda)$ for which the distributions of ξ_n and ξ equal $\lambda_1 \circ \eta_n^{-1}$ and $\lambda_1 \circ \eta^{-1}$, where λ_1 is Lebesgue measure on $[0,1]$, and $\eta_n(s) \to \eta(s)$ in $L^0(\lambda)$ for almost every s. Hence the functions $|\eta_n(s)(t) - \eta(s)(t)|$ tend to zero in measure $\lambda \otimes \lambda_1$. Since the integral of $|\eta_n(s)(t)|^p$ against $\lambda \otimes \lambda_1$ is $\mathbb{E}\, \|\xi_n\|_p^p$, we obtain convergence to zero of the integrals of $|\eta_n(s)(t) - \eta(s)(t)|^r$ whenever $r < p$, i..e, the processes η_n converge to η in $L^r(\lambda \otimes \lambda_1)$, which gives weak convergence of their distributions in $L^r(\lambda)$. □

For $r = p$ the conclusion can fail, but if the processes ξ_n are uniformly bounded, then it remains true for all $r \in [1, +\infty)$. Let us draw the reader's attention to the difference between the hypotheses of Theorem 2.7.11 and Theorem 2.7.14.

By the assumed separability of the space L^0 weak convergence of finite-dimensional distributions yields the uniform tightness of $\{\mu_n\}$ in L^0. If the set $\mathcal{F}$ is measurable in L^0 with respect to all Borel measures (say, is a Souslin space, as is the case for $C[0,1]$), then the measures μ_n are uniformly tight not only in L^0, but also in $\mathcal{F}_d$.

We recall that, by the Skorohod theorem, Corollary 2.7.15 enables us to obtain random processes η_n and η for which $P_{\xi_n} = P_{\eta_n}$, $P_\xi = P_\eta$ and $\eta_n(\cdot, \omega) \to \eta(\cdot, \omega)$ in measure λ for almost all ω. A similar assertion is also true in the situation of Corollary 2.7.16, which is clear from its justification.

Sometimes one has to consider measures on path spaces not on the Borel σ-algebra, but on smaller σ-algebras. For example, it is known (see Borovkov, Sakhanenko [**110**]) that the Wiener measure on the space of continuous paths on $[0, +\infty)$ with finite norm $\sup_t |x(t)/\alpha(t)|$ cannot be defined on the whole Borel σ-algebra associated with this norm due to the nonseparability of the obtained space. In such cases it is usual to employ smaller σ-algebras, say, the σ-algebra $\mathcal{B}_S(X)$ generated by balls in a metric space. Other approaches are possible; for example, for the Wiener measure one can take the separable subspace of paths with finite norm $\sup_t |x(t)/\beta(t)|$ consisting of functions for which $x(t)/\beta(t) \to 0$ as $t \to \infty$. Certainly, this norm may be not appropriate for some problems. An advantage of the metric of convergence in measure is its separability under broad conditions.

2.7(vi). The Skorohod space

Let (E, ϱ) be a metric space. The Skorohod space $D_1(E)$ is defined to be the space of mappings $x\colon [0,1] \to E$ that are right continuous and have left limits at all points $t > 0$; it is equipped with the metric

$$d(x,y) = \inf\bigl\{\varepsilon > 0 \mid \exists\, h \in \Lambda[0,1]\colon |t - h(t)| \leqslant \varepsilon, \varrho\bigl(x(t), y(h(t))\bigr) \leqslant \varepsilon\bigr\},$$

where $\Lambda[0,1]$ is the set of homeomorphisms h of the interval $[0,1]$ with $h(0) = 0$, $h(1) = 1$. This space was introduced by Skorohod [**584**] and is frequently used in the theory of random processes. If the space E is Polish, then so is $D_1(E)$ (for

$E = \mathbb{R}^1$ the proof can be found in Billingsley [**67**], the general case is similar). For a complete space E, the space $D_1(E)$ is not always complete with respect to the metric d, but it is complete with respect to the following metric generating the same topology:

$$d_0(x,y) = \inf\bigl\{\varepsilon > 0 \mid \ \exists\, h \in \Lambda[0,1]\colon \\ \sup_{t>s}\bigl|\ln\bigl([h(t)-h(s)]/(t-s)\bigr)\bigr| \leqslant \varepsilon, \varrho\bigl(x(t), y(h(t))\bigr) \leqslant \varepsilon\bigr\}.$$

Similarly one defines the Skorohod space of mappings $D(E)$ on the half-line. In the case $E = \mathbb{R}^1$ a thorough discussion of the Skorohod space can be found in Billingsley [**67**]. For a separable metric space E the Borel σ-algebra of $D_1(E)$ is generated by the mappings $x \mapsto x(t)$, $t \in [0,1]$, although they are discontinuous (for $E = \mathbb{R}$). Various conditions for uniform tightness of measures on $D_1(E)$ are known (see [**67**, Chapter 3]). Let us mention a typical result. For any trajectory $x \in D_1(\mathbb{R})$ let us set $\omega_D(x,\delta) := \inf \max_{i\leqslant k} \sup_{t,s\in[t_{i-1},t_i)} |x(t)-x(s)|$, where inf is taken over all finite collections of points with $t_i - t_{i-1} > \delta$, $t_0 = 0$.

2.7.17. THEOREM. *A sequence of measures $\mu \in \mathcal{P}\bigl(D_1(\mathbb{R})\bigr)$ is uniformly tight precisely when for every $\varepsilon > 0$ one can find $R > 0$, $\delta \in (0,1)$ and $N \in \mathbb{N}$ such that $\mu_n\bigl(x\colon \sup_t |x(t)| > R\bigr) \leqslant \varepsilon$, $\mu_n\bigl(x\colon \omega_D(x,\delta) \geqslant \varepsilon\bigr) \leqslant \varepsilon$ for all $n \geqslant N$.*

2.7(vii). Gaussian measures

We have already encountered Gaussian measures as the most important representatives of classes of limit distributions for some specially arranged sequences of random variables. Gaussian measures are of obvious independent interest and there is a vast literature on this subject, see Bogachev [**78**], [**82**], [**83**], Fernique [**229**], Kuo [**401**], Lifshits [**431**], and Vakhania, Tarieladze, Chobanyan [**629**]. Here we present only some basic facts about Gaussian measures which directly related to weak convergence (unlike other topics of this chapter, Gaussian measures are discussed here in the setting of locally convex spaces, because no specific features of metric spaces are essential in this discussion). Some additional information about the central limit theorem in infinite-dimensional spaces is presented in § 4.8.

2.7.18. DEFINITION. A Radon probability measure γ on a locally convex (for example, Banach) space X is called Gaussian if, for every $f \in X^*$, the measure $\gamma\circ f^{-1}$ is Gaussian; if all such measures have zero mean, then γ is called centered.

It is easy to show that the Fourier transform of a Gaussian measure γ, i.e., the function on X^* defined as the integral of $\exp(if)$ with respect to the measure γ (in more detail the Fourier transform in infinite-dimensional spaces is considered in § 4.6), has the form

$$\int_X \exp(if)\,d\gamma = \exp\bigl(if(a) - Q(f,g)/2\bigr), \quad f(a) = \int_X f\,d\gamma$$

with some vector $a \in X$, where

$$Q(f,g) = \int_X [f - f(a)]\,[g - g(a)]\,d\gamma = \int_X f^2\,d\gamma - \Bigl(\int_X f\,d\gamma\Bigr)^2.$$

The vector a is called the mean of the measure γ and the function Q is called the covariance of the measure γ.

On $\mathbb{R}^n$ the covariance has the form (Ky, y), where K is a nonnegative definite symmetric operator on $\mathbb{R}^n$. In the eigenbasis of K the Fourier transform of the measure γ has the form

$$e^{ia_1y_1 - k_1y_1^2/2} \cdots e^{ia_ny_n - k_ny_n^2/2}.$$

Thus, in this basis the measure γ is the product of n one-dimensional Gaussian measures. If the operator K is invertible, then the measure γ is given by density

$$(2\pi k_1)^{-1/2}e^{-(x_1-a_1)^2/(2k_1)} \cdots (2\pi k_n)^{-1/2}e^{-(x_n-a_n)^2/(2k_n)},$$

which is the product of Gaussian densities on the real line. On the other hand, for every vector $a \in \mathbb{R}^n$ and nonnegative definite symmetric operator A, the function $\exp\bigl(i(y,a) - (Ay,y)/2\bigr)$ is the Fourier transform of a Gaussian measure.

Weak convergence of Gaussian measures γ_j on $\mathbb{R}^n$ to a Gaussian measure γ is equivalent to convergence of their means a_j to the mean a of γ and convergence of their covariance operators A_j to the covariance operator A of γ.

The situation is similar (although a bit more complicated) in case of a separable Hilbert space H. As before, the Fourier transform of the measure γ is given by the formula (in which the dual space is identified with H itself)

$$\exp\bigl(i(y,a) - (Ky,y)/2\bigr),$$

where K is a bounded symmetric operator on H with a nonnegative quadratic form. However, now not every such operator corresponds to a measure on H. A necessary and sufficient condition is that K be nuclear.

The mean a is the vector Bochner integral of x with respect to the measure γ (see §2.5). For a centered measure, the covariance operator is calculated on the basis of the equality

$$(Ku, v) = \int_H (u,x)(v,x)\,\gamma(dx).$$

If we take an orthonormal basis $\{e_n\}$ consisting of eigenvectors of the covariance operator K of the Gaussian measure γ, then we can identify H with l^2 and write the Fourier transform of the measure γ in the form

$$\exp\biggl(\sum_{n=1}^{\infty}(iy_na_n - k_ny_n^2/2)\biggr).$$

Extending the measure γ to $\mathbb{R}^\infty$, we obtain that this is precisely the countable product of the one-dimensional Gaussian measures with means a_n and variances k_n. If such a product is originally defined on $\mathbb{R}^\infty$, then the condition $\sum_{n=1}^\infty k_n < \infty$ is needed exactly for obtaining the equality $\gamma(l^2) = 1$, which enables us to restrict the measure γ to l^2. In this sense Gaussian measures on Hilbert spaces are still products of one-dimensional Gaussian measures. A Gaussian measure γ with mean a coincides with the shift of the centered Gaussian measure γ_0 with the same covariance by the the vector a: $\gamma(B) = \gamma_0(B - a)$.

An important characteristic of a Radon Gaussian measure γ is its Cameron–Martin subspace $H(\gamma)$, which in the case of a centered measure is defined as the set of all vectors with finite norm

$$|h|_H := \sup\biggl\{f(h)\colon\ f \in X^*,\ \int_X f^2\,d\gamma \leqslant 1\biggr\}.$$

This space is separable Hilbert and its closed unit ball is compact in X.

For a measure with a nonzero mean, the Cameron–Martin subspace is the same as for the corresponding centered measure. For the countable power of the standard Gaussian measure on $\mathbb{R}$ the Cameron–Martin subspace is the usual space l^2.

In the theory of Gaussian measures the following result of Fernique plays an important role (see [**78**, Theorem 2.8.5]). Let γ be a centered Radon Gaussian measure on a locally convex space X and let q be a γ-measurable seminorm on some measurable linear subspace in X of measure 1. Suppose that $\gamma(q \leqslant \tau) = c > 1/2$ for some $\tau > 0$. Then there exist numbers $\alpha = \alpha(\tau, c) > 0$ and $I(\tau, c)$ such that

$$\int_X \exp(\alpha q^2)\, d\gamma \leqslant I(\tau, c). \tag{2.7.2}$$

For example, the classical Wiener measure on $C[0,1]$, defined usually as the distribution of the Wiener process (see Bulinskii, Shiryaev [**129**], Wentzel [**653**]), can be also obtained as the image of the countable power of the standard Gaussian measure on $\mathbb{R}$ under the mapping

$$Tx(t) = \sum_{n=1}^{\infty} x_n \int_0^t \varphi_n(s)\, ds,$$

where $\{\varphi_n\}$ is an arbitrary orthonormal basis in $L^2[0,1]$. The Wiener process itself is usually defined as a process w_t with independent increments for which $w_t - w_s$ with $t \geqslant s$ is a Gaussian random variable with zero mean and variance $t - s$.

General Radon Gaussian measure have some product structure, although in a weaker form than in Hilbert spaces. The celebrated Tsirelson theorem (generalizing the Itô–Nisio theorem for Banach spaces) asserts that for every centered Radon Gaussian measure γ on a locally convex space X there exists a Borel injective linear mapping T with values in X defined on a Borel linear subspace $L \subset \mathbb{R}^\infty$ of measure 1 with respect to the standard Gaussian measure γ_0 (the countable power of the one-dimensional standard Gaussian measure) such that $\gamma = \gamma_0 \circ T^{-1}$. This mapping can be defined constructively: if $\{e_n\}$ is an orthonormal basis in the Cameron–Martin space $H(\gamma)$, then the series $\sum_{n=1}^\infty x_n e_n$ converges in X for γ_0-almost all $x = (x_n)$ and its sum can be taken for Tx.

The limit of a weakly convergent sequence of Gaussian measures is a Gaussian measure (Exercise 2.7.49).

2.7.19. PROPOSITION. (i) *A family Γ of Radon Gaussian measures on a locally convex (for example, Banach) space X is uniformly tight precisely when the set of their means is contained in a compact set and the family of the corresponding centered measures is uniformly tight. In this case for every continuous seminorm q there exists $\alpha > 0$ such that*

$$\sup_{\gamma \in \Gamma} \int_X \exp(\alpha q^2)\, d\gamma < \infty.$$

(ii) *If a sequence of Radon Gaussian measures γ_n converges weakly to a Radon measure γ, then, for every continuous seminorm q, there exists $\alpha > 0$ such that*

$$\lim_{n\to\infty} \int_X \exp(\alpha q^2)\, d\gamma_n = \int_X \exp(\alpha q^2)\, d\gamma.$$

In addition, for every $r > 0$ one has

$$\lim_{n\to\infty} \int_X q^r\, d\gamma_n = \int_X q^r\, d\gamma.$$

Finally, the means of the measures γ_n converge to the mean of the measure γ.

PROOF. (i) It is clear that the shifts of measures from a uniformly tight family by vectors from a compact set constitute a uniformly tight family. For the proof of the converse it suffices to find a compact sets containing the means a_γ of the measures $\gamma \in \Gamma$. Let Γ_1 and Γ_2 denote the family of measures obtained from Γ by the mappings $x \mapsto x/\sqrt{2}$ and $x \mapsto -x/\sqrt{2}$. The measure with the Fourier transform $\exp\bigl(if(a) - Q(f)/2\bigr)$ is taken by these transformations to the measures γ_1 and γ_2 with the Fourier transforms

$$\exp\bigl(if(a)/\sqrt{2} - Q(f)/4\bigr) \quad \text{and} \quad \exp\bigl(-if(a)/\sqrt{2} - Q(f)/4\bigr),$$

respectively. It is easy to verify (in § 4.6 this is done for general measures) that the set of convolutions of the measures $\gamma_1 * \gamma_2$ of the indicated form is also uniformly tight. The Fourier transform of the convolution equals the product of the Fourier transforms of the measures γ_1 and γ_2, i.e., this is $\exp\bigl(Q(f)/2\bigr)$. Hence we have the uniform tightness of the set Γ_0 of the centered measures obtained from measures of the family Γ. Thus, there exists a compact set K of measure greater than $3/4$ with respect to all measures in Γ and Γ_0. Hence $\gamma(K) > 3/4$, $\gamma(K + a_\gamma) > 3/4$ for all $\gamma \in \Gamma$, whence we obtain that a_γ belongs to the compact set $K - K$. The last assertion in (i) follows from (2.7.2).

(ii) If $\gamma_n \Rightarrow \gamma$, then, taking $c > 0$ such that $\gamma(q < c) > 3/4$, due to weak convergence and the condition that $\{q < c\}$ is open we obtain that $\gamma_n(q < c) > 3/4$ for all sufficiently large n. Hence by (2.7.2) for some $\alpha_0 > 0$ the integrals of $\exp(\alpha_0 q^2)$ with respect to these measures are uniformly bounded. If $\alpha < \alpha_0$, then we obtain convergence of the integrals of $\exp(\alpha q^2)$, which yields also convergence of the integrals of q^r (see Theorem 2.7.1 and its topological version in Theorem 4.3.14).

It remains to prove convergence of the means a_n of the measures γ_n to the mean a of the measure γ. By applying the shift $x \mapsto x - a$ we can assume that $a = 0$. Clearly, one has weak convergence. In the case of a Banach space these measures are uniformly tight, hence by (i) the vectors a_n belong to a compact set, which gives their convergence to a. In the case of a general locally convex space one can pass to its completion and assume that X is complete. Then (taking into account weak convergence) it suffices to show that the sequence $\{a_n\}$ is totally bounded, i.e., can be covered by finitely many shifts of every absolutely convex neighborhood of zero V. To this end we find a compact set K with $\gamma(K) > 3/4$. Let us observe that the centered measures μ_n obtained from γ_n also weakly converge to γ, because they are obtained as the convolutions $\nu_n * \sigma_n$ of weakly convergent images of the measures γ_n under the two homotheties indicated above. Hence for all n sufficiently large we have $\gamma_n(K + V/4) > 3/4$, $\gamma_n(K + V/4 + a_n) > 3/4$. Therefore, $K + V/4$ and $K + V/4 + a_n$ are disjoint, whence we have $a_n \in (K - K) + V/2$. The compact set $K - K$ is covered by finitely many neighborhoods $k_i + V/2$, which gives a cover of $\{a_n\}$ by the sets $k_i + V$. □

Identification of conditions for weak convergence of Gaussian measures on Banach spaces is relatively simple in the Hilbert case, although even in this case, as compared to the case of $\mathbb{R}^n$, a description of convergence in terms of means and covariance operators undergoes a nontrivial change requiring to strengthen the mode of convergence of covariances. For Gaussian measures γ_n with a common covariance K and means a_n weak convergence reduces to convergence of a_n in norm (this is clear from § 2.5). Let us consider the case of different covariances.

2.7.20. LEMMA. *A family Γ of centered Gaussian measures on a separable Hilbert space H is uniformly tight precisely when for some (and then for every) orthonormal basis $\{e_i\}$ in H the covariance operators K_γ of the measures $\gamma \in \Gamma$ satisfy the following condition: the series $\sum_{n=1}^\infty (K_\gamma e_n, e_n)$ are uniformly bounded and converge uniformly in $\gamma \in \Gamma$.*

PROOF. Let Γ be uniformly tight. Take a compact set S such that $\gamma(S) > 3/4$ for all $\gamma \in \Gamma$. Since any compact set is contained in a compact ellipsoid, we can assume that S is an ellipsoid of the form $\{x\colon q^2 \leqslant 1\}$, $q^2 := \sum_{n=1}^\infty C_n^2 (x, e_n)^2$, with some $C_n \uparrow +\infty$. Applying (2.7.2) to the norm q defining this ellipsoid on its linear span, we obtain the uniform boundedness of the integrals of q^2 with respect to the measures from the family Γ, i.e., the uniform boundedness of the series $S_\gamma = \sum_{n=1}^\infty C_n^2 (K_\gamma e_n, e_n)$. This yields both indicated properties. The converse is clear from the results in § 2.5 for general measures. □

We now obtain a characterization of weak convergence of Gaussian measures.

2.7.21. THEOREM. *Gaussian measures γ_n on a separable Hilbert space with means a_n and covariance operators K_n converge weakly to a Gaussian measure γ with mean a and covariance operator K precisely when*

$$\|a_n - a\| \to 0, \quad \|\sqrt{K_n} - \sqrt{K}\|_{_{HS}} \to 0.$$

The second condition is also equivalent to convergence $\|K_n - K\|_1 \to 0$. In addition, it is equivalent to convergence in the weak operator topology along with convergence of traces $\operatorname{tr} K_n \to \operatorname{tr} K$.

PROOF. First we consider centered measures. Let $\gamma_n \Rightarrow \gamma$. Convergence of the Fourier transforms yields convergence of covariances, which implies convergence $(K_n u, v) \to (Ku, v)$. Proposition 2.7.19 yields convergence of the integrals of (x, x), which enables us to apply Proposition 2.5.2 giving convergence of K_n in the nuclear norm. By Lemma 2.5.1 this is equivalent to convergence of the square roots of K_n in the Hilbert–Schmidt norm.

Conversely, if one of the indicated forms of operator convergence holds, then we obtain from Proposition 2.5.2 that the sequence $\{\gamma_n\}$ is uniformly tight on H. Due to convergence of the Fourier transforms of γ_n this means weak convergence. The general case follows by the previous results. □

2.7(viii). The invariance principle and the Brownian bridge

To the early classics of the theory of weak convergence belongs Donsker's result (see Donsker [**175**]) called the *invariance principle* and asserting that the distribution P_W of the Wiener process is a weak limit in $C[0, 1]$ of the distributions P_n of random piece-wise linear functions defined by the formula

$$X_n(t) = n^{-1/2} S_{[nt]} + (nt - [nt]) n^{-1/2} \xi_{[nt]+1},$$

where $[r]$ is the integer part of the number r, $S_n = \xi_1 + \cdots + \xi_n$, and ξ_i are independent random variables with an identical distribution with zero mean and unit variance. This result is also called a functional central limit theorem (the term "the invariance principle" says that the limit is independent of the distribution of ξ_1 similarly to the usual central limit theorem). The trajectory $X_n(t)$ is piece-wise linear with nods $(i/n, n^{-1/2} S_i)$. Convergence of finite-dimensional distributions can be easily derived from the central limit theorem in $\mathbb{R}^d$. For example, convergence of the distributions of $X_n(t)$ to the centered Gaussian measure γ_t with variance t is seen from the fact that the second term in $X_n(t)$ tends to zero in probability (and in L^2), while the distribution of the first term tends to γ_t by the one-dimensional central limit theorem, since $[nt]/n \to t$.

Weak convergence of k-dimensional distributions defined by points $t_1 < \cdots < t_k$ is verified similarly: the variables $X_n(t_1), X_n(t_2) - X_n(t_1), \ldots, X_n(t_k) - X_n(t_{k-1})$ are independent and the vector with these components is obtained from the vector with components $X_n(t_i)$ by a nondegenerate linear transformation in $\mathbb{R}^k$ independent of n, and the same transformation takes the distribution of the vector $(W_{t_1}, \ldots, W_{t_k})$ to the distribution of the vector $(W_{t_1}, \ldots, W_{t_k} - W_{t_{k-1}})$ with independent components.

However, for weak convergence in $C[0,1]$ we also need the uniform tightness of the measures P_n in $C[0,1]$. Let us verify the conditions of Corollary 2.7.13. The first condition is trivially fulfilled. For verification of the second one we need the following Kolmogorov estimate for the distribution of the maximum of $|S_n|$ (see Billingsley [**67**, p. 70]):):

$$P\bigl(\max_{i\leqslant n} |S_i| \geqslant rn^{1/2}\bigr) \leqslant 2P\bigl(|n^{-1/2}S_n| \geqslant r/2\bigr), \quad r > 2\sqrt{2}.$$

The right-hand side tends to $2P(\xi \geqslant r/2)$, where ξ is the standard Gaussian random variable. Hence, for every $\varepsilon > 0$, whenever $r \geqslant r_\varepsilon$, we have

$$\limsup_{n\to\infty} 2P(\max_{i\leqslant n} |S_i| \geqslant rn^{1/2}) \leqslant \varepsilon^3 r^{-2}.$$

For $\delta = \varepsilon^2 r_\varepsilon^{-2}$ we obtain $\limsup_{n\to\infty} 2P(\max_{i\leqslant [n\delta]} |S_i| \geqslant r[n\delta]^{1/2}) \leqslant \varepsilon\delta$. Let $t \in [0,1]$. Pick k and j such that $k/n \leqslant t < (k+1)/n$, $(j-1)/n \leqslant t + \delta/2 < j/n$. By the piece-wise linearity of the function X_n we have

$$\sup_{s\in[t,t+\delta/2]} |X_n(s) - X_n(t)| \leqslant 2 \max_{i\leqslant j-k} |S_{k+i} - S_k| n^{-1/2}.$$

Whenever $n \geqslant 4/\delta$ we have $j - k < n\delta$, so the right-hand side of the last estimate does not decrease if we take the maximum in $i \leqslant n\delta$. Thus, we have $\mu_n(x\colon\ \sup_{s\in[t,t+\delta/2]} |x(s) - x(t)| \geqslant \varepsilon) \leqslant \varepsilon\delta/2$, since $\varepsilon n^{1/2}[n\delta]^{-1/2} \geqslant \varepsilon\delta^{-1/2} = r_\varepsilon$. The conditions of Corollary 2.7.13 are fulfilled.

With the aid of distributions of random polygonal functions one can also obtain the distribution of the process called the Brownian bridge on $[0,1]$ and defined by the formula $W_t^0 = W_t - tW_1$, where W_t is the Wiener process. The distribution P_0 of this process in $C[0,1]$ is obtained by the linear mapping $Ax(t) = x(t) - tx(1)$ from the Wiener measure P_W. It can be also obtained as the weak limit (as $\varepsilon \to 0$) of the measures $P_\varepsilon(B) := P_W(B \cap S_\varepsilon)/P_W(S_\varepsilon)$, where $S_\varepsilon = \{x\colon\ |x(1)| \leqslant \varepsilon\}$. We observe that $\mathbb{E}\, W_t W_s = \min(t,s)$, therefore, $\mathbb{E}\, W_t^0 W_s^0 = \min(t,s) - ts$. In addition, we have $\mathbb{E}\, W_t^0 = 0$.

Indeed, let Z be a closed set in $C[0,1]$. For any fixed $\delta > 0$, whenever $\varepsilon < \delta$ we have

$$P_W(Z \cap S_\varepsilon) \leqslant P_W(x \in S_\varepsilon\colon\ Ax \in Z^\delta)$$

since $\|x - Ax\| \leqslant \varepsilon < \delta$ whenever $x \in S_\varepsilon$, because $x(t) - Ax(t) = tx(1)$. Hence

$$P_\varepsilon(Z) \leqslant P_W(x \in S_\varepsilon\colon\ Ax \in Z^\delta)/P_W(S_\varepsilon).$$

We now observe that

$$P_W(x \in S_\varepsilon\colon\ Ax \in B)/P_W(S_\varepsilon) = P_W(x\colon\ Ax \in B) = P_0(B)$$

for every Borel set $B \subset C[0,1]$. This follows from the fact that both sides of the equality above are Borel measures and are equal on all cylinders B of the form $B = \{x\colon (x(t_1),\dots,x(t_n)) \in C\}$, where $t_1 < \cdots < t_n$ and $C \in \mathcal{B}(\mathbb{R}^n)$. The latter is obvious from the independence of W_1 and the random vector with components $W_{t_i} - t_i W(1)$, which holds, since their joint distribution is Gaussian and the expectations of $W_{t_i} W_1 - t_i W(1)^2$ are zero. Thus, $\limsup_{\varepsilon\to 0} P_\varepsilon(Z) \leqslant P_0(Z^\delta)$. As $\delta \to 0$, we obtain $\limsup_{\varepsilon\to 0} P_\varepsilon(Z) \leqslant P_0(Z)$, hence $P_\varepsilon \Rightarrow P_0$.

2.7(ix). Extensions of mappings

The Tietze–Urysohn theorem on extensions of continuous functions mentioned in § 2.1 was generalized by Dugundji [**196**] as follows.

2.7.22. THEOREM. *Let Z be a closed subset of a metric space M, let E be a locally convex space (for example, a normed space), and let $f\colon Z \to E$ be a continuous mapping. Then there exists a continuous mapping $\widetilde{f}$ from the space M to the convex hull of $f(Z)$ that coincides with f on Z.*

A detailed proof can be found in Borsuk [**114**, p. 77].

Klee [**367**] obtained the following result (needed in Chapter 3), the included proof of which is borrowed from Toruńczyk [**618**].

2.7.23. THEOREM. *Let X and Y be normed spaces, let $A \subset X$ and $B \subset Y$ be closed sets, and let $f\colon A \to B$ be a homeomorphism. Then, there exists a homeomorphism $g\colon X\times Y \to X\times Y$ such that $g(x,0) = (0, f(x))$ for all $x \in A$.*

PROOF. By the Dugundji theorem the mapping f extends to a continuous mapping $f_1\colon X \to Y$. Let us set $h_1(x,y) = (x, y + f_1(x))$, which, as can be easily verified, gives a homeomorphism of the space $X \times Y$. Similarly one can extend f^{-1} to a continuous mapping $f_2\colon Y \to X$ and obtain the homeomorphism $h_2(x,y) = (x + f_2(y), y)$. Finally, take $g = h_2^{-1}\circ h_1$. □

Exercises

2.7.24. Prove that on every noncompact metric space there exists an unbounded continuous function.

HINT: take a sequence $\{x_n\}$ without limit points, consider a function f_n equal to n at x_n and vanishing outside the ball of radius r_n centered at x_n by taking r_n such that the ball of the doubled radius contains no other points of this sequence; consider the series of f_n.

2.7.25. Let (X,d) be a metric space.

(i) Let f be a bounded function on a set $A \subset X$ such that $|f(x) - f(y)| \leqslant d(x,y)$ for all $x, y \in A$. Let

$$g(x) := \max\bigl\{\sup_{y\in A}\bigl(f(y) - d(x,y)\bigr), \inf_A f\bigr\}.$$

Show that $g(x) = f(x)$ if $x \in A$, $\sup_{y\in X} |g(y)| = \sup_{x\in A} |f(x)|$ and $|g(x) - g(y)| \leqslant d(x,y)$ for all $x, y \in X$.

(ii) Prove that all bounded uniformly continuous functions on X are uniformly approximated by Lipschitz functions.

HINT: derive (ii) from (i); to this end, given $\varepsilon > 0$, take $\delta > 0$ such that the given function f satisfies the condition $|f(x) - f(y)| < \varepsilon$ whenever $d(x,y) < \delta$; then use Zorn's lemma to find a maximal (by inclusion) set of points A with pairwise distances from the interval $[\delta/2, \delta]$; observe that the function f is Lipschitz on A.

2.7.26. (Talagrand [**606**]) Show that there exists a nonseparable metric space the Borel σ-algebra of which is generated by balls.

2.7.27. (Michael [**461**], Toruńczyk [**618**]) Show that every metric space is isometric to a closed part of some normed space (in Proposition 1.3.1 this occurs only for a complete space).

HINT: modify the proof of Proposition 1.3.1 as follows: let Ω be the set of all finite subsets of a nonempty metric space X; an embedding $j\colon X \to B(\Omega)$ is defined by an analogous formula $j(x)(S) = \operatorname{dist}(x, S) - \operatorname{dist}(x_0, S)$, $S \in \Omega$, where $x_0 \in X$ is a fixed point; as in Proposition 1.3.1, the mapping j preserves distances, but now $j(X)$ turns out to be a closed set in its linear span L. Indeed, the set $j(A)$ is closed in L for every closed set A from X, since if elements $j(a_n)$, where $a_n \in A$, converge to an element $y = c_1 j(x_1) + \cdots + c_m j(x_m) \in L$, then, letting $S = \{x_0, x_1, \ldots, x_m\}$, we obtain that $\lim\limits_{n\to\infty} \operatorname{dist}(a_n, S) = \lim\limits_{n\to\infty} j(a_n)(S) = 0$, whence it follows that $\{a_n\}$ contains a subsequence converging to some point x_i in S, then $x_i \in A$ and $y = j(x_i) \in j(A)$.

2.7.28. Show that the σ-algebra generated by all first category sets consists of the first category sets and their complements.

2.7.29. Prove that every Borel measure on the real line is concentrated on a countable union of compact sets without inner points.

HINT: first consider the case of a probability measure μ without points of positive measure and for fixed $\varepsilon > 0$ remove the union of open intervals U_k centered at rational points r_k picked such that $\mu(U_k) < \varepsilon 2^{-k}$; in the general case use points outside the set of atoms of μ.

2.7.30. Let μ and ν be two Radon measures on a metric space X assigning equal integrals to all functions in some algebra of functions $\mathcal{F} \subset C_b(X)$ separating points in X and containing constants. Prove the equality $\mu = \nu$.

HINT: it suffices to show that $\mu(K) = \nu(K)$ for all compact sets K; for every $\varepsilon > 0$ there exists a compact set K_ε with $|\mu|(X\backslash K_\varepsilon) < \varepsilon$, $|\nu|(X\backslash K_\varepsilon) < \varepsilon$; next, there exists a continuous function $f\colon X \to [-1, 1]$ equal to 1 on K for which the integrals with respect to both measures differ from the values of the measures on K less than by ε; by the Stone–Weierstrass theorem there exists a function $g \in \mathcal{F}$ for which $|f(x) - g(x)| \leqslant \varepsilon$ for all $x \in K_\varepsilon$; by a limit procedure it is easy to verify that both measure assign equal integrals to the function $h = \max\bigl(-1 - \varepsilon, \min(g, 1 + \varepsilon)\bigr)$, for which we also have $|f(x) - h(x)| \leqslant \varepsilon$ for all $x \in K_\varepsilon$; it remains to compare the integrals of f.

2.7.31. Show that the Rao Theorem 2.2.8 does not extend to uniformly bounded nets of signed measures even if Γ is a uniformly Lipschitz uniformly bounded family.

HINT: by the non coincidence of the weak topology with the norm topology on the unit ball U in l^1, there exists a net of elements $\mu_\alpha \in U$ that converges weakly to zero and does not converge in norm. Consider μ_α as measures on $\mathbb{N}$ and observe that the set Γ of all functions f on $\mathbb{N}$ with $\sup|f| \leqslant 1$ is uniformly Lipschitz with constant 2. In addition, $\|\mu_\alpha\|$ is the supremum of the integrals of functions $f \in \Gamma$ with respect to the measure μ_α.

2.7.32. (i) Suppose that a sequence of Borel measures μ_n on a metric space X converges weakly to a Radon measure μ and, in addition, is uniformly tight. Let a family $\Gamma \subset C_b(X)$ be uniformly bounded and pointwise equicontinuous. Show that

$$\lim_{n\to\infty} \sup_{f\in\Gamma} \left| \int f\, d(\mu_n - \mu) \right| = 0.$$

(ii) Show that if (X, d) is a separable metric space and $\Gamma \subset C_b(X)$ is uniformly bounded and satisfies the condition

$$\lim_{\varepsilon\to 0} \sup\bigl\{|f(x) - f(y)|\colon\ f \in \Gamma, d(x, y) \leqslant \varepsilon\bigr\} = 0,$$

then the indicated equality is true for every weakly convergent sequence of Borel measures (not necessarily uniformly tight).

HINT: (i) one can assume that $|f| \leqslant 1$ for all $f \in \Gamma$ and $\|\mu_n\| \leqslant 1$. Suppose that we have a number $\varepsilon > 0$ and a sequence $\{f_n\} \subset \Gamma$ such that

$$\Bigl|\int_X f_n \, d(\mu_n - \mu)\Bigr| > \varepsilon.$$

Find a compact set K such that $|\mu|(X\backslash K) + |\mu_n|(X\backslash K) < \varepsilon/4$ for all n. By the Ascoli–Arzelà theorem the sequence $\{f_n\}$ contains a subsequence uniformly converging on K to some function f. We can assume that this is the whole sequence $\{f_n\}$. There is a function $g \in C_b(X)$ with $g|_K = f|_K$ and $|g| \leqslant 1$. For all n sufficiently large we obtain that $\sup_{x\in K} |g(x) - f_n(x)| \leqslant \varepsilon/4$ and the absolute value of the integral of g with respect to $\mu_n - \mu$ does not exceed $\varepsilon/4$, which leads to a contradiction. Assertion (ii) follows from (i), since on the completion of X the sequence $\{\mu_n\}$ is uniformly tight and the functions from Γ can be extended by continuity to the completion, moreover, the family of extensions satisfies the indicated condition.

2.7.33. (Rao [**541**], Bhattacharia, Ranga Rao [**64**, Theorem 1.11, p. 22, 23]) Let $\mathcal{V}$ be the class of all convex Borel sets in $\mathbb{R}^d$. Let $\mu_n \in \mathcal{P}(\mathbb{R}^d)$, $\mu_n \Rightarrow \mu$, and let μ be absolutely continuous. Then $\sup_{V\in\mathcal{V}} |\mu_n(V) - \mu(V)| \to 0$.

2.7.34. (Billingsley, Topsøe [**68**], Bhattacharia, Ranga Rao [**64**, p. 8, Theorem 2.4]) Let X be a separable metric space, let $\mu \in \mathcal{P}(X)$, and let $\mathcal{F}$ be a uniformly bounded family of Borel functions. Then the following conditions are equivalent:

(i) $\lim\limits_{\varepsilon\to 0} \sup_{f\in\mathcal{F}} \mu\bigl(x\colon \sup_{u,v\in U(x,\varepsilon)} |f(u) - f(v)| > \delta\bigr) = 0 \ \ \forall\,\delta > 0$;

(ii) if $\mu_n \Rightarrow \mu$, where $\mu_n \in \mathcal{P}(X)$, then the integrals of functions $f \in \mathcal{F}$ with respect to μ_n converge to the integral of f with respect to μ uniformly in f.

2.7.35. Derive Theorem 2.3.9 from Corollary 2.3.5 for complete separable spaces.

2.7.36. Let (X, d) be a noncompact metric space. Show that one can find a metric $\widetilde{d}$ on X generating the original topology such that there exists a sequence of signed Radon measures μ_n and a Radon measure μ with the following property: for every function f that is bounded and uniformly continuous in the metric $\widetilde{d}$, its integrals with respect to the measures μ_n converge to the integral of f with respect to measure μ, but the measures μ_n do not converge weakly to μ. The original metric possesses this property if there exist two sequences $\{x_n\}$ and $\{y_n\}$ with $x_n \neq y_n$ which have no limit points and the distance between x_n and y_n tends to zero.

HINT: if the latter condition is fulfilled, it suffices to consider the signed measures $\mu_n = \delta_{x_n} - \delta_{y_n}$, the integrals with respect to which of every uniformly continuous function tend to zero. Each point x_n has a neighborhood V_n in which there are no other elements of both sequences. There is a bounded continuous function f such that $f(x_n) = 1$ for all n and $f = 0$ outside $\bigcup_{n=1}^\infty V_n$. Hence there is no weak convergence of $\{\mu_n\}$ to zero. In the general case one can pass to a metric d_0 generating the original topology such that $d_0(x, y) \leqslant 1$ for all x, y. In this case either X contains a Cauchy sequence $\{x_n\}$ that does not converge, i.e., the indicated condition is fulfilled, or there is a countable set of points x_n with mutual distances separated from zero. Hence it suffices to consider the case where for some $r > 0$ there are no points x with $d_0(x, x_n) \leqslant r$. Now we define a new metric on X as follows: $\widetilde{d}(x, y) = d_0(x, y)$ if $x, y \notin \{x_n\}$, $\widetilde{d}(x, x_n) = r + 1$ if $x \notin \{x_n\}$, $\widetilde{d}(x_n, x_k) := r|1/n - 1/k|$. See also Varadarajan [**635**, Part 2, Theorem 4].

2.7.37. Let (X, d) be a separable metric space. Prove that for weak convergence of a net of measures $\mu_\alpha \in \mathcal{P}(X)$ to a measure $\mu \in \mathcal{P}(X)$ it suffices to have convergence of the integrals with respect to these measures for all functions

$$f(x) = \max\bigl(c_0, c_1 - d(x, y_1), \ldots, c_n - d(x, y_n)\bigr),$$

where $c_0, \dots, c_n$ are real numbers, $y_1, \dots, y_n \in X$. The condition of separability of X can be replaced by the τ-additivity of these measures.

HINT: the indicator function of any closed set can be represented as a limit of a decreasing net of functions of the indicated form.

2.7.38. Suppose that vectors v_n in a Hilbert space H converge weakly to a vector v and that $\|v_n\| \to \|v\|$. Prove that $\|v - v_n\| \to 0$.

2.7.39. (i) In the situation of Theorem 2.4.9 prove that the standard Gaussian measure on $\mathbb{R}^\infty$ is a weak limit of the normalized Lebesgue measures on the balls of radius $\sqrt{n}$ centered at the origin in $\mathbb{R}^n$.

(ii) Prove that the normalized Lebesgue measures on the unit balls in $\mathbb{R}^n$ and the normalized surface measures on the unit spheres in $\mathbb{R}^n$ converge weakly to the Dirac measure at the origin in $\mathbb{R}^\infty$.

HINT: (i) let ν_n be the normalized Lebesgue measure on the ball $n^{1/2}U_n$ and let σ_n be the normalized spherical measure on the boundary of $n^{1/2}U_n$, where U_n is the unit ball in $\mathbb{R}^n$; let $\varepsilon_n = n^{-1/2}/\ln(n+1)$, $r_n = 1 - \varepsilon_n$; then $r_n^n \to 0$, whence $\nu_n(r_n n^{1/2}U_n) \to 0$, i.e., $\|\nu_n - \nu_n|_{K_n}\| \to 0$, where K_n is the ring $n^{1/2}U_n \backslash r_n n^{1/2}U_n$; finally, one can verify that for every Lipschitz function f on $\mathbb{R}^d$ the difference of the integrals of f with respect to the measures σ_n and $\nu_n|_{K_n}$ is estimated in absolute value by $Cn^{-1/2}$; to this end, one can write the integral with respect to the second measure in polar coordinates and observe that $|f(x) - f(r_n x)| \leqslant C\varepsilon_n|x|$ and the integrals of $x_1^2 + \dots + x_d^2$ with respect to the considered measures with $n \geqslant d$ does not exceed d, because the integral of $x_1^2 + \dots + x_n^2$ does not exceed n, and the integrals of the summands are equal; (ii) the integral of x_i^2 with $i \leqslant n$ with respect to each of the indicated measures does not exceed $1/n$, since the integral of $x_1^2 + \dots + x_n^2$ does not exceed 1 and the integrals of the summands are equal; hence for every fixed d the integrals of the norm in $\mathbb{R}^d$ tend to zero, which gives convergence of integrals of any Lipschitz function f on $\mathbb{R}^d$ to $f(0)$. See also Ledoux [**420**] and Zorich [**676**] about the phenomenon of measure concentration.

2.7.40. (Borovkov [**111**]) Let $B_p^n(r)$ be the closed ball centered at the origin with radius r with respect to the l^p-norm on $\mathbb{R}^n$, let $S_p^n(r)$ be the boundary of $B_p^n(r)$, and let U_n^p and Q_n^p be the uniform distributions on $B_p^n(r)$ and $S_p^n(r)$, respectively. Let $U_{n,k}^p(r)$ and $Q_{n,k}^p(r)$ be the distributions of subvectors x^k of the vectors x^n with distribution $U_n^p(r)$ and $Q_n^p(r)$, let P_k^p be the distribution of the vector $Y = (Y_1, \dots, Y_k)$, where random variables Y_i are independent and equally distributed with density $c_p \exp(-|x|^p/p)$, where $c_p^{-1} = 2\Gamma(1/p)p^{1-1/p}$. Then for all $n > k + p$ we have

$$\|Q_{n,k}^p(r) - P_k^p(rn^{-1/p})\| \leqslant \frac{k+2p}{n-k-p}, \quad p \geqslant 0,$$

$$\|U_{n,k}^p(r) - P_k^p(rn^{-1/p})\| \leqslant \frac{k+2p}{n-k-p}, \quad p \geqslant 1.$$

See also Diaconis, Freedman [**167**], Khokhlov [**360**], Klartag [**366**].

2.7.41. (i) (Dembski [**164**]) Let X be a separable metric space. A class $\mathcal{D}$ of Borel sets is called determining weak convergence if, for Borel probability measures μ_n and μ on X, convergence $\mu_n(D) \to \mu(D)$ for all $D \in \mathcal{D}$ such that $\mu(\partial D) = 0$ yields weak convergence of μ_n to μ. Prove that, for every class $\mathcal{D}$ determining weak convergence and every Borel probability measure ν, the subclass $\mathcal{D}^\nu$ in $\mathcal{D}$ consisting of sets with boundary of ν-measure zero is also a class determining weak convergence.

(ii) Show that for every Borel probability measure ν, the class $\mathcal{B}(X)^\nu = \Gamma_\nu$ of all Borel sets with boundary of ν-measure zero is a class determining weak convergence. In particular, if $\mu_n, \mu \in \mathcal{P}(X)$ and $\lim\limits_{n\to\infty} \mu_n(D) = \mu(D)$ for all sets $D \in \mathcal{B}(X)^\nu$, then $\mu_n \Rightarrow \mu$.

(iii) Let $\mathcal{D}$ be the class of all compact sets in $[0,1]$ with boundary of positive Lebesgue measure. Prove that if Borel probability measures μ and μ_n, where $n \in \mathbb{N}$, on $[0,1]$ satisfy the condition $\lim_{n\to\infty} \mu_n(D) = \mu(D)$ for all $D \in \mathcal{D}$, then the measures μ_n converge weakly to μ, although the class $\mathcal{D}$ is not determining weak convergence.

HINT: (i) let

$$\lim_{n\to\infty} \mu_n(D) = \mu(D) \text{ if } D \in \mathcal{D}^\nu \text{ and } \mu(\partial D) = 0.$$

Then we obtain $\lim_{n\to\infty} \frac{1}{2}(\mu_n + \nu)(D) = \frac{1}{2}(\mu+\nu)(D)$ if $D \in \mathcal{D}$ is such that $(\mu+\nu)(\partial D) = 0$. Hence the measures $(\mu_n+\nu)/2$ converge weakly to $(\mu+\nu)/2$, which gives weak convergence of $\{\mu_n\}$ to μ. Clearly, (i) yields (ii). (iii) Suppose that Borel probability measures μ_n and μ on $[0,1]$ are such that $\lim_{n\to\infty} \mu_n(D) = \mu(D)$ for all $D \in \mathcal{D}$. We have to show that $\lim_{n\to\infty} F_{\mu_n}(t) = F_\mu(t)$ for every continuity point t of the distribution function F_μ of the measure μ. If there exists $\varepsilon > 0$ such that $F_{\mu_n}(t) > F_\mu(t) + \varepsilon$ for infinitely many n, then one can find $s > t$ such that $F_\mu(s) < F_\mu(t) + \varepsilon/2$. It is clear that (t, s) contains a compact set K with boundary of positive Lebesgue measure. Then $D = [0, t] \cup K \in \mathcal{D}$, and we obtain a contradiction, since $\mu_n(D) \geqslant F_{\mu_n}(t)$. If $F_{\mu_n}(t) < F_\mu(t) - \varepsilon$ for infinitely many n, then there exists $s < t$ such that $F_\mu(s) > F_\mu(t) - \varepsilon/2$. Again we find a set $D \in \mathcal{D}$ of the form $D = [0, s] \cup K$, $K \subset (s, t)$, which gives $\mu_n(D) \leqslant F_{\mu_n}(t) \leqslant \mu(D) - \varepsilon/2$, since one has the bound $\mu(D) \geqslant F_\mu(s)$.

2.7.42. (Hartman, Marczewski [**311**]) Let $(X, \mathcal{A}, \mu)$ be a probability space, let (Y, d) be a separable metric space, and let f and f_n, where $n \in \mathbb{N}$, be Borel measurable mappings from X to Y. Prove that $f_n \to f$ in measure, i.e.,

$$\lim_{n\to\infty} \mu\Bigl(d\bigl(f_n(x), f(x)\bigr) > \varepsilon\Bigr) = 0 \quad \text{for all } \varepsilon > 0,$$

precisely when $\lim_{n\to\infty} \mu\bigl(f_n^{-1}(E) \triangle f^{-1}(E)\bigr) = 0$ for every set $E \in \mathcal{B}(Y)$ such that the boundary of E has measure zero with respect to $\mu\circ f^{-1}$.

2.7.43. Let X be a separable Banach space and let μ be a Borel probability measure on X. For every compact set $K \subset X$, the concentration function $C_\mu(K)$ is defined by the formula $C_\mu(K) = \sup_{x\in X} \mu(K + x)$. Prove that the following conditions are equivalent for every sequence of Borel probability measures μ_n on X:

(i) $\sup_{K\in\mathcal{K}} \inf_n C_{\mu_n}(K) = 1$, where $\mathcal{K}$ is the family of all compact sets in X,

(ii) every subsequence in $\{\mu_n\}$ contains a further subsequence $\{\nu_n\}$ for which there are vectors $x_n \in X$ such that the sequence of measures $\nu_n(\,\cdot\, + x_n)$ is uniformly tight.

HINT: see Hengartner, Teodorescu [**317**, Chapter 5], where one can find additional information about concentration functions.

2.7.44. (Slutsky [**587**]) Let $(\Omega, \mathcal{A}, P)$ be a probability space, let E be a separable Banach space, and let ξ_n, ξ and η_n be $\bigl(\mathcal{A}, \mathcal{B}(E)\bigr)$-measurable mappings such that the measures $P\circ\xi_n^{-1}$ converge weakly to the measure $P\circ\xi^{-1}$ and $\eta_n \to 0$ a.e. Show that the measures $P\circ(\xi_n + \eta_n)^{-1}$ converge weakly to $P\circ\xi^{-1}$ as well.

HINT: apply the Egorov theorem to $\{\eta_n\}$ (see § 2.1).

2.7.45. (Wichura [**659**]) Suppose that (X, d) is a metric space, (Ω, P) is a probability space, $\xi_n, \xi\colon \Omega \to X$ are measurable mappings, $T_n\colon X \to X$ are Borel mappings such that, for each n, the measures $P\circ(T_n\circ\xi_k)^{-1}$ converge weakly to $P\circ(T_n\circ\xi)^{-1}$ as $k\to\infty$, the sequence $d(\xi, T_n\circ\xi)$ converges to 0 in probability and for every $\varepsilon > 0$ one has

$$\lim_{n\to\infty} \limsup_{k\to\infty} P\bigl(d(\xi_k, T_n\circ\xi_k) \geqslant \varepsilon\bigr) = 0.$$

Prove that the measures $P\circ\xi_k^{-1}$ converge weakly to the measure $P\circ\xi^{-1}$.

2.7.46. Let T be an infinite set. Prove that the sets in the σ-algebra $\sigma(\mathbb{R}^T)$ generated by cylinders $\bigl\{x\colon \bigl(x(t_1), \ldots, x(t_n)\bigr) \in B\bigr\}$, $B \in \mathcal{B}(\mathbb{R}^n)$, have the following form: $\bigl\{x\colon \bigl(x(t_i)\bigr)_{i=1}^\infty \in E\bigr\}$ for some countable set of points $t_i \in T$ and some set $E \in \mathcal{B}(\mathbb{R}^\infty)$.

HINT: verify that all sets of the indicated form constitute a σ-algebra.

2.7.47. Let f_n be measurable mappings from a probability space (X, μ) to a separable metric space S converging in measure to a mapping f. Let $\Psi\colon S \to M$ be a continuous mapping with values in a metric space (M, d). Show that the mappings $\Psi\circ f_n$ converge in measure to $\Psi\circ f$.

HINT: show that the integrals of $\min\bigl(1, d(\Psi\circ f_n, \Psi\circ f)\bigr)$ converge to zero by using that for every subsequence in $\{f_n\}$ there exists a further subsequence converging to f almost everywhere.

2.7.48. Prove that on the nonseparable Banach space l^∞ of bounded sequences $x = (x_n)$ with norm $\|x\|_\infty = \sup_n |x_n|$ a Borel measure μ is a weak limit of its finite-dimensional projections under the mappings $P_n x = (x_1, \ldots, x_n, 0, 0, \ldots)$ precisely when it is concentrated on the subspace c_0 of sequences tending to zero.

HINT: for a measure on c_0 weak convergence of its projections follows from the fact that $P_n x \to x$ on c_0; the converse follows from the fact that c_0 is closed in l^∞, hence for a measure μ with $|\mu|(l^\infty\backslash c_0) > 0$ one can find a function $f \in C_b(l^\infty)$ vanishing on c_0 with a nonzero integral.

2.7.49. Prove that the limit of any weakly convergent sequence of Gaussian measures is a Gaussian measure.

HINT: consider the case of $\mathbb{R}$ with the aid of the Fourier transform.

2.7.50. Construct an example of a sequence of centered Gaussian measures μ_n on $\mathbb{R}^\infty$ equivalent to the countable power γ of the standard Gaussian measure on the real line such that $\lim\limits_{n\to\infty} \mu_n(B) = \gamma(B)$ for every Borel set B, but there is no convergence in variation.

HINT: consider the measure μ_n that differs from γ by replacing the nth copy of the standard Gaussian measure in the product by the centered Gaussian measure with variance 2; observe that the measures μ_n converge to γ setwise, because there is convergence on cylindrical sets and the Radon–Nikodym densities $d\mu_n/d\gamma$ are uniformly bounded, however, the variation distance $\|\mu_n - \gamma\|$ is constant.

2.7.51. Construct an example of a sequence of signed Borel measures μ_n on $\mathbb{R}^\infty$ such that, for every fixed k, the projections of the measures μ_n on $\mathbb{R}^k$ converge weakly to zero, but the measures μ_n are not uniformly bounded in the variation norm.

HINT: consider $\mu_n = n\delta_{e_{n+1}} - n\delta_{-e_{n+1}}$, where $e_{n+1} \in \mathbb{R}^{n+1}$ is orthogonal to $\mathbb{R}^n$ and $\|e_{n+1}\| = 1$.

2.7.52. (i) A Radon probability measure μ on a Fréchet space X (or a locally convex space) is called convex or logarithmically concave if its finite-dimensional projections are convex (see p. 40). This property is equivalent (see Bogachev [**81**, § 7.14]) to the inequality

$$\mu\bigl(tA + (1-t)B\bigr) \leqslant \mu(A)^t \mu(B)^{1-t} \quad \forall\, A, B \in \mathcal{B}(X),\ t \in [0,1].$$

Deduce from Theorem 1.7.4 that on $\mathbb{R}^\infty$ and l^2 every convex measure is the weak limit of a sequence of uniform distributions on finite-dimensional convex compact sets.

(ii) Prove that the product, the image under a continuous affine mapping and the convolution of convex measures are convex.

(iii) Prove that the limit of a weakly convergent sequence of convex measures is a convex measure.

HINT: (ii) use the definition and the characterization mentioned in (i) and also that the convolution is an affine image of the product; (iii) reduces to finite-dimensional projections, hence to the case of $\mathbb{R}^d$, where it suffices to verify the inequality from (i) for open sets A and B with boundaries of zero measure with respect to the limiting measure μ.

CHAPTER 3

Metrics on spaces of measures

In this chapter we start a discussion of those aspects of weak convergence that are connected with topological and geometric properties of spaces of measures. In the previous chapter weak convergence of sequences or even uncountable nets of Borel measures on metric spaces was discussed, but so far it was not clarified whether there is a metric or topology behind this convergence. Here we introduce the weak topology on spaces of measures (further topological issues are the subject of Chapter 5) and also a number of classical metrics connected with this topology. Spaces of measures on a given space equipped with such metrics turn out to be a very interesting object at the junction of different areas of mathematics.

3.1. The weak topology and the Prohorov metric

Before discussing weak topologies and the corresponding metrics it is useful to recall that not every reasonable convergence is generated by a norm, a metric or at least a topology. For example, convergence in measure on the space of measurable (or continuous) functions on an interval is generated by a metric, but not by a norm. The pointwise convergence of continuous functions on an interval is generated by a topology, but not by a metric. Convergence almost everywhere on the set of continuous functions cannot be defined even by a topology (Exercise 3.5.14).

However, behind weak convergence of measures is obviously the following *weak topology*, making the space $\mathcal{M}(X)$ locally convex.

3.1.1. DEFINITION. Let X be a metric space. The weak topology on the space $\mathcal{M}(X)$ of Borel measures on X is the topology $\sigma\bigl(\mathcal{M}(X), C_b(X)\bigr)$, i.e., the topology generated by seminorms

$$p_f(\mu) = \left|\int_X f\,d\mu\right|, \quad f \in C_b(X).$$

A base of the weak topology consists of the sets

$$U_{f_1,\ldots,f_n,\varepsilon}(\mu) = \left\{\nu\colon \ \left|\int_X f_i\,d\mu - \int_X f_i\,d\nu\right| < \varepsilon, \ i = 1,\ldots,n\right\}, \tag{3.1.1}$$

where $\mu \in \mathcal{M}(X)$, $f_i \in C_b(X)$, $\varepsilon > 0$. Sets of such a form are called basic neighborhoods of the measure μ in the weak topology.

The weak topology is actually the weak-$*$ topology in the terminology of functional analysis (however, following tradition, we call it the *weak topology*, which does not match the terminology introduced in § 1.3 for Banach spaces!). The considered weak topology on $\mathcal{M}(X)$ is actually the topology of duality $\sigma\bigl(\mathcal{M}(X), C_b(X)\bigr)$, where a bounded continuous function f on X is regarded as the functional on $\mathcal{M}(X)$

acting by the formula

$$\mu \mapsto \int_X f\,d\mu.$$

Various subsets of $\mathcal{M}(X)$, including $\mathcal{M}_r(X)$, $\mathcal{M}_\tau(X)$, $\mathcal{P}(X)$, and $\mathcal{P}_r(X)$, are also equipped with the weak topology. As usual, once a topology appears, the questions arise about its metrizability (on the whole space or on its parts) and its properties such as conditions for compactness. More systematically such questions are studied in Chapter 5, where analogous weak topologies are introduced on sets of measures on general spaces, but here we concentrate on the metric side.

First of all, let us observe that the weak topology on the whole space $\mathcal{M}(X)$ in nontrivial cases (that is, for infinite X) is nonmetrizable. This follows from the explanation in § 1.3 of nonmetrizability of the topology $\sigma(E, F)$ with a space of functionals F that is not the linear span of a countable set, since the space $C_b(X)$ is always Banach, and hence by Baire's category theorem it can be the linear span of a countable set only in case it is finite-dimensional.

For example, for $X = \mathbb{N}$ we obtain for $\mathcal{M}(X)$ the space l^1, and its weak topology from the point of view of the space of measures coincides with its weak topology of a Banach space, since $C_b(\mathbb{N})$ coincides with l^∞. If the weak topology on l^1 were metrizable at least on the unit ball, then it would coincide with the norm topology, since weakly convergent sequences in l^1 converge in norm (Exercise 1.7.15). However, this is false, since in an infinite-dimensional normed space there are no bounded nonempty sets open in the weak topology: the basic open sets in the weak topology indicated in § 1.3 must contain nontrivial linear subspaces in the infinite-dimensional case. In particular, every weak neighborhood of zero must contain points of the unit sphere, so the ball of radius $1/2$ centered at zero cannot contain a weak neighborhood of zero.

It turns out, however, that for a separable space X the weak topology is metrizable on the cone of nonnegative measures, hence also on the set of probability measures. There are several interesting metrics which can be used for this. They are considered below, starting from the Prohorov metric. We see that the space $\mathcal{M}(X)$, nonmetrizable in the weak topology, is the algebraic sum of two metrizable cones $\mathcal{M}^+(X)$ and $-\mathcal{M}^+(X)$.

A base of the weak topology on the set of probability measures can be defined by means of sets, not functions. Let us consider the following two classes of sets in the space $\mathcal{P}(X)$ of probability measures:

$$W_{F_1,\ldots,F_n,\varepsilon}(\mu) = \Bigl\{\nu \in \mathcal{P}(X)\colon\ \nu(F_i) < \mu(F_i) + \varepsilon,\ i = 1, \ldots, n\Bigr\},$$

$$F_i = f_i^{-1}(0),\ f_i \in C(X),\ \varepsilon > 0, \tag{3.1.2}$$

$$W_{G_1,\ldots,G_n,\varepsilon}(\mu) = \Bigl\{\nu \in \mathcal{P}(X)\colon\ \nu(G_i) > \mu(G_i) - \varepsilon,\ i = 1, \ldots, n\Bigr\},$$

$$G_i = X \backslash f_i^{-1}(0),\ f_i \in C(X),\ \varepsilon > 0. \tag{3.1.3}$$

Since we deal with a metric space, the sets F_i represent arbitrary closed sets and G_i represent arbitrary open sets. However, F_i and G_i are intentionally represented by means of continuous functions, since in this way the weak topology will be introduced for general topological spaces in Chapter 4.

3.1.2. THEOREM. *The bases introduced above generate the weak topology on the set of probability measures $\mathcal{P}(X)$. On the set of all nonnegative measures $\mathcal{M}^+(X)$*

a base of the weak topology is formed by all neighborhoods of the form (3.1.2) *or* (3.1.3) *along with neighborhoods of the form* $\{\nu\colon\ |\mu(X)-\nu(X)|<\varepsilon\}$.

PROOF. The coincidence of the bases (3.1.2) and (3.1.3) is obvious from the defining expressions. Let U be a neighborhood of the form (3.1.1). We can assume that $0<f_i<1$. Let us fix a number $k\in\mathbb{N}$ with $k^{-1}<\varepsilon/4$. For each $i=1,\ldots,n$ there exist points $c_{i,j}\in[0,1]$ such that $0=c_{i,0}<\cdots<c_{i,m}=1$, $c_{i,j+1}-c_{i,j}<\varepsilon/4$ and $\mu\bigl(f_i^{-1}(c_{i,j})\bigr)=0$. Let $A_{i,1}=\{0\leqslant f_i<c_{i,1}\}$, ..., $A_{i,m}=\{c_{i,m-1}\leqslant f_i<c_{i,m}\}$. We show that there is a neighborhood V of the form (3.1.2) such that for all i and j and $\nu\in V$ we have $|\mu(A_{i,j})-\nu(A_{i,j})|<\delta$, where $\delta=(4m)^{-1}\varepsilon$. Then $V\subset U$ by the inequality $\bigl|f_i-\sum_{j=1}^m c_{i,j}I_{A_{i,j}}\bigr|<\varepsilon/4$. Indeed,

$$\left|\int_X f_i\,d\mu-\int_X f_i\,d\nu\right|\leqslant\sum_{j=1}^m c_{i,j}|\mu(A_{i,j})-\nu(A_{i,j})|+\varepsilon/2<\varepsilon.$$

For the required neighborhood V of the measure μ we take the intersection of the neighborhood V_1 of the form (3.1.2) with δ in place of ε, generated by the closed sets $F_{i,j}=\{c_{i,j-1}\leqslant f_i\leqslant c_{i,j}\}$, where $i\leqslant n$, $j\leqslant k$, and an analogous neighborhood V_2 of the form (3.1.3) with the open sets $G_{i,j}=\{c_{i,j-1}<f_i<c_{i,j}\}$. Clearly, for all $\nu\in V_2$ we have

$$\nu(A_{i,j})\geqslant\nu(G_{i,j})>\mu(G_{i,j})-\delta=\mu(A_{i,j})-\delta,$$

since $\mu\bigl(f_i^{-1}(c_{i,j})\bigr)=0$. Similarly, $\nu(A_{i,j})<\mu(A_{i,j})+\delta$ if $\nu\in V_1$.

We now show that every neighborhood of the form (3.1.2) contains a neighborhood from the weak topology. It suffices to consider neighborhoods defined by a single closed set F_1. We can assume that $F_1=f_1^{-1}(0)$, where $0\leqslant f_1\leqslant 1$. Let us find $c>0$ such that $\mu\bigl(\{0<f_1<c\}\bigr)<\varepsilon/2$. Let $\zeta\in C_b(\mathbb{R})$, $\zeta(t)=1$ if $t\leqslant 0$, $\zeta(t)=0$ if $t\geqslant c$, $0<\zeta(t)<1$ if $t\in(0,c)$. Set $f=\zeta(f_1)$. It remains to observe that $\nu(F_1)<\mu(F_1)+\varepsilon$ whenever $\displaystyle\int_X f\,d\nu<\int_X f\,d\mu+\varepsilon/2$. Indeed, $\nu(F_1)\leqslant\displaystyle\int_X f\,d\nu$, since $f=1$ on F_1. On the other hand,

$$\int_X f\,d\mu\leqslant\mu(\{f=1\})+\mu\bigl(\{0<f<1\}\bigr)=\mu(F_1)+\mu\bigl(\{0<f_1<c\}\bigr),$$

which is smaller than $\mu(F_1)+\varepsilon/2$. The case of $\mathcal{M}^+(X)$ is similar. □

3.1.3. EXAMPLE. The set of all measures on X of the form $\sum_{j=1}^n c_j\delta_{x_j}$, where $c_j\in\mathbb{R}^1$, $x_j\in X$, is everywhere dense in $\mathcal{M}(X)$ with the weak topology.

Indeed, suppose we have a neighborhood U of the form (3.1.1). We can assume that $\|\mu\|\leqslant 1$. There exist simple functions g_i for which $\sup|f_i-g_i|<\varepsilon/4$ for all $i=1,\ldots,n$. Hence in order to show that U contains a measure ν of the required form with $\|\nu\|\leqslant 1$, it suffices to find a measure ν of the required form such that the functions g_i have the same integrals with respect to ν and μ. Thus, everything reduces to choosing points x_j and numbers c_j such that, for a given finite partition of X into disjoint Borel sets A_i, $i=1,\ldots,k$, one has the equality $\nu(A_i)=\mu(A_i)$. It remains to take a point x_i in each set A_i and put $c_i=\mu(A_i)$.

The established property is not equivalent to another one that every measure can be obtained as a limit of a sequence of finite linear combinations of Dirac measures (which is true only for τ-additive measures, see Example 2.2.7). However, for separable X the space $\mathcal{M}(X)$ with the weak topology is separable as well.

Let us now introduce the *Prohorov metric* (or *Lévy–Prohorov metric*) on the set of Borel probability measures $\mathcal{P}(X)$ by the formula

$$d_P(\mu,\nu) = \tag{3.1.4}$$
$$= \inf\Bigl\{\varepsilon > 0\colon\ \nu(B) \leqslant \mu(B^\varepsilon) + \varepsilon,\ \mu(B) \leqslant \nu(B^\varepsilon) + \varepsilon \ \ \forall B \in \mathcal{B}(X)\Bigr\}.$$

3.1.4. THEOREM. *The weak topology on the set $\mathcal{P}_\tau(X)$ of probability τ-additive measures is generated by the Prohorov metric. In particular, if X is separable, then the weak topology on the whole set $\mathcal{P}(X)$ is generated by the metric d_P and the space $\bigl(\mathcal{P}(X), d_P\bigr)$ is separable.*

If $\mathcal{P}(X) \neq \mathcal{P}_\tau(X)$, then the weak topology cannot be generated by a metric on $\mathcal{P}(X)$.

PROOF. Let us verify that d_P is a metric on $\mathcal{P}(X)$. Clearly, $d_P(\mu,\nu) = d_P(\nu,\mu)$. If $d_P(\mu,\nu) = 0$, then $\mu(B) = \nu(B)$ for every closed set B and hence $\mu = \nu$. Let

$$\nu(B) \leqslant \mu(B^\varepsilon) + \varepsilon, \quad \mu(B) \leqslant \nu(B^\varepsilon) + \varepsilon,$$
$$\mu(B) \leqslant \eta(B^\delta) + \delta, \quad \eta(B) \leqslant \mu(B^\delta) + \delta$$

for all $B \in \mathcal{B}(X)$. Then

$$\nu(B) \leqslant \eta(B^{\varepsilon+\delta}) + \varepsilon + \delta, \quad \eta(B) \leqslant \nu(B^{\varepsilon+\delta}) + \varepsilon + \delta,$$

whence $d_P(\nu,\eta) \leqslant d_P(\nu,\mu) + d_P(\mu,\eta)$.

Let us show that every neighborhood W of the form (3.1.2) contains a ball centered at μ of positive radius in the metric d_P. Let us pick $\delta \in (0, \varepsilon/2)$ such that $\mu(F_i^\delta) < \mu(F_i) + \varepsilon/2$ for all $i = 1,\ldots,n$. If $d_P(\mu,\nu) < \delta$, then we have the inequalities $\nu(F_i) < \mu(F_i^\delta) + \delta < \mu(F_i) + \varepsilon$, i.e., ν belongs to the neighborhood W. So far we have not used the τ-additivity of measures.

We now show that every ball in $\mathcal{P}_\tau(X)$ in the Prohorov metric with center μ and radius ε contains a neighborhood of the form (3.1.2). Let us take $\delta < \varepsilon/3$. We can cover the separable support of μ by a countable family of open balls V_n of diameter less than δ having boundaries of μ-measure zero (recall that the τ-additivity means the separability of support). This is possible, because the spheres of different radii with a common center are disjoint, so only countably many of them can have a nonzero measure. Let us construct pairwise disjoint sets A_n with boundary of μ-measure zero covering the support of μ. For such sets one can take $A_n = \bigcup_{i=1}^n V_i \backslash \bigcup_{i=1}^{n-1} V_i$, where $A_1 = V_1$. Let us find k such that

$$\mu\Bigl(\bigcup_{i=1}^k A_i\Bigr) > 1 - \delta. \tag{3.1.5}$$

By Theorem 2.4.1 there is a neighborhood W of the form (3.1.2) such that

$$|\mu(A) - \nu(A)| < \delta \tag{3.1.6}$$

for all $\nu \in W$ and every set A that is the union of some of the sets $A_1,\ldots, A_k$. We verify that $d_P(\mu,\nu) < \varepsilon$ for all $\nu \in W$. Let $B \in \mathcal{B}(X)$. Consider the set A that is the union of those sets $A_1,\ldots,A_k$ that intersect B. Then $B \subset A \bigcup \bigcup_{i=k+1}^\infty A_i$ and $A \subset B^\delta$, since the diameter of each A_i is smaller than δ. Whenever $\nu \in W$, from (3.1.5) and (3.1.6) we obtain

$$\mu(B) < \mu(A) + \delta < \nu(A) + 2\delta \leqslant \nu(B^\delta) + 2\delta.$$

Since (3.1.5) and (3.1.6) yield that $\nu\Bigl(\bigcup_{i=1}^k A_i\Bigr) > 1-2\delta$, similarly we obtain $\nu(B) < \mu(B^\delta) + 3\delta$. Thus, $d_{\mathrm{P}}(\mu,\nu) < 3\delta < \varepsilon$. The separability of $\bigl(\mathcal{P}(X), d_{\mathrm{P}}\bigr)$ for separable X is clear from the example above.

Finally, if the weak topology is metrizable on $\mathcal{P}(X)$, then, according to Example 3.1.3, every measure $\mu \in \mathcal{P}(X)$ is the limit of a sequence of measures μ_n with finite supports and hence has a separable support. Thus, if $\mathcal{P}(X)$ is broader than $\mathcal{P}_\tau(X)$, then the Prohorov metric generates on $\mathcal{P}(X)$ a strictly stronger topology than the weak topology (and the latter is not metrizable). □

The Prohorov metric admits another representation due to Strassen [**597**]. Given a pair of measures $\mu,\nu \in \mathcal{P}_r(X)$, let $\Pi(\mu,\nu)$ denote the set of all measures $\sigma \in \mathcal{P}_r(X^2)$ for which the projections on the first and second factors coincide with μ and ν, respectively. For example, we have $\mu\otimes\nu \in \Pi(\mu,\nu)$. Measures in $\Pi(\mu,\nu)$ are called couplings of the pair (μ,ν) and the measures μ and ν are called *marginals* or *marginal distributions* of measures in $\Pi(\mu,\nu)$. If μ and ν are distributions of random elements ξ and η, then random elements with distributions in $\Pi(\mu,\nu)$ are called couplings of the pair (ξ,η); the distributions of their components, but not the joint distributions of components, coincide with the distributions of the given elements.

3.1.5. THEOREM. *Let X be a complete separable metric space. Then*

$$(3.1.7)\quad d_{\mathrm{P}}(\mu,\nu) = \inf\Bigl\{\varepsilon > 0\colon \exists\sigma \in \Pi(\mu,\nu) \ \ \text{with} \ \ \sigma\bigl((x,y) \in X\times X\colon\ d(x,y) > \varepsilon\bigr) \leqslant \varepsilon\Bigr\}.$$

In addition, there exists a measure $\sigma_0 \in \Pi(\mu,\nu)$ for which the estimate above holds with $\varepsilon = d_{\mathrm{P}}(\mu,\nu)$.

PROOF. Suppose that for some $\varepsilon > 0$ there exists a measure $\sigma \in \Pi(\mu,\nu)$ satisfying the inequality above. Then for every $B \in \mathcal{B}(X)$ we have

$$\nu(B) = \sigma\bigl((x,y)\colon\ y \in B\bigr) \leqslant \sigma\bigl((x,y)\colon\ x \in B^\varepsilon\bigr) + \varepsilon = \mu(B^\varepsilon) + \varepsilon.$$

The same inequality is true when we interchange our measures, hence we have the estimate $d_{\mathrm{P}}(\mu,\nu) \leqslant \varepsilon$, i.e., the left-hand side of (3.1.7) is not greater than the right-hand side.

The opposite inequality is more difficult and can hardly be verified directly. This can be done by using the Hahn–Banach theorem or passing to the discrete case and applying rather nontrivial combinatorial arguments, see Dudley [**193**, § 11.6]. A shorter way is based on the Kantorovich duality method (see Villani [**648**, p. 44]). We apply not the duality method itself, but a theorem due to Strassen from the cited paper, obtained by means of this method, which gives the desired estimate at once. Certainly, on this way the main difficulty of the proof is moved to justification of the following assertion, which can be found in the original paper [**597**] or in Rachev, Rüschendorf [**532**, Theorem 5.4.1] and which we present here not in the most general form: if X is a complete separable metric space, $\mu,\nu \in \mathcal{P}(X)$ and $F \subset X\times X$ is a closed set such that for some $\varepsilon > 0$ for all $A, B \in \mathcal{B}(X)$ with $(A\times B)\cap F = \varnothing$ we have

$$\mu(A) + \nu(B) \leqslant 1 + \varepsilon,$$

then there exists a measure $\sigma \in \Pi(\mu,\nu)$ for which $\sigma(F) \geqslant 1-\varepsilon$.

We observe that it suffices to verify the indicated condition only for closed sets A, B. Suppose now that for each closed set $A \subset X$ we have $\mu(A) \leqslant \nu(A^\varepsilon) + \varepsilon$. Let us apply the cited theorem to the set $F = \{(x,y)\colon d(x,y) \leqslant \varepsilon\}$, which is obviously closed. If $A, B \subset X$ are closed and $(A \times B) \cap F = \varnothing$, then $B \subset X \backslash A^\varepsilon$, whence $\nu(B) \leqslant 1 - \nu(A^\varepsilon)$. Hence $\mu(A) + \nu(B) \leqslant \mu(A) + 1 - \nu(A^\varepsilon) \leqslant 1 - \varepsilon$, which gives a measure $\sigma \in \Pi(\mu, \nu)$ with $\sigma(F) \geqslant 1 - \varepsilon$, i.e., the right-hand side of (3.1.7) does not exceed ε, hence is not greater than the left-hand side. The existence of σ_0 is clear from the cited theorem, but follows also from the Prohorov theorem, since $\Pi(\mu, \nu)$ is uniformly tight, so one can find a weakly convergent sequence of measures σ_j for the numbers $\varepsilon_j = d_{\mathrm{P}}(\mu, \nu) + 1/j$. □

Equality (3.1.7) is true even without completeness of X, see [**193**, § 11.6]. The Strassen theorem can be restated in the following form.

3.1.6. THEOREM. *Let μ and ν be two Radon probability measures on a metric space (X, d). Then there exists a probability space $(\Omega, \mathcal{F}, P)$ and measurable mappings ξ and η from Ω to X such that $\mu = P \circ \xi^{-1}$, $\nu = P \circ \eta^{-1}$ and $d_{\mathrm{P}}(\mu, \nu) = K(\xi, \eta)$, where $K(\xi, \eta)$ is the Ky Fan semimetric defined by the formula*

$$K(\xi, \eta) := \inf\bigl\{\varepsilon > 0\colon\ P\bigl(d(\xi, \eta) > \varepsilon\bigr) \leqslant \varepsilon\bigr\}.$$

The Ky Fan semimetric defines convergence in probability for random elements with values in X. It has already been noted that every measure can be obtained as the distribution of a random element. However, two essentially different random elements can have the same distribution. For example, the standard Gaussian random variable ξ has the same distribution as $-\xi$. Hence in the stated theorem one cannot take arbitrary random elements with given distributions.

3.1.7. REMARK. (i) Let Y be a closed subset of X. Then the weak topology on $\mathcal{M}(Y)$ coincides with the weak topology induced from $\mathcal{M}(X)$ when $\mathcal{M}(Y)$ is regarded as the subset of the space $\mathcal{M}(X)$ consisting of measures concentrated on Y. This follows from the fact that every bounded continuous function on Y extends to a bounded continuous function on X.

(ii) If Y is an arbitrary Borel set, then the weak topology on $\mathcal{M}(Y)$ induced from $\mathcal{M}(X)$ can be weaker than the topology $\sigma\bigl(\mathcal{M}(Y), C_b(Y)\bigr)$. Even the set of convergent sequences in it can be smaller. For example, if $X = [0,1]$ and $Y = \{1/n\}$, then the measures $\delta_{1/n} - \delta_{1/(n+1)}$ on the interval converge weakly to zero, but as measures on the countable space Y they do not converge, since the function $f(1/n) = (-1)^n$ is continuous on Y. A similar example can be constructed for every nonclosed set in a metric space. It is important here that we deal with signed measures. For probability τ-additive measures such problems do not arise, since the Prohorov distance between probability measures μ and ν on a Borel set Y in X as an independent metric space coincides with the Prohorov distance between them as measures on all of X. Indeed, if for some $\varepsilon > 0$ we have $\nu(B) \leqslant \mu(B^\varepsilon) + \varepsilon$ and $\mu(B) \leqslant \nu(B^\varepsilon) + \varepsilon$ for all Borel sets $B \subset X$, then these inequalities remain also valid for $B \in \mathcal{B}(Y)$, since the ε-neighborhood of such a set B in Y has the form $B^\varepsilon \cap Y$. Hence the distance does not decrease when we pass to X. It does not increase, because the validity of the indicated inequalities for all sets $B \in \mathcal{B}(Y)$ implies their validity for all $B \in \mathcal{B}(X)$, since $\nu(B) = \nu(B \cap Y)$, $(B \cap Y)^\varepsilon \subset B^\varepsilon$, whence $\mu\bigl((B \cap Y)^\varepsilon\bigr) \leqslant \mu(B^\varepsilon)$, and similarly for the second inequality.

The Prohorov metric can be used for metrization of the whole cone $\mathcal{M}^+(X)$, but in the next section we do this in a more natural way. We now show that on the whole space $\mathcal{M}(X)$ of signed measures the weak topology is nonmetrizable if X is infinite, and in case of a noncompact X the weak topology is nonmetrizable even on the unit ball in $\mathcal{M}_r(X)$. We recall that for compact X bounded sets in $\mathcal{M}(X)$ with the weak topology are metrizable by the metrizability (and compactness) of the ball in the dual space to a separable Banach space with the weak-$*$ topology.

3.1.8. PROPOSITION. *If X is an infinite metric space, then the weak topology on the space $\mathcal{M}_r(X)$ of signed Radon measures is not metrizable. If X is noncompact, then the weak topology is not metrizable on the ball $\{\mu \in \mathcal{M}_r(X)\colon \|\mu\| \leqslant 1\}$.*

PROOF. The lack of metrizability of the infinite-dimensional space $\mathcal{M}(X)$ with the weak topology has already been explained (see p. 102), and the same reasoning works for $\mathcal{M}_r(X)$.

If X is complete and noncompact, then it contains a sequence of points x_n with mutual distances uniformly separated from zero. Then the set $S = \{x_n\}$ is closed in X. The same is true if X is incomplete and noncompact, since in this case one can find a Cauchy sequence $\{x_n\}$ that has no limit. Such a sequence is a closed set in X. Hence by the previous remark the restriction of the weak topology $\mathcal{M}(X)$ to $\mathcal{M}(S)$ coincides with the weak topology of the space $\mathcal{M}(S)$. The latter can be identified with l^1 with the weak topology (associating the measure $\sum_{n=1}^\infty c_n\delta_{x_n}$ with the element (c_n)). It has already been explained that the ball in l^1 with the weak topology is nonmetrizable. □

Let us now give a more topological version of the Prohorov theorem from § 2.3 also connected with metrization of sets of signed measures.

3.1.9. THEOREM. *Let X be a complete metric space. Then every weakly compact subset $\mathcal{M}_r(X)$ is uniformly tight. In addition, such a subset is metrizable in the weak topology.*

Conversely, every bounded uniformly tight set in $\mathcal{M}_r(X)$ has compact closure in the weak topology.

PROOF. Suppose that $M \subset \mathcal{M}_r(X)$ is weakly compact, but is not uniformly tight. Let us consider the functions f_j and measures μ_n constructed in the proof of Theorem 2.3.4 (their construction employed only the lack of uniform tightness, the Radon property of the measures of the given family and completeness of X). Now, however, we have only the relative weak compactness of $\{\mu_n\}$, which does not mean the existence of convergent subsequences. Nevertheless, by the relative weak compactness of $\{\mu_n\}$ the sequence $a_n := (a_n^i) \in l^1$, where a_n^i is the integral of f_i against μ_n, is relatively weakly compact in l^1. Indeed, the mapping from $\mathcal{M}_r(X)$ to l^1 associating to a measure μ the sequence of the integrals of the functions f_i with respect to μ, is continuous when $\mathcal{M}_r(X)$ and l^1 are equipped with the weak topologies. The latter is seen from the fact that, as noted in the proof of Theorem 2.3.4, for every element $\lambda = (\lambda_i)_{i=1}^\infty$ of the space $l^\infty = (l^1)^*$ the function $f^\lambda = \sum_{i=1}^\infty \lambda_i f_i$ is continuous and bounded, which gives the continuity of the functional taking μ to the integral of f^λ against μ. Hence the image of M under the indicated mapping is weakly compact in l^1, whence it follows that $a_n^n \to 0$ and we again arrive at a contradiction.

Let us show that M is metrizable in the weak topology. It follows from our condition and the first assertion that $\sup_{\mu\in M} \|\mu\| \leqslant C$ for some $C < \infty$ and that for every n there is a compact set K_n with $|\mu|(X\backslash K_n) \leqslant 1/n$ for all measures $\mu \in M$. We can assume that $C = 1$. It is easy to see that there exists a countable family of functions $f_k \in C_b(X)$ such that $|f_k| \leqslant 1$ and the family of the restrictions of the functions f_k to K_n is everywhere dense in the unit ball of the space $C_b(K_n)$ for each n. We define the metric d_M on the set M by means of the norm

$$q(\mu) = \sum_{k=1}^{\infty} 2^{-k} \left| \int_X f_k \, d\mu \right|$$

on the linear span of M. We observe that this is a norm. Indeed, if $\mu \neq 0$, then there exists a function $f \in C_b(X)$ with $|f| \leqslant 1$ and the integral $\alpha > 0$ with respect to the measure μ. Since μ belongs to the linear span of M, there exists K_n with $|\mu|(X\backslash K_n) < \alpha/8$. Then there is a function f_k with $\sup_{x \in K_n} |f(x) - f_k(x)| < \alpha/8$. The integrals of the functions f and f_k over the complement of K_n are less than $\alpha/8$, while their integrals over K_n differ by at most $\alpha/8$. Hence the integral of f_k over X cannot vanish, so $q(\mu) > 0$.

Let us verify that M with the norm q is compact. It follows from the Prohorov theorem that every sequence $\{\mu_n\}$ in M contains a weakly convergent subsequence $\{\mu_{n_i}\}$, and by the weak compactness of M its limit μ must also belong to M. Since the integrals of every function f_k with respect to the measures μ_{n_i} tend to the integral of f_k with respect to μ and all these integrals do not exceed 1 in absolute value, we obtain that $q(\mu - \mu_{n_i}) \to 0$, which we wanted to verify. In addition, the topology on M generated by the norm q is majorized by the weak topology, which yields the coincidence of both topologies on M due to compactness of M with respect to them (see Exercise 3.5.19). One could also apply a more general assertion on metrizability of compact sets in locally convex spaces (see Proposition 4.5.12). To this end, it is necessary to have a countable set of functions in $C_b(X)$ separating measures in M; though, for separable X this is obvious.

Conversely, if a set $M \subset \mathcal{M}_r(X)$ is bounded and uniformly tight, then by the Prohorov theorem the sets M^+ and M^- of positive and negative parts from the Hahn–Jordan decomposition are contained in compact sets in the weak topology. Hence their difference is also contained in a compact set. □

In the next section we introduce certain classical norms which could be used in place of q in this proof (see Remark 3.2.5).

3.1.10. COROLLARY. *Let X be a complete metric space and let $M \subset \mathcal{M}_r(X)$ be a weakly compact set. Then the sets $M^+, M^-, |M|$ consisting, respectively, of the positive parts, negative parts and total variations of measures in M have compact closures in the weak topology.*

In addition, the sets $M, M^+, M^-, |M|$ are contained in an absolutely convex weakly compact set. Finally, the same is true for the set of measures $\nu \in \mathcal{M}_r(X)$ such that there exists a measure $\mu \in M$ with $|\nu(B)| \leqslant |\mu|(B)$ for all $B \in \mathcal{B}(X)$.

PROOF. The first assertion is clear from what has been said above, since the set $|M|$ is uniformly tight. The assertion about inclusion of these sets to an absolutely convex weakly compact set follows from the fact that the closed absolutely convex hull of a metrizable compact set in a sequentially complete locally convex space is metrizable and compact (see Bogachev, Smolyanov [**97**, Theorem 5.6.10]). This fact applies to the space of Radon measures by Theorem 2.3.9 and completeness of X, due to which any weakly convergent sequence of Radon measures is concentrated on a separable subspace of X and its limit is a Radon measure. The last assertion follows from the already proven results. □

In Chapter 4 we discuss analogous results for general spaces, including incomplete metric spaces.

3.2. The Kantorovich and Fortet–Mourier metrics

In this section X is a metric space with a metric ϱ. As already noted in § 3.1, excepting the case of finite X, the weak topology on $\mathcal{M}(X)$ cannot be generated by a metric, hence by a norm. However, $\mathcal{M}(X)$ can be equipped with a norm such that the generated topology coincides with the weak topology on the set of τ-additive nonnegative measures (hence also on the set of τ-additive probability measures). We shall consider even two such (equivalent) norms.

Let us consider on the space $\mathcal{M}(X)$ the following *Kantorovich–Rubinshtein norm*:

$$\|\mu\|_{\mathrm{KR}} = \sup\biggl\{ \int_X f\,d\mu\colon\ f \in \mathrm{Lip}_1(X),\ \sup_{x\in X} |f(x)| \leqslant 1 \biggr\},$$

where

$$\mathrm{Lip}_1(X) := \bigl\{ f\colon\ X \to \mathbb{R},\ |f(x) - f(y)| \leqslant \varrho(x, y)\ \ \forall\, x, y \in X \bigr\}.$$

Clearly, $\|\mu\|_{\mathrm{KR}} \leqslant \|\mu\|$. The metric corresponding to the introduced norm is called the *Kantorovich–Rubinshtein metric*.

An equivalent norm is called the *Fortet–Mourier norm* and is given by the formula

$$\|\mu\|_{\mathrm{FM}} = \sup\biggl\{ \int_X f\,d\mu\colon\ f \in \mathrm{BL}(X),\ \|f\|_{BL} \leqslant 1 \biggr\},$$

where $\mathrm{BL}(X)$ is the class of all bounded Lipschitz functions on X with the norm

$$\|f\|_{BL} := \sup_x |f(x)| + \mathrm{Lip}(f), \quad \mathrm{Lip}(f) = \sup_{x\neq y} \frac{|f(x) - f(y)|}{\varrho(x, y)}.$$

The equivalence of the Kantorovich–Rubinshtein norm and the Fortet–Mourier norm is seen from the fact that the former actually employs a different (but equivalent) norm on $\mathrm{BL}(X)$ of the form $\max\bigl(\sup_x |f(x)|, \mathrm{Lip}(f)\bigr)$. This norm is not greater than the norm $\|f\|_{BL}$, but is not smaller than the half of the norm $\|f\|_{BL}$. Hence

$$\|\mu\|_{\mathrm{FM}} \leqslant \|\mu\|_{\mathrm{KR}} \leqslant 2\|\mu\|_{\mathrm{FM}}.$$

The Kantorovich–Rubinshtein and Fortet–Mourier norms have the form

$$\|\mu\|_{\mathcal{H}} = \sup\biggl\{ \biggl| \int_X h\,d\mu \biggr|\colon\ h \in \mathcal{H} \biggr\},$$

where $\mathcal{H}$ is some class of functions. Below we encounter also other norms of such a type. Certainly, on every normed space the norm can be written in a similar way due to the Hahn–Banach theorem, however, in concrete examples the specific features of different classes $\mathcal{H}$ play an important role.

Let $\mathcal{M}^1(X)$ denote the set of measures μ in $\mathcal{M}(X)$ such that for some $x_0 \in X$ the function $x \mapsto \varrho(x, x_0)$ is integrable. We recall that the integrability with respect to a signed measure means the integrability with respect to its total variation. Our choice of the point x_0 plays no role, since for every other point x_1 we have $\varrho(x, x_1) \leqslant \varrho(x, x_0) + \varrho(x_0, x_1)$. Measures in $\mathcal{M}^1(X)$ are called measures with *finite first moment.* A measure μ is contained in $\mathcal{M}^1(X)$ precisely when $\mathrm{Lip}_1(X) \subset L^1(\mu)$. It is clear that $\mathcal{M}^1(X)$ is a linear subspace in $\mathcal{M}(X)$.

On the linear subspace $\mathcal{M}_0^1(X)$ in $\mathcal{M}^1(X)$ consisting of all measures σ with $\sigma(X) = 0$ we define the *Kantorovich norm* by the formula

$$\|\sigma\|_{\mathrm{K}} = \sup\biggl\{ \int_X f\,d\sigma\colon\ f \in \mathrm{Lip}_1(X)\biggr\},$$

The condition $\sigma(X) = 0$ is needed to obtain a finite norm (because for f one can take any constant). This norm can be extended to the subspace $\mathcal{M}^1(X)$ by the formula

$$\|\mu\|_{\mathrm{K}} := |\mu(X)| + \sup\biggl\{ \int_X f\,d\mu\colon\ f \in \mathrm{Lip}_1(X),\ f(x_0) = 0\biggr\},$$

where x_0 is a fixed point of the space (common for all measures). Certainly, the extension depends on our choice of a fixed point x_0. It is clear that $\|\mu\|_{\mathrm{KR}} \leqslant \|\mu\|_{\mathrm{K}}$, since $f = f - f(x_0) + f(x_0)$.

If the diameter of X does not exceed 1, then for any measure σ with $\sigma(X) = 0$ we have $\|\sigma\|_{\mathrm{KR}} = \|\sigma\|_{\mathrm{K}}$, since $|f(x)| \leqslant 1$ for all $f \in \mathrm{Lip}_1(X)$ with $f(x_0) = 0$. For general bounded spaces the equality can fail, but the equivalence of the two norms remains in force.

Let us introduce the metrics generated by the indicated norms:

$$d_{\mathrm{KR}}(\mu,\nu) = \|\mu - \nu\|_{\mathrm{KR}},\ d_{\mathrm{FM}}(\mu,\nu) = \|\mu - \nu\|_{\mathrm{FM}},\ d_{\mathrm{K}}(\mu,\nu) = \|\mu - \nu\|_{\mathrm{K}}.$$

A remarkable property of the Kantorovich metric is that the set of Dirac measures in $\mathcal{M}^1(X)$ with the Kantorovich norm is isometric to the space X itself. Indeed, for any $x, y \in X$ we have

$$\varrho(x,y) = \|\delta_x - \delta_y\|_{\mathrm{K}},$$

since the difference of the integrals of the function $f \in \mathrm{Lip}_1(X)$ with respect to the Dirac measures δ_x and δ_y is not greater than $\varrho(x,y)$ and for $f(u) = \varrho(u,y)$ one has precisely $\varrho(x,y)$. For the Kantorovich–Rubinshtein metric this equality is true for all points at the distance at most 1.

On the real line the Kantorovich distance can be easily expressed through the distribution functions (Dall'Aglio [**155**, p. 42], Vallander [**630**]).

3.2.1. Example. For any probability measures μ and ν on the real line with finite first moments and the distribution functions Φ_μ and Φ_ν, the following equality holds:

$$d_{\mathrm{K}}(\mu,\nu) = \int_{-\infty}^{+\infty} |\Phi_\mu(t) - \Phi_\nu(t)|\,dt.$$

Indeed, if $f \in \mathrm{Lip}_1(\mathbb{R})$, then by the integration by parts formula

$$\int_{\mathbb{R}} f\,d(\mu - \nu) = \int_{-\infty}^{+\infty} f'(t)\bigl(\Phi_\nu(t) - \Phi_\mu(t)\bigr)\,dt,$$

where f' exists almost everywhere and takes values in $[-1,1]$. Hence the left-hand side is not greater than $\|\Phi_\mu - \Phi_\nu\|_{L^1(\mathbb{R}^1)}$. On the other hand, for every measurable function g with values in $[-1,1]$ vanishing outside some interval, the function

$$f(x) = \int_{-\infty}^{x} g(t)\,dt$$

is 1-Lipschitz and $f'(t) = g(t)$ almost everywhere. Since $\|\Phi_\mu - \Phi_\nu\|_{L^1(\mathbb{R}^1)}$ coincides with the exact upper bound of the integrals of $g(\Phi_\mu - \Phi_\nu)$ with respect to such functions g, we see that $\|\Phi_\mu - \Phi_\nu\|_{L^1(\mathbb{R}^1)}$ is not greater than $d_{\rm K}(\mu,\nu)$. A more general fact is presented in Exercise 3.5.25.

There is a number of less obvious relations between the introduced metrics.

3.2.2. THEOREM. (i) *For every two Borel probability measures μ and ν on a metric space X, there hold the following relations between the Prohorov, Fortet–Mourier and Kantorovich–Rubinshtein metrics*:

$$\frac{2d_{\rm P}(\mu,\nu)^2}{2+d_{\rm P}(\mu,\nu)} \leqslant \|\mu-\nu\|_{\rm FM} \leqslant \|\mu-\nu\|_{\rm KR} \leqslant 3d_{\rm P}(\mu,\nu). \tag{3.2.1}$$

In addition, $\|\mu-\nu\|_{\rm FM} \leqslant 2d_{\rm P}(\mu,\nu)$.

(ii) *On the set $\mathcal{P}_\tau(X)$ all indicated metrics generate the weak topology.*

(iii) *If X is complete, then the space $\mathcal{P}_r(X) = \mathcal{P}_\tau(X)$ is complete as well with any of the indicated metrics. If X is separable, then $\mathcal{P}(X)$ is also separable with these metrics.*

(iv) *If X is compact, then $\mathcal{P}(X)$ is compact with any of the indicated metrics.*

PROOF. (i) Let $d_{\rm P}(\mu,\nu) > r > 0$. It follows from the definition of the metric $d_{\rm P}$ that there exists a closed set B with

$$\mu(B) > \nu(B^r) + r \quad (\text{or with } \nu(B) > \mu(B^r) + r).$$

There is a function f on X such that $|f(x) - f(y)| \leqslant 2\varrho(x,y)/r$, $|f(x)| \leqslant 1$, $f = 1$ on B and $f = -1$ outside B^r; for example, one can take $f(x) = \theta\bigl(\operatorname{dist}(x,B)\bigr) - 1$, $\theta(t) = 2(1-t/r)I_{[0,r]}(t)$. Then

$$\begin{aligned}(1+2/r)\|\mu-\nu\|_{\rm FM} &\geqslant \int_X f\,d(\mu-\nu)\\ &= \int_X (f+1)\,d(\mu-\nu) \geqslant 2\mu(B) - 2\nu(B^r) \geqslant 2r.\end{aligned}$$

Therefore, $\|\mu-\nu\|_{\rm FM} \geqslant 2r^2/(2+r)$, whence the first inequality in (3.2.1) follows, since $r < d_{\rm P}(\mu,\nu)$ was arbitrary.

Let now $f \in \mathrm{Lip}_1(X)$, $|f| \leqslant 1$ and $\displaystyle\int_X f\,d(\mu-\nu) > 3r > 0$. Let us consider the distribution functions $\Phi_\mu(t) = \mu(f < t)$, $\Phi_\nu(t) = \nu(f < t)$. Integrating by parts, taking into account the equalities $\Phi_\mu(1+) = \Phi_\nu(1+) = 1$ and applying formula (1.2.1), we obtain

$$\int_{-1}^1 \bigl[\Phi_\nu(t) - \Phi_\mu(t)\bigr]\,dt = \int_{-1}^1 t\,d(\Phi_\mu - \Phi_\nu)(t) = \int_X f\,d(\mu-\nu) > 3r. \tag{3.2.2}$$

Let us show that there exists $\tau \in [-1,1]$ such that

$$\Phi_\nu(\tau) > \Phi_\mu(\tau + r) + r. \tag{3.2.3}$$

Indeed, otherwise $\Phi_\nu(t) \leqslant \Phi_\mu(t+r) + r$ for all t. Integrating over $[-1,1]$, we have

$$\int_{-1}^1 \Phi_\nu(t)\,dt \leqslant \int_{-1+r}^{1+r} \Phi_\mu(t)\,dt + 2r.$$

Since $\Phi_\mu(t) = 1$ whenever $t > 1$, from the previous inequality we obtain

$$\int_{-1}^{1} \Phi_\nu(t)\,dt \leqslant \int_{-1}^{1} \Phi_\mu(t)\,dt + 3r$$

contrary to (3.2.2). Let $B := f^{-1}\bigl([-1,\tau)\bigr)$. Then

$$B^r \subset f^{-1}\bigl([-1,\tau+r)\bigr),$$

since $f \in \mathrm{Lip}_1(X)$. Hence (3.2.3) gives $\nu(B) > \mu(B^r)+r$, i.e., we have $d_{\mathrm{P}}(\mu,\nu) \geqslant r$. The established bound yields the last estimate in (3.2.1) by choosing $3r$ sufficiently close to $\|\mu-\nu\|_{\mathrm{KR}}$. The estimate $\|\mu-\nu\|_{\mathrm{FM}} \leqslant 2d_{\mathrm{P}}(\mu,\nu)$ is proved similarly taking into account that in case $\|f\|_{\mathrm{BL}} = 1$ the function f is Lipschitz with the Lipschitz constant $1-\sup_x |f(x)|$.

(ii) It follows from (i) that the Kantorovich–Rubinshtein and Fortet–Mourier metrics define the same topology as the Prohorov metric. Hence on $\mathcal{P}_\tau(X)$ they generate the weak topology.

(iii) The assertion about completeness for the Kantorovich–Rubinshtein metric follows from Corollary 2.3.5, since any Cauchy sequence of Radon probability measures in this metric is weakly convergent, so it converges to its limit in the given metric too. The same is true for the Fortet–Mourier metric and by (i) also for the Prohorov metric. The assertion about the separability follows from the separability of $\mathcal{P}(X)$ in the weak topology and the mutual bounds for these metrics.

(iv) For compact X, the compactness of $\mathcal{P}(X)$ with the indicated metrics is clear from the Banach–Alaoglu theorem as well as from the Prohorov theorem on weak compactness. □

If X contains an infinite Cauchy sequence, then $\mathcal{M}(X)$ with the norms $\|\cdot\|_{\mathrm{KR}}$ and $\|\cdot\|_{\mathrm{FM}}$ is not complete: otherwise the bound $\|\mu\|_{\mathrm{KR}} \leqslant \|\mu\|$ and the classical Banach theorem would imply the equivalence of the norms $\|\cdot\|_{\mathrm{KR}}$ and $\|\cdot\|$ (see Dunford, Schwartz [**197**, p. 58]), which does not take place (Exercise 3.5.17).

Let us now consider the following example on the real line demonstrating that for signed measures even with unit total variation norms convergence in the Kantorovich norm does not imply weak convergence. Note, however, that the situation is different for the unit ball of the space of measures on $\mathbb{N}$ (see Exercise 3.5.16) and obviously in the case of a compact metric space.

3.2.3. Example. We construct a sequence of signed measures μ_n on the real line of the form $\mu_n = \varrho_n\,dx$, where ϱ_n is a bounded function with support in the interval $[n,n+1]$, such that $\|\mu_n\| = 1$, but for the Kantorovich norm we have the bound $\|\mu_n\|_{\mathrm{K}} \leqslant 2^{-n}$. The sequence $\{\mu_n\}$ converges to zero in the Kantorovich norm, but is not uniformly tight, hence does not converge weakly.

To this end we partition $[n,n+1]$ into 2^n equal intervals I_k of length 2^{-n} and set $\varrho_n = (-1)^k$ on I_k, $\varrho_n = 0$ outside $[n,n+1]$. Let

$$F_n(x) := \int_0^x \varrho_n(t)\,dt.$$

Then the following relations are obvious:

$$\int_n^{n+1} |\varrho_n(t)|\,dt = 1, \quad \int_{-\infty}^{+\infty} |F_n(x)|\,dx \leqslant 2^{-n}.$$

For every function f that is Lipschitz with constant 1 on the real line the equality $F_n(n) = F_n(n+1) = 0$ yields

$$\left|\int_{-\infty}^{+\infty} f(t)\varrho_n(t)\,dt\right| = \left|\int_n^{n+1} f'(t)F_n(t)\,dt\right| \leqslant 2^{-n},$$

as required.

The next example shows that any measure can be approximated in the Kantorovich norm by its normalized restrictions to sets of an arbitrarily small fixed measure.

3.2.4. EXAMPLE. If X is a separable metric space, $\mu \in \mathcal{P}(X)$ has no atoms and $\alpha \in (0,1)$, then there exist sets $B_n \in \mathcal{B}(X)$ with $\mu(B_n) = \alpha$ such that for the measures $\mu_n := \alpha^{-1} I_{B_n} \cdot \mu$ we have $\|\mu_n - \mu\|_{\mathrm{KR}} \to 0$, and if $\mu \in \mathcal{P}^1(X)$, then also $\|\mu_n - \mu\|_{\mathrm{K}} \to 0$ as $n \to \infty$.

Indeed, for every n we can split X into disjoint Borel parts $E_{n,i}$ of diameter less than $1/n$ with $\mu(E_{n,i}) > 0$. The absence of atoms provides Borel sets $B_{n,i} \subset E_{n,i}$ with $\mu(B_{n,i}) = \alpha\mu(E_{n,i})$. Set $B_n := \bigcup_{i=1}^{\infty} B_{n,i}$. Let $f \in \mathrm{Lip}_1(X)$, $\sup_x |f(x)| \leqslant 1$. Pick arbitrary points $x_{n,i} \in B_{n,i}$. Then $\sup_{x \in E_{n,i}} |f(x) - f(x_{n,i})| \leqslant n^{-1}$, hence for every measure ν we have

$$\left|f(x_{n,i})\nu(E_{n,i}) - \int_{E_{n,i}} f\,d\nu\right| \leqslant n^{-1}\nu(E_{n,i}).$$

In addition, $\mu_n(E_{n,i}) = \mu(E_{n,i})$. Hence the sum $\sum_{i\geqslant 1} f(x_{n,i})\mu(E_{n,i})$ differs in absolute value from the integrals of the function f with respect to the measures μ and μ_n by at most $1/n$. Therefore, the absolute value of the difference between these integrals does not exceed $2/n$, whence $\|\mu_n - \mu\|_{\mathrm{KR}} \leqslant 2/n$. Similarly, for $\mu \in \mathcal{P}^1(X)$ we obtain the estimate $\|\mu_n - \mu\|_{\mathrm{K}} \leqslant 2/n$.

3.2.5. REMARK. In Theorem 3.1.9 we proved the metrizability of any weakly compact set M of Radon measures on a complete space X. For this purpose we used some explicitly constructed norm q. Having defined the Fortet–Mourier norm in this section, it is natural to ask whether it can be used for the same purpose. The answer is positive: the set M is also compact with respect to the norms $\|\cdot\|_{\mathrm{KR}}$ and $\|\cdot\|_{\mathrm{FM}}$, which generated on it the weak topology. Indeed, as in the proof of the cited theorem, from every sequence $\{\mu_n\}$ in M we can choose a subsequence that converges weakly to some measure $\mu \in M$. One can assume that this is the whole original sequence. The sequence of measures μ_n^+ from the Hahn–Jordan decomposition also contains a subsequence that converges weakly to some measure ν_1. Next we take a further subsequence $\{n_i\}$ giving convergence of the negative parts $\mu_{n_i}^-$ to some measure ν_2. Then $\mu = \nu_1 - \nu_2$ and

$$\|\mu_{n_i} - \mu\|_{\mathrm{FM}} \leqslant \|\mu_{n_i}^+ - \nu_1\|_{\mathrm{FM}} + \|\mu_{n_i}^- - \nu_2\|_{\mathrm{FM}} \to 0,$$

which proves our assertion.

Note also that if K is a compact metric space, then the closed balls in $\mathcal{M}(K)$ with the weak topology are compact and metrizable with the Kantorovich or Fortet–Mourier metric.

The Kantorovich metric has two non-trivial representations discovered by Kantorovich himself. First we give a simpler one, called the duality formula. Let $\mathcal{P}^1(X)$ denote the set of Borel probability measures on X with finite first moments.

3.2.6. PROPOSITION. *For all $\mu,\nu\in\mathcal{P}^1(X)$ we have*

$$\|\mu-\nu\|_{\mathrm{K}} = \widehat{W}(\mu,\nu) := \sup\Bigl\{\int_X f\,d\mu + \int_X g\,d\nu\colon \\ f,g\in C(X), f(x)+g(y)\leqslant \varrho(x,y)\Bigr\}. \tag{3.2.4}$$

PROOF. On the one hand, $\|\mu-\nu\|_{\mathrm{K}}\leqslant \widehat{W}(\mu,\nu)$, since $f(x)-f(y)\leqslant\varrho(x,y)$ for $f\in\mathrm{Lip}_1(X)$ and one can take $g(y)=-f(y)$. On the other hand, given continuous functions f and g such that $f(x)+g(y)\leqslant\varrho(x,y)$, we set

$$h(x)=\inf_y\,[\varrho(x,y)-g(y)].$$

Then $f\leqslant h\leqslant -g$, hence for all x,x' we have

$$h(x)-h(x')\leqslant \sup_y\,[\varrho(x,y)-\varrho(x',y)]\leqslant\varrho(x,x'),$$

i.e., $h\in\mathrm{Lip}_1(X)$. In addition,

$$\int_X f\,d\mu+\int_X g\,d\nu\leqslant\int_X h\,d(\mu-\nu),$$

which gives (3.2.4). □

The next important result due to Kantorovich gives another expression for the norm $\|\mu-\nu\|_K$. We recall that $\Pi(\mu,\nu)$ is the set of all Radon probability measures on $X\times X$ whose projections on the first and second factors are μ and ν.

3.2.7. THEOREM. *Let μ and ν be Radon probability measures of class $\mathcal{P}^1(X)$. Then the Kantorovich distance $\|\mu-\nu\|_{\mathrm{K}}$ can be represented in the form*

$$\|\mu-\nu\|_{\mathrm{K}} = W(\mu,\nu) := \inf_{\lambda\in\Pi(\mu,\nu)}\int_{X\times X}\varrho(x,y)\,\lambda(dx,dy). \tag{3.2.5}$$

In addition, there exists a measure $\lambda_0\in\Pi(\mu,\nu)$ at which the value $W(\mu,\nu)$ is attained; it is called an optimal plan.

PROOF. We observe that $\|\mu-\nu\|_{\mathrm{K}}\leqslant W(\mu,\nu)$, since for all $\lambda\in\Pi(\mu,\nu)$ and every function $f\in\mathrm{Lip}_1(X)$ we have

$$\int_X f\,d(\mu-\nu)=\int_{X\times X}[f(x)-f(y)]\,\lambda(dx,dy)\leqslant\int_{X\times X}\varrho(x,y)\,\lambda(dx,dy).$$

Let us consider the case of a finite space X. We show that $\widehat{W}(\mu,\nu)=W(\mu,\nu)$. We have $\widehat{W}(\mu,\nu)\leqslant W(\mu,\nu)$. On the linear space L of all functions of the form $\varphi(x,y)=f(x)+g(y)$ on $X\times X$ we consider the functional

$$l(\varphi)=\int_X f\,d\mu+\int_X g\,d\nu.$$

It is readily seen that the functional l is well-defined. The set

$$U=\bigl\{\varphi\in C(X\times X)\colon\ \varphi(x,y)<\varrho(x,y)\bigr\}$$

is convex and open in the space $C(X\times X)$ and l is bounded on $U\cap L$. By the Hahn–Banach theorem l extends to a linear functional l_0 on $C(X\times X)$ satisfying the equality $\sup_U l_0=\sup_{U\cap L} l$. In addition, $l_0(u)\geqslant 0$ whenever $u\geqslant 0$, since $\varrho-1-cu\in U$ for all $c>0$ and $\sup_{c>0} l(\varrho-1-cu)<\infty$. Hence there exists a nonnegative measure λ on $X\times X$ defining l_0. Since $l_0=l$ on L, the measures λ and

μ assign equal integrals to $f(x)$ and the measures λ and ν assign equal integrals to $g(y)$, i.e., $\lambda \in \Pi(\mu, \nu)$. It is readily seen that

$$\widehat{W}(\mu, \nu) = \int_{X\times X} \varrho(x, y)\, \lambda(dx\, dy),$$

which completes the proof for finite X.

In the general case we take two sequences of probability measures μ_n and ν_n with finite supports X_n that converge weakly to μ and ν, respectively, such that both sequences are uniformly tight. One can assume that all sets X_n contain a fixed point a. Let $\lambda_n \in \Pi(\mu_n, \nu_n)$ be a probability measure on $X_n \times X_n$ for which

$$\int_{X\times X} \varrho(x, y)\, \lambda_n(dx\, dy) = \|\mu_n - \nu_n\|_{\mathrm{K}}.$$

The sequence of measures λ_n with uniformly tight projections is also uniformly tight on X. Indeed, if K_1 and K_2 are compact sets such that the projections of the measure λ_n assign to them the values larger than $1 - \varepsilon$, then on the compact set $K_1 \times K_2$ the value of the measure λ_n is greater than $1 - 2\varepsilon$, since its complement belongs to the union of the complements of the sets $K_1 \times X$ and $X \times K_2$. Passing to a subsequence, one can assume that the measures λ_n converge weakly to a measure λ on $X \times X$. It is clear that $\lambda \in \Pi(\mu, \nu)$, since the projections of weakly convergent measures converge weakly to the projections of the limit measure. By the Radon property of the measures μ and ν we can assume that the space X is separable. For every n there is a function f_n on X_n such that it is Lipschitz with constant 1, $f_n(a) = 0$ (where $a \in X$ is the common point fixed above) and

$$\begin{aligned} \int_X f_n\, d(\mu_n - \nu_n) &= \|\mu_n - \nu_n\|_{\mathrm{K}} = W(\mu_n, \nu_n) \\ &= \int_{X\times X} \varrho(x, y)\, \lambda_n(dx\, dy). \end{aligned} \tag{3.2.6}$$

The functions f_n can be extended to the whole space X with the same Lipschitz constant (see Exercise 2.7.25). Denoting the extension again by f_n we pick in $\{f_n\}$ a subsequence converging on a countable everywhere dense set. By the uniform Lipschitzness, this subsequence, denoted again by $\{f_n\}$, converges at every point. It is clear that the limit f of this subsequence is Lipschitz with constant 1 and that $f(a) = 0$. By Theorem 2.2.8 we obtain

$$\lim_{n\to\infty} \int_X f_n\, d(\mu_n - \nu_n) = \int_X f\, d(\mu - \nu). \tag{3.2.7}$$

By the continuity of the function ϱ we have

$$\int_{X\times X} \varrho(x, y)\, \lambda(dx\, dy) \leqslant \liminf_{n\to\infty} \int_{X\times X} \varrho(x, y)\, \lambda_n(dx\, dy),$$

since this is true for all bounded functions $\min(\varrho, k)$. Therefore, on account of (3.2.6) and (3.2.7) we obtain

$$W(\mu, \nu) \leqslant \int_{X\times X} \varrho(x, y)\, \lambda(dx\, dy) \leqslant \int_X f\, d(\mu - \nu) \leqslant \|\mu - \nu\|_{\mathrm{K}}.$$

However, $\|\mu - \nu\|_{\mathrm{K}} \leqslant W(\mu, \nu)$, therefore, there are equalities here. □

If X is bounded, then $\mathcal{P}^1(X) = \mathcal{P}(X)$. Note that for every metric space (X, ϱ) the metric $d := \varrho/(\varrho + 1)$ or $d := \min(\varrho, 1)$ generates the original topology, but is bounded, so the function W from (3.2.5) metrizes the weak topology on $\mathcal{P}_r(X)$. By using a homeomorphic embedding of a separable space into $[0, 1]$, one can even find a totally bounded metric on X generating the same topology.

In many cases the measure λ_0 at which the infimum is attained can be obtained in the form $\mu\circ\Psi^{-1}$ with some measurable mapping $\Psi\colon X \to X\times X$, moreover, under appropriate conditions (but not always, of course) the mapping Ψ can be even written in the form $\Psi(x) = \bigl(x, F(x)\bigr)$ with some mapping $F\colon X \to X$. This connects this problem with the study of transformations of measures, because we obviously have $\nu = \mu\circ F^{-1}$. Finding F is the classical Monge problem of mass transportation (see also Comments, p. 252), while finding λ_0 is the Kantorovich problem. The Monge–Kantorovich problem is also posed for more general functions h in place of the distance function ϱ. Under broad assumptions F turns out to be a sufficiently regular mapping (for example, the differential or the subdifferential of a convex function). Unfortunately, it is impossible to discuss here in detail this interesting direction at the junction of measure theory, calculus of variations, and nonlinear differential equations. The interested reader can consult the survey works Ambrosio, Gigli [**12**], Ambrosio, Gigli, Savaré [**13**], Bogachev, Kolesnikov [**88**], Rachev, Rüschendorf [**532**], and Villani [**648**], [**649**].

Let us mention an interesting result which will be proved below in Chapter 5 in Theorem 5.1.7.

3.2.8. THEOREM. *If E is a complete separable metric space, then the weak topology on the unit sphere in $\mathcal{M}(E)$ consisting of all signed measures μ with the total variation norm $\|\mu\| = 1$ can be defined by a metric with respect to which this sphere will be a complete separable metric space.*

The proof of this theorem is rather involved, which is not surprising taking into account that the ball in $\mathcal{M}(E)$ is not metrizable in the weak topology for infinite E and also that the Kantorovich–Rubinshtein and Fortet–Mourier norms generating the weak topology on the cone of nonnegative measures are not suitable for a complete metrization of the sphere with the weak topology!

3.2.9. EXAMPLE. The unit sphere S in the space of signed measures on the interval $[0, 1]$ is not complete with respect to the Kantorovich–Rubinshtein norm and also with respect to the Kantorovich norm that coincides with the latter in this case (certainly, also with respect to the equivalent Fortet–Mourier norm). Indeed, one can find a sequence of measures μ_n in S converging to zero in the Kantorovich norm. The measures μ_n can be defined by densities φ_n with respect to Lebesgue measure on $[0, 1]$ constructed as follows: the function φ_n assumes the value -1 on the left half of every interval obtained by diving $[0, 1]$ into n equal parts, and on the right half of such an interval it assumes the value 1. The function φ_n coincides (except for the points of division) with the derivative of the Lipschitz function Φ_n that is affine on each of the $2n$ halves of the indicated intervals and equals zero at the points $0, 1/n, 2/n, \dots, 1$. Hence for every function $f \in \mathrm{Lip}_1$ we have

$$\int_0^1 f(t)\varphi_n(t)\,dt = -\int_0^1 f'(t)\Phi_n(t)\,dt \leqslant \max_t |\Phi_n(t)| \leqslant (2n)^{-1},$$

as required. A similar example can be constructed in every metric space where there is an infinite Cauchy sequence (Exercise 3.5.17).

On the unit sphere of the space l^1 the weak topology coincides with the norm topology (Exercise 3.5.16), so for $E = \mathbb{N}$ the weak topology on the sphere in $\mathcal{M}(E)$ is generated by the Kantorovich–Rubinshtein norm.

One should bear in mind that although the weak topology on $\mathcal{P}_\tau(X)$ or on $\mathcal{P}_r(X)$ is generated by some metrics d (say, by the Prohorov metric and the Kantorovich–Rubinshtein metric), the collections of Cauchy sequences in this topology and such metrics need not coincide. Certainly, a similar phenomenon is observed even on the real line, which can be metrized by the metric $|\arctan x - \arctan y|$ that have different Cauchy sequences as compared with the usual metric. However, the collection of weakly fundamental sequences of measures does not depend on the choice of the metric, since it is determined entirely by the space of continuous functions that depends only on the given topology. As already noted above and will be proven in Chapter 5, any weakly fundamental sequence of measures must have a weak limit independently of the completeness of the space. Hence weakly fundamental sequences always converge in metrics generating the weak topology. Therefore, at the level of measures non equivalence of being weakly fundamental and fundamental in metric happens in the following situations.

If a complete metric space is noncompact, then it can be equipped with a metric generating the original topology, but such that with respect to it the space will no longer be complete (Exercise 3.5.15). Hence there exists a sequence of Dirac measures that is fundamental in the Kantorovich metric generated by this new metric, but is not weakly fundamental, since otherwise it would converge weakly to some measure (due to the sequential weak completeness of the space of measures) that would be also Dirac's measure at a point, and then the given sequence would converge to this point (see Example 2.2.3).

If our metric space X does not admit a complete metric generating the same topology (as an example one can take $X = \mathbb{Q}$), then there also exists a sequence of Dirac measures that is fundamental with respect to d, but has no limit. Certainly, this sequence is not fundamental in the weak topology.

For a compact metric space X, the space $\mathcal{P}(X)$ is compact in the weak topology, hence all different metrics on $\mathcal{P}(X)$ generating the weak topology possess the same supply of fundamental sequences.

3.3. The Kantorovich metric of order p

Similarly to the Kantorovich metric $W(\mu,\nu)$ considered above one can introduce the *Kantorovich L^p-metric* (or *Kantorovich p-metric*). The quantity

$$W_p(\mu,\nu) = \inf_{\sigma\in\Pi(\mu,\nu)} \left(\int_{X\times X} \varrho(x,y)^p\,\sigma(dx\,dy)\right)^{1/p}$$

is a metric for each $p \geqslant 1$ on the space $\mathcal{P}_r^p(X)$ of Radon probability measures μ such that the function $x \mapsto \varrho(x,x_0)$ belongs to $L^p(\mu)$ for some (and then for all) $x_0 \in X$; in case $0 < p < 1$ the function W_p^p is a metric. The integrability of $\varrho(x,y)^p$ with respect to $\sigma \in \Pi(\mu,\nu)$ follows from the inequality

$$\varrho(x,y)^p \leqslant 2^p\varrho(x,x_0)^p + 2^p\varrho(y,x_0)^p$$

and the assumed integrability of the terms in the right-hand side with respect to μ and ν, respectively. The metric W_p is also called the *Kantorovich p-distance* and a minimizing measure is called an *optimal plan* for ϱ^p.

Let $\mathcal{M}_r^p(X)$ denote the linear space of all Radon measures μ such that the function $\varrho(\,\cdot\,,x_0)^p$ is integrable with respect to the measure $|\mu|$.

By Hölder's inequality $\|\cdot\|_q \leqslant \|\cdot\|_p$ we obtain

$$W_q(\mu,\nu) \leqslant W_p(\mu,\nu) \quad \text{if } 1 \leqslant q \leqslant p.$$

Clearly, $W_p(\mu,\nu) = W_p(\nu,\mu)$; it is easy to see that $W_p(\mu,\nu) = 0$ precisely when $\mu = \nu$; the equality $W_p(\mu,\mu) = 0$ holds, since for σ one can take the image of the measure μ under the mapping $x \mapsto (x,x)$ concentrated on the diagonal. Unlike most of practically encountered metrics, the verification of the triangle inequality for W_p is not completely obvious. If we are given three measures $\mu_1,\mu_2,\mu_3 \in \mathcal{P}_r^p(X)$, then how can we estimate $W_p(\mu_1,\mu_3)$ by $W_p(\mu_1,\mu_2) + W_p(\mu_2,\mu_3)$? If $\sigma_{1,2}$ and $\sigma_{2,3}$ are optimal plans for the pairs (μ_1,μ_2) and (μ_2,μ_3) and ϱ^p, then it suffices to have a probability measure η on $X\times X\times X$ whose projection on the product of the first two factors is $\sigma_{1,2}$ and the projection on the product of the last two factors is the measure $\sigma_{2,3}$. Then $W_p(\mu_1,\mu_3)$ is estimated as follows:

$$\begin{aligned}\biggl|\int \varrho(x_1,x_3)^p\,\sigma_{1,3}(dx_1dx_3)\biggr|^{1/p} &= \biggl|\int \varrho(x_1,x_3)^p\,\eta(dx_1dx_2dx_3)\biggr|^{1/p} \\ &\leqslant \biggl|\int [\varrho(x_1,x_2)+\varrho(x_2,x_3)]^p\;\eta(dx_1dx_2dx_3)\biggr|^{1/p} \\ &\leqslant \biggl|\int \varrho(x_1,x_2)^p\,\eta(dx_1dx_2dx_3)\biggr|^{1/p} + \biggl|\int \varrho(x_2,x_3)^p\,\eta(dx_1dx_2dx_3)\biggr|^{1/p},\end{aligned}$$

which coincides with $\|\varrho\|_{L^p(\sigma_{1,2})} + \|\varrho\|_{L^p(\sigma_{2,3})} = W_p(\mu_1,\mu_2) + W_p(\mu_2,\mu_3)$. However, the existence of a measure η requires a justification. Let us give it.

3.3.1. LEMMA. *Let X_1, X_2, X_3 be three metric spaces. Suppose that $X_1\times X_2$ and $X_2\times X_3$ are equipped with Radon probability measures $\sigma_{1,2}$ and $\sigma_{2,3}$ with equal projections on X_2. Then on $X_1\times X_2\times X_3$ there is a Radon probability measure η with projections $\sigma_{1,2}$ and $\sigma_{2,3}$ on $X_1\times X_2$ and $X_2\times X_3$, respectively .*

PROOF. Suppose first that these spaces are compact. Then on the linear subspace in $C(X_1\times X_2\times X_3)$ of all functions of the form

$$\varphi(x_1,x_2,x_3) = f(x_1,x_2) + g(x_2,x_3), \; f \in C(X_1\times X_2), \; g \in C(X_2\times X_3),$$

one can define a linear functional

$$l(\varphi) = \int_{X_1\times X_2} f\, d\sigma_{1,2} + \int_{X_1\times X_3} g\, d\sigma_{1,3}.$$

The assumption of the equality of projections on X_2 guarantees that it is well-defined: if we have $f(x_1,x_2)+g(x_2,x_3) = f_0(x_1,x_2)+g_0(x_2,x_3)$, then the difference of $f(x_1,x_2) - f_0(x_1,x_2)$ does not depend on x_1 and hence has the form $\psi(x_2)$, and the measures $\sigma_{1,2}$ and $\sigma_{2,3}$ assign the same integral to $\psi(x_2)$. The norm of l equals 1, since $l(1) = 1$ and $l(f+g) \leqslant 1$ if $f(x_1,x_2) + g(x_2,x_3) \leqslant 1$, which is seen from the representation

$$f(x_1,x_2) + g(x_2,x_3) = f(x_1,x_2) + \max_{x_3} g(x_2,x_3) + g(x_2,x_3) - \max_{x_3} g(x_2,x_3),$$

where $f(x_1,x_2) + \max_{x_3} g(x_2,x_3) \leqslant 1$, $g(x_2,x_3) - \max_{x_3} g(x_2,x_3) \leqslant 0$.

By the Hahn–Banach theorem l extends to a linear functional on the space $C(X_1{\times}X_2{\times}X_3)$ with unit norm, which by the Riesz theorem is given as the integral with respect to some Radon measure ν with $\|\nu\| = 1$. Since $l(1) = 1$, the measure ν

is a probability measure (if $\nu^+(X)+\nu^-(X)=\nu^+(X)-\nu^-(X)=1$, then $\nu^+(X)=1$ and $\nu^-(X)=0$), and its has the required projections (it suffices to substitute $g=0$ and then $f=0$).

In the general case three proofs can be given. First, one can derive the general assertion from the case of finite measures. Second, one can reduce it to the case of compact sets, passing to the Stone–Čech compactifications (see p. 140) and taking into account that by the Radon property of the considered measures we can assume that all the three spaces are countable unions of compact sets, due to which they are Borel subsets of their Stone–Čech compactifications.

Finally, for Souslin spaces (to which the case of Radon measures reduces) the desired measure can explicitly expressed via conditional measures (see § 2.1):

$$\eta(dx_1dx_2dx_3)=\sigma_{1,2}(dx_1|x_2)\sigma_{2,3}(dx_3|x_2)\pi(dx_2),$$

where π is the common projection of $\sigma_{1,2}$ and $\sigma_{2,3}$ on X_2, $\sigma_{1,2}(\,\cdot\,|x_2)$ and $\sigma_{2,3}(\,\cdot\,|x_2)$ are the corresponding conditional measures, i.e.,

$$\sigma_{1,2}(dx_1dx_2)=\sigma_{1,2}(dx_1|x_2)\pi(dx_2),\quad \sigma_{2,3}(dx_2dx_3)=\sigma_{2,3}(dx_3|x_2)\pi(dx_2).$$

This means that we have the following equality:

$$\int f\,d\eta=\int_{X_2}\int_{X_3}\int_{X_1} f(x_1,x_2,x_3)\,\sigma_{1,2}(dx_1|x_2)\,\sigma_{2,3}(dx_3|x_2)\,\pi(dx_2).$$

For example, if f depends only on x_2,x_3, then the first integral gives $f(x_2,x_3)$, which yields the integral of f with respect to the measure $\sigma_{2,3}$. If f does not depend on x_3, then we obtain the integral of f with respect to the measure $\sigma_{1,2}$, which means that the required conditions on projections hold. We observe that the last way of reasoning applies, certainly, without any preliminary consideration of compact or finite spaces. □

3.3.2. LEMMA. *For every two measures $\mu,\nu\in\mathcal{P}_r^p(X)$, there exists a measure $\sigma\in\Pi(\mu,\nu)$ at which the quantity $W_p(\mu,\nu)^p$ is attained.*

PROOF. Let us take measures $\sigma_n\in\Pi(\mu,\nu)$ with respect to which the integrals of ϱ^p tend to $W_p(\mu,\nu)^p$. By the Radon property of the measures μ and ν, the sequence of measures σ_n is uniformly tight, hence one can assume that it converges weakly to some Radon probability measure σ. It is clear that $\sigma\in\Pi(\mu,\nu)$. For every fixed $k\in\mathbb{N}$, the integral of the function $\min(k,\varrho^p)$ with respect to the measure σ equals the limit of its integrals with respect to the measures σ_n, hence it is not greater that the limit of the integrals of ϱ^p, which equals $W_p(\mu,\nu)^p$. Hence the integral of ϱ^p with respect to the measure σ does not exceed $W_p(\mu,\nu)^p$, but then it is equal to this number. □

The metric W_p is estimated via the variation as follows.

3.3.3. PROPOSITION. *For all $p\geqslant 1$, $x_0\in X$ one has*

$$W_p(\mu,\nu)^p\leqslant 2^{p-1}\|\varrho(\,\cdot\,,x_0)^p(\mu-\nu)\|,\quad \mu,\nu\in\mathcal{P}_r^p(X). \tag{3.3.1}$$

PROOF. Let us consider the measure $\mu\wedge\nu=\mu-(\mu-\nu)^+$. If μ and ν are given by densities f and g with respect to a common probability measure λ (say, their half-sum), then $\mu\wedge\nu=\min(f,g)\cdot\lambda$. Let us set $\alpha:=(\mu-\nu)^+(X)=(\mu-\nu)^-(X)$ and introduce on $X\times X$ the measure

$$\pi=(\mu\wedge\nu)\circ J^{-1}+\alpha^{-1}(\mu-\nu)^+\otimes(\mu-\nu)^-,$$

where $J: X \to X \times X$, $x \mapsto (x, x)$. The first term is concentrated on the diagonal and its mass equals $1-\alpha$, hence π is a probability measure. The projection of π on the first factor is μ, since it equals the sum of the measures $\mu \wedge \nu$ and $(\mu-\nu)^+$. The projection on the second factor is the sum of the measures $\mu \wedge \nu$ and $(\mu-\nu)^-$, i.e., $\mu-(\mu-\nu)^+ + (\mu-\nu)^- = \mu-(\mu-\nu) = \nu$. Hence $\pi \in \Pi(\mu, \nu)$. Since the part of the measure concentrated on the diagonal gives zero contribution to the integral of ϱ, we have

$$W_p(\mu,\nu)^p \leqslant \int_{X\times X} \varrho(x,y)^p\, \pi(dx\, dy) = \frac{1}{\alpha}\int_X\int_X \varrho(x,y)^p\, (\mu-\nu)^+(dx)(\mu-\nu)^-(dy),$$

which by the triangle inequality for the metric ϱ and the elementary inequality $(a+b)^p \leqslant \max(1, 2^{p-1})(a^p+b^p)$ for $a, b \geqslant 0$ is estimated by

$$\frac{\max(1,2^{p-1})}{\alpha}\int_X\int_X \bigl[\varrho(x,x_0)^p + \varrho(x_0,y)^p\bigr]\, (\mu-\nu)^+(dx)(\mu-\nu)^-(dy),$$

which coincides with the quantity

$$\max(1,2^{p-1})\int_X \varrho(x,x_0)^p\, \bigl((\mu-\nu)^+ + (\mu-\nu)^-\bigr)(dx) \\ = \max(1,2^{p-1})\int_X \varrho(x,x_0)^p\, |\mu-\nu|(dx),$$

since $(\mu-\nu)^+(X) = (\mu-\nu)^-(X) = \alpha$. □

Let us note the following simple fact. Suppose we are given a sequence of functions $\theta_j \in \mathrm{Lip}_1(X)$, $0 \leqslant \theta_j \leqslant 1$ such that $\theta_j(x) = 1$ for all x in the ball $U(x_0, j)$ of radius j with some common center x_0 and $\theta_j(x) = 0$ if $x \notin U(x_0, j+1)$. It is clear that, for every measure μ, the measures $\theta_j \cdot \mu$ converge weakly (and even in variation) to μ. The same is true for the metric W_p. Let $\xi_j(\mu)$ denote the probability measure $C_{n,j}\theta_j \cdot \mu$, $C_{n,j} = (\theta_j \cdot \mu(X))^{-1}$ if $\theta_j \cdot \mu(X) > 0$, and if $\theta_j \cdot \mu(X) = 0$, then we set $\xi_j(\mu) = \mu$.

3.3.4. LEMMA. *For every $\mu \in \mathcal{P}_r^p(X)$ we have $\lim_{j\to\infty} W_p(\xi_j(\mu), \mu) = 0$. Moreover,*

$$\lim_{j\to\infty} \sup_n W_p\bigl(\xi_j(\mu_n), \mu_n\bigr) = 0$$

if measures $\mu_n, \mu \in \mathcal{P}_r^p(X)$ are such that the measures $\bigl(1+\varrho(\,\cdot\,, x_0)^p\bigr)\mu_n$ converge weakly to the measure $\bigl(1+\varrho(\,\cdot\,, x_0)^p\bigr)\mu$.

PROOF. The first assertion follows from (3.3.1). For the proof of the second one we have to verify that the relation $\|\varrho(\,\cdot\,, x_0)(\xi_j(\mu_n)-\mu_n)\| \to 0$ as $j \to \infty$ is fulfilled uniformly in n. Weak convergence of the indicated measures yields that for any given $\delta \in (0, 1/2)$ there exists $R > 0$ such that the integrals of $1+\varrho(x,x_0)^p$ over the complement of the ball $U(R, x_0)$ with respect to all measures μ_n and μ are less than δ. The integrals over the whole space are bounded by some number C. We can consider only those n for which

$$\mu_n\bigl(U(R, x_0)\bigr) > 1-\delta.$$

Whenever $j > R$ we have $\theta_j(x) = 1$ for all $x \in U(R, x_0)$, hence the restriction of the measure $\xi_j(\mu_n)$ to the ball $U(R, x_0)$ equals the restriction of the measure $C_{n,j}\mu_n$, where $1 \leqslant C_{n,j} \leqslant 1+2\delta$. In addition, $\xi_j(\mu_n) \leqslant 2\mu_n$. Hence the norm $\|\varrho(\,\cdot\,, x_0)^p(\xi_j(\mu_n)-\mu_n)\|$ is estimated by $2C\delta + 3\delta$. □

3.3.5. COROLLARY. *The set of measures with bounded support, hence also the set of measures with finite support, is everywhere dense in the space* $(\mathcal{P}_r^p(X), W_p)$.

There is the following connection between the metric W_p with $p>1$ and weak convergence.

3.3.6. THEOREM. *A sequence of measures* $\mu_n \in \mathcal{P}_r^p(X)$ *converges to a measure* $\mu \in \mathcal{P}_r^p(X)$ *in the metric* W_p *precisely when* $\{\mu_n\}$ *converges to* μ *weakly and there holds the equality*

$$\lim_{n\to\infty} \int_X \varrho(x,x_0)^p\,\mu_n(dx) = \int_X \varrho(x,x_0)^p\,\mu(dx) \tag{3.3.2}$$

for some (and then for each) element $x_0 \in X$. *This is also equivalent to weak convergence of the bounded measures* $\bigl(1+\varrho(x,x_0)^p\bigr)\mu_n$ *to the measure* $\bigl(1+\varrho(x,x_0)^p\bigr)\mu$.

PROOF. Weak convergence of $\{\mu_n\}$ to μ and (3.3.2) follow from weak convergence of the measures $\nu_n = \bigl(1+\varrho(x,x_0)^p\bigr)\mu_n$ to the measure $\nu = \bigl(1+\varrho(x,x_0)^p\bigr)\mu$, since the function $f(x)\bigl(1+\varrho(x,x_0)^p\bigr)^{-1}$ is bounded and continuous if $f\in C_b(X)$. In turn, they imply weak convergence, since, under the condition $\nu_n(X)\to\nu(X)$, for weak convergence it suffices to have convergence of the integrals of functions $f\in C_b(X)$ with bounded support, but for such functions the multiplier $1+\varrho(x,x_0)^p$ is bounded.

Let $W_p(\mu_n,\mu)\to 0$. Then $W(\mu_n,\mu)\to 0$, hence we have weak convergence. Let us verify (3.3.2). It follows from weak convergence and consideration of cut-offs $\min\bigl(\varrho(x,x_0)^p, N\bigr)$ that the right-hand side is not greater than the lim inf of the integrals of $\varrho(x,x_0)^p$ against μ_n. On the other hand, the triangle inequality $\varrho(x_0,x)\leqslant \varrho(x,y)+\varrho(x_0,y)$ yields that for every $q>1$ there is $C>0$ such that $\varrho(x,x_0)^p \leqslant C\varrho(x,y)^p + q\varrho(x_0,y)^p$. Taking a measure $\sigma_n\in\Pi(\mu_n,\mu)$ the integral with respect to which of the function $\varrho(x,y)^p$ equals $W_p(\mu_n,\mu)^p$ (it was shown above that such measures exist) and integrating with respect to it the previous equality, we obtain the estimate

$$\int_X \varrho(x,x_0)^p\,\mu_n(dx) \leqslant CW_p(\mu_n,\mu)^p + q\int_X \varrho(x,x_0)^p\,\mu(dx).$$

Since $W_p(\mu_n,\mu)\to 0$, we see that the lim sup of the left-hand side of this estimate is not greater than the product of q and the right-hand side of (3.3.2). Letting $q\to 1$ we obtain an estimate completing the proof of (3.3.2).

Finally, we show that weak convergence of the measures ν_n to ν implies convergence $W_p(\mu,\mu_n)\to 0$. Lemma 3.3.4 reduces everything to the case of measures with support in some ball, hence to the case of a bounded metric. In the latter case our assertion is true by virtue of the results in § 3.2 (see Theorem 3.2.2). □

For $p>1$ the metric W_p on the simplex of probability measures cannot be the restriction of a norm on the space of measures due to the lack of convexity on intervals: even for $X=\{0,1\}$ we have $W_p(\delta_0,(1-t)\delta_0+t\delta_1)=t^{1/p}$. However, the following fact is true.

3.3.7. COROLLARY. *On the set* $\mathcal{P}_r^p(X)$ *the topology generated by the metric* W_p *is also generated by the norm*

$$K_p(\mu) := \|(1+\varrho(\,\cdot\,,x_0)^p)\mu\|_K,\quad \mu\in\mathcal{M}_r^p(X).$$

PROOF. We know that convergence of sequences in the space $\mathcal{P}_r^p(X)$ with the indicated norm is equivalent to weak convergence after multiplication by the function $1+\varrho(\,\cdot\,,x_0)^p$, which is also equivalent to convergence in the metric W_p. The norm K_p can fail to generate the metric W_p on $\mathcal{P}_r^p(X)$, it generates just the same topology. □

3.4. Gromov metric triples

Here we briefly discuss the notion of a Gromov metric triple (or Gromov–Vershik triple). The main object is a triple (X,d,μ) consisting of a metric space (X,d) with a Borel probability measure μ whose topological support is the whole space X, i.e., the measure μ does not vanish on open balls. Two such triples (X,d,μ) and (X',d',μ') are called isomorphic if there exists an isometry $j\colon X_0\to X_0'$ of subsets $X_0\subset X$, $X_0'\subset X'$ of measure 1 under which the measure μ is taken to the measure μ'. However, the condition with full support is not essential and can be omitted (especially when one deals with equivalence classes in place of triples).

For every metric triple (X,d,μ) we can define a mapping

$$X^\infty\to M_\infty,\quad x=(x_i)\mapsto \bigl(d(x_i,x_j)\bigr)_{i,j\geqslant 1}$$

on the countable power of X with values in the space of infinite symmetric matrices; this space can be regarded just as a subset of the space of two-index sequences equipped with coordinate-wise convergence, or, equivalently, with the metric

$$d_\infty(u,v)=\sum_{i,j=1}^\infty 2^{-i-j}\min\bigl(|u_{ij}-v_{ij}|,1\bigr).$$

The measure μ^∞ (the countable power of the measure μ) is taken by this mapping to the measure P_μ on M_∞ called the *matrix distribution* of the measure μ.

The following assertion about a classification of metric triples is true (see Gromov [**301**] and Vershik [**640**]).

3.4.1. THEOREM. *Two metric triples with complete separable metric spaces are isomorphic precisely when they have equal matrix distributions.*

PROOF. We shall assume that separable metric spaces are realized as subsets of the universal separable space $C=C[0,1]$. Let us consider the aforementioned mapping

$$j\colon C^\infty\to M_\infty,\quad x=(x_i)\mapsto \bigl(d(x_i,x_j)\bigr)_{i,j\geqslant 1}.$$

For every triple (X,d,μ), we consider the mapping

$$\mu\mapsto \mu^\infty\circ j^{-1}.$$

Suppose that $\mu^\infty\circ j^{-1}=\nu^\infty\circ j^{-1}$ for two triples (X,d_1,μ) and (Y,d_2,ν). Then there exist points $x=(x_i)$ and $y=(y_i)$ with equal images under the mapping j such that the sequence $\{x_i\}$ is everywhere dense in X and the sequence $\{y_i\}$ is everywhere dense in Y (Exercise 3.5.18). Thus, $d_1(x_i,x_j)=d_2(y_i,y_j)$ for all i,j. Then we obtain an isometry $x_i\mapsto y_i$ of countable everywhere dense sets that extends by continuity to an isometry of X and Y. Moreover, x and y can be taken in such a way that $n^{-1}\sum_{i=1}^n\delta_{x_i}\Rightarrow\mu$, $n^{-1}\sum_{i=1}^n\delta_{y_i}\Rightarrow\nu$ (Exercise 3.5.18). Comparing the integrals of functions of the form (2.2.5), we conclude that μ is taken to ν. □

For the study of isometric embeddings it is useful to know the concept of the Urysohn universal metric space U (constructed in [**623**]): this is a complete separable metric space containing an isometric copy of every separable metric space and having the property that U is homogeneous in the sense that every two finite isometric sets in U can be taken one to the other by an isometry of the whole space U (the latter property of homogeneity fails for the best known universal metric space $C[0,1]$, and due to this property U is actually unique up to an isometry).

Our next step is to endow the set of triples with the structure of a metric space. The *Hausdorff distance* between sets A and B in a common metric space is defined as the maximum of the numbers $\sup_{a\in A} \operatorname{dist}(a,B)$, $\sup_{b\in B} \operatorname{dist}(b,A)$. It is easy to see that this distance (which can be infinite) satisfies the triangle inequality and is a metric on the set of bounded closed sets. The distance between sets with equal closures equals zero. The space $\mathfrak{M}(X)$ of closed bounded subsets of a complete metric space with the Hausdorff metric is complete; for compact X the space $\mathfrak{M}(X)$ is compact as well, see Burago, Burago, Ivanov [**130**, § 7.3].

Next, the *Gromov–Hausdorff distance* $d_{GH}(X,Y)$ between two metric spaces X and Y is the infimum of numbers r for which there is an isometric embedding of X and Y into a common metric space Z such that the Hausdorff distance between the images X and Y is not larger than r. The distance between bounded spaces is finite. Certainly, for obtaining a metric (not a semimetric) it is necessary to pass to equivalence classes similarly to metrics on classes of integrable functions.

Now in this construction measures must be somehow taken into account. One way is to complement the restriction on the number r above by the requirement that the distance between the images of the given measures on X and Y under their isometric embedding into a common space Z will not exceed r. To this end, we have to a choose a metric on $\mathcal{P}(Z)$. It is customary to use the Prohorov metric; then the obtained distance between triples is called the *Gromov–Hausdorff–Prohorov distance.* However, one can also take the Kantorovich–Rubinshtein distance (and for bounded spaces the Kantorovich distance). There is another possibility for metrization of triples: the distance between two triples is defined as the Prohorov distance between the images of the measures in these triples under the aforementioned mapping j. We shall now introduce two more metrics (as above, these metrics are obtained on equivalence classes, not on triples themselves).

The *Gromov–Prohorov distance (metric)* $d_{GP}(\mathcal{X},\mathcal{Y})$ between two such metric triples $\mathcal{X}=(X,d_X,\mu)$ and $\mathcal{Y}=(Y,d_Y,\nu)$ with probability measures is defined by the formula

$$d_{GP}(\mathcal{X},\mathcal{Y}) := \inf d_{\mathrm{P}}(\mu\circ f^{-1}, \nu\circ g^{-1}),$$

where inf is taken over all isometric embeddings $f\colon X\to Z$, $g\colon Y\to Z$ to metric spaces Z. Certainly, in place of the Prohorov metric one can use other metrics, say, the Kantorovich W_p-metric (the case $p=2$ is considered in Sturm [**600**]).

Let us fix $\lambda>0$. For two functions d_1,d_2 on $[0,1]^2$ we denote by $\square_\lambda(d_1,d_2)$ the infimum of numbers $\varepsilon>0$ such that there exists a measurable set $S\subset[0,1]$ of Lebesgue measure at least $\lambda\varepsilon$ such that $|d_1(t,s)-d_2(t,s)|\leqslant\varepsilon$ for all $s,t\in S$.

The Gromov box distance $\underline{\square}_\lambda(\mathcal{X},\mathcal{Y})$ between two metric triples $\mathcal{X}=(X,d_X,\mu)$ and $\mathcal{Y}=(Y,d_Y,\nu)$ with probability measures is introduced (see Gromov [**301**]) as the infimum of numbers

$$\square_\lambda(d_X\circ\varphi, d_Y\circ\psi)$$

over all Borel mappings $\varphi\colon [0,1] \to X$ and $\psi\colon [0,1] \to Y$ for which $\mu = \lambda_1 \circ \varphi^{-1}$ and $\nu = \lambda_1 \circ \psi^{-1}$, where λ_1 is Lebesgue measure on $[0,1]$.

The following fact is proved in Löhr [**439**] (see also [**438**]).

3.4.2. THEOREM. *We have the equality* $\underline{\Box}_{1/2} = 2d_{GP}$. *In addition, for every* $\lambda > 0$ *we have the inequality*

$$\min(2, 1/\lambda) \cdot d_{GP} \leqslant \underline{\Box}_\lambda \leqslant \max(2, 1/\lambda) \cdot d_{GP}.$$

PROOF. Let $\mathcal{X}_i = (X_i, d_i, \mu_i)$, $i = 1, 2$. Suppose that $d_{GP}(\mathcal{X}_1, \mathcal{X}_2) < r$ for some $r > 0$. Then the spaces (X_i, d_i) can be regarded as subsets of a common separable metric space (X, d) in which $d_P(\mu_1, \mu_2) < r$. According to the representation of the Prohorov metric from Theorem 3.1.5, there exists a measure $\sigma \in \mathcal{P}(X^2)$ with projections μ_1 and μ_2 such that

$$\sigma(Y_r) \leqslant r, \quad Y_r := \{(x, y) \in X^2\colon d(x, y) \geqslant r\}.$$

Let us take a Borel mapping $f\colon [0,1] \to X^2$ taking Lebesgue measure to σ. Let π_1 and π_2 be the projection mappings from X^2 on the factors. Then the mapping $f_i = \pi \circ f$ takes Lebesgue measure to μ_i. We show that $\Box_{1/2}(d \circ f_1, d \circ f_2) \leqslant 2r$. Indeed, $\lambda_1\bigl(f^{-1}(Y_r)\bigr) = \sigma(Y_r) \leqslant r$. Whenever $s, t \in [0,1]\backslash f^{-1}(Y_r)$, we have $d\bigl(f_1(s), f_2(s)\bigr) \leqslant r$ by the definition of Y_r. So

$$d\bigl(f_1(s), f_1(t)\bigr) \leqslant d\bigl(f_1(s), f_1(t)\bigr) + 2r.$$

Hence we obtain

$$\underline{\Box}_{1/2}(\mathcal{X}_1, \mathcal{X}_2) \leqslant \Box_{1/2}(d \circ f_1, d \circ f_2) \leqslant 2r.$$

This proves the right bound in the desired inequality for $\lambda = 1/2$, because r can be taken as close to $d_{GP}(\mathcal{X}_1, \mathcal{X}_2)$ as we wish.

Let now $\underline{\Box}_{1/2}(\mathcal{X}_1, \mathcal{X}_2) < 2r$ and let $f_i\colon [0,1] \to X_i$ be parametrizations of measures μ_i with $\Box_{1/2}(d_1 \circ f_1, d_2 \circ f_2) < 2r$. There is a Borel set $S \subset [0,1]$ with $\lambda_1(S) > 1 - r$ and $|d_1 \circ f_1 - d_2 \circ f_2| < 2r$ on S^2. On the disjoint union $X := X_1 \cup X_2$ of copies of the given spaces we define a metric d by the formula

$$d|_{X_i^2} = d_i, \ d(x, y) = \inf_{s \in S} \Bigl[d_1\bigl(x, f_1(s)\bigr) + d_2\bigl(f_2(s), y\bigr)\Bigr] + r, \quad x \in X_1, \ y \in X_2.$$

Below we verify that this is a metric. Extend the measures μ_i to measures on X by zero on the other component. Let us estimate the distance between them in the Prohorov metric on (X, d). Let $F \subset X$ be closed. By construction $d\bigl(f_1(s), f_2(s)\bigr) = r$ for all $s \in S$. Hence for every $r_0 > r$ we have

$$f_2\bigl(f_1^{-1}(F) \cap S\bigr) \subset F^{r_0}, \quad F^{r_0} = \{x \in X\colon \ d(x, F) < r_0\}.$$

Therefore,

$$\begin{aligned} \mu_1(F) = \lambda_1\bigl(f_1^{-1}(F)\bigr) &\leqslant \lambda_1\bigl(f_1^{-1}(F) \cap S\bigr) + r \\ &\leqslant \mu_2\bigl(f_2\bigl(f_1^{-1}(F) \cap S\bigr)\bigr) + r \leqslant \mu_2(F^{r_0}) + r. \end{aligned}$$

Since $r_0 > r$ was arbitrary, we have $d_{\mathrm{P}}(\mu_1, \mu_2) < r$, whence $d_{GP}(\mathcal{X}_1, \mathcal{X}_2) < r$. Hence the left bound holds for $\lambda = 1/2$. The case of any $\lambda > 0$ is similar.

It remains to verify the triangle inequality for d. For all triples $x, z \in X_1, y \in X_2$ we have

$$d(x, y) \leqslant \inf_{s \in S} \Bigl[d_1(x, z) + d_1\bigl(z, f_1(s)\bigr) + d_2\bigl(f_2(s), y\bigr)\Bigr] + r = d(x, z) + d(z, y)$$

by the equality $d_1(x,z) = d(x,z)$. For all $x, y \in X_1$, $z \in X_2$ we have

$$\begin{aligned} d(x,y) &\leqslant \inf_{s,t\in S} \Big[d_1\big(x, f_1(s)\big) + d_1\big(f_1(s), f_1(t)\big) + d_1\big(f_1(t), y\big)\Big] \\ &\leqslant \inf_{s,t\in S} \Big[d_1\big(x, f_1(s)\big) + d_2\big(f_2(s), f_2(t)\big) + d_1\big(f_1(t), y\big)\Big] + 2r \\ &\leqslant \inf_{s,t\in S} \Big[d_1\big(x, f_1(s)\big) + d_2\big(f_2(s), z)\big) + r \\ &\quad + \inf_t d_2\big(z, f_2(t)\big) + d_1\big(f_1(t), y\big)\Big] + r = d(x,z) + d(z,y). \end{aligned}$$

Other cases are obtained by symmetry. □

Fukaya [**247**] introduced an interesting "measurable Hausdorff topology" on the space of compact metric spaces (K, d, μ) with probability measures (that is, on the space of triples), in which convergence of a net of triples $(K_\alpha, d_\alpha, \mu_\alpha)$ to a triple (K, d, μ) means the existence of Borel mappings $\psi_\alpha\colon X_\alpha \to X$ and numbers $\varepsilon_\alpha > 0$ for which $\varepsilon_\alpha \to 0$, $\mu_\alpha \circ \psi_\alpha^{-1} \Rightarrow \mu \circ \psi^{-1}$, and also $\big(\psi_\alpha(X_\alpha)\big)^{\varepsilon_\alpha} = X$ and $d\big(\psi_\alpha(x), \psi_\alpha(y)\big) < \varepsilon_\alpha$ for all $x, y \in X_\alpha$. In Evans, Winter [**212**] a complete metric is defined generating this topology. Gromov–Hausdorff type metrics were first considered in Edwards [**199**]. On Gromov and Gromov–Prohorov metrics and related questions, see also Abraham, Delmas, Hoscheit [**1**], Athreya, Löhr, Winter [**25**], Blumberg, Gal, Mandell, Pancia [**73**], Evans [**210**], Evans, Molchanov [**211**], Funano [**248**], Gadgil, Krishnapur [**249**], Greven, Pfaffelhuber, Winter [**292**], Melleray, Petrov, Vershik [**456**], Ozawa [**497**], Vershik [**639**]–[**643**], Vershik, Zatitskii, Petrov [**647**], and the book Shioya [**580**]. In Depperschmidt, Greven, Pfaffelhuber [**165**] and Kliem, Löhr [**368**], a modification of Gromov triples is considered ("marked metric measure spaces") in which measures are defined not on metric spaces X, but on products $X \times I$, where I is some space of indices.

3.5. Complements and exercises

3.5(i). Zolotarev metrics

Let $\mathcal{P}^s(\mathbb{R}^d)$ be the space of probability measures on $\mathbb{R}^d$ having finite moments of order s. Let us consider the metric introduced by Zolotarev [**670**], [**671**] on the space $\mathcal{P}^{p+k}(\mathbb{R}^d)$, $p \geqslant 1$, $k \in \mathbb{N}$. The weighted Zolotarev metric is defined by

$$Z_{k,p}(\mu, \sigma) = \sup_{\mathcal{F}_{k,p}} \int_{\mathbb{R}^d} f\, d(\mu - \sigma),$$

where $\mathcal{F}_{k,p}$ is the set of functions $f \in C^{k-1}(\mathbb{R}^d)$ such that $f^{(i)}(0) = 0$ whenever $0 \leqslant i \leqslant k-1$ and

$$\|f^{(k-1)}(x) - f^{(k-1)}(y)\| \leqslant |x-y|(1 + |x|^{p-1} + |y|^{p-1}) \quad \forall\, x, y \in \mathbb{R}^d.$$

In addition, $Z_{0,p}(\mu, \sigma) = \|(1 + |x|^{p-1})(\mu - \sigma)\|$.

For $d = 1$, Rio [**551**], [**552**] obtained estimates for the Kantorovich metric via the Zolotarev metric of the following form with some constants c_k:

$$W_k \leqslant c_k (Z_{k,1})^{1/k}.$$

Bogachev, Doledenok, Shaposhnikov [**85**] proved the following estimate: for any numbers $d, k \in \mathbb{N}$ and $p \geqslant 1$, there exists a number $\gamma = \gamma(d,k,p) > 0$ such that for every $\varepsilon > 0$ one has

$$Z_{k,p}(\mu,\sigma) \leqslant \gamma(1+\varepsilon^{p-1})\big(\varepsilon Z_{k-1,p}(\mu,\sigma) + \varepsilon^{-1} Z_{k+1,p}(\mu,\sigma)\big).$$

In addition, for all $p \geqslant 1$ and $\varepsilon \in (0,1)$ one has

$$W_p^p(\mu,\sigma) \leqslant 2pZ_{1,p}(\mu,\sigma) \leqslant \gamma_{d,p}\varepsilon\|(1+|x|^{p-1})(\mu-\sigma)\|_{TV} + \gamma_{d,p}\varepsilon^{-1}Z_{2,p}(\mu,\sigma).$$

In the cited paper also some weaker analog of the aforementioned one-dimensional estimate from [**551**], [**552**] is obtained. Let $s > 1$. Then

$$W_p(\mu,\sigma) \leqslant \gamma(d,k,s,p)\big(Z_{k,1}(\mu,\sigma)\big)^{1/(skp)} \times \\ \times \Big(\int |x|^{\frac{sp-1}{s-1}}\,\mu(dx) + \int |x|^{\frac{sp-1}{s-1}}\,\sigma(dx)\Big)^{(s-1)/(sp)}.$$

For $p = 1$ we have $Z_{1,1} \leqslant \gamma_{d,k}(Z_{k,1})^{1/k}$ (see [**671**]), $d_{\rm P} \leqslant \gamma_{d,k}(Z_{k,1})^{1/k}$ (see Yamukov [**664**]), and one also has the estimate (see [**85**])

$$Z_{k,1} \leqslant \gamma_{d,k}(Z_{k-1,1}Z_{k+1,1})^{1/2}, \quad (Z_{k,1})^{1/k} \leqslant \gamma_{d,k}(Z_{k+1})^{1/(k+1)}.$$

3.5(ii). Lower bounds for the Kantorovich norm in the Nikolskii–Besov classes

Here we give an estimate of the variation norm of a measure on $\mathbb{R}^d$ via its Kantorovich norm. Certainly, in the general case this is impossible, but for measures with sufficiently regular densities such estimates are available in a rather compact form by using Sobolev type norms (including fractional norms). The estimates presented below are obtained in papers Bogachev, Shaposhnikov [**95**], Bogachev, Wang, Shaposhnikov [**98**], Bogachev, Zelenov, Kosov [**100**], Bogachev, Kosov, Zelenov [**90**] (see also Kohn, Otto [**373**], Seis [**574**]), concerned with multidimensional analogs of the Hardy–Landau–Littlewood inequality (see Hardy, Landau, Littlewood [**310**])

$$\|f'\|_1^2 \leqslant C\|f\|_1\|f''\|_1 \tag{3.5.1}$$

for all integrable functions on the real line with integrable first and second derivatives. The minimal possible constant C in this inequality equals 2. An immediate corollary of this inequality is the following estimate for the distance in L^1 between probability densities f and g on the real line via the variation of their difference and the Kantorovich norm:

$$\|f-g\|_1^2 \leqslant 2\mathrm{Var}(f-g)\|f-g\|_{\rm K}. \tag{3.5.2}$$

In case of absolutely continuous densities f and g we have

$$\mathrm{Var}(f-g) = \|f'-g'\|_1.$$

Inequality (3.5.2) is deduced from (3.5.1) in the following way. It suffices to prove inequality (3.5.2) for smooth probability densities with compact support. In this case we apply (3.5.1) to the difference of the functions

$$F(x) = \int_0^x f(y)\,dy, \quad G(x) = \int_0^x g(y)\,dy.$$

This difference has compact support and, as we know, $\|F-G\|_1 = \|f-g\|_{\rm K}$. In addition, $\|F'-G'\|_1 = \|f-g\|_1$ and $\|F''-G''\|_1 = \|f'-g'\|_1 = \mathrm{Var}(f-g)$.

Let us recall (see Adams, Fournier [**3**], Besov, Il'in, Nikolskii [**63**], Nikolskii [**488**]) that the Nikolskii–Besov class $B^{\alpha}_{1,\infty}(\mathbb{R}^k)$ or order $\alpha \in (0,1)$ consists of all functions $\varrho \in L^1(\mathbb{R}^k)$ for which

$$\|\varrho(\cdot + h) - \varrho\|_{L^1} \leqslant C(\varrho)|h|^{\alpha} \quad \forall\, h \in \mathbb{R}^k$$

for some number $C(\varrho)$. This class is a special case of the class $H^{\alpha}_p(\mathbb{R}^k)$ defined similarly with the L^p-norm in place of the L^1-norm. Below we use the shortened notation $B^{\alpha}(\mathbb{R}^k)$, moreover, this notation will also be used for $\alpha = 1$, when the indicated condition gives the class $BV(\mathbb{R}^k)$ of functions of bounded variation (but not the broader Nikolskii–Besov with $\alpha = 1$ defined by means of the difference $\varrho(\cdot + h) + \varrho(\cdot - h) - 2\varrho$). The class $B^1(\mathbb{R}^k)$ contains the Sobolev class $W^{1,1}(\mathbb{R}^k)$ of integrable functions possessing integrable first order generalized partial derivatives. It will be more convenient to regard functions in these classes as densities of Borel measures on $\mathbb{R}^k$. Let ν be a bounded Borel measure on $\mathbb{R}^k$ and let ν_h denote its shift by the vector h:

$$\nu_h(A) = \nu(A - h).$$

Let $0 < \alpha \leqslant 1$. Then the Nikolskii–Besov class $B^{\alpha}(\mathbb{R}^k)$ coincides with the class of densities of all bounded Borel measures ν on $\mathbb{R}^k$ such that for some number C_ν we have

$$\|\nu_h - \nu\|_{\mathrm{TV}} \leqslant C_\nu |h|^{\alpha} \quad \forall\, h \in \mathbb{R}^k,$$

where $\|\sigma\|_{\mathrm{TV}} = \|\sigma\|$ is the variation of the measure σ (the index TV is added here for greater clarity). The norm $\|\cdot\|_{\mathrm{TV}}$ defines the variation distance $d_{\mathrm{TV}}(\mu,\nu)$ between Borel measures μ,ν on $\mathbb{R}^k$. It will be more convenient to deal with measures possessing densities in these classical spaces and not with functions themselves. So we shall identify measures with their densities and in this sense speak of membership of measures in $B^{\alpha}(\mathbb{R}^k)$.

We shall need the following norm on the space $B^{\alpha}(\mathbb{R}^k)$:

$$\|\nu\|_{B^{\alpha}} := \inf\{C\colon\ \|\nu - \nu_h\|_{\mathrm{TV}} \leqslant C|h|^{\alpha}\}.$$

It is readily seen that this is indeed a norm. However, the space $B^{\alpha}(\mathbb{R}^k)$ is incomplete with this norm: its standard Banach norm is given by $\|\nu\|_{\mathrm{TV}} + \|\nu\|_{B^{\alpha}}$. The latter is greater than $\|\nu\|_{B^{\alpha}}$, moreover, these norms are not equivalent. The situation is similar with Sobolev spaces when only the norm of the gradient is used. For the classes $B^{\alpha}(\mathbb{R}^k)$ one has the following embedding (see [**488**, § 6.3]):

$$B^{\alpha}(\mathbb{R}^k) \subset H^{\beta}_p(\mathbb{R}^k) \subset L^p(\mathbb{R}^k), \quad \beta = \kappa\alpha,\ \kappa = 1 - \frac{k(p-1)}{\alpha p}.$$

Hence all measures in $B^{\alpha}(\mathbb{R}^k)$ have densities of class $L^p(\mathbb{R}^k)$ for all $p < k/(k-\alpha)$. The indicated embeddings to L^p composed with restricting functions to balls are compact operators.

Let us give a sufficient (as recently shown in Bogachev, Kosov, Popova [**89**], also necessary) condition for membership in the class $B^{\alpha}(\mathbb{R}^k)$.

3.5.1. PROPOSITION. *Let $\alpha \in (0,1]$ and let ν be a Borel measures on $\mathbb{R}^k$ such that for every function $\varphi \in C_b^{\infty}(\mathbb{R}^k)$ and every unit vector $e \in \mathbb{R}^k$ one has*

$$\int_{\mathbb{R}^k} \partial_e\varphi(x)\,\nu(dx) \leqslant C\|\varphi\|_{\infty}^{\alpha}\,\|\partial_e\varphi\|_{\infty}^{1-\alpha}.$$

Then the following estimate holds:

$$\|\nu_h - \nu\|_{\mathrm{TV}} \leqslant 2^{1-\alpha} C |h|^\alpha \quad \forall\, h \in \mathbb{R}^k,$$

i.e., $\nu \in B^\alpha(\mathbb{R}^k)$ *and* $\|\nu\|_{B^\alpha} \leqslant 2^{1-\alpha}C$. *In particular, the density of the measure* ν *belongs to all* $L^p(\mathbb{R}^k)$ *whenever* $p < k/(k-\alpha)$.

PROOF. Let $e = |h|^{-1}h$. We have

$$\begin{aligned}\|\nu_h - \nu\|_{\mathrm{TV}} &= \sup_{\varphi \in C_b^\infty(\mathbb{R}^k),\ \|\varphi\|_\infty \leqslant 1} \int_{\mathbb{R}^k} \varphi(x)\,(\nu_h - \nu)(dx) \\ &= \sup_{\varphi \in C_b^\infty(\mathbb{R}^k),\ \|\varphi\|_\infty \leqslant 1} \int_{\mathbb{R}^k} [\varphi(x+h) - \varphi(x)]\,\nu(dx) \\ &= \sup_{\varphi \in C_b^\infty(\mathbb{R}^k),\ \|\varphi\|_\infty \leqslant 1} \int_{\mathbb{R}^k}\int_0^{|h|} \partial_e\varphi(x+se)\,ds\,\nu(dx).\end{aligned}$$

Let $\varphi \in C_b^\infty(\mathbb{R}^k)$ and $\|\varphi\|_\infty \leqslant 1$. Set

$$\Phi(x) = \int_0^{|h|} \varphi(x+se)\,ds.$$

It is clear that $\sup_{x\in\mathbb{R}^k} |\Phi(x)| \leqslant |h|$ and

$$|\partial_e \Phi(x)| = \left| \int_0^{|h|} \partial_e\varphi(x+se)\,ds \right| = |\varphi(x+h) - \varphi(x)| \leqslant 2.$$

By the hypotheses of the theorem

$$\int_{\mathbb{R}^k} \partial_e \Phi(x)\,\nu(dx) \leqslant C|h|^\alpha 2^{1-\alpha},$$

hence

$$\int_{\mathbb{R}^k}\int_0^{|h|} \partial_e\varphi(x+se)\,ds\,\nu(dx) \leqslant C2^{1-\alpha}|h|^\alpha,$$

as desired. □

Let us now establish a multidimensional fractional analog of the aforementioned Hardy–Landau–Littlewood estimate (for $\alpha = 1$ we obtain an analog of (3.5.2)). For simplicity we consider the case of $\mathbb{R}^k$, although in the papers cited above more general results are obtained for Riemannian manifolds and even for some measures on infinite-dimensional spaces.

3.5.2. THEOREM. *Let* $\nu, \sigma \in B^\alpha(\mathbb{R}^k)$ *be Borel probability measures on* $\mathbb{R}^k$. *Then*

$$\|\sigma - \nu\|_{\mathrm{TV}} \leqslant C(k,\alpha)\|\sigma - \nu\|_{B^\alpha}^{1/(1+\alpha)} d_{\mathrm{K}}(\sigma,\nu)^{\alpha/(1+\alpha)}, \tag{3.5.3}$$

where

$$C(k,\alpha) = 1 + \int_{\mathbb{R}^k} |x|^\alpha\,\gamma_k(dx).$$

PROOF. Let γ_k^ε be the centered Gaussian measure on $\mathbb{R}^k$ with covariance matrix $\varepsilon^2\mathrm{I}$, i.e., with the scaled density $(2\pi\varepsilon^2)^{-k/2}\exp\bigl(-|x|^2/(2\varepsilon^2)\bigr)$. By the triangle inequality

$$\|\sigma - \nu\|_{\mathrm{TV}} \leqslant \|(\sigma-\nu) - (\sigma-\nu) * \gamma_k^\varepsilon\|_{\mathrm{TV}} + \|\sigma * \gamma_k^\varepsilon - \nu * \gamma_k^\varepsilon\|_{\mathrm{TV}}. \tag{3.5.4}$$

For every function $\varphi \in C_b^\infty(\mathbb{R}^k)$ with $\|\varphi\|_\infty \leqslant 1$ we have (all integrals are taken over the whole space $\mathbb{R}^k$):

$$\int \varphi\, d(\sigma * \gamma_k^\varepsilon - \nu * \gamma_k^\varepsilon) = \int \varphi(x) \int (2\pi\varepsilon^2)^{-k/2} \exp\Bigl(-\frac{|y-x|^2}{2\varepsilon^2}\Bigr)\,(\nu-\sigma)(dy)\,dx$$
$$= \int \Bigl(\int \varphi(x)(2\pi\varepsilon^2)^{-k/2} \exp\Bigl(-\frac{|y-x|^2}{2\varepsilon^2}\Bigr)\,dx \Bigr)(\nu-\sigma)(dy).$$

Let us consider the function

$$\Phi(y) := \int \varphi(x)(2\pi\varepsilon^2)^{-k/2} \exp\Bigl(-\frac{|y-x|^2}{2\varepsilon^2}\Bigr)\,dx.$$

Then

$$\nabla\Phi(y) = \varepsilon^{-1} \int \varphi(y+\varepsilon z)(2\pi)^{-k/2} z \exp\Bigl(-\frac{|z|^2}{2}\Bigr)\,dz,$$

hence $|\Phi(y)| \leqslant 1$, $|\nabla\Phi(y)| \leqslant \varepsilon^{-1}$. Therefore,

$$\|\sigma * \gamma_k^\varepsilon - \nu * \gamma_k^\varepsilon\|_{\mathrm{TV}} \leqslant \varepsilon^{-1} d_{\mathrm{K}}(\sigma,\nu). \tag{3.5.5}$$

We now estimate the remaining term in the right-hand side of (3.5.4):

$$\|(\sigma-\nu) - (\sigma-\nu) * \gamma_k^\varepsilon\|_{\mathrm{TV}}$$
$$= \sup_{\|\varphi\|_\infty \leqslant 1} \int \Bigl((2\pi\varepsilon^2)^{-k/2} \exp\Bigl(-\frac{|y|^2}{2\varepsilon^2}\Bigr) \int \varphi(x)\,\bigl((\sigma-\nu) - (\sigma_y - \nu_y)\bigr)(dx) \Bigr)\,dy$$
$$\leqslant \|\sigma-\nu\|_{B^\alpha} \int (2\pi\varepsilon^2)^{-k/2} \exp\Bigl(-\frac{|y|^2}{2\varepsilon^2}\Bigr) |y|^\alpha\,dy$$
$$= \varepsilon^\alpha \|\sigma-\nu\|_{B^\alpha} (2\pi)^{-k/2} \int |y|^\alpha \exp\Bigl(-\frac{|y|^2}{2}\Bigr)\,dy.$$

Therefore,

$$\|\sigma-\nu\|_{\mathrm{TV}} \leqslant \varepsilon^{-1} d_{\mathrm{K}}(\sigma,\nu) + \varepsilon^\alpha \|\sigma-\nu\|_{B^\alpha} \int |x|^\alpha\, \gamma_k(dx).$$

Taking $\varepsilon = \bigl(\|\sigma-\nu\|_{\mathrm{K}}/\|\sigma-\nu\|_{B^\alpha}\bigr)^{1/(1+\alpha)}$, we obtain (3.5.3). □

3.5.3. Remark. An analogous estimate is also true for the Fortet–Mourier metric:

$$\|\sigma-\nu\|_{\mathrm{TV}} \leqslant C(k,\alpha)\|\sigma-\nu\|_{B^\alpha}^{1/(1+\alpha)} d_{\mathrm{FM}}(\sigma,\nu)^{\alpha/(1+\alpha)} + d_{\mathrm{FM}}(\sigma,\nu)$$
$$\leqslant \bigl(C(k,\alpha)\|\sigma-\nu\|_{B^\alpha}^{1/(1+\alpha)} + 2^{1/(1+\alpha)}\bigr) d_{\mathrm{FM}}(\sigma,\nu)^{\alpha/(1+\alpha)},$$

where $C(k,\alpha)$ is the same as above. For the proof in place of (3.5.5) we write $\|\sigma * \gamma_k^\varepsilon - \nu * \gamma_k^\varepsilon\|_{\mathrm{TV}} \leqslant \bigl(\varepsilon^{-1}+1\bigr) d_{\mathrm{FM}}(\sigma,\nu)$ and continue as above.

For $\alpha = 1$ we obtain that for probability measures μ and ν with densities ϱ_μ and ϱ_ν from the Sobolev class $W^{1,1}(\mathbb{R}^k)$ there holds the inequality

$$\|\mu-\nu\|_{\mathrm{TV}}^2 \leqslant C(k)\|\nabla\varrho_\mu - \nabla\varrho_\nu\|_1 \|\mu-\nu\|_{\mathrm{K}}. \tag{3.5.6}$$

Let us mention the following elegant inequality established by Krugova [**396**] for convex measures (see p. 40) on general locally convex spaces (presented here in the case of $\mathbb{R}^n$):

$$\|\mu - \mu_h\| \geqslant 2 - \exp(-\|\partial_h \varrho\|/2),$$

where ϱ is the density of μ; earlier she proved in [**395**] that $\partial_h\varrho$ is automatically a measure of bounded variation (in other words, the density of an absolutely continuous convex measure is of class $BV(\mathbb{R}^n)$). It follows from this estimate combined with Theorem 1.7.3 that if convex measures $\mu_j = \varrho_j\,dx$ converge weakly to an absolutely continuous convex measure μ on $\mathbb{R}^n$, then $\sup_j \|\partial_{x_i}\varrho_j\| < \infty$ for all $i \leqslant n$. Indeed, otherwise we can arrive at the situation where $\|\partial_{x_i}\varrho_j\| \to \infty$ for some i. Hence $\|\mu_j - (\mu_j)_{se_i}\| \to 2$ for any $s > 0$. Since we know that there is also convergence in variation, we obtain $\|\mu - \mu_{se_i}\| = 2$, which means that μ and μ_{se_i} are mutually singular for all $s > 0$. This is impossible, since μ is absolutely continuous. Furthermore, from (3.5.6) we obtain

$$\|\mu - \mu_j\| \leqslant C\sqrt{d_{\mathrm{K}}(\mu, \mu_j)},$$

where the number C depends on μ and the whole sequence $\{\mu_j\}$.

Since many statistics popular in applications are polynomials or expressed via polynomials, it is of interest to have conditions for convergence of distributions of polynomials in random variables. Distributions of polynomials on spaces with Gaussian measures are considered in Bogachev, Kosov, Zelenov [**90**], Bogachev, Zelenov [**99**], Bogachev, Zelenov, Kosov [**100**], Nourdin, Nualart, Poly [**491**], Nourdin, Peccati [**492**], Nourdin, Poly [**493**], Peccati, Taqqu [**506**], see also the survey Bogachev [**84**]. Let γ be the countable power of the standard Gaussian measure on the real line and let $\mathcal{P}_d(\gamma)$ be the class of γ-measurable polynomials of degree d, i.e., the set of limits of almost everywhere convergent sequences of polynomials of degree d in finitely many variables. As shown in Sevastyanov [**579**] and Arcones [**20**], the set of distributions of polynomials from $\mathcal{P}_2(\gamma)$ is closed in the weak topology. It is unknown whether this is true for the classes $\mathcal{P}_d(\gamma)$ with $d > 2$. However, weak convergence of the distributions of polynomials $f_n \in \mathcal{P}_d(\gamma)$ such that their limit is not a Dirac measure yields their convergence in variation (see [**491**], [**493**]). Moreover, it is proved in Bogachev, Kosov, Zelenov [**90**] that the distribution of any nonconstant polynomial $f(\xi_1, \ldots, \xi_n)$ of degree d in standard Gaussian random variables $\xi_1, \ldots, \xi_n$ belongs to the Nikolskii–Besov class $B^{1/d}(\mathbb{R})$ of order $1/d$, which does not depend on the number of variables. It is also shown there that for every $a > 0$ there exists $C(a) > 0$ such that if f and g are polynomials of degree d and for both the supremum of the quantities $\|\partial_h f\|_{L^2(\gamma_n)}$ and $\|\partial_h g\|_{L^2(\gamma_n)}$, where γ_n is the standard Gaussian measure on $\mathbb{R}^n$, over unit vectors h is not less than a, then

$$\|\gamma_n\circ f^{-1} - \gamma_n\circ g^{-1}\|_{\mathrm{TV}} \leqslant C(a)\|\gamma_n\circ f^{-1} - \gamma_n\circ g^{-1}\|_{\mathrm{K}}^{1/(d+1)}.$$

Similar results hold for k-dimensional distributions of random vectors with polynomial coefficients, i.e., images of the measure γ under mappings of the form $f = (f_1, \ldots, f_k)$, where all f_i are polynomials of degree $d > 1$. The measures $\gamma_n\circ f^{-1}$ belong to the class $B^\alpha(\mathbb{R}^k)$ with any $\alpha < \bigl(4k(d-1)\bigr)^{-1}$ if this measure is absolutely continuous, and the latter is true precisely when there is no nonconstant polynomial Q such that $Q(f_1, \ldots, f_k)$ is a constant.

3.5(iii). Bounds in terms of Fourier transforms

Let μ and ν be two Borel probability measures on the real line with distribution functions F_μ and F_ν and Fourier transforms (characteristic functionals) $\widetilde{\mu}$ and $\widetilde{\nu}$, respectively. Let us mention several useful inequalities for various distances in terms of Fourier transforms. A survey of this direction is given in Bobkov [**74**], where one can find proofs, comments and references; see also Goudon, Junca, Toscani [**285**].

The Kolmogorov distance is the uniform distance between $\widetilde{\mu}$ and $\widetilde{\nu}$. Proposition 1.6.1 gives the estimate (if the integral is finite)

$$\|F_\mu - F_\nu\|_\infty \leqslant \frac{1}{2\pi}\int_{-\infty}^{+\infty}\Bigl|\frac{\widetilde{\mu}(y)-\widetilde{\nu}(y)}{y}\Bigr|\,dy.$$

There is also the following bound established in Bentkus, Götze [**53**]:

$$\|F_\mu - F_\nu\|_\infty \leqslant \frac{1}{2\pi}\int_{-T}^{T}\Bigl|\frac{\widetilde{\mu}(y)-\widetilde{\nu}(y)}{y}\Bigr|\,dy + \frac{1}{T}\int_{-T}^{T}\bigl(|\widetilde{\mu}(y)|+|\widetilde{\nu}(y)|\bigr)\,dy.$$

As shown in Faĭnleĭb [**219**], there is an absolute constant c such that

$$c\|F_\mu - F_\nu\|_\infty \leqslant \int_{-T}^{T}\Bigl|\frac{\widetilde{\mu}(y)-\widetilde{\nu}(y)}{y}\Bigr|\,dy + \frac{1}{T}\int_{-T}^{T}|\widetilde{\nu}(y)|\,dy.$$

Let us also mention the following bound due to Esseen valid for all $T>0$ and $b>1/(2\pi)$ in the case where F_ν is Lipschitz with constant L:

$$\|F_\mu - F_\nu\|_\infty \leqslant b\int_{-T}^{T}\Bigl|\frac{\widetilde{\mu}(y)-\widetilde{\nu}(y)}{y}\Bigr|\,dy + c(b)\frac{L}{T},$$

where $c(b)$ depends only on b.

Let us now turn to the Lévy metric L (the Lévy–Prohorov metric is its natural generalization, in the case of the real line $L(\mu,\nu)\leqslant d_{\rm P}(\mu,\nu)$) given by

$$L(\mu,\nu)=\inf\{h\geqslant 0\colon F_\nu(x-h)-h\leqslant F_\mu(x)\leqslant F_\nu(x+h)+h\ \ \forall x\in\mathbb{R}\}. \tag{3.5.7}$$

In general, $L(\mu,\nu)\leqslant\|F_\mu-F_\nu\|_\infty$, and if F_ν is Lipschitz with constant L (as in Esseen's bound), then $\|F_\mu-F_\nu\|_\infty\leqslant(1+L)L(\mu,\nu)$. The following estimate was proved in Bohman [**102**]:

$$\frac{1}{2}L(\mu,\nu)\leqslant\sup_y\Bigl|\frac{\widetilde{\mu}(y)-\widetilde{\nu}(y)}{y}\Bigr|.$$

Zolotarev [**667**] proved the bound

$$L(\mu,\nu)\leqslant\frac{1}{2\pi}\int_{-T}^{T}\Bigl|\frac{\widetilde{\mu}(y)-\widetilde{\nu}(y)}{y}\Bigr|\,dy + 2e\frac{\ln T}{T},\quad T>1.3.$$

For the total variation distance one has

$$\|\mu-\nu\|^4\leqslant\int_{-\infty}^{+\infty}|\widetilde{\mu}(y)-\widetilde{\nu}(y)|\,dy\int_{-\infty}^{+\infty}|\widetilde{\mu}'(y)-\widetilde{\nu}'(y)|\,dy.$$

It is also worth noting that for any $p\in[2,+\infty)$ and $q=p/(p-1)$ one has

$$\|F_\mu-F_\nu\|^q_{L^p(\mathbb{R})}\leqslant\frac{1}{2\pi}\int_{-\infty}^{+\infty}\Bigl|\frac{\widetilde{\mu}(y)-\widetilde{\nu}(y)}{y}\Bigr|^q\,dy.$$

This inequality does not extend to the case $p<2$. For $p=1$ there is the following bound:

$$\|F_\mu-F_\nu\|_{L^1(\mathbb{R})}\leqslant\frac{1}{2}\int_{-\infty}^{+\infty}\Bigl|\frac{\widetilde{\mu}(y)-\widetilde{\nu}(y)}{y}\Bigr|^2\,dy+\frac{1}{2}\int_{-\infty}^{+\infty}\Bigl|\frac{d}{dy}\frac{\widetilde{\mu}(y)-\widetilde{\nu}(y)}{y}\Bigr|^2\,dy.$$

Finally, let us mention a lower bound obtained in Bobkov [**74**] for any signed measure σ:

$$\|F_\sigma\|_\infty \geqslant \frac{1}{3T}\left|\int_0^T \widetilde{\sigma}(y)(1-yT^{-1})\,dy\right|, \quad T>0.$$

3.5(iv). Discrete approximations

Here we briefly discuss certain quantitative problems connected with approximations of measures on $\mathbb{R}^d$ by measures with finite supports. A thorough discussion of this direction (called "quantization" in the literature) can be found in Graf, Luschgy [**286**].

For every $n\in\mathbb{N}$, let $\mathcal{F}_n$ be the family of all Borel mappings from $\mathbb{R}^d$ to $\mathbb{R}^d$ with n values and let $\mathcal{P}_n$ be the set of all probability measures on $\mathbb{R}^d$ concentrated at n points. We know that every probability measure can be approximated by discrete measures, but the rate of best possible approximations in a suitable metric by measures from $\mathcal{P}_n$ can be quite different for different measures (similarly to different rates of approximation of smooth functions by polynomials of degree n).

Let $\|\cdot\|$ be a norm on $\mathbb{R}^d$, say, its standard norm.

Let μ be a Borel probability measure on $\mathbb{R}^d$ with a finite moment of order r. Set

$$V_{n,r}(\mu) = \inf_{f\in\mathcal{F}_n}\int_{\mathbb{R}^d}\|x-f(x)\|^r\,\mu(dx).$$

In particular,

$$V_{1,r}(\mu) = \inf_{a\in\mathbb{R}^d}\int_{\mathbb{R}^d}\|x-a\|^r\,\mu(dx).$$

Approximation by discrete measures involves certain problems of independent interest related to approximation of sets. Let A be a locally finite subset of $\mathbb{R}^d$. For each $a\in A$, the *Voronoi region* generated by a is the set

$$W(a|A) = \{x\in\mathbb{R}^d\colon \|x-a\| = \min_{b\in A}\|x-b\|\}.$$

The family of sets $\{W(a|A)\colon a\in A\}$ is called the *Voronoi diagram* of A. It is not difficult to verify that this family covers $\mathbb{R}^d$.

3.5.4. LEMMA. *One has*

$$V_{n,r}(\mu) = \inf_{\mathrm{Card}A\leqslant n}\int_{\mathbb{R}^d}\min_{a\in A}\|x-a\|^r\,\mu(dx).$$

In addition,

$$V_{n,r}(\mu) = \inf_{A_1,\ldots,A_n}\sum_{i=1}^n V_{1,r}\bigl(\mu(A_i)^{-1}\mu|_{A_i}\bigr)\mu(A_i),$$

where inf *is taken over all partitions of* $\mathbb{R}^d$ *into disjoint Borel parts* $A_1,\ldots,A_n$.

PROOF. For any mapping f with values $a_1,\ldots,a_n$ the integral of $\|x-f(x)\|^r$ against μ equals the sum of the integrals of $\|x-a_i\|^r$ over the sets $A_i=f^{-1}(a_i)$, hence is not less than the sum of the integrals of $\min_{b\in A_i}\|x-b\|^r$ over A_i, which is the right-hand side of our equality. Conversely, given a finite set $A=\{a_1,\ldots,a_n\}$, we take its Voronoi partition $A_i,\ldots,A_n$ of $\mathbb{R}^d$ and define $f\in\mathcal{F}_n$ by $f|_{A_i}=a_i$. It is straightforward to verify that the integral of $\|x-f(x)\|^r$ against μ equals the integral of $\min_{a\in A}\|x-a\|^r$. □

The next result connects $V_{n,r}$ with the Kantorovich distance W_r.

3.5.5. THEOREM. *The following equalities hold:*

$$V_{n,r}(\mu) = \inf_{f\in\mathcal{F}_n} W_r(\mu, \mu\circ f^{-1}) = \inf_{\nu\in\mathcal{P}_n} W_r(\mu,\nu).$$

PROOF. Let $f \in \mathcal{F}_n$ and let σ_f be the image of the measure μ under the mapping $x \mapsto \bigl(x, f(x)\bigr)$. Then $\sigma_f \in \Pi(\mu, \mu\circ f^{-1})$, hence

$$\int_{\mathbb{R}^d} \|x - f(x)\|^r\, \mu(dx) = \int_{\mathbb{R}^d\times\mathbb{R}^d} \|x-y\|^r\, \sigma_f(dx\,dy) \geqslant W_r(\mu,\mu\circ f^{-1})^r.$$

Hence the left-hand side of the equality to be proven is not less than the right-hand side. Now let $\nu \in \mathcal{P}_n$ be concentrated on a set $A = \{a_1, \ldots, a_n\}$. For any measure $\sigma \in \Pi(\mu,\nu)$ the integral of $\|x-y\|^r$ against σ reduces to the integral over $\mathbb{R}^d \times A$, hence is not less than the integral of $\min_{A\in A} \|x-a\|^r$ against σ, which equals the integral of $\min_{A\in A}\|x-a\|^r$ against μ. By the lemma above, the latter integral is $V_{n,r}(\mu)$. Hence the right-hand side of the desired equality is not less than the left-hand side. □

For a bounded Borel set A of positive Lebesgue measure (denoted by λ_d) let

$$M_{n,r}(A) = V_{n,r}(U_A)\lambda_d(A)^{-r/d},$$

where $U_A = \lambda_d(A)^{-1}\lambda_d|_A$ is the uniform distribution on A. It is not difficult to verify that $\inf_{n\geqslant 1} n^{r/d} M_{n,r}([0,1]^d) > 0$.

3.5.6. THEOREM. *Let $r \geqslant 1$, $\delta > 0$, and let μ be a Borel probability measure on $\mathbb{R}^d$ with a finite moment of order $r+\delta$. Let ϱ_a be the density of the absolutely continuous component of μ with respect to Lebesgue measure. Then*

$$\lim_{n\to\infty} n^{r/d} V_{n,r}(\mu) = \|\varrho_a\|_{L^{d/(d+r)}(\mathbb{R}^d)} \inf_{n\geqslant 1} n^{r/d} M_{n,r}([0,1]^d).$$

For a bounded set A let

$$e_{n,\infty}(A) = \inf_{\operatorname{Card} A\leqslant n} \max_{x\in A}\min_{a\in A} \|x-a\|.$$

One can show that the number $c_{d,n} := e_{n,\infty}([0,1]^d)$ is positive.

3.5.7. THEOREM. *Let A be a nonempty compact set in $\mathbb{R}^d$ with boundary of Lebesgue measure zero. Then* $\lim\limits_{n\to\infty} n^{1/d} e_{n,\infty}(A) = c_{d,n}\lambda_d(A)^{1/d}$.

For proofs, see [**286**, Theorem 6.2, Theorem 10.7]. On this topic, see also Dereich, Scheutzow, Schottstedt [**166**], Graf, Luschgy, Pagés [**287**], Kloeckner [**371**], Kreitmeier [**392**], and Exercise 3.5.33.

3.5(v). Extensions of metrics

First we present a result on extension of metrics going back to Hausdorff (the included proof is borrowed from Toruńczyk [**618**], where one can find some related bibliographic references).

3.5.8. THEOREM. *Let (X,d) be a metric space and let Y be its closed subset such that Y is equipped with a new metric d_Y with the same supply of convergent sequences (i.e., generating the induced topology). Then this metric can be extended to a metric on X generating the original topology. In the case where X is complete and the new metric on Y is complete, the extended metric on X can be also made complete.*

PROOF. According to Exercise 2.7.27 we can assume that the space (X, d) is a closed subset of a normed space E (which is complete in case of complete X) and that (Y, d_Y) is a subset of another normed space F (also complete in case of a complete metric d_Y). The space $E \times F$ is equipped with its natural norm $\|(x, y)\| = \|x\|_E + \|y\|_F$. The set Y in E is homeomorphic to the same set considered in F (with the metric d_Y), hence by Theorem 2.7.23 the natural homeomorphism can be extended to a homeomorphism h of the whole space $E \times F$. The mapping $x \mapsto h(x, 0)$ defines a homeomorphism of X and a closed subset in $E \times F$, and this mapping on Y with the metric d_Y gives an isometric embedding of (Y, d_Y) into the space $E \times F$. For the required metric we take $\varrho(x_1, x_2) = \|h(x_1, 0) - h(x_2, 0)\|$.

It is clear that we have obtained a metric on X. For all $x_1, x_2 \in Y$ the points $h(x_1, 0), h(x_2, 0)$ belong to (Y, d_Y) regarded as a subset of F (because h extends the natural homeomorphism of Y with the induced metric and the new metric), hence $\varrho(x_1, x_2) = d_Y(x_1, x_2)$. It follows from what has been said above that the metric ϱ generates on X the original topology. In case of complete (X, d) and (Y, d_Y), the metric ϱ is also complete. □

It is clear from this general fact that every metric generating weak convergence on the set of probability measures extends to a metric on the set of nonnegative measures generating the weak topology (if we consider separable spaces or measures).

As we have seen, the simplex of probability measures $\mathcal{P}(X)$ on a complete separable metric space X is a complete separable metrizable subset of the whole locally convex space of measures $\mathcal{M}(X)$ with the weak topology, which is not metrizable in nontrivial cases. The linear span of this simplex is the whole space $\mathcal{M}(X)$, on which there is the Kantorovich–Rubinshtein norm, by means of which one can metrize this simplex, but not the whole space. It turns out that this is a special case of the following very general assertion.

3.5.9. THEOREM. (i) (*Drewnowski* [**181**]) *Let E be a separated locally convex space and let A be its subset that is separable and metrizable in the induced topology. Then on the linear span of A there is a metrizable locally convex topology that is majorized by the original topology and the restriction of which to A coincides with the original topology.*

(ii) (*Larman, Rogers* [**413**]) *If, in addition, A is locally bounded, i.e., for every $a \in A$ there is a neighborhood of zero U such that $(a + U) \cap A$ is bounded, then on the linear span of A there is a norm generating on A the original topology (but the topology generated by this norm is not always majorized by the original topology on the linear span of the set A).*

It is interesting to investigate which position is occupied by the metrics and norms discussed above among metrics and norms from this theorem when we apply it to the space of measures $\mathcal{M}(X)$ with the weak topology.

3.5(vi). Merging sequences

In relation to weak convergence we mention the notion of merging sequences of measures.

3.5.10. DEFINITION. We say that two sequences of Borel measures $\{\mu_n\}$ and $\{\nu_n\}$ on a metric space X are weakly merging if $\mu_n - \nu_n \Rightarrow 0$.

If $\{\mu_n\}$ and $\{\nu_n\}$ are weakly merging sequences of Borel probability measures on a separable metric space (X,d), then we have $\|\mu_n-\nu_n\|_{\rm KR}\to 0$ and $d_{\rm P}(\mu_n,\nu_n)\to 0$ according to Exercise 2.7.32. However, the condition that $d_{\rm P}(\mu_n,\nu_n)\to 0$ (or, equivalently, $\|\mu_n-\nu_n\|_{\rm KR}\to 0$) does not imply that $\{\mu_n\}$ and $\{\nu_n\}$ are weakly merging. For example, let μ_n be the Dirac measure at the point n on the real line and let ν_n be the Dirac measure at the point $n+1/n$. Clearly, $\|\mu_n-\nu_n\|_{\rm KR}\to 0$, but the measures $\delta_n-\delta_{n+1/n}$ do not converge weakly. See Dudley [**193**, § 11.7] for the proof of the following assertion.

3.5.11. PROPOSITION. *Let (X,d) be a separable metric space. Let $\{\mu_n\}$ and $\{\nu_n\}$ be two sequences in $\mathcal{P}(X)$. The following conditions are equivalent:*

(a) $d_{\rm P}(\mu_n,\nu_n)\to 0$,

(b) $\|\mu_n-\nu_n\|_{\rm KR}\to 0$,

(c) *one can find a probability space (Ω,P) and two sequences of measurable mappings $\xi_n,\eta_n\colon\Omega\to X$ such that $P\circ\xi_n^{-1}=\mu_n$, $P\circ\eta_n^{-1}=\nu_n$ and $d(\xi_n,\eta_n)\to 0$ a.e.*

In D'Aristotile, Diaconis, Freedman [**157**] in the case of a separable metric space X a stronger merging of measures, called F-merging, is studied: it is required that the relation $\varrho(\mu_n,\nu_n)\to 0$ hold for every metric ϱ on $\mathcal{P}(X)$ generating weak convergence. In order to see that this is indeed a stronger condition, let us consider the following example. Let the measure μ_n on the real line assign the values $1/n$ to the points $1,\ldots,n$. Let $\nu_n=\mu_{n+1}$. Then the measures $\mu_n-\nu_n$ converge to zero even in variation, but are not F-merging, since the sets $\{\mu_{2n}\}$ and $\{\nu_{2n}\}$ are closed in $\mathcal{P}(\mathbb{R})$ and disjoint, which gives a function $\Phi\in C_b\bigl(\mathcal{P}(\mathbb{R})\bigr)$ such that the numbers $\Phi(\mu_n)-\Phi(\nu_n)$ do not tend to zero (then one can take the metric $d_{\rm P}(\mu,\nu)+|\Phi(\mu)-\Phi(\nu)|$ on $\mathcal{P}(X)$). In [**157**], among other things, the following result is proved.

3.5.12. THEOREM. *Suppose that X is a separable metric space and $\{\mu_n\},\{\nu_n\}$ are two sequences in $\mathcal{P}(X)$. The following conditions are equivalent:*

(a) *the sequences $\{\mu_n\}$ and $\{\nu_n\}$ are F-merging,*

(b) *for all functions $\Phi\in C_b\bigl(\mathcal{P}(X)\bigr)$ we have $\Phi(\mu_n)-\Phi(\nu_n)\to 0$,*

(c) *for every function $\Psi\in C_b\bigl(\mathcal{P}(X)\times\mathcal{P}(X)\bigr)$ vanishing on the diagonal we have $\Psi(\mu_n,\nu_n)\to 0$.*

If $\mu_n\neq\nu_n$ for all n, then yet another equivalent condition is this:

(d) *every subsequence in $\{\mu_n\}$ has a further subsequence that converges weakly and the corresponding subsequence in $\{\nu_n\}$ converges weakly to the same limit.*

Finally, if X is complete, then (d) *is equivalent to the condition that both sequences are uniformly tight and weakly merging.*

The weak merging is equivalent to F-merging precisely when X is compact. Analogous questions are considered in [**157**] for nets.

Bergin [**54**] proved the following fact. It would be interesting to extend it to more general topological spaces.

3.5.13. THEOREM. *Suppose that X and Y are separable metric spaces and $\mu\in\mathcal{P}(X)$, $\nu\in\mathcal{P}(Y)$, $\eta\in\mathcal{P}(X\times Y)$ are such that the projections of η on X and Y equal μ and ν. Suppose that sequences $\{\mu_n\}\subset\mathcal{P}(X)$ and $\{\nu_n\}\subset\mathcal{P}(Y)$ converge weakly to μ and ν, respectively. Then, there exists a sequence of measures $\eta_n\in\mathcal{P}(X\times Y)$ weakly converging to η such that for every n the projections of the measures η_n on the spaces X and Y equal μ_n and ν_n, respectively.*

Exercises

3.5.14. Prove that on the set of continuous functions on the interval $[0,1]$ there is no topology such that convergence of a sequence of functions in this topology coincides with convergence almost everywhere.

HINT: find a sequence of continuous functions f_n on $[0,1]$ that converges at no point, but converges to zero in measure; use the Riesz theorem, according to which every its subsequence contains a further subsequence converging almost everywhere; observe that in the presence of a topology with the stated properties the sequence $\{f_n\}$ would converge to zero in this topology.

3.5.15. Prove that on every noncompact complete metric space there is a metric generating the same topology, but such that with respect to it the space will not be complete.

HINT: assuming that the given space (X,d) is complete, take a sequence $\{x_n\}$ with $d(x_n,x_k)\geqslant r>0$ for $n\neq k$ and construct a continuous function f for which $f(x_n)=n$ (see Exercise 2.7.24); consider the metric $\arctan\bigl(d(x,y)+|f(x)-f(y)|\bigr)$. One can also use Theorem 3.5.8.

3.5.16. Prove that on the unit sphere of the space l^1 the weak topology coincides with the norm topology.

HINT: show that every ball of radius $r>0$ centered at a point a with $\|a\|_1=1$ contains the intersection of the sphere with a weak neighborhood of a; for this take m such that $\sum_{i=1}^m |a_i|>1-r/4$, then $U=\{x\in l^1\colon\ |x_i-a_i|<r/(4m),\ i=1,\ldots,m\}$ is the desired neighborhood. If now $x\in U$ and $\|x\|_1=1$, then

$$\|x-a\|_1\leqslant\sum_{i=1}^m|x_i-a_i|+\sum_{i>m}(|a_i|+|x_i|)\leqslant r/4+r/4+r/2,$$

since $\sum_{i>m}|x_i|=1-\sum_{i=1}^m|x_i|\leqslant 1-\sum_{i=1}^m|a_i|+r/4<r/2$.

3.5.17. Suppose that a metric space X contains an infinite Cauchy sequence. Construct a sequence of signed Borel measures μ_n on X for which $\|\mu_n\|=1$ and simultaneously $\|\mu_n\|_{\mathrm{K}}\to 0$ as $n\to\infty$.

HINT: take different points x_i in the given sequence such that $d(x_i,x_{i+1})\leqslant 2^{-i}$, consider the signed discrete measures $\mu_n=n^{-1}(\delta_{x_n}-\delta_{x_{n+1}}+\cdots-\delta_{x_{2n-1}})$.

3.5.18. Let μ be a Borel probability measure on a separable metric space X such that its topological support is X (i.e., the measure μ is positive on all open balls). Let μ^∞ be the countable power of the measure μ defined on X^∞. Prove that for μ^∞-almost every point $x=(x_i)$ the sequence $\{x_i\}$ is everywhere dense in X and $(\delta_{x_1}+\cdots+\delta_{x_n})/n\Rightarrow\mu$.

HINT: it suffices to verify that μ^∞-almost every sequence hits every ball U of a rational radius centered at a point of a countable everywhere dense set in X, and for this it is enough to show that almost every sequence hits any fixed ball of this type; the set of sequences not hitting U belongs to the set $(X\backslash U)^\infty$ of zero μ^∞-measure, since $\mu(X\backslash U)<1$. Concerning weak convergence, see a more general fact in Exercise 5.8.42.

3.5.19. Let $f\colon\ X\to Y$ be a continuous one-to-one mapping of compact spaces X and Y. Prove that f^{-1} is also continuous.

HINT: use the compactness of the images of closed sets.

3.5.20. Let X be a Banach space, $\mu\in\mathcal{M}_r(X)$, $\sigma\in\mathcal{P}_r(X)$. Prove the following estimate for convolutions (see Definitions 1.1.4 and 4.6.2):

$$\|\mu*\sigma\|_{\mathrm{KR}}\leqslant\|\mu\|_{\mathrm{KR}},\quad \|\mu*\sigma\|_{\mathrm{K}}\leqslant\|\mu\|_{\mathrm{K}}.$$

3.5.21. Let $\mu,\nu\in\mathcal{P}(X_1)$ and $\sigma\in\mathcal{P}(X_2)$, where (X_1,d_1) and (X_2,d_2) are two metric spaces. Suppose that the product $X_1\times X_2$ is equipped with the standard metric $d\bigl((x_1,y_1),(x_2,y_2)\bigr)=d_1(x_1y_1)+d_2(x_2,y_2)$. Prove that $d_{\mathrm{P}}(\mu\otimes\sigma,\nu\otimes\sigma)=d_{\mathrm{P}}(\mu,\nu)$.

HINT: by (3.1.4) the left-hand side is not smaller than the right-hand side, since $F^\varepsilon \times X_2 = (F \times X_2)^\varepsilon$; on the other hand, if $E \subset X_1 \times E_2$ is a closed subset such that $\nu\otimes\sigma(E) > \mu\otimes\sigma(E^\varepsilon) + \varepsilon$ for some $\varepsilon > 0$, then by Fubini's theorem there exists $y \in X_2$ such that for the sections $E_y = \{x\colon (x,y) \in E\}$ and $(E^\varepsilon)_y = \{x\colon (x,y) \in E^\varepsilon\}$ we have $\nu(E_y) > \mu\bigl((E^\varepsilon)_y\bigr) + \varepsilon$, where $(E_y)^\varepsilon \subset (E^\varepsilon)_y$.

3.5.22. (Pachl [**499**]) Let X be a complete metric space and let $BUC(X)$ be the set of all bounded uniformly continuous functions on X. (i) Prove that the space $\mathcal{M}_r(X)$ of all Radon measures on X is sequentially complete in the topology $\sigma\bigl(\mathcal{M}_r(X), BUC(X)\bigr)$, which means that every fundamental sequence in this topology is convergent.

(ii) Prove that for every bounded set M in $\mathcal{M}_r(X)$ the following conditions are equivalent: (a) M has compact closure in the Kantorovich–Rubinshtein norm; b) M is relatively countably compact in the topology $\sigma\bigl(\mathcal{M}_r(X), BUC(X)\bigr)$, i.e., every sequence in it has a limit point in $\mathcal{M}_r(X)$. Therefore, if a sequence is bounded in variation and converges in this topology, then it converges with respect to the norm $\|\cdot\|_{\mathrm{KR}}$. For measures $\mu_n, \nu_n \in \mathcal{P}_r(X)$ this yields the equivalence of convergence $\mu_n - \nu_n \to 0$ in this topology to the condition $\|\mu_n - \nu_n\|_{\mathrm{KR}} \to 0$ or $d_{\mathrm{P}}(\mu_n, \nu_n) \to 0$ (see Davydov, Rotar [**160**]).

3.5.23. (Bartoszyński [**42**]) Let (X,d) be a complete separable metric space, let a sequence $\{f_i\} \subset C_b(X)$ be uniformly bounded, and let

$$d^*(x,y) = \sup_n |f_n(x) - f_n(y)|, \quad \pi_k = (f_1, \ldots, f_k)\colon X \to \mathbb{R}^k.$$

Let $\mu, \mu_n \in \mathcal{P}(X)$, $n \in \mathbb{N}$. (i) Show that if functions f_i are equicontinuous at every point (i.e., convergence $d(x_n, x) \to 0$ implies convergence $d^*(x_n, x) \to 0$) and $\mu_n \Rightarrow \mu$, then

$$\lim_{n\to\infty} \sup_k d_{\mathrm{P}}(\mu_n \circ \pi_k^{-1}, \mu \circ \pi_k^{-1}) = 0,$$

where in the calculation of d_{P} on $\mathbb{R}^k$ the norm $\|y\|_\infty = \max_{i\leqslant k} |y_i|$ is used in place of the standard Euclidean norm. (ii) Show that if convergence $d^*(x_n, x) \to 0$ implies convergence $d(x_n, x) \to 0$, then the equality in (i) implies convergence $\mu_n \Rightarrow \mu$. Thus, in the case where d^* is a metric equivalent to d this equality is equivalent to weak convergence.

HINT: (i) apply Theorem 2.2.8 to functions of the form $\varphi(f_1, \ldots, f_k)$, where φ is a 1-Lipschitz function on $\mathbb{R}^k$ with the norm $\|\cdot\|_\infty$ and $|\varphi| \leqslant 1$.

3.5.24. (Choquet [**137**], Fremlin, Garling, Haydon [**245**]) Let X be a metric space. Then every countable set in the space $\mathcal{M}_r^+(X)$ that is compact in the weak topology is uniformly tight.

3.5.25. (Major [**449**]) If F and G are distribution functions on $\mathbb{R}$, then

$$\int_0^1 \psi\bigl(F^{-1}(t) - G^{-1}(t)\bigr)\, dt = \inf \mathbb{E}\, \psi(\xi - \eta)$$

for any convex function ψ, where inf is taken over all pairs of random variables ξ and η with the distribution functions F and G, $F^{-1}(t) := \sup\{x\colon F(x) \leqslant t\}$. In particular, this holds for $\psi(s) = |s|^p$, $p \geqslant 1$.

3.5.26. Suppose that Borel probability measures μ_n on a metric space converge weakly to a measure μ and that their topological supports converge in the sense of Hausdorff (see p. 123) to a closed set S. Prove that the support of μ is contained in S.

3.5.27. Let X be a compact metric space. Show that the mapping $S\colon \mu \mapsto \mathrm{supp}(\mu)$ from $\mathcal{P}(X)$ to the space $\mathcal{F}(X)$ of closed subsets of X with the Hausdorff metric is Borel. For noncompact X this can be false, but S is measurable with respect to the Effros σ-algebra on $\mathcal{F}(X)$ generated by sets of the form $B = \{F \in \mathcal{F}(X)\colon F \subset E\}$, $E \in \mathcal{F}(X)$.

HINT: the second assertion is obvious, since $S^{-1}(B)$ is closed. For the first assertion it suffices to verify that $S^{-1}(B)$ is also Borel for open E, which follows from the fact that for compact X the set E is the union of increasing compact sets E_n such that every closed set in E is contained in some E_n (this property fails, e.g., for the open unit ball in l^2; one can show that the set of probability measures on l^2 with support in the open unit ball is not Souslin by using results in Christensen [**139**, Chapter 3]).

3.5.28. (Kechris [**356**, Theorem 29.26]) Let X and Y be Polish space, let $A \subset X \times Y$ be a Souslin set, and let $A_x = \{y \in Y : (x,y) \in A\}$. Then

$$\{(\mu, x, \alpha) \in \mathcal{P}(Y) \times X \times [0,1] : \mu(A_x) > \alpha\}$$

is a Souslin set provided that $\mathcal{P}(Y)$ is equipped with the weak topology.

3.5.29. Let (X,d) be a separable metric space. Prove that the dual space to the space $\mathcal{M}_0^1(X)$ of Borel measures with zero value on X equipped with the Kantorovich norm coincides with the space of Lipschitz functions on X vanishing at a fixed point x_0 equipped with the norm $\|f\|_{\rm Lip} = \sup_{x \neq y} |f(x) - f(y)|/d(x,y)$, in addition, the general form of a continuous functional on $\mathcal{M}_0^1(X)$ is the integral of a Lipschitz function.

HINT: the integral of $f \in \mathrm{Lip}_1(X)$ is a functional with norm $\|f\|_{\rm Lip}$; for a functional L on $\mathcal{M}_0^1(X)$ with norm 1, the function $f(x) := L(\delta_x - \delta_{x_0})$ is Lipschitz with norm 1 and the functional generated by it coincides with L on the linear span of the differences $\delta_x - \delta_{x_0}$, which is everywhere dense in $\mathcal{M}_0^1(X)$.

3.5.30. (Melleray, Petrov, Vershik [**456**], Vershik [**644**], and Vershik, Petrov, Zatitskiy [**646**]). (i) Let (X,d) be a separable metric space and let $V_0(X)$ be the linear span of the differences $\delta_x - \delta_y$, where $x, y \in X$. A norm q on $V_0(X)$ is called compatible with d if $q(\delta_x - \delta_y) = d(x,y)$ for all $x, y \in X$. Prove that every compatible norm q has the form

$$q(\mu) = \sup_{f \in \mathcal{F}} \Bigl|\sum_{k=1}^n c_k f(z_k)\Bigr|, \quad \mu = \sum_{k=1}^n c_k \delta_{z_k}$$

with some class of functions $\mathcal{F} \subset \mathrm{Lip}_1(X)$. Deduce from this that the Kantorovich norm is the maximal compatible norm on $V_0(X)$.

(ii) Let $\mathcal{F}$ be the class of all functions of the form $f_{x,y}(z) = \bigl(d(z,x) - d(z,y)\bigr)/2$ and let q_{dp} be the corresponding norm defined by the above formula and called two-point. The space X is called linearly rigid in [**456**] if the norm q_{dp} coincides with the Kantorovich norm $\|\cdot\|_{\rm K}$. Show that a finite space with at least three points and $[0,1]$ are not linearly rigid. The universal Urysohn space is linearly rigid (see [**456**] and Holmes [**330**]).

HINT: (i) the space $V_0(X)$ with the norm q is separable by the separability of X, hence $q(\mu) = \sup_j l_j(\mu)$ for some countable collection of functionals l_j with unit norm; let $f_j(x) = l(\delta_x - \delta_{x_0})$, where δ_{x_0} is a fixed point; then $l_j(\mu)$ coincides with the integral of f_j with respect to every measure $\mu \in V_0(X)$; observe that $f_j \in \mathrm{Lip}_1(X)$.

3.5.31. Show that for the Lévy metric L on $\mathcal{P}(\mathbb{R})$ defined by (3.5.7) one has the inequality $L(\mu,\nu) \leqslant d_{\rm P}(\mu,\nu)$ and that it also generates the weak topology on $\mathcal{P}(\mathbb{R})$.

3.5.32. Show that for a space (X,d) of diameter at most $r \leqslant 1$ the Prohorov metric does not exceed r and $d_{\rm P}(\delta_x, \delta_y) = d(x,y)$.

3.5.33. (Tikhonov, Shaposhnikov, Sheipak [**609**]) Use Theorem 3.5.6 to prove that for an almost everywhere differentiable function $f \in L^{r+\delta}[0,1]$, where $1 \leqslant r < \infty$ and $\delta > 0$, the following conditions are equivalent: (i) $f' = 0$ a.e., (ii) there is a sequence of measurable functions f_k with $N(k)$ values such that $\lim\limits_{k\to\infty} N(k)\|f - f_k\|_{L^r[0,1]} = 0$.

In addition, for a bounded measurable function f on $[0,1]$ the following conditions are equivalent: (i) the image of Lebesgue measure under f has compact support of zero Lebesgue measure; (ii) there is a sequence of measurable functions f_k with $N(k)$ values such that $\lim\limits_{k\to\infty} N(k)\|f - f_k\|_\infty = 0$.

3.5.34. (Givens, Shortt [**277**]) For Gaussian measures μ_1 and μ_2 on $\mathbb{R}^n$ with means m_i and covariance matrices Q_i one has

$$W_2(\mu_1,\mu_2)^2 = |m_1 - m_2|^2 + \mathrm{tr}\,(Q_1 + Q_2) - 2\mathrm{tr}\,\bigl[\bigl(Q_1^{1/2} Q_2 Q_1^{1/2}\bigr)^{1/2}\bigr].$$

CHAPTER 4

Convergence of measures on topological spaces

In this chapter we proceed to the study of weak convergence of measures on topological spaces. All of the fundamental facts known from Chapter 2 remain valid also in this generality. This especially concerns the results of A.D. Alexandroff, which are most naturally formulated in this maximal generality. In addition to using more general spaces, an important feature of this chapter is the involvement of nets in many problems related to convergence of measures. Unlike Chapter 2, here this becomes crucially important, since a considerable part of the information cannot be expressed in terms of sequences. This chapter contains also important and useful results specifically for sequences of measures.

4.1. Borel, Baire and Radon measures

Below we consider only Hausdorff (separated) topological spaces. Most of the facts about topological spaces needed below can be found not only in standard books on general topology (see Arkhangelskii, Ponomarev [**23**], Engelking [**203**], Fedorchuk, Filippov [**217**], and Kuratowski [**402**]), but also in some texts on functional analysis, see Bogachev, Smolyanov [**96**], Edwards [**201**], Kolmogorov, Fomin [**379**], and Reed, Simon [**545**].

As above, the symbol $C_b(X)$ denotes the set of bounded continuous functions on a topological space X. It is a Banach space with the norm $\|f\| = \sup_{x\in X} |f(x)|$.

A *compact space* is a Hausdorff space such that every family of open sets, the union of which is the whole space, contains a finite subfamily also covering this space. A set in a topological space is called *compact* if it is compact as an independent space.

Recall that the product $\prod_t X_t$ of a collection of nonempty topological spaces X_t is equipped with the *product topology* (or the *Tychonoff product topology*) in which nonempty open sets are arbitrary unions of products of the form $\prod_t U_t$, where the sets U_t are open in X_t and only finitely many of them are different from the whole of X_t. The latter restriction is very important (if it is forgotten, then we obtain the so-called *box topology*, which in typical cases is stronger than the Tychonoff topology and is much less frequently used).

A very important fact for topological measure theory is that by the classical Tychonoff theorem the product of compact spaces with the Tychonoff topology is always compact (this is false for the box topology).

There exist nonmetrizable compacta (i.e., their topology cannot be generated by a metric). As an example one can take the product of the continuum of compact intervals. A criterion of metrizability for a compact space K is the existence of a countable collection of continuous functions separating its points (Exercise 4.8.27). This is also equivalent to the existence of a countable everywhere dense set in the Banach space of continuous functions $C(K)$ with the norm $\sup_{x\in K} |f(x)|$.

A Hausdorff topological space is called completely regular (or *Tychonoff*) if every nonempty open set contains a *functionally open* set, i.e., a set of the form $\{x\colon f(x) > 0\}$, where f is a continuous function. It is clear that without loss of generality one can assume that such a function takes values in $[0,1]$. Any set of the form $f^{-1}(U)$, where f is a continuous function and U is an open set on the real line, is also functionally open. This follows from the fact that on the real line every open set can be written in the form $\{\psi > 0\}$ by means of some continuous function ψ.

A set is *functionally closed* if it has the form $f^{-1}(0)$, where $f \in C(X)$. Equivalently, the functionally closed sets can be defined as sets of the form $f^{-1}(Z)$, where Z is a closed set on the real line and f is a continuous function. The next lemma is obvious from our previous discussion and the fact that every closed set on the real line has the form $f^{-1}(0)$, $f \in C(\mathbb{R})$. Sometimes functionally closed sets are called zero sets and their complements are called cozero sets.

4.1.1. LEMMA. *A set $U \subset X$ is functionally open precisely when it has the form $U = \varphi^{-1}(W)$, where $\varphi \in C(X)$ and $W \subset \mathbb{R}$ is open. A set Z is functionally closed precisely when it has the form $Z = \psi^{-1}(S)$, where $S \subset \mathbb{R}$ is a closed set and $\psi \in C(X)$.*

4.1.2. LEMMA. *Let Z_1 and Z_2 be two disjoint functionally closed sets in a topological space X. Then there exists a function $f \in C_b(X)$ with values in $[0,1]$ such that $Z_1 = f^{-1}(0)$, $Z_2 = f^{-1}(1)$.*

PROOF. The sets Z_i have the form $Z_i = \psi_i^{-1}(0)$, where $\psi_i \in C_b(X)$ and $0 \leqslant \psi_i \leqslant 1$. We can take $f = \psi_1/(\psi_1 + \psi_2)$. □

The complete regularity of a space does not mean that all open sets in it are functionally open or that all closed sets are functionally closed. The spaces with the latter stronger property are called *perfectly normal.* All compact spaces are completely regular.

An important example of a compact space that is not perfectly normal is the product of the continuum of copies of the interval $[0,1]$. A simple closed set in it that is not functionally closed is any singleton (on this space every continuous function depends only on countably many coordinates, which is seen from the Stone–Weierstrass theorem giving uniform approximations by functions in finitely many variables).

To every completely regular space X there corresponds the so-called *Stone–Čech compactification* βX: that is, a compact space containing an everywhere dense set homeomorphic to X with the property that every bounded continuous function on X after this embedding extends to a continuous function on βX. The existence of this compactification is not obvious at all (see Bogachev, Smolyanov [**96**, Chapter 1], Engelking [**203**, § 3.6]). One construction is this: in the space of all functions from $C_b(X)$ to $[0,1]$ with the topology of pointwise convergence, one takes the closure of the set of functions of the form $\Psi_x(\varphi) = \varphi(x)$, where $x \in X$, $\varphi\colon C_b(X) \to [0,1]$.

Čech complete spaces are those completely regular spaces which are G_δ-sets in their Stone–Čech compactifications, i.e., countable intersections of open sets. For metric spaces this is equivalent to the existence of a complete metric generating the original topology. It is clear that countable unions of compact sets (the so called σ-compact spaces) are Čech complete. A less obvious example is that *locally compact* spaces (i.e., spaces possessing a topology base consisting of sets with compact closures) are always Čech complete. The set of irrational numbers with its usual topology is not locally compact, but it is Čech complete. About these spaces, see Arkhangelskii, Ponomarev [**23**], Engelking [**203**].

A Hausdorff space is *Lindelöf* if every cover of this space by open sets contains an at most countable subcover. For metric spaces this is equivalent to separability. It is obvious that compact spaces are Lindelöf.

4.1.3. DEFINITION. The Borel σ-algebra in $\mathcal{B}(X)$ is the smallest σ-algebra containing all open sets in X. The Baire σ-algebra $\mathcal{B}a(X)$ is the smallest σ-algebra containing all functionally open sets, i.e., the smallest σ-algebra with respect to which all continuous functions are measurable.

In general completely regular spaces the Borel σ-algebra is often broader than the Baire one. In a metric space these two σ-algebras coincide, since all open sets there are functionally open. In a compact space the coinciding of the Borel and Baire σ-algebras is equivalent to the space being perfectly normal.

In general a closed Baire set need not be functionally closed, but this is true if the space is completely regular and is a Baire set in its Stone–Čech compactification (say, is compact).

4.1.4. DEFINITION. Let X be a topological space.

(i) A countably additive measure on the Borel σ-algebra $\mathcal{B}(X)$ is called a Borel measure on X.

(ii) A countably additive measure on the Baire σ-algebra $\mathcal{B}a(X)$ is called a Baire measure on X.

(iii) A Borel measure μ on X is called Radon if, for every set $B \in \mathcal{B}(X)$ and every $\varepsilon > 0$, there exists a compact set $K_\varepsilon \subset B$ such that $|\mu|(B \backslash K_\varepsilon) < \varepsilon$.

(iv) A Borel measure μ on X is called regular if, for every set $B \in \mathcal{B}(X)$ and every $\varepsilon > 0$, there is a closed set $F \subset B$ with $|\mu|(B \backslash F) < \varepsilon$.

As in the case of metric spaces, one introduces the notion of τ-additivity (now not equivalent to the separability of the topological support). All Radon measures are τ-additive.

4.1.5. DEFINITION. A Borel measure μ on a topological space X is called τ-additive (or τ-regular, or τ-smooth) if, for every increasing net of open sets $(U_\lambda)_{\lambda\in\Lambda}$ in X, one has the equality

$$|\mu|\Bigl(\bigcup\nolimits_{\lambda\in\Lambda} U_\lambda\Bigr) = \lim_{\lambda} |\mu|(U_\lambda). \tag{4.1.1}$$

If (4.1.1) is fulfilled for all nets with $\bigcup_\lambda U_\lambda = X$, then μ is called τ_0-additive (or weakly τ-additive).

Equivalently, for a decreasing net of closed sets Z_λ we have $\mu(Z_\lambda) \to \mu(\bigcap_\lambda Z_\lambda)$.

One can introduce weaker properties of τ-additivity and τ_0-additivity considering only functionally open sets. However, this is not of practical interest, because on a completely regular space every τ_0-additive Baire measure in the sense of functionally open sets possesses a unique extension to a Borel τ-additive measure (see Bogachev [**81**, Corollary 7.3.3]).

As we know, even on subsets of the interval $[0, 1]$ there exist non-Radon τ-additive measures. However, on compact spaces the τ-additivity of a Borel measure is equivalent to the Radon property, moreover, every tight τ-additive measure is

Radon (see Bogachev [**81**, Proposition 7.2.2]). It is worth noting here that there exist also Borel measures on compact spaces that are not Radon (hence are not regular).

4.1.6. LEMMA. *Let μ be a regular τ-additive (for example, Radon) measure on a topological space X and let $\{f_\alpha\}$ be an increasing net of lower semicontinuous nonnegative functions such that the function $f = \lim\limits_{\alpha} f_\alpha$ is bounded. Then*

$$\lim_{\alpha} \int_X f_\alpha(x)\,\mu(dx) = \int_X f(x)\,\mu(dx).$$

PROOF. We can assume that μ is nonnegative, since the general case is obtained from the Hahn decomposition. In addition, we can assume that $f < 1$. Set now

$$f_{\alpha,n} = \frac{1}{n}\sum_{k=1}^{n} I_{\{f_\alpha > (k-1)/n\}}, \quad f_n = \frac{1}{n}\sum_{k=1}^{n} I_{\{f > (k-1)/n\}}.$$

By the lower semicontinuity of the functions f_α the function f is also lower semicontinuous. Thus, the sets $\{f_\alpha > (k-1)/n\}$ are open and for each fixed n and k they form a net increasing to the open set $\{f > (k-1)/n\}$. By the τ-additivity $\lim\limits_{\alpha} \mu\bigl(f_\alpha > (k-1)/n\bigr) = \mu\bigl(f > (k-1)/n\bigr)$. Hence for every n we have $\lim\limits_{\alpha} \int_X f_{\alpha,n}\,d\mu = \int_X f_n\,d\mu$. Due to the estimates $|f_{\alpha,n} - f_\alpha| \leqslant 1/n$, $|f_n - f| \leqslant 1/n$ this completes the proof. □

4.1.7. COROLLARY. *If μ is a regular τ-additive measure on a topological space X and $\{f_\alpha\} \subset C_b(X)$ is a net decreasing to zero, then*

$$\lim_{\alpha} \int_X f_\alpha(x)\,\mu(dx) = 0.$$

PROOF. Let α_0 be any fixed index. We observe that the net $f_{\alpha_0} - f_\alpha$, where $\alpha \geqslant \alpha_0$, is increasing to f_{α_0} and consists of nonnegative functions. It remains to apply the previous lemma and the additivity of the integral. □

The monotonicity of the net is important (see Exercise 4.8.25).

For measures with a domain of definition $\mathfrak{A}$ more narrow than $\mathcal{B}(X)$ there is a reasonable analog of the property of tightness. A set function $m \geqslant 0$ on $\mathfrak{A}$ is called *tight* if, for every $\varepsilon > 0$, there is a compact set K_ε such that $m(A) \leqslant \varepsilon$ for all $A \in \mathfrak{A}$ with $K_\varepsilon \cap A = \varnothing$. If $\mathfrak{A}$ is a σ-algebra, then this means that $m^*(K_\varepsilon) \geqslant m(X) - \varepsilon$. For example, the property of tightness of a Baire measure cannot be defined by means of the measure of $X\backslash K_\varepsilon$, since compact sets need belong to $\mathcal{B}a(X)$; say, in $\mathcal{B}a(\mathbb{R}^{[0,1]})$ there are no nonempty compact sets. A signed Baire measure μ is called tight if so is $|\mu|$. The following result (see [**81**, Corollary 7.3.5] for its proof) explains why this notion is useful.

4.1.8. THEOREM. *Suppose that the space X is completely regular and a family $\Gamma \subset C(X)$ separates points. Then, for every measure μ on the σ-algebra $\sigma(\Gamma)$ generated by Γ such that the measure $|\mu|$ is tight on $\sigma(\Gamma)$, there is a unique extension to a Radon measure on X.*

For example, for Γ one can take $C_b(X)$, in which case $\sigma(\Gamma)$ is the whole Baire σ-algebra, and for a locally convex space X one can take the space of continuous linear functionals $\Gamma = X^*$ and the corresponding σ-algebra denoted by the symbol

$\sigma(X)$ and called the *cylindric* σ-algebra. The latter σ-algebra is generated by all cylinders of the form

$$\{x \in X\colon\ (f_1(x), \dots, f_n(x)) \in B\},$$

where $f_i \in X^*$, $B \in \mathcal{B}(\mathbb{R}^n)$. It is also generated by standard basic neighborhoods in the weak topology $\sigma(X, X^*)$.

It is worth noting that the σ-algebra $\sigma(X)$ in a locally convex space X also coincides with the Baire σ-algebra with respect to the weak topology (see Bogachev [**81**, Theorem 6.10.6]), but it can be more narrow than the Borel σ-algebra for the weak topology (and also smaller than the Baire and Borel σ-algebras of the original topology).

Notation: let $\mathcal{M}(X)$ be the set of all Borel measures on X, $\mathcal{M}_\tau(X)$ the set of all τ-additive Borel measures on X, $\mathcal{M}_r(X)$ the set of all Radon measures on X, $\mathcal{M}_\sigma(X)$ the set of all Baire measures on X, and $\mathcal{M}_t(X)$ the set of all tight Baire measures on X. The symbols like $\mathcal{M}^+(X)$ and $\mathcal{P}(X)$ (with the corresponding indices) will denote the subsets consisting of nonnegative and probability measures.

It is clear from what has been said above that for a completely regular space (which will be assumed in most of the results below) one can identify the spaces $\mathcal{M}_t(X)$ and $\mathcal{M}_r(X)$; in addition, in the property of τ-additivity it is sufficient to consider only Baire sets.

Measures from the classes introduced above correspond to continuous linear functionals on the Banach space $C_b(X)$ of bounded continuous functions. For measures of class $\mathcal{M}_\sigma(X)$ and functionally open sets the exact analog of Lemma 2.1.11 holds with the same proof. Theorem 2.1.15 has the following extension.

4.1.9. THEOREM. *Let X be a topological space.*

(i) *A continuous linear functional L on $C_b(X)$ has the form*

$$L(f) = \int_X f\, d\mu$$

with a measure $\mu \in \mathcal{M}_\sigma(X)$ precisely when $L(f_n) \to 0$ for every sequence of functions $f_n \in C_b(X)$ pointwise decreasing to zero (note that such a sequence is automatically uniformly bounded).

(ii) *Suppose that X is completely regular. A continuous linear functional L on $C_b(X)$ has the indicated form with a Radon measure μ precisely when for every $\varepsilon > 0$ there exists a compact set $K_\varepsilon \subset X$ such that*

$$|L(f)| \leqslant \varepsilon \sup_{x\in X} |f(x)|$$

for every function $f \in C_b(X)$ vanishing on K_ε.

(iii) *Suppose that X is completely regular. A continuous linear functional L on $C_b(X)$ has the indicated form with a τ-additive measure μ precisely when $L(f_\alpha) \to 0$ for every uniformly bounded net of functions $f_\alpha \in C_b(X)$ pointwise decreasing to zero. If X is compact, then this is true for each $L \in C_b(X)^*$ (the Riesz theorem). In all these cases one has $\|L\| = \|\mu\|$.*

In Chapter 2 we have already mentioned Souslin sets. Let us recall some basic facts about them (see Bogachev [**81**, Chapters 6, 7]).

4.1.10. DEFINITION. A Souslin space is a Hausdorff space that is the image of a complete separable metric space under a continuous mapping.

A Luzin space is a Hausdorff space that is the image of a complete separable metric space under an injective continuous mapping.

In the next theorem we have collected some necessary facts about Souslin spaces, including those related to measures.

4.1.11. THEOREM. (i) *In any complete separable metric space all Borel sets are Souslin, but in any uncountable space there are always non-Borel Souslin sets.*

(ii) *The image of a Souslin set under a Borel mapping of Souslin spaces is a Souslin set.*

(iii) *On any Souslin space all Borel measures are Radon. In addition, each compact Souslin space is metrizable, hence any Borel measure on a Souslin space is concentrated on a countable union of metrizable compact sets.*

The last assertion implies that every Borel probability measure on a Souslin space can be obtained as the image of Lebesgue measure on $[0,1]$ under a Borel mapping.

For the questions we discuss, an important advantage of Souslin sets as compared with Borel sets is the fact that the image of a Souslin set under a continuous or just Borel mapping is a Souslin set (the image of a Borel set need not be Borel even for infinitely differentiable function on the real line).

4.1.12. THEOREM. *Suppose that X and Y are Souslin spaces and $f\colon X \to Y$ is a surjective Borel mapping. Then, for every measure $\nu \in \mathcal{M}(Y)$, there exists a measure $\mu \in \mathcal{M}(X)$ such that $\nu = \mu\circ f^{-1}$ and $\|\mu\| = \|\nu\|$. If $\nu \in \mathcal{P}(Y)$, then μ can be found in $\mathcal{P}(X)$.*

PROOF. Suppose first that $\nu \in \mathcal{P}(Y)$. By the measurable choice theorem (see, e.g., Bogachev [**81**, Chapter 9]) there exists a mapping $g\colon Y \to X$ measurable with respect to all Borel measures on Y such that $g(Y)$ is a Souslin set in X and $f\bigl(g(y)\bigr) = y$ for all $y \in Y$. Hence for μ one can take $\nu\circ g^{-1}$. In the general case we consider the decomposition $\nu = \nu^+ - \nu^-$ and the corresponding disjoint decomposition $Y = Y^+ \cup Y^-$ into Borel parts. Then $g(Y^+) \cap g(Y^-) = \varnothing$, hence one can take the measures $\mu^+ := \nu^+\circ g^{-1}$, $\mu^- := \nu^-\circ g^{-1}$, $\mu := \mu^+ - \mu^-$. $\square$

In the consideration of weak convergence of measures on completely regular spaces it is sometimes useful to extend measures to the Stone–Čech compactifications. In this procedure even a non-Radon measure gains a Radon extension to the compactification.

4.1.13. LEMMA. *If X is a completely regular space with the Stone–Čech compactification βX, then, for every measure $\mu \in \mathcal{M}_\sigma(X)$, there exists a unique measure $\widehat{\mu} \in \mathcal{M}_r(\beta X)$ for which*

$$\int_X f\,d\mu = \int_{\beta X} \widehat{f}\,d\widehat{\mu} \quad \forall\, f \in C_b(X),$$

where $\widehat{f}$ is the extension of f to βX by continuity. Furthermore, for a measure $\nu \in \mathcal{M}_r(\beta X)$ one can find a measure $\mu \in \mathcal{M}_\tau(X)$ with $\nu = \widehat{\mu}$ precisely when $|\nu|(K) = 0$ for every compact subset K of the set $\beta X\backslash X$.

PROOF. The first assertion is obvious from the Riesz theorem, since the left-hand side defines a continuous linear functional on $C_b(\beta X)$ on account of the equality $\widehat{f}|_X = f$. It suffices to verify the second assertion for probability measures. Let

$\mu \in \mathcal{P}_\tau(X)$, let $\widehat{\mu} = \nu \in \mathcal{M}_r(\beta X)$, and let K be a compact set in $\beta X \backslash X$. Suppose that $\widehat{\mu}(K) > 0$. Let us consider the set

$$\mathcal{F} := \{f \in C_b(\beta X)\colon\ 0 \leqslant f \leqslant 1,\ f|_K = 0\}.$$

This set is partially ordered by the pointwise comparison and is directed, since if $f, g \in \mathcal{F}$, then there exists $h \in \mathcal{F}$ with $h \geqslant f$, $h \geqslant g$. For h one can take $h = \max(f, g)$. It is readily seen that $\sup_{f\in\mathcal{F}} \widehat{f}(x) = I_U(x)$, where $U = \beta X\backslash K$. On the set X we have $\sup_{f\in\mathcal{F}} f(x) = 1$. By Lemma 4.1.6 the τ-additivity of the measure μ yields the equality

$$\mu(X) = \sup_{f\in\mathcal{F}} \int_X f\,d\mu = \sup_{f\in\mathcal{F}} \int_{\beta X} \widehat{f}\,d\widehat{\mu} \leqslant \widehat{\mu}(U) < 1.$$

The obtained contradiction shows that $\widehat{\mu}(K) = 0$. Conversely, assume that $\widehat{\mu}(K) = 0$ for every compact set $K \subset \beta X\backslash X$. This means that $\widehat{\mu}^*(X) = 1$. Hence by the general construction of restriction to a set of full outer measure (Example 1.1.2) the measure $\widehat{\mu}$ can be restricted to a Borel probability measure μ on X by the formula

$$\mu(B \cap X) = \widehat{\mu}(B), \quad B \in \mathcal{B}(\beta X).$$

We have to verify the τ-additivity of the measure μ. Suppose we are given a decreasing family of closed sets Z_α in X with empty intersection. We have to show that their measures decrease to zero. There are compact sets K_α in βX with $Z_\alpha = K_\alpha \cap X$. Hence $K := \bigcap_\alpha K_\alpha \subset \beta X\backslash X$. Then we obtain $\widehat{\mu}(K) = 0$. Therefore, $\mu(Z_\alpha) = \widehat{\mu}(K_\alpha) \to 0$, as required. □

For Radon measures there is a useful generalization of the classical *Luzin theorem* on connections between measurability and continuity.

4.1.14. THEOREM. *Let μ be a Radon measure on a completely regular space X, let Y be a separable metric space, and let $f\colon X \to Y$ be a mapping measurable with respect to μ, i.e., all sets $f^{-1}(B)$, where $B \in \mathcal{B}(Y)$, are μ-measurable. Then, for every $\varepsilon > 0$, there exists a compact set K_ε in X such that $|\mu|(X\backslash K_\varepsilon) < \varepsilon$ and $f\colon K_\varepsilon \to Y$ is continuous. If Y is a separable Fréchet space, then there exists a continuous mapping $g\colon X \to Y$ for which $g|_{K_\varepsilon} = f|_{K_\varepsilon}$.*

4.2. The weak topology

Let $\{\mu_\alpha\}$ be a net (for example, a countable sequence) of finite measures defined on the Baire σ-algebra $\mathcal{B}a(X)$ of a topological space X. In this section we introduce one of the most important types of convergence of such nets, but the definition is identical with the one used for metric spaces (see (2.2.1)).

4.2.1. DEFINITION. A net $\{\mu_\alpha\} \subset \mathcal{M}_\sigma(X)$ is called weakly convergent to a measure $\mu \in \mathcal{M}_\sigma(X)$ if, for every bounded continuous real function f on X, we have the equality

$$\lim_\alpha \int_X f(x)\,\mu_\alpha(dx) = \int_X f(x)\,\mu(dx). \tag{4.2.1}$$

Notation: $\mu_\alpha \Rightarrow \mu$.

We shall say that a sequence of Baire measures μ_n on the space X is *weakly fundamental* if, for every bounded continuous function f on X, the sequence of integrals $\displaystyle\int_X f\,d\mu_n$ is fundamental (hence converges).

For Borel measures weak convergence is understood as weak convergence of their Baire restrictions. In the last section we discuss another natural convergence of Borel measures (convergence in the A-topology), which in the general case is not equivalent to weak convergence, but is closely connected with it.

In the same way as in the metric case, weak convergence can be defined by a topology.

4.2.2. DEFINITION. The weak topology of the space $\mathcal{M}_\sigma(X)$ of Baire measures on a topological space X is the topology $\sigma\bigl(\mathcal{M}_\sigma(X), C_b(X)\bigr)$, i.e., a base of the weak topology consists of the sets

$$U_{f_1,\ldots,f_n,\varepsilon}(\mu) = \biggl\{\nu\colon \biggl|\int_X f_i\,d\mu - \int_X f_i\,d\nu\biggr| < \varepsilon,\ i = 1,\ldots,n\biggr\}, \tag{4.2.2}$$

where $\mu \in \mathcal{M}_\sigma(X)$, $f_i \in C_b(X)$, $\varepsilon > 0$. Sets of such a form are called basic neighborhoods of the measure μ in the weak topology.

Precisely in the same way the weak topology is defined on other classes of measures introduced above (Borel, Radon, τ-additive, nonnegative, probability, etc.) by regarding them as subsets of $\mathcal{M}_\sigma(X)$.

On the space $\mathcal{M}_\sigma(X)$ the weak topology is obviously always Hausdorff, but for the spaces $\mathcal{M}(X)$ and $\mathcal{M}_r(X)$ this can be false (for example, for the trivial reason of absence of nonconstant continuous functions on X). If X is completely regular, then the weak topology is Hausdorff on $\mathcal{M}_\tau(X)$ and on $\mathcal{M}_r(X)$, but not always on $\mathcal{M}(X)$ (even on a compact space a Baire measure can have different Borel extensions, see Bogachev [**81**, Example 7.1.3]).

As already noted in Chapter 3, the weak topology is actually the weak-$*$ topology in the terminology of functional analysis. Convergence in this topology is called sometimes *w^*-convergence* (or, even more rarely, *narrow convergence*). Random elements are called *convergent in distribution* if their distributions converge weakly.

4.2.3. EXAMPLE. If a net of measures μ_α converges in variation to a measure μ, then it converges to μ weakly. More generally, if there exists α_1 such that $\sup_{\alpha\geqslant\alpha_1}\|\mu_\alpha\| < \infty$, and has the equality $\lim\limits_{\alpha}\mu_\alpha(B) = \mu(B)$ for every $B \in \mathcal{B}a(X)$ or at least for every set B of the form $B = \{f < c\}$, where $f \in C_b(X)$ and $|\mu|(\{f = c\}) = 0$, then $\mu_\alpha \Rightarrow \mu$.

PROOF. Justification does not differ from the one-dimensional case and a countable sequence (see Example 1.4.2). It suffices to prove the last assertion.

Let $\|\mu_\alpha\| \leqslant C$ and $\|\mu\| \leqslant C$, $f \in C_b(X)$ and $\varepsilon > 0$. One can assume that $|f| < 1$. There are $c_i \in [-1,1]$, $i = 1,\ldots,n$, such that

$$0 < c_{i+1} - c_i < \varepsilon,\ c_1 = -1,\ c_n = 1,\ |\mu|(\{f = c_i\}) = 0.$$

Let $g(x) = c_i$ if $c_i \leqslant f(x) < c_{i+1}$. Then we have $|f(x) - g(x)| < \varepsilon$. Starting from some index α_0, the absolute value of the difference between the integrals of g with respect to the measures μ and μ_α is less than ε, since

$$\lim_\alpha \mu_\alpha(\{c_i \leqslant f < c_{i+1}\}) = \mu(\{c_i \leqslant f < c_{i+1}\})$$

by our hypothesis and the equality $\{c_i \leqslant f < c_{i+1}\} = \{f < c_{i+1}\}\backslash\{f < c_i\}$. Hence, whenever $\alpha \geqslant \alpha_0$, the absolute value of the difference between the integrals of f with respect to the measures μ and μ_α does not exceed $(2C+1)\varepsilon$. □

Here is a version of Example 2.2.3 (see also Exercise 4.8.28).

4.2.4. EXAMPLE. A net $\{x_\alpha\}$ of elements of a completely regular space X converges to an element $x \in X$ if and only if the Dirac measures δ_{x_α} converge weakly to δ_x (recall that $\delta_x(A) = 1$ if $x \in A$, $\delta_x(A) = 0$ if $x \notin A$).

Indeed, if the net $\{x_\alpha\}$ does not converge to x, then there exists its subnet $\{x'_\alpha\}$ such that $f(x'_\alpha) = 0$ and $f(x) = 1$ for some function $f \in C_b(X)$. To this end, we take a subnet belonging to the complement of some neighborhood of x.

An analog of Example 3.1.3 is this (with the same proof).

4.2.5. EXAMPLE. The set of all measures of the form $\sum_{j=1}^n c_j\delta_{x_j}$, where $c_j \in \mathbb{R}$, $x_j \in X$, is everywhere dense in $\mathcal{M}_\sigma(X)$ in the weak topology.

As in the case of metric spaces, the Banach–Steinhaus theorem implies the following fact.

4.2.6. PROPOSITION. *Let $\mathcal{M} \subset \mathcal{M}_\sigma(X)$ be a family of measures such that*

$$\sup_{\mu\in\mathcal{M}} \int_X f\,d\mu < \infty \quad \text{for all } f \in C_b(X).$$

Then $\sup_{\mu\in\mathcal{M}} \|\mu\| < \infty$. In particular, every weakly convergent sequence of Baire measures is bounded in variation.

An analogous assertion is true, of course, also for complex measures, one just needs to consider the absolute values of integrals (in the real case this gives an equivalent condition, since in place of f one can take $-f$).

A simple sufficient (but not necessary) condition for weak convergence is given in the following generalization of Theorem 2.3.10, which is proved by the same reasoning.

4.2.7. THEOREM. *Let $\{\mu_\alpha\}$ be a uniformly bounded in variation and uniformly tight net (see Definition 1.4.10 or Definition 4.5.1 below) of Radon measures on a topological space X and let μ be a Radon measure on X such that*

$$\int_X f\,d\mu = \lim_\alpha \int_X f\,d\mu_\alpha$$

for all functions f from some subalgebra in $C_b(X)$ containing 1 and separating points. Then $\mu_\alpha \Rightarrow \mu$.

4.3. The case of probability measures

Similarly to the case of metric spaces, a base of the weak topology on the set of probability measures on a topological space can be defined by means of sets in place of functions. Let us consider the following two classes of sets in the space $\mathcal{P}_\sigma(X)$ of Baire probability measures:

$$\begin{aligned} W_{F_1,\ldots,F_n,\varepsilon}(\mu) &= \Bigl\{\nu \in \mathcal{P}_\sigma(X)\colon\ \nu(F_i) < \mu(F_i) + \varepsilon,\ i = 1,\ldots,n\Bigr\}, \\ & F_i = f_i^{-1}(0),\ f_i \in C(X),\ \varepsilon > 0, \end{aligned} \tag{4.3.1}$$

$$\begin{aligned} W_{G_1,\ldots,G_n,\varepsilon}(\mu) &= \Bigl\{\nu \in \mathcal{P}_\sigma(X)\colon\ \nu(G_i) > \mu(G_i) - \varepsilon,\ i = 1,\ldots,n\Bigr\}, \\ & G_i = X\backslash f_i^{-1}(0),\ f_i \in C(X),\ \varepsilon > 0. \end{aligned} \tag{4.3.2}$$

In the case of a metrizable space, arbitrary closed and open sets can be written as F_i and G_i, respectively, but in the general case this is not true.

An analog of Theorem 3.1.2 is now the following fact with the same proof (the only difference is that now we consider functionally open and functionally closed sets in place of arbitrary open and closed ones).

4.3.1. THEOREM. *The bases indicated above generate the weak topology on the set of probability measures $\mathcal{P}_\sigma(X)$. On the set of all nonnegative measures $\mathcal{M}_\sigma^+(X)$ a base of the weak topology is formed by neighborhoods of the form* (4.3.1) *or* (4.3.2) *along with neighborhoods of the form* $\{\nu\colon |\mu(X)-\nu(X)|<\varepsilon\}$.

The next classical characterization of weak convergence found by A.D. Alexandroff [**9**] was discussed in Chapter 1 for the real and in Chapter 2 for metric spaces. It follows directly from Theorem 4.3.1. Alternatively, one can use the same reasoning as in the case of the real line.

4.3.2. THEOREM. *Let $\{\mu_\alpha\}$ be a net of Baire probability measures on a topological space X and let μ be a Baire probability measure on X. The following conditions are equivalent:*

(i) *the net $\{\mu_\alpha\}$ converges weakly to μ;*

(ii) *for every functionally closed set F one has*

$$\limsup_\alpha \mu_\alpha(F) \leqslant \mu(F); \tag{4.3.3}$$

(iii) *for every functionally open set U one has*

$$\liminf_\alpha \mu_\alpha(U) \geqslant \mu(U). \tag{4.3.4}$$

In case of measures μ_α, $\mu\in\mathcal{M}_\sigma^+(X)$, condition (i) *is equivalent to either of conditions* (ii) *and* (iii) *complemented by the equality $\lim\limits_\alpha \mu_\alpha(X)=\mu(X)$.*

Since a Baire measure can fail to have Borel extensions (or can have several different Borel extensions), the consideration of of (4.3.3), (4.3.4) and their validity for arbitrary closed sets F and open sets U require additional conditions. Certainly, no additional conditions are needed if all closed sets are represented as the sets of zeros of continuous functions (i.e., if X is perfectly normal).

4.3.3. COROLLARY. (a) *If X is perfectly normal, then condition* (i) *is equivalent to condition* (ii) *with every closed set F and condition* (iii) *with every open set U. This is also true if X is completely regular, the measures μ_α are Borel and the measure μ is τ-additive (for example, is Radon).*

(b) *If X is completely regular and the limit measure μ is τ_0-additive, then condition* (i) *implies condition* (ii) *for all closed Baire sets F (not necessarily functionally closed) and condition* (iii) *for all open Baire sets U. In particular, this is true if the measure μ is tight.*

PROOF. The first assertion in (a) is obvious. The second one follows from the fact that in case of a completely regular space X the value of a τ-additive measure μ on every open set U equals the supremum of measures of functionally open sets contained in U. For the proof of assertion (b) it suffices to apply the theorem on existence of a τ-additive extension of μ (mentioned in § 4.1) and assertion (a). □

Repeating the reasoning from the proof of Corollary 2.2.6 we obtain the following assertion.

4.3.4. COROLLARY. *Suppose that a net of Borel probability measures μ_α on a completely regular space X converges weakly to a Borel probability measure μ that is τ-additive (for example, is Radon). If f is a bounded upper semicontinuous function, then*

$$\limsup_\alpha \int_X f\,d\mu_\alpha \leqslant \int_X f\,d\mu.$$

If f is a bounded lower semicontinuous function, then

$$\liminf_\alpha \int_X f\,d\mu_\alpha \geqslant \int_X f\,d\mu.$$

We have already seen in Chapter 2 that weak convergence of probability measures ensures convergence on some "sufficiently regular" sets. Let us consider similar results for topological spaces, giving proofs only when some new features appear.

4.3.5. COROLLARY. *A net $\{\mu_\alpha\}$ of Baire probability measures on a topological space X converges weakly to a Baire probability measure μ precisely when*

$$\lim_\alpha \mu_\alpha(E) = \mu(E) \tag{4.3.5}$$

for every set $E \in \mathcal{B}a(X)$ with the following property: one can find a functionally open set W and a functionally closed set F in X such that $W \subset E \subset F$ and $\mu(F\backslash W) = 0$.

PROOF. In case of weak convergence we have

$$\limsup_\alpha \mu_\alpha(E) \leqslant \limsup_\alpha \mu_\alpha(F) \leqslant \mu(F) = \mu(E).$$

Similarly, $\liminf_\alpha \mu_\alpha(E) \geqslant \mu(E)$, which yields (4.3.5). Suppose now that (4.3.5) holds. Let $U = \{f > 0\}$, where $f \in C(X)$, and let $\varepsilon > 0$. It is readily seen that there exists $c > 0$ such that $\mu(U) < \mu(\{f > c\}) + \varepsilon$ and $\mu(\{f > c\}) = \mu(\{f \geqslant c\})$. Then for $E = \{f > c\}$ we have (4.3.5), since one can take the sets $W = E$ and $F = \{f \geqslant c\}$, the first of which is functionally open and the second one is functionally closed. Thus, $\liminf_\alpha \mu_\alpha(U) \geqslant \mu(U) - \varepsilon$, which yields (4.3.4), since ε was arbitrary. □

It is clear that in the case where X is a metric space, the sets E with the aforementioned property are precisely the Borel sets whose boundary has μ-measure zero. Let us formulate an analogous assertion for Borel measures. Let μ be a nonnegative Borel measure on a topological space X. Denote by Γ_μ the class of Borel sets $E \subset X$ having boundary of μ-measure zero. The boundary ∂E of a set E is defined as the closure E without the interior of E, hence is a Borel set for any E. As in the metric case, sets from Γ_μ are called *continuity sets of the measure* μ. They have two important properties.

4.3.6. PROPOSITION. (i) *Γ_μ is a subalgebra in $\mathcal{B}(X)$.*
(ii) *If X is completely regular, then Γ_μ contains a topology base.*

The proof of Proposition 2.4.2 given earlier also works in this case (actually it was the complete regularity of X that was needed).

4.3.7. LEMMA. *Let μ be a Radon probability measure on a completely regular space X and let $U \subset X$ be an open set. Then, for every $\varepsilon > 0$, there exists an open set V such that $\overline{V} \subset U$, $\mu(\partial V) = 0$ and $\mu(U \setminus V) < \varepsilon$. The same is true for every Borel probability measure if X is a perfectly normal space. Finally, this assertion is* *also true for every Baire measure on an arbitrary space if U is a functionally open set in this space.*

PROOF. There is a compact set $K \subset U$ for which $\mu(U\backslash K) < \varepsilon$. Since X is completely regular, one can find a continuous function on X such that $f|_K = 1$ and $f = 0$ outside U. Then we can find $c \in (0,1)$ such that the set $\{f = c\}$ has μ-measure zero. We can take $V = \{f > c\}$, since $\partial V \subset \{f = c\}$. In the case where the set U has the form $\{f > 0\}$ for some continuous function f, the same reasoning applies to c sufficiently small. □

In particular, the assertion of Lemma 4.3.7 is true for every Borel measure on a universally measurable subspace of a completely regular Souslin space.

4.3.8. THEOREM. *Let $\{\mu_\alpha\}$ be a net of Borel probability measures on a topological space X and let μ be a Borel probability measure on X.*

(i) *If*

$$\lim_\alpha \mu_\alpha(E) = \mu(E) \quad \text{for all sets} \quad E \in \Gamma_\mu, \tag{4.3.6}$$

then the net $\{\mu_\alpha\}$ converges weakly to μ.

(ii) *Let X be completely regular. If a net $\{\mu_\alpha\}$ converges weakly to μ and μ is τ-additive, then* (4.3.6) *holds. If X is perfectly normal (say, metrizable), then the τ-additivity of μ is not needed.*

PROOF. For the proof of (i) we observe that every set E with the property indicated in Corollary 4.3.5 is contained in Γ_μ. Assertion (ii) follows from Corollary 4.3.3 and the reasoning in the proof of Corollary 4.3.5. Note that for weak convergence of signed measures (4.3.6) is sufficient, but not necessary (see Example 4.2.3 and Example 1.4.8, and also Corollary 4.4.2). □

An immediate corollary of the obtained results is the following assertion.

4.3.9. COROLLARY. *Let X be perfectly normal. Then the following conditions are equivalent:*

(i) *a net $\{\mu_\alpha\}$ of Borel probability measures converges weakly to a Borel probability measure μ;*

(ii) $\limsup_\alpha \mu_\alpha(F) \leqslant \mu(F)$ *for every closed set F;*

(iii) $\liminf_\alpha \mu_\alpha(U) \geqslant \mu(U)$ *for every open set U;*

(iv) $\lim_\alpha \mu_\alpha(E) = \mu(E)$ *for all $E \in \Gamma_\mu$.*

These conditions remain equivalent for an arbitrary completely regular space X if the measure μ is τ-additive (for example, is Radon).

The result of Kolmogorov and Prohorov from [**380**] presented in Theorem 2.4.10 for metric spaces is true with the same justification for topological spaces.

4.3.10. THEOREM. *Let $\{\mu_\alpha\}$ be a net of Borel probability measures on a topological space X and let μ be a τ-additive probability measure on X. Suppose that the equality*

$$\lim_\alpha \mu_\alpha(U) = \mu(U)$$

holds for all elements U of some topology base $\mathcal{O}$ that is closed with respect to finite intersections. Then the net $\{\mu_\alpha\}$ converges weakly to μ.

See also Exercise 4.8.30, where there is a stronger assertion.

Let us now extend the Ranga Rao theorem (Theorem 2.2.8) to the case of topological spaces. For completeness, we include a detailed proof, although it would be possible just to indicate some minor changes needed in the general case. We give the formulation for Baire measures, which, certainly, makes this theorem applicable to Borel measures.

4.3.11. THEOREM. *Suppose that a net $\{\mu_\alpha\}$ of Baire probability measures on a completely regular Lindelöf space X converges weakly to a Baire probability measure μ. If a family of functions $\Gamma \subset C_b(X)$ is uniformly bounded and pointwise equicontinuous (i.e., for every x and $\varepsilon > 0$, there exists a neighborhood U of x with $|f(x) - f(y)| < \varepsilon$ for all $y \in U$ and $f \in \Gamma$), then*

$$\lim_{\alpha} \sup_{f\in\Gamma} \left| \int_X f\,d\mu_\alpha - \int_X f\,d\mu \right| = 0. \tag{4.3.7}$$

PROOF. We can assume that the measures μ_α and μ are Borel and τ-additive, since by the Lindelöf property of X they are τ-additive in the sense of functionally open sets, hence by the result mentioned in § 4.1 they possess unique τ-additive extensions. We can also assume that $|f| \leqslant 1$ for all $f \in \Gamma$. Let $\varepsilon > 0$. By the complete regularity of X and our hypotheses, every point x has a functionally open neighborhood U_x such that $\mu(\partial U_x) = 0$ and $|f(x) - f(y)| < \varepsilon$ for all $y \in U_x$ and $f \in \Gamma$. Since X is Lindelöf, some countable collection of sets U_{x_n} covers X. Let $V_n = U_{x_n} \backslash \bigcup_{i=1}^{n-1} V_i$, $V_1 = U_{x_1}$. It is readily verified that the pairwise disjoint sets V_n cover the whole space X and $\mu(\partial V_n) = 0$. Let

$$\nu = \sum_{n=1}^{\infty} \mu(V_n)\delta_{x_n}, \quad \nu_\alpha = \sum_{n=1}^{\infty} \mu_\alpha(V_n)\delta_{x_n}.$$

We observe that

$$\lim_{\alpha} \sup_{f\in\Gamma} \left| \int_X f\,d\nu_\alpha - \int_X f\,d\nu \right| \leqslant \lim_{\alpha} \sum_{n=1}^{\infty} |\mu_\alpha(V_n) - \mu(V_n)| = 0. \tag{4.3.8}$$

The last equality in (4.3.8) follows from the equality

$$\lim_{\alpha} \mu_\alpha(V_n) = \mu(V_n)$$

for every fixed n (which holds according to Theorem 4.3.8(ii)) and the equality $\sum_{n=1}^{\infty} \mu_\alpha(V_n) = \sum_{n=1}^{\infty} \mu(V_n) = 1$. Next,

$$\begin{aligned}
&\left| \int_X f\,d\mu_\alpha - \int_X f\,d\mu \right| \\
&\leqslant \left| \int_X f\,d\mu_\alpha - \int_X f\,d\nu_\alpha \right| + \left| \int_X f\,d\nu_\alpha - \int_X f\,d\nu \right| + \left| \int_X f\,d\nu - \int_X f\,d\mu \right| \\
&\leqslant \sum_{n=1}^{\infty} \int_{V_n} |f(x) - f(x_n)|\,(\mu_\alpha + \mu)(dx) + \left| \int_X f\,d\nu_\alpha - \int_X f\,d\nu \right| \\
&\leqslant 2\varepsilon + \left| \int_X f\,d\nu_\alpha - \int_X f\,d\nu \right|,
\end{aligned}$$

because $|f(x) - f(x_n)| \leqslant \varepsilon$ for all $x \in V_n$, as $V_n \subset U_{x_n}$. Since ε was arbitrary, equality (4.3.7) follows from (4.3.8). □

Concerning signed measures, see Exercise 2.7.31 and Exercise 4.8.32.

Let us now present general topological versions of a number of results of Chapter 2, including proofs only in those cases where some differences occur.

4.3.12. THEOREM. *Suppose that a net of Baire measures μ_α on a topological space X converges weakly to a measure μ. Then the following assertions are valid.*

(i) *For every continuous mapping $F\colon X \to Y$ to a topological space Y, the net of measures $\mu_\alpha \circ F^{-1}$ converges weakly to the measure $\mu \circ F^{-1}$.*

(ii) *Let the space X be completely regular, let the measures μ_α be nonnegative Borel, and let the measure μ be τ-additive. If a Borel mapping F from X to a topological space Y is continuous μ-almost everywhere, then $\mu_\alpha \circ F^{-1} \Rightarrow \mu \circ F^{-1}$.*

4.3.13. COROLLARY. *Let $\{\mu_\alpha\}$ be a net of Borel probability measures on a completely regular space X and let μ be a τ-additive probability measures. Then $\{\mu_\alpha\}$ converges weakly to μ precisely when the equality*

$$\lim_\alpha \int_X f\, d\mu_\alpha = \int_X f\, d\mu$$

is valid for every bounded Borel function f that is μ-almost everywhere continuous.

Let us give a modification of Theorem 2.7.1 proved for metric spaces.

4.3.14. THEOREM. *If a net $\{\mu_\alpha\}$ of Baire probability measures on a topological space X converges weakly to a Baire measure μ, then, for every continuous function f on X satisfying the condition*

$$\lim_{R\to\infty} \sup_\alpha \int_{|f|\geqslant R} |f|\, d\mu_\alpha = 0,$$

one has $\lim\limits_\alpha \int_X f\, d\mu_\alpha = \int_X f\, d\mu$.

If X is completely regular, the measures μ_α and μ are Radon and for every $\varepsilon > 0$ there is a compact set K_ε with $\sup_\alpha \mu_\alpha(X\backslash K_\varepsilon) < \varepsilon$, then it suffices to have the continuity of f only on each K_ε.

PROOF. The reasoning from the justification of Theorem 2.7.1 applies for the proof of the first assertion. Let us explain a minor change in this reasoning needed for the second assertion. Set $A = \{|f| \leqslant R\}$. By the closedness of A and our assumption, there exists a compact set $K \subset A$ such that the function f is continuous on it and

$$\mu_\alpha(A\backslash K) + \mu(A\backslash K) < \varepsilon R^{-1}$$

for all α. The function $f|_K$ can be extended from K to the whole space to a continuous function g for which $|g| \leqslant R$. □

The next result follows from the last assertion in Corollary 4.3.9.

4.3.15. PROPOSITION. *Suppose that a net $\{\mu_\alpha\}$ of Borel probability measures on a completely regular space X converges weakly to a τ-additive Borel probability measure μ and a Borel set $X_0 \subset X$ is equipped with the induced topology. Then the induced measures μ_α^0 on X_0 converge weakly to the measure μ^0 induced by μ in either of the following cases:* (i) *X_0 has measure 1 for all measures μ_α and μ;*

(ii) *X_0 is either open or closed and* $\lim\limits_\alpha \mu_\alpha(X_0) = \mu(X_0)$.

4.3.16. PROPOSITION. *Suppose that a net of Borel probability measures μ_α on a completely regular space X converges weakly to a τ-additive Borel probability* *measure* μ *and a bounded Borel function* f *is continuous at* μ*-almost all points. Then the measures* $f\cdot\mu_\alpha$ *converge weakly to the measure* $f\cdot\mu$.

Let us recall that the product of two τ-additive measures on completely regular spaces X and Y extends to a τ-additive measure on $X\times Y$ and the product of Radon measures has a Radon extension (the question about extensions arises, because the classical product is defined on the product of the Borel σ-algebras, which can be smaller than the Borel σ-algebra of the product); see details in Bogachev [**81**, Chapter 7].

4.3.17. THEOREM. *Let* $\{\mu_\alpha\}$ *and* $\{\nu_\alpha\}$ *be two nets of* τ*-additive probability measures on completely regular spaces* X *and* Y *converging weakly to* τ*-additive measures* μ *and* ν*, respectively. Then the* τ*-additive extensions of the measures* $\mu_\alpha\otimes\nu_\alpha$ *converge weakly to the* τ*-additive extension of the measure* $\mu\otimes\nu$.

PROOF. Denote by $\mathcal{U}_\mu$ and $\mathcal{U}_\nu$ the classes of open sets in X and Y with boundaries of zero measure with respect to μ and ν, respectively. By Proposition 4.3.6 these classes form topology bases in X and Y. Then $\mathcal{U}=\{U\times V\colon U\in\mathcal{U}_\mu, V\in\mathcal{U}_\nu\}$ is a topology base in $X\times Y$. The class $\mathcal{U}$ is closed with respect to finite intersections, since this property holds, as one can easily see, for $\mathcal{U}_\mu$ and $\mathcal{U}_\nu$. Since for all $U\in\mathcal{U}_\mu$ and $V\in\mathcal{U}_\nu$ we have

$$\lim_\alpha \mu_\alpha\otimes\nu_\alpha(U\times V)=\lim_\alpha \mu_\alpha(U)\lim_\alpha \nu_\alpha(V)=\mu\otimes\nu(U\times V),$$

by Theorem 4.3.10 our claim follows. □

The proven fact extends immediately to finite products, which gives also the case of arbitrary products of probability measures by Theorem 4.3.10 and the definition of the product topology, the base of which is formed by cylinders (see p. 140).

Here is an analog of Proposition 2.7.8 for signed measures.

4.3.18. PROPOSITION. *Let* $\{\mu_n\}$ *and* $\{\nu_n\}$ *be two sequences of Radon measures on completely regular spaces* X *and* Y *converging weakly to Radon measures* μ *and* ν*, respectively. Suppose that both sequences are uniformly tight. Then the Radon extensions of the measures* $\mu_n\otimes\nu_n$ *on* $X\times Y$ *converge weakly to the Radon extension of the measure* $\mu\otimes\nu$.

PROOF. The hypothesis yields the uniform tightness of the Radon extensions of the measures $\mu_n\otimes\nu_n$. It remains to observe that the functions of the form $f_1(x)g_1(y)+\cdots+f_k(x)g_k(y)$, where $f_i\in C_b(X)$, $g_i\in C_b(Y)$, separate points in the product $X\times Y$, hence we can apply Theorem 4.2.7. □

It would be interesting to study whether the uniform tightness can be omitted. We recall that even on the real line weak convergence of measures does not imply any convergence of their densities with respect to a common dominating measure (see Exercise 1.7.16). However, we note the following simple fact.

4.3.19. PROPOSITION. *Let* $\{\mu_n\}$ *be a sequence of Baire probability measures on a topological space* X *converging weakly to a Baire probability measure* μ*. Let* ν *be a Baire probability measure on* X *such that* $\mu_n\ll\nu$ *and* $\mu\ll\nu$*, i.e.,* $\mu_n=\varrho_n\cdot\nu$ *and* $\mu=\varrho\cdot\nu$*. Then the functions* $\varrho_n I_{\{\varrho=0\}}$ *converge to* 0 *in measure* ν.

In particular, if a Baire probability measure λ *on* X *is mutually singular with* μ*, then the densities of absolutely continuous components of the measures* μ_n *with respect to* λ *converge to* 0 *in measure* λ.

PROOF. Given $\varepsilon > 0$, we find a functionally closed set $F \subset E := \{\varrho = 0\}$ such that $\nu(E\backslash F) < \varepsilon$. Since $\mu(F) = 0$, we have $\mu_n(F) \to 0$, i.e., $\|\varrho_n I_F\|_{L^1(\nu)} \to 0$. Hence $\varrho_n I_F \to 0$ in measure ν, which gives the first assertion. The second one follows by choosing a measure ν such that $\mu_n \ll \nu$, $\mu \ll \nu$ and $\lambda \ll \nu$. □

Let us establish one more result related to densities of weakly convergent measures and giving a sufficient condition for the absolute continuity of the limit measure.

4.3.20. PROPOSITION. *Suppose that a sequence of Baire probability measures μ_n on a topological space X converges weakly to a Baire probability measure μ and that $\mu_n = f_n \cdot \nu$, where ν is some Baire probability measures. Let*

$$\sup_n \int_X \Psi(f_n)\, d\nu \leqslant C < \infty,$$

where Ψ is a convex function on $[0, +\infty)$ with $\lim_{t\to\infty} \Psi(t)/t = +\infty$. Then $\mu \ll \nu$ and

$$\int_X \Psi(f)\, d\nu \leqslant C, \quad \text{where } f = d\mu/d\nu.$$

PROOF. By the known theorem of Komlós (see, for example , Bogachev [**81**, Theorem 4.7.24]) one can find a subsequence $\{f_{n_k}\}$ such that the functions

$$g_k := (f_{n_1} + \cdots + f_{n_k})/k$$

converge almost everywhere to some function f. By the convexity of Ψ we have

$$\sup_k \int_X \Psi(g_k)\, d\nu \leqslant C.$$

Hence $g_k \to f$ in $L^1(\nu)$ (see, for example, Theorem 1.3.11). By Fatou's theorem the integral of $\Psi(f)$ with respect to the measure ν is not larger than C. The measures $g_k \cdot \nu$ converge to $f \cdot \nu$ in variation, hence also weakly. Since they converge weakly to μ, we obtain the equality $\mu = f \cdot \nu$. □

4.4. Results of A.D. Alexandroff

This section contains in a revised form the results of the fundamental series of papers by A.D. Alexandroff [**9**] on weak convergence of measures on general spaces. A.D. Alexandroff [**9**, § 15] obtained the following criterion of weak convergence. Let $\mathcal{Z}$ be the class of all functionally closed sets and let $\mathcal{G}$ be the class of all functionally open sets in a given space.

4.4.1. THEOREM. *A sequence of Baire measures μ_n is fundamental in the weak topology precisely when it is bounded in variation and, for every $Z \in \mathcal{Z}$ and $U \in \mathcal{G}$ with $U \supset Z$ and every $\varepsilon > 0$, there is N such that for all $n, k > N$ we have*

$$\inf\bigl\{|\mu_n(V) - \mu_k(V)|\colon\ V \in \mathcal{G},\ Z \subset V \subset U\bigr\} < \varepsilon.$$

Weak convergence of μ_n to μ is equivalent to the condition that $\{\mu_n\}$ is bounded in variation and, for every $Z \in \mathcal{Z}$ and $U \in \mathcal{G}$ with $U \supset Z$, we have

$$\lim_{n\to\infty} \inf\bigl\{|\mu_n(V) - \mu(V)|\colon\ V \in \mathcal{G},\ Z \subset V \subset U\bigr\} = 0.$$

Finally, if measures $\mu_n \geqslant 0$ converge weakly to μ, then, whenever $Z \in \mathcal{Z}$, $U \in \mathcal{G}$ and $U \supset Z$, there exists a set $V \in \mathcal{G}$ such that $Z \subset V \subset U$ and $\lim_{n\to\infty} \mu_n(V) = \mu(V)$. In the case of signed measures the latter condition is sufficient for weak convergence.

PROOF. Suppose that a sequence $\{\mu_n\}$ is fundamental in the weak topology. Lemma 4.1.2 gives a function $f \in C_b(X)$ such that $Z = f^{-1}(0)$, $X\backslash U = f^{-1}(1)$. For a set $V \in \mathcal{G}$ with $Z \subset V \subset U$ such that $|\mu_n(V) - \mu_k(V)| < \varepsilon$ for all n and k sufficiently one can take some of the sets $\{f < t\}$ with $t \in (0,1)$. This follows from the second condition in Theorem 1.4.7 and the fact that the sequence of measures $\mu_n \circ f^{-1}$ on $[0,1]$ is fundamental, hence converges weakly (because the dual space to $C[0,1]$ is the space of measures). If the measures μ_n are nonnegative and the distribution function of $\mu \circ f^{-1}$ is continuous at the point t, then again by Theorem 1.4.7 we obtain the equality $\lim\limits_{n\to\infty} \mu_n(V) = \mu(V)$.

Conversely, suppose that the condition indicated in the theorem is fulfilled. We can assume that $\|\mu_n\| \leqslant 1$. We observe that the sequence $\mu_n(X)$ converges, since one can take $Z = U = X$. Let $\varphi \in C_b(X)$, $0 \leqslant \varphi < 1$ and $\varepsilon = 1/p$, $p \in \mathbb{N}$. Let us consider the sets

$$U_j = \{\varphi < \varepsilon j\}, \quad Z_j = \{\varphi \leqslant \varepsilon(j-1)\}, \quad j = 1, \ldots, p.$$

By our hypothesis there is N such that, for every $j \leqslant p$ and every $n, k > N$, there exists a functionally open set $V_{j,n,k}$ for which $Z_j \subset V_{j,n,k} \subset U_j$ and

$$|\mu_n(V_{j,n,k}) - \mu_k(V_{j,n,k})| < \varepsilon p^{-2}.$$

We can also assume that $|\mu_n(X) - \mu_k(X)| < \varepsilon p^{-2}$ whenever $n, k > N$. For any fixed n and k, the sets

$$W_{1,n,k} := V_{1,n,k}, W_{2,n,k} := V_{2,n,k}\backslash V_{1,n,k}, \ldots, W_{p+1,n,k} := X\backslash V_{p,n,k}$$

form a partition of X. It is easy to see that the values of the measures μ_n and μ_k on these sets differ in absolute value by at most ε/p; for example,

$$|\mu_n(V_{1,n,k}) - \mu_k(V_{1,n,k})| < \varepsilon p^{-2}, \ |\mu_n(V_{2,n,k}\backslash V_{1,n,k}) - \mu_k(V_{2,n,k}\backslash V_{1,n,k})| < 2\varepsilon p^{-2},$$

and so on. Finally, it remains to observe that

$$\left| \int_X \varphi \, d\mu_n - \sum_{j=1}^{p+1} (j-1)p^{-1}\mu_n(W_{j,n,k}) \right| \leqslant \varepsilon$$

and $\bigl|\sum_{j=1}^{p+1}(j-1)p^{-1}\bigl(\mu_n(W_{j,n,k}) - \mu_k(W_{j,n,k})\bigr)\bigr| \leqslant \varepsilon(p+1)/p$. The assertion about convergence to μ is proved similarly. The last assertion follows from the proof. □

4.4.2. COROLLARY. *If $\mu_n(V) \to \mu(V)$ for all functionally open sets V of class Γ_P for some measure $P \in \mathcal{P}_\sigma(X)$, then $\mu_n \Rightarrow \mu$.*

PROOF. For the function f from the proof there exists $t \in (0,1)$ such that $P\bigl(f^{-1}(t)\bigr) = 0$ and $Z \subset \{f < t\} \subset U$. □

We need two lemmas due to A.D. Alexandroff **[9]** on functionally closed sets.

4.4.3. LEMMA. *Let Z_n be disjoint functionally closed sets in a topological space X.*

(i) *Suppose that every union of sets Z_n is functionally closed. Then the sets Z_n possess pairwise disjoint functionally open neighborhoods U_n.*

(ii) *If X is normal, then the usual closedness of all unions of the sets Z_n yields their functional closedness.*

PROOF. (i) Sets U_n are constructed by induction. We find disjoint functionally open neighborhoods U_1 and V_1 of the functionally closed sets Z_1 and $\bigcup_{n=2}^{\infty} Z_n$.

Next, in V_1 we find disjoint functionally open neighborhoods of the sets Z_2 and $\bigcup_{n=3}^\infty Z_n$ and so on.

(ii) In a normal space X every pair of disjoint closed sets can be separated by disjoint functionally open sets. Hence for the disjoint closed sets $Z=\cup_{n=1}^\infty Z_n$ and $S=X\backslash\bigl(\bigcup_{n=1}^\infty U_n\bigr)$, where U_n are the disjoint functionally open neighborhoods of the sets Z_n constructed in (i), we can find a continuous function $\theta\colon X\to[0,1]$ such that $\theta|_Z=1$, $\theta|_S=0$. In addition, according to (i) there exist continuous functions $h_n\colon X\to[0,1]$ for which $Z_n=h_n^{-1}(1)$, $h_n|_{X\backslash U_n}=0$. The function $f=\theta\sum_{n=1}^\infty h_n$ is continuous and $f^{-1}(1)=Z$. Indeed, $f|_Z=1$ and $f<1$ outside Z. The continuity of f at all points of U_n is obvious. Let $x\in S$. Then $f(x)=0$. For every $\varepsilon>0$ the set $V=\theta^{-1}\bigl((-\varepsilon,\varepsilon)\bigr)$ is a neighborhood of x and $f<\varepsilon$ on V, since we have $f\leqslant\theta$. □

In Exercise 4.8.50 an example is described showing that in (ii) one cannot omit the condition of normality of the space (mistakenly omitted in Exercise 6.10.79(i) in Bogachev [**81**]).

4.4.4. LEMMA. (i) *Let F_n be functionally closed sets in a topological space X and $F_{n+1}\subset F_n$, $n\in\mathbb{N}$. Then, there exist functionally open sets G_n such that $F_n\subset G_n$ and $G_{n+1}\subset G_n$ for all n and, in addition, $\bigcap_{n=1}^\infty G_n=\bigcap_{n=1}^\infty F_n$.*

(ii) *If in (i) we have the equality $\bigcap_{n=1}^\infty F_n=\varnothing$, then the sets $Z_n:=F_n\backslash G_{n+1}$ are disjoint and $\bigcup_{n=1}^\infty Z_n$ is functionally closed. Furthermore, the sets G_n can be found with the following additional property: the closedness of a set in X is equivalent to the closedness of its intersections with the sets $X\backslash G_n$, hence the continuity of a function on X is equivalent to the continuity of its restrictions to the sets $X\backslash G_n$.*

PROOF. (i) Let us take $f_n\in C(X)$ such that $0\leqslant f_n\leqslant 1$, $F_n=f_n^{-1}(0)$. Set $h_n:=f_1+\cdots+f_n$ and $G_n:=\{x\colon h_n(x)<1/n\}$. Then $F_n\subset G_n$, $G_{n+1}\subset G_n$. If $x\notin\bigcap_{n=1}^\infty F_n$, then there is n such that $h_n(x)>0$. Hence there is $m>n$ such that $h_m(x)>1/m$, i.e., we have $x\notin\bigcap_{n=1}^\infty G_n$.

(ii) The disjointness of Z_n is obvious, since $Z_n\cap F_{n+1}=\varnothing$. We show first that there exist decreasing functionally open sets $W_n\supset F_n$ with empty intersection such that the closedness of the intersections with the sets $X\backslash W_n$ implies the closedness in X. Let $F_n=f_n^{-1}(0)$, where $f_n\in C_b(X)$, $0\leqslant f_n\leqslant 1$. Set

$$g_n=\max(f_1,\dots,f_n).$$

Then $g_n\in C_b(X)$, $0\leqslant g_n\leqslant 1$, $g_n\leqslant g_{n+1}$, $F_n=g_n^{-1}(0)$. The sets

$$S_n=g_n^{-1}\bigl([1/n,+\infty)\bigr)$$

are functionally closed, the sets $V_n=g_n^{-1}\bigl((1/(n+1),+\infty)\bigr)$ are functionally open, and we have $S_n\subset V_n\subset S_{n+1}$ due to increasing of $\{g_n\}$. Moreover, $X=\bigcup_{n=1}^\infty S_n$, since for every $x\in X$ there is n such that $x\in V_n$ (otherwise $g_n(x)\leqslant 1/(n+1)$ for all n, whence $f_k(x)=0$ for all k, i.e., $x\in F_k$ for all k). If a set A is such that the sets $A\cap S_n$ are closed, then A is closed. Indeed, if a is a limit point of A, then $a\in V_n$ for some n. Hence a is a limit point of $A\cap S_{n+1}$ by the inclusion $V_n\subset S_{n+1}$, hence $a\in A\cap S_{n+1}\subset A$.

We show that $Z=\bigcup_{k=1}^\infty Z_n$ is functionally closed. By using the complementary sets W_n constructed above we find by induction two sequences of functionally open sets U_n and U_n' with the following properties: $Z_n\subset U_n\subset G_n\cap W_n$, $U_n\cap U_n'=\varnothing$, $F_{n+1}\subset U_n'$, $U_{n+1}\subset U_n'$. To this end, we include Z_1 and F_2 in disjoint functionally open sets U_1 and U_1' contained in $G_1 \cap W_1$. Next we take a functionally open set $G_2 \cap W_2 \cap U_1'$ containing disjoint functionally closed sets Z_2 and F_3, and so on. The sets U_n are disjoint. There are functions $h_n \in C_b(X)$ such that $0 \leqslant h_n \leqslant 1$, $Z_n = h_n^{-1}(1)$, $X\backslash U_n \subset h_n^{-1}(0)$. Let $h = \sum_{n=1}^{\infty} h_n$. Clearly, $0 \leqslant h \leqslant 1$, $h^{-1}(1) = Z$. The function h is continuous, since for every n the functions h_k with $k > n$ vanish on the set $X\backslash W_n \subset X\backslash(G_n \cap W_n)$. □

4.4.5. REMARK. A.D. Alexandroff [**9**, § 17] introduced the following terminology (he considered Borel measures on normal spaces). A set M of Baire measures on a topological space X has an *eluding load* equal to a number $a \neq 0$ if M contains an infinite sequence of measures μ_n such that one has $\mu_n(Z_n)/a \geqslant 1$ for some sequence of pairwise disjoint functionally closed sets Z_n with the property that the union of every subfamily in $\{Z_n\}$ is functionally closed (A.D. Alexandroff called such sequences *divergent*; for a normal space it suffices to have the closedness of these unions by Lemma 4.4.3).

If for some $a \neq 0$ the set M has eluding load equal to a, then we say that M *possesses eluding load. The absence of eluding load* is equivalent to the property that

$$\lim_{n\to\infty} \sup_{\mu\in M} |\mu|(Z_n) = 0$$

for every divergent sequence of functionally closed sets Z_n. Indeed, suppose that $|\mu_n|(Z_n) \geqslant a > 0$. Passing to a subsequence, we can assume that $\mu_n^+(Z_n) \geqslant a/2$ (or $\mu_n^-(Z_n) \leqslant -a/2$). Then there exist functionally closed sets $F_n \subset Z_n \cap X_n^+$, where $X = X_n^+ \cup X_n^-$ is the Hahn decomposition for μ_n, such that $\mu_n(F_n) \geqslant a/4$, i.e., $\{F_n\}$ is divergent.

A very important property of weak convergence is described in the following result due to A.D. Alexandroff [**9**, § 18].

4.4.6. PROPOSITION. *Suppose that a sequence of Baire measures μ_n on a topological space X converges weakly to a measure μ. Then this sequence has no eluding load, that is,*

$$\lim_{n\to\infty} \sup_{k} |\mu_k|(Z_n) = 0 \qquad (4.4.1)$$

for every sequence of pairwise disjoint functionally closed sets Z_n with the property that the union of every subfamily in $\{Z_n\}$ is functionally closed.

PROOF. Otherwise, passing to a subsequence, we can assume that there is $c > 0$ such that $|\mu_n(Z_n)| \geqslant c > 0$ (see the remark above). By Lemma 4.4.3 there are pairwise disjoint functionally open sets U_n with $Z_n \subset U_n$ and $|\mu_n|(U_n\backslash Z_n) \leqslant c/2$. We show that there are functions $f_n \in C_b(X)$ such that $0 \leqslant f_n \leqslant 1$, $f_n = 0$ outside U_n and

$$\left|\int_X f_n \, d\mu_n\right| \geqslant c/2, \qquad (4.4.2)$$

and, in addition, for every bounded sequence $\{c_n\}$ the function $\sum_{n=1}^{\infty} c_n f_n$ is bounded and continuous. This will lead to a contradiction, since by the hypothesis the sequence of the integrals of any such function with respect to the measures μ_n converges, i.e., the sequence

$$l_n := \left\{\int f_k \, d\mu_n\right\}_{k=1}^{\infty}$$

of elements of l^1 converges weakly, which contradicts (4.4.2) by Corollary 1.3.7. For constructing functions f_n we take (by Lemma 4.1.2 and the assumed functional closedness of the union of Z_n) a continuous function f such that $0 \leqslant f \leqslant 1$, $f = 1$ on $\bigcup_{n=1}^\infty Z_n$ and $f = 0$ outside $\bigcup_{n=1}^\infty U_n$. Set $f_n = f$ on U_n and $f_n = 0$ outside U_n. The nonnegative function f_n is continuous, since for each $c \geqslant 0$ we have

$$\{f_n > c\} = \{f > c\} \cap U_n, \quad \{f_n < c\} = \{f < c\} \cup \Bigl(\bigcup\nolimits_{k\neq n} U_k\Bigr),$$

and the sets in the right-hand sides are open. For the same reason every function $h = \sum_{n=1}^\infty c_n f_n$, where $|c_n| < 1$, is continuous, since for each $c < 0$ we have

$$\{h > c\} = \bigcup\nolimits_{n:\ c_n<0} (U_n \cap \{f < c/c_n\}) \cup \{f < |c|\} \cup \bigl(\bigcup\nolimits_{n:\ c_n\geqslant 0} U_n\bigr)$$

and $\{h > c\} = \bigcup_{n:\ c_n>0} \bigl(U_n \cap \{f > c/c_n\}\bigr)$ for all $c \geqslant 0$. Similarly we prove that the set $\{h < c\}$ is open. □

If the measures μ_n are Radon and the space is completely regular, the following fact is true, which is to be compared with the seemingly close equality from Theorem 5.6.13, where the stronger setwise convergence is considered, but there is no condition of local finiteness of $\{U_n\}$.

4.4.7. PROPOSITION. *If a sequence of Radon measures μ_n on a completely regular space X converges weakly to a Baire measure μ, then*

$$\lim_{n\to\infty} \sup_k |\mu_k|(U_n) = 0 \tag{4.4.3}$$

for every locally finite sequence of disjoint open sets U_n.

PROOF. Here we can apply the same reasoning which proves the previous result of A.D. Alexandroff (in this case it becomes even simpler). Arguing from the contradiction and passing to a subsequence, we obtain compact sets $K_n \subset U_n$ such that $|\mu_n|(K_n) \geqslant c > 0$. There are continuous functions $\psi_n\colon X \to [-1,1]$ such that $\psi_n = 0$ outside U_n and the integral of ψ_n with respect to the measure μ_n is not smaller than $c/2$.

For every bounded sequence of numbers c_k, the function $\varphi_{\{c_k\}} := \sum_{k=1}^\infty c_k\psi_k$ is bounded on X and continuous by the local finiteness of $\{U_n\}$. By assumption, the sequence of integrals of these functions converges. This means that the sequence of vectors $v^n = (v_1^n, v_2^n, \ldots)$, where

$$v_k^n = \int_X \psi_k\, d\mu_n,$$

is weakly fundamental in the Banach space l^1. Indeed, $v^n \in l^1$ by the disjointness of supports of the functions ψ_k. The value on v^n of the functional on l^1 given by the bounded sequence $\{c_k\}$ is

$$\sum_{k=1}^\infty c_k \int_X \psi_k\, d\mu_n = \int_X \varphi_{\{c_k\}}\, d\mu_n.$$

Since weakly fundamental sequences in l^1 converge in norm, there is a number N such that $|v_n^n| < c/4$ for all $n > N$, which contradicts our assumption. □

Note that the limit measure μ is not supposed to be Radon (and can be non-Radon indeed).

The following result was also obtained by A.D. Alexandroff (see [**9**, §18]). The notation $Z_n \downarrow \varnothing$ means that $Z_{n+1} \subset Z_n$ and $\bigcap_{n=1}^\infty Z_n = \varnothing$.

4.4.8. PROPOSITION. *A family $\mathcal{M}$ of Baire measures on a topological space X has no eluding load precisely when for every sequence of functionally closed sets Z_n with $Z_n \downarrow \varnothing$ we have*

$$\lim_{n\to\infty} \sup_{\mu\in\mathcal{M}} |\mu|(Z_n) = 0. \tag{4.4.4}$$

PROOF. Suppose that $\mathcal{M}$ has no eluding load and Z_n are increasing functionally closed sets with empty intersection. If (4.4.4) is false, then, passing to a subsequence, we can assume that we have measures $\mu_n \in \mathcal{M}$ with

$$|\mu_n|(Z_n) > c > 0.$$

Passing to a subsequence once again, we reduce everything to the case where $\mu_n^+(Z_n) > c/2$. By Lemma 4.4.4 there exist decreasing functionally open sets G_n with empty intersection and $Z_n \subset G_n$. In addition, by the last assertion of that lemma we can ensure that the sets $Z_{n_k}\backslash G_{n_{k+1}}$, then also the sets F_k, form a divergent sequence in the sense of Remark 4.4.4.

There is n_1 with $\mu_1^+(G_{n_1}) < c/4$. Then $\mu_1^+(Z_1\backslash G_{n_1}) > c/4$. Next we find $n_2 > n_1$ with $\mu_{n_1}^+(G_{n_2}) < c/4$, whence $\mu_{n_1}^+(Z_{n_1}\backslash G_{n_2}) > c/4$. By induction we obtain increasing numbers n_k with

$$\mu_{n_k}^+(Z_{n_k}\backslash G_{n_{k+1}}) > c/4.$$

By the definition of $\mu_{n_k}^+$, for every k one can find a functionally closed set F_k in $Z_{n_k}\backslash G_{n_{k+1}}$ such that

$$\mu_{n_k}(F_k) > \mu_{n_k}^+(Z_{n_k}\backslash G_{n_{k+1}}) - c/8 > c/8.$$

Thus, $\mathcal{M}$ has an eluding load, a contradiction.

Suppose that the family $\mathcal{M}$ possesses an eluding load. Then there exist a divergent sequence of functionally closed sets F_n, measures $\mu_n \in \mathcal{M}$ and a number $a \neq 0$ with $\mu_n(F_n)/a \geqslant 1$. The sets $Z_n := \bigcup_{k=n}^{\infty} F_k$ are functionally closed (by the definition of a divergent sequence) and decrease to the empty set. In addition, $|\mu_n|(Z_n) \geqslant |\mu_n(F_n)| \geqslant |a|$, so (4.4.4) fails. □

4.4.9. REMARK. For every sequence of Radon measures μ_n on a normal space property (4.4.3) is equivalent to (4.4.1) for every sequence of pairwise disjoint closed sets Z_n with the property that the union of every subfamily in $\{Z_n\}$ is closed.

Indeed, if (4.4.1) is true and $\{U_n\}$ is a locally finite sequence of disjoint open sets for which there are indices k_i and n_i with $|\mu_{k_i}|(U_{n_i}) > \varepsilon > 0$, then there exist compact sets $K_{n_i} \subset U_{n_i}$ with $|\mu_{k_i}|(K_{n_i}) > \varepsilon > 0$ such that all their unions are closed. Conversely, suppose that for every sequence of locally finite disjoint open sets U_n we have property (4.4.3).

It suffices to verify (4.4.1) for compact sets. Let $\{Z_n\}$ be a sequence of disjoint compact sets all unions of which are closed. Suppose that $\mu_n(Z_n) \geqslant c > 0$. By induction we construct pairwise disjoint open sets $U_n \supset Z_n$. For example, to find U_1 we separate the compact set Z_1 and the closed (by assumption) set $\bigcup_{n=2}^{\infty} Z_n$ by open sets U_1 and V_1; next, in V_1 we take a neighborhood U_2 of the compact set Z_2 which does not intersect a neighborhood of the closed set $\bigcup_{n=3}^{\infty} Z_n$, and so on. Since the space is normal, there is a continuous function $f\colon X \to [0,1]$ equal to 1 on the closed union of Z_n and 0 outside the open union of U_n. Now we set $\psi_n = f$ on U_n and $\psi_n = 0$ outside U_n. It is straightforward to verify the continuity of ψ_n and $\varphi_{\{c_n\}}$ (see the proof of Proposition 4.4.6).

Hence for perfectly normal spaces (where closed sets are functionally closed) and Radon measures property (4.4.1) for general closed sets and property (4.4.3) are equivalent.

The following fact is due to A.D. Alexandroff [**9**] and Varadarajan [**635**]. For the definition of a paracompact space, see Exercise 4.8.35 or Engelking [**203**]; all metric spaces are paracompact.

4.4.10. COROLLARY. *Suppose that X is a paracompact space and τ-additive measures μ_n, where $n \in \mathbb{N}$, converge weakly to a Baire measure μ. Then, for every net of open sets U_α increasing to X, we have* $\lim_{\alpha} \sup_n |\mu_n|(X\backslash U_\alpha) = 0$.

PROOF. According to Exercise 4.8.36 the measure μ is also τ-additive and there exists a Lindelöf closed subspace $Z \subset X$ with $|\mu|(X\backslash Z) = |\mu_n|(X\backslash Z) = 0$ for all n. The restrictions of the measures μ_n to Z converge weakly to the restriction of the measure μ (since every continuous function on Z extends to a continuous function on X and these measures are concentrated on Z). Hence everything reduces to a Lindelöf space, and, on account of the complete regularity of X, reduces further to the case of a countable increasing sequence of functionally open sets U_k, where we can apply Proposition 4.4.8. □

4.5. Weak compactness

In Chapters 2 and 3 we already discussed conditions for weak compactness of families of Borel measures on metric spaces, i.e., compactness in the weak topology $\sigma\bigl(\mathcal{M}, C_b(X)\bigr)$ generated by duality with the space of bounded continuous functions. In this section we continue a discussion of this question for general topological spaces. An important difference is that now compactness is no longer determined in terms of sequences. For this reason another type of compactness (also important for applications) arises, which is called the sequential compactness and is defined as the existence of convergent subsequences in every countable sequence. As in the metric case, the key role will be played by the notion of uniform tightness of families of measures.

4.5.1. DEFINITION. A family $\mathcal{M}$ of Radon measures on a topological space X is called uniformly tight if, for every $\varepsilon > 0$, there exists a compact set K_ε such that $|\mu|(X\backslash K_\varepsilon) < \varepsilon$ for all $\mu \in \mathcal{M}$.

A family $\mathcal{M}$ of Baire measures on X is called uniformly tight if, for every $\varepsilon > 0$, there exists a compact set K_ε such that $|\mu|_*(X\backslash K_\varepsilon) < \varepsilon$ for all $\mu \in \mathcal{M}$, i.e., $|\mu|(C) < \varepsilon$ for every set $C \in \mathcal{B}a(X)$ such that $C \cap K_\varepsilon = \varnothing$.

Uniformly tight families are also called *tight families* for brevity.

In the case of a completely regular space, the uniform tightness of a family of Baire measures is equivalent to the existence of uniformly tight Radon extensions of these measures.

In the previous chapters we have already seen the importance of this concept. Let us consider reflexive Banach spaces.

4.5.2. EXAMPLE. A family $\mathcal{M}$ of Borel probability measures on a separable reflexive Banach space X is uniformly tight on X with the weak topology precisely when there exists a function $V\colon X \to [0, \infty)$ continuous in the norm topology such that the following conditions are fulfilled:

$$\lim_{\|x\|\to\infty} V(x) = \infty \quad \text{and} \quad \sup_{\mu\in\mathcal{M}} \int_X V(x)\,\mu(dx) < \infty.$$

PROOF. The justification of Example 2.3.7 can be easily modified in order to use the compactness of balls in a reflexive Banach space with the weak topology. Here the function V can be taken in the form $V(x) = f(\|x\|)$ with some concave positive continuous function f on the half-line increasing to infinity. $\square$

We now prove the following reinforced version of one implication in the Prohorov theorem.

4.5.3. THEOREM. *Suppose that a family $\mathcal{K} \subset \mathcal{M}_r(X)$ of Radon measures on a completely regular space X is uniformly bounded in variation and uniformly tight. Then $\mathcal{K}$ has compact closure in the weak topology. If, in addition, for every $\varepsilon > 0$ there exists a metrizable compact set K_ε such that $|\mu|(X\backslash K_\varepsilon) < \varepsilon$ for all $\mu \in \mathcal{K}$, then every sequence in $\mathcal{K}$ contains a weakly convergent subsequence. In particular, this is true if all compact subsets of X are metrizable.*

PROOF. Let us consider $\mathcal{K}$ as a subset of the dual space to the space $C_b(X)$ equipped with the weak-$*$ topology. By the Banach–Alaoglu theorem (applicable due to the norm boundedness of $\mathcal{K}$) the set $\mathcal{K}$ has compact closure in the indicated topology. However, we have to verify that every limit point of F for the set $\mathcal{K}$ is representable as the integral with respect to a Radon measure (since not all continuous linear functionals on $C_b(X)$ have this property). Here we need uniform tightness. We can assume that $\|\mu\| \leqslant 1$ for all $\mu \in \mathcal{K}$. Let $\varepsilon > 0$ and let K_ε be a compact set such that $|\mu|(X\backslash K_\varepsilon) < \varepsilon$ for all measures $\mu \in \mathcal{K}$. If $f \in C_b(X)$, $f = 0$ on K_ε and $|f| \leqslant 1$, then

$$|F(f)| \leqslant \sup_{\mu\in\mathcal{K}} \left| \int_X f\,d\mu \right| \leqslant \varepsilon.$$

By Theorem 4.1.9 the functional F is given by some Radon measure ν, which is a limit point of $\mathcal{K}$ in the weak topology. The second assertion was actually obtained in the proof of the Prohorov theorem, since we used only the metrizability of compact sets K_ε on which the considered sequence of measures was uniformly concentrated. $\square$

For many spaces important in applications, uniform tightness turns out to be a necessary condition for weak compactness of families of measures. We discuss such spaces in § 4.7, but here we only recall that if X is a complete metric space, then every weakly compact subset of $\mathcal{M}_r(X)$ is uniformly tight, which was proved in Theorem 3.1.9.

In the general case, unlike the case of a complete metric space, the hypothesis of Theorem 4.5.3 is not necessary: even on a countable nonmetrizable space a weakly convergent sequence of probability measures need not be uniformly tight, as the following example shows.

4.5.4. EXAMPLE. Let $X = \mathbb{N} \cup \{\infty\}$, where the points of $\mathbb{N}$ are open and the neighborhoods of ∞ have the form $U \cup \{\infty\}$, where U is a subset of $\mathbb{N}$ with density 1, i.e., $\lim\limits_{n\to\infty} N(U,n)/n = 1$, where $N(U,n)$ is the number of points in U not exceeding n. Then the sequence $n^{-1}\sum_{i=1}^n \delta_i$ of the arithmetic means of the

Dirac measures at the points i converges weakly to the Dirac measure δ_∞, but is not uniformly tight.

PROOF. We observe that $n^{-1}\sum_{i=1}^n \delta_i(U) \to 1$ for every open neighborhood U of the point ∞, for all other open sets $\delta_\infty(U) = 0$. All compact sets in the considered topology are finite. Indeed, any infinite sequence $\{n_k\}$ contains an infinite subsequence $\{n_{k_i}\}$ with the property that the set $U := X\backslash\{n_{k_i}\}$ is open in our topology. Then U and the points n_{k_i} form an open cover of $\{n_k\}\cup\{\infty\}$ that contains no finite subcover. □

In applications it is often useful to have various additional special conditions for weak compactness oriented towards specific features of the studied problems. For example, for distributions of random processes in functional spaces such conditions can be formulated in terms of covariance functions, sample moduli of continuity, etc., and for measures on linear spaces there are efficient conditions in terms of Fourier transforms (see §4.6). The following fact will be useful in Example 4.7.8, where we establish the Prohorov property of strict inductive limits.

4.5.5. EXAMPLE. Let $X = \bigcup_{n=1}^\infty X_n$ be a locally convex space that is the strict inductive limit of an increasing sequence of its closed linear subspaces X_n, i.e., every X_n is a proper closed linear subspace in a locally convex space X_{n+1} and convex neighborhoods of zero in X are convex sets V such that $V\cap X_n$ is a neighborhood of zero in X_n (the inductive limit topology is the strongest locally convex topology in X for which all embeddings $X_n \to X$ are continuous, see Bogachev, Smolyanov [**97**, Chapter 2]). If a sequence $\{\mu_i\}$ of nonnegative τ-additive (for example, Radon) measures on X converges weakly to a τ-additive measure μ, then, for every $\varepsilon > 0$, there exists an index $n \in \mathbb{N}$ such that $\mu_i(X\backslash X_n) < \varepsilon$ for all $i \in \mathbb{N}$.

Moreover, for every family $\{\mu_\alpha\}$ of nonnegative τ-additive measures on X that has compact closure in the weak topology in space $\mathcal{M}_\tau(X)$ and every $\varepsilon > 0$, there exists $n \in \mathbb{N}$ such that $\mu_\alpha(X\backslash X_n) < \varepsilon$ for all α.

PROOF. Without loss of generality we can assume that μ_i and μ are probability measures (if $\mu_i(X) \to 0$, the assertion is trivial). If our assertion is false, then, for every $n \in \mathbb{N}$, there exists $i(n) \in \mathbb{N}$ with $\mu_{i(n)}(X_n) < 1-\varepsilon$. Passing to a new sequence of measures, we can assume that $i(n) = n$. Pick $m \in \mathbb{N}$ such that $\mu(X_m) > 1-\varepsilon/2$. Set $k_1 := m$. Find a number $k_2 > m$ for which $\mu_m(X_{k_2}) > 1-\varepsilon/2$. Next we find a convex symmetric open set U_1 in the space X_{k_2} such that $X_m \subset U_1$ and $\mu_m(U_1) < 1-\varepsilon$. Such a set U_1 exists.

Indeed, by the Hahn–Banach theorem the subspace X_m is the intersection of all closed hyperplanes containing it. By the τ-additivity of μ_m there exists a finite collection of closed hyperplanes $L_1, \ldots, L_p$ in X_{k_2} such that $X_m \subset \bigcap_{i=1}^p L_i$ and $\mu_m\bigl(\bigcap_{i=1}^p L_i\bigr) < 1-\varepsilon$. Then $L_i = l_i^{-1}(0)$ for some $l_i \in X_{k_2}^*$, and the set $\bigcap_{i=1}^p l_i^{-1}(-\delta,\delta)$ can be taken for U_1 for sufficiently small $\delta > 0$. Next we take $k_3 \geqslant k_2$ with $\mu_{k_2}(X_{k_3}) > 1-\varepsilon/2$. There is a convex symmetric neighborhood of zero $W \subset X_{k_3}$ such that $W\cap X_{k_2} = U_1$ (see Bogachev, Smolyanov [**97**, Lemma 1.3.12] or Schaefer [**566**, II.6.4]). As above, there exists a convex symmetric open set V in the space X_{k_3} such that $X_{k_2} \subset V$ and $\mu_{k_2}(V) < 1-\varepsilon$. Set $U_2 := W\cap V$. Continuing the described process by induction, we obtain an increasing sequence of indices $k_n \geqslant n$ such that each space $X_{k_{n+1}}$ contains a convex symmetric open set U_n with the following properties: 1) $U_n \cap X_{k_n} = U_{n-1}$, 2) $\mu_{k_n}(U_n) < 1-\varepsilon$, $\mu_{k_n}(X_{k_{n+1}}) > 1-\varepsilon/2$.

According to the aforementioned definition of a strict inductive limit, the set $U=\bigcup_{n=1}^{\infty} U_n$ is a neighborhood of zero in X. By construction for every n we have the estimate

$$\mu_{k_n}(U) < \mu_{k_n}(U\cap X_{k_{n+1}}) + \varepsilon/2 = \mu_{k_n}(U_n) + \varepsilon/2 < 1-\varepsilon/2,$$

which contradicts weak convergence (see Corollary 4.3.9), since $\mu(U) > 1-\varepsilon/2$.

For a relatively weakly compact family $\{\mu_\alpha\}$ the reasoning is similar. As above, we pick a sequence $\{\mu_{\alpha(n)}\}$ and denote by μ its weak limit point. The previous choice of U leads again to a contradiction with Corollary 4.3.9, since there exists a subnet $\{\mu_\beta\}$ in $\{\mu_{\alpha(n)}\}$ convergent weakly to μ. □

Another justification can be obtained from Theorem 4.8.10 below.

The next result from Hoffmann-Jørgensen [**325**] extends some properties of weakly compact sets of nonnegative measures on metric spaces to completely regular spaces that admit injective embeddings into metric spaces. Actually, it deals with completely regular spaces on which there are weaker metrizable topologies.

4.5.6. THEOREM. *Suppose that a completely regular space X has a continuous injective mapping into a metric space (for example, let X be a completely regular Souslin space). Then, for every set M in the space $\mathcal{M}_t^+(X)$ with the weak topology, the following conditions are equivalent:*

(i) *every infinite sequence in M has a limit point in $\mathcal{M}_t^+(X)$;*

(ii) *every infinite sequence in M has a convergent subsequence in $\mathcal{M}_t^+(X)$;*

(iii) *the closure of M is compact;*

(iv) *the closure of M is compact and metrizable.*

PROOF. It suffices to show that (i) implies (iv). Let $h\colon X\to Y$ be continuous and injective, where Y is a metric space. The mapping

$$\widehat{h}\colon \mathcal{M}_t^+(X)\to\mathcal{M}_t^+(Y), \quad \mu\mapsto\mu\circ h^{-1}$$

is also continuous in the weak topology and injective (because measures from $\mathcal{M}_t(X)$ have unique Radon extensions and on compact sets h is a homeomorphism). Since $\mathcal{M}_t^+(Y)$ with the weak topology is metrizable, our assertion follows from Exercise 4.8.46. □

A useful technical result characterizing weak compactness for nonnegative measures was obtained in Topsøe [**614**].

4.5.7. THEOREM. *Let X be a completely regular space. A set $M\subset\mathcal{M}_t^+(X)$ has compact closure in the weak topology precisely when it is bounded in variation and, for every $\varepsilon>0$ and each collection $\mathcal{U}$ of open sets with the property that every compact set is contained in a set from $\mathcal{U}$, there exist $U_1,\ldots,U_n\in\mathcal{U}$ such that*

$$\min\{\mu(X\backslash U_i)\colon\ 1\leqslant i\leqslant n\}<\varepsilon \quad \forall\,\mu\in M.$$

PROOF. For simplicity we consider the case of probability measures. Let M be a compact set in $\mathcal{P}_t(X)$, $\mathcal{U}$ let be a class of open sets the subsets of which contain all compact sets in X, and let $\varepsilon>0$. We shall deal with Radon extensions of the measures in M. For every measure $\mu\in M$ we can find a compact set K_μ with $\mu(K_\mu)>1-\varepsilon/2$ and take an open set $U_\mu\in\mathcal{U}$ containing K_μ. Let us take a neighborhood of the measure μ in the weak topology of the form

$$W_\mu=\{\nu\in\mathcal{P}_r(X)\colon\ \nu(U_\mu)>\mu(U_\mu)-\varepsilon/2\}.$$

There is a finite subcover of M by sets $W_{\mu_1}, \ldots, W_{\mu_n}$. This means that every measure μ in M satisfies the condition $\mu(U_{\mu_i}) > \mu_i(U_{\mu_i}) - \varepsilon/2 > 1 - \varepsilon$ for some $i \leqslant n$, i.e., $\mu(X \backslash U_{\mu_i}) < \varepsilon$, as required.

Suppose now that the indicated condition is fulfilled. We extend all measures in M to the Stone-Čech compactification βX of the space X and take a limit point $\mu_0 \in \mathcal{P}_r(\beta X)$ of the obtained set of measures on this compact space (which exists by the weak compactness of the space of probability measures on any compact space). We show that, for every $\varepsilon > 0$, there exists a compact set $K_\varepsilon \subset X$ with $\mu_0(K_\varepsilon) > 1 - \varepsilon$. Suppose that this is false. Then for some $q < 1$ we have $\mu_0(K) \leqslant q$ for all compact sets $K \subset X$. Now for every compact set $K \subset X$ we take a functionally open set $U_K \subset \beta X$ for which $\mu_0(\partial U_K) = 0$ and $\mu_0(U_K) < q + \delta$, where $\delta = (1-q)/2$. By the assumption of the theorem for the number $\delta/4$ there exists a finite collection of the constructed sets $U_{K_1}, \ldots, U_{K_n}$ such that for every measure $\mu \in M$ there is a number $i \leqslant n$ with $\mu(X \backslash U_{K_i}) < \delta/4$. Denote by M_i the subset of M corresponding to the number i (these n subsets can have intersections). Since the measure μ_0 is a limit point of the set M, it is a limit point of some set M_i, i.e., the limit of some net of measures $\mu_\alpha \in M_i$. Then the equality $\mu_0(\partial U_{K_i}) = 0$ must yield the equality $\mu_0(U_{K_i}) = \lim\limits_{\alpha} \mu_\alpha(U_{K_i})$, which leads to a contradiction, since we have $\mu_0(U_{K_i}) < q + \delta = 1 - \delta/2$, $\mu_\alpha(U_{K_i}) \geqslant 1 - \delta/4$. □

4.5.8. Corollary. (i) *If a set $\mathcal{M} \subset \mathcal{M}_t^+(X)$ has compact closure in the weak topology in $\mathcal{M}_t^+(X)$, then, for every closed set $Y \subset X$, the family of restrictions of measures in $\mathcal{M}$ on Y has compact closure in the weak topology in $\mathcal{M}_t^+(Y)$.*

(ii) *If Y is completely regular, then $\mathcal{M} \subset \mathcal{M}_t^+(X \times Y)$ has compact closure in the weak topology if the projections of $\mathcal{M}$ to X and Y have compact closures.*

Proof. (i) Suppose that we are given a collection $\mathcal{V}$ of open sets in Y for which every compact set in Y is contained in some $V \in \mathcal{V}$. For every $V \in \mathcal{V}$ there is an open set U_V in X for which $V = U_V \cap Y$. Let us consider the collection of open sets $(X \backslash Y) \cup U_V$. Every compact subset of X is contained in a set from the obtained collection $\mathcal{U}$ of open sets, since its intersection with Y is compact (by the closedness of Y) and is contained in a set from $\mathcal{V}$. By the previous theorem, for every $\varepsilon > 0$, there are sets $U_1, \ldots, U_n \in \mathcal{U}$ for which we have the estimate indicated in the theorem. We have to find analogous sets in $\mathcal{V}$ for the restrictions of the measures $\mu \in M$ to Y. If $U_i = (X \backslash Y) \cup U_{V_i}$, where $V_i \in \mathcal{V}$, then we take such a set V_i. In this case we have $\mu(Y \backslash V_i) = \mu(X \backslash U_i)$. Hence the hypothesis of the theorem is fulfilled. Justification of (ii) is delegated to Exercise 4.8.40. □

Assertion (i) of the corollary is rather unexpected (although for Polish spaces it is obvious from the Prohorov criterion and for normal spaces it follows from Theorem 4.5.10 below), since weak convergence does not imply convergence on closed sets. In particular, the limit of the restrictions of measures from a weakly convergent sequence to a closed set need not coincide with the restriction of the limit of this sequence (see Example 1.4.3). In case of a complete metric space the previous corollary remains also valid for signed measures by Theorem 3.1.9. However, to signed measures on general spaces this corollary does not extend even in the case of a weakly convergent sequence of measures. The next example employs ordinal numbers, see Bogachev, Smolyanov [**96**], Kolmogorov, Fomin [**379**].

4.5.9. EXAMPLE. Let $X = \bigl([0,\omega_1]\times[0,\omega_0]\bigr)\backslash(\omega_1,\omega_0)$, where ω_0 is the ordinal corresponding to $\mathbb{N}$, ω_1 is the first uncountable ordinal, and the intervals of ordinal numbers are equipped with the natural order topology (generated by intervals of ordinal numbers). Set

$$Y = \{(\omega_1, 2n)\}_{n=1}^{\infty}, \quad M = \{\delta(\omega_1, 2n) - \delta(\omega_1, 2n+1)\}_{n=1}^{\infty} \cup \{0\}.$$

The set M is weakly compact in the space $\mathcal{M}_t(X)$, since the sequence of measures $\mu_n := \delta(\omega_1, 2n) - \delta(\omega_1, 2n+1)$ converges weakly to zero.

Indeed, one can verify (Exercise 4.8.47) that every continuous function g on $[0,\omega_1)$ is constant on some interval $[t,\omega_1)$, i.e., on $[0,\omega_1)$ there are no continuous functions like $\sin\bigl(1/(x-1)\bigr)$. This is due to the fact that for every sequence of ordinal numbers $\tau_k < \omega_1$ there is an upper bound smaller than ω_1. Hence for every function $f \in C_b(X)$ and for every k there exists $t_k \in [0,\omega_1)$ such that $f(\omega,k) = f(t_k,k)$ for all $\omega \geqslant t_k$. Then $\tau = \sup_k t_k < \omega_1$, hence $f(\omega,k) = f(\tau,k)$ for all $\omega \geqslant \tau$ and all $k \in \mathbb{N}$. Since $k \to \omega_0$ in $[0,\omega_0]$, we have $f(\tau,k) \to f(\tau,\omega_0)$. Therefore, $f(\omega_1,k) \to f(\tau,\omega_0)$, because the function $\omega \mapsto f(\omega,k)$ is constant on $[\tau,\omega_1]$. Thus, $f(\omega_1,2n) - f(\omega_1,2n+1) \to 0$, which means weak convergence of the measures μ_n to zero.

The set Y (homeomorphic to the set of even natural numbers with its usual topology) is closed in X and discrete (has no limit points), and the restrictions of the measures from M to Y form a discrete set of Dirac measures in $\mathcal{M}_t(Y)$ that has no limit points, hence obviously is not contained in a compact set. Note also that in this example the set of total variations of measures from M is not contained in a compact set.

Let us give a simple criterion of the relative weak compactness in the space of nonnegative Baire measures on an arbitrary space X going back to A.D. Alexandroff.

A sequence of functionally closed sets Z_n in a topological space X is called *regular* if $Z_n \subset Z_{n+1}$, $X = \bigcup_{n=1}^{\infty} Z_n$, and there exist functionally open sets U_n such that $Z_n \subset U_n \subset Z_{n+1}$.

4.5.10. THEOREM. *A bounded set $M \subset \mathcal{M}_\sigma^+(X)$ has compact closure in the weak topology precisely when*

$$\lim_{n\to\infty} \sup_{\mu\in M} \int_X f_n\,d\mu = 0$$

for every sequence of functions $f_n \in C_b(X)$ pointwise decreasing to 0. An equivalent condition:

$$\lim_{n\to\infty} \sup_{\mu\in M} \mu(X\backslash Z_n) = 0$$

for every regular sequence of functionally closed sets Z_n.

PROOF. Suppose that the first of these conditions is fulfilled. The bounded set M has compact closure M' in the space $C_b(X)^*$. In addition, every element $\mu \in M'$ belongs to $\mathcal{M}_\sigma^+(X)$ by Theorem 4.1.9, since if a sequence $\{f_n\} \subset C_b(X)$ is pointwise decreasing to zero and for a given number $\varepsilon > 0$ the integrals of the functions f_n with respect to the measures from M are smaller than ε in absolute value for all $n \geqslant n_\varepsilon$, then the same estimate is true for the values of functional μ on the functions f_n. Note that this reasoning applies to signed measures as well.

Conversely, if M has compact closure M' in the weak topology, then the measures in M' are nonnegative and the functions $\mu \mapsto \int_X f_n \, d\mu$ on M' decrease to zero. By Dini's theorem (Exercise 4.8.26) they converge to zero uniformly on M'.

If $\{Z_n\}$ is a regular sequence, then there exists a sequence $\{f_n\} \subset C_b(X)$ pointwise decreasing to zero such that $f_n = 1$ on $X \backslash Z_n$ (see Lemma 4.1.2). Then

$$\mu(X \backslash Z_n) \leqslant \int_X f_n \, d\mu \quad \forall \, \mu \in M.$$

Conversely, if functions $f_n \in C_b(X)$ decrease to 0, then for any given $\varepsilon > 0$ we set $U_n = \{f_n < \varepsilon\}$. It is readily verified that there exists a regular sequence of functionally closed sets Z_n for which $Z_n \subset U_n$ (one can take sets of the form $Z_n := \{\min(f_n, \varepsilon) \leqslant \varepsilon - 1/n\}$). Then the integral of f_n with respect to the measure μ does not exceed $\varepsilon\mu(X) + \mu(X \backslash Z_n)$, which shows the equivalence of the indicated conditions. □

An analogous assertion (with uniform convergence of integrals) is also true for signed measures, but the proof of one implication is more complicated, for this reason it is postponed to Theorem 4.8.3 below.

4.5.11. PROPOSITION. *Let X be a locally compact space. Suppose that measures $\mu \in \mathcal{M}_r(X)$ and $\mu_n \in \mathcal{M}_r(X)$, where $n \in \mathbb{N}$, satisfy the following condition: for every continuous function φ with compact support*

$$\lim_{n\to\infty} \int_X \varphi \, d\mu_n = \int_X \varphi \, d\mu.$$

Suppose that $\|\mu_n\| \to \|\mu\|$. Then the sequence $\{\mu_n\}$ is uniformly tight and converges weakly to μ.

PROOF. Let $\varepsilon > 0$. We find a compact set K and a number n_ε such that $|\mu|(K) > \|\mu\| - \varepsilon$ and $\|\mu_n\| < \|\mu\| + \varepsilon$ for all $n \geqslant n_\varepsilon$. By the local compactness of the space there exists a continuous function f with compact support S containing K (and contained in a neighborhood of K with compact closure) such that $|f| \leqslant 1$ and $\int_X f \, d\mu \geqslant |\mu|(K) - \varepsilon$. We now take $N \geqslant n_\varepsilon$ such that

$$\int_X f \, d\mu_n \geqslant \int_X f \, d\mu - \varepsilon \quad \text{for all } n \geqslant N.$$

Then $|\mu_n|(S) \geqslant \|\mu_n\| - 4\varepsilon$ for all $n \geqslant N$. Thus, we have established the tightness of the given sequence of measures, whence its weak convergence to the measure μ follows at once. □

Let us give a simple condition for metrizability of compact sets in locally convex spaces. With its help one can verify the metrizability of compact sets in the space of measures with the weak topology by means of an appropriate choice of countable collections of bounded continuous functions the integrals of which distinguish measures of a given family.

4.5.12. PROPOSITION. *A compact set K in a locally convex space is metrizable precisely when there is a countable collection of continuous linear functionals separating points in K.*

PROOF. If a sequence $\{f_n\}$ from $C(K)$ separates points of K, then the mapping $f\colon K \to \mathbb{R}^\infty$, $x \mapsto \bigl(f_n(x)\bigr)_{n=1}^{\infty}$ is continuous and injective. Hence $f(K)$ is compact in the metrizable space $\mathbb{R}^\infty$ and f is a homeomorphism between K and $f(K)$. Hence K is metrizable. Conversely, if K is metrizable, then the space $C(K)$ is separable (Exercise 1.7.8). Hence there is a countable everywhere dense set in its subspace consisting of the restrictions to K of continuous linear functions. This set separates points in K. □

4.6. The Fourier transform and weak convergence

In this section we obtain a characterization of weak convergence and weak compactness in terms of Fourier transforms (characteristic functionals). We first recall the Lévy theorem proved in Chapter 1. This theorem asserts that a sequence $\{\mu_j\}$ of probability measures on $\mathbb{R}^d$ converges weakly precisely when the sequence of their characteristic functionals $\widetilde{\mu}_j$ converges at every point and the function $\varphi(x) := \lim\limits_{j\to\infty} \widetilde{\mu}_j(x)$ is continuous at the origin; in this case φ is the characteristic functional of the probability measure μ to which the measures μ_j converge. In addition, a family M of probability measures on $\mathbb{R}^d$ is uniformly tight precisely when the family of functions $\{\widetilde{\mu}\}_{\mu\in M}$ is equicontinuous on $\mathbb{R}^d$ (or at least at the origin). Here we consider some analogs of this theorem for infinite-dimensional spaces. First we extend to the setting of locally convex spaces the Fourier transform and convolution (for technical details, see Bogachev [**81**, Chapter 7]).

For a locally convex space X let X^* denote the topological dual, i.e., the set of all continuous linear functionals on X. The symbol $\sigma(X)$ denotes the σ-algebra generated by all elements of X^*. It is called the *cylindrical* σ-algebra, since it is generated by the algebra of cylinders of the form $C = \bigl\{x\in X\colon \bigl(f_1(x),\dots,f_n(x)\bigr)\in B\bigr\}$, where $f_i \in X^*$, $B \in \mathcal{B}(\mathbb{R}^n)$.

4.6.1. DEFINITION. For a measure μ on the σ-algebra $\sigma(X)$ in a locally convex space the Fourier transform (the characteristic functional) is defined by the formula

$$\widetilde{\mu}\colon X^* \to \mathbb{C}, \quad \widetilde{\mu}(f) = \int_X \exp(if)\, d\mu.$$

The results in § 1.6 for $\mathbb{R}^n$ yield the equality of two measures with equal Fourier transforms, since such measures coincide on all cylinders.

4.6.2. DEFINITION. The convolution of two measures μ and ν on $\sigma(X)$ is defined as the image of the product $\mu\otimes\nu$ under the mapping $(x,y) \mapsto x+y$, i.e., it is the measure defined by the formula

$$\mu * \nu(B) = \int_X \mu(B-x)\,\nu(dx).$$

One can verify that the function $x \mapsto \mu(B-x)$ is measurable with respect to $\sigma(X)$, and if both measures are Radon, then this function is Borel measurable for every Borel set B, which enables us to define the convolution as a Radon measure by the same formula. For Radon measures the equality of the Fourier transforms also yields the equality of measures. Fubini's theorem gives a useful formula

$$\widetilde{\mu * \nu} = \widetilde{\mu}\widetilde{\nu}.$$

Therefore, the convolution is commutative: $\mu * \nu = \nu * \mu$.

For Dirac's measure at the origin the equality $\mu*\delta_0 = \mu$ holds for every measure. Let us note the following fact.

4.6.3. LEMMA. *Let μ and ν be Radon probability measures on a locally convex space X such that $\mu * \nu = \mu$. Then ν is Dirac's measure at the origin.*

PROOF. For the Fourier transforms we have $\widetilde{\mu}\cdot\widetilde{\nu} = \widetilde{\mu}$, therefore, the same equality holds for convolutions of all one-dimensional projections, so that our assertion reduces to the one-dimensional case. In this case either $\widetilde{\nu} \equiv 1$, i.e., $\nu = \delta_0$, or there exists a sequence of points $y_n \to 0$ with $\widetilde{\nu}(y_n) \neq 1$ (if $\widetilde{\nu} = 1$ in a neighborhood of zero, then $\widetilde{\nu} \equiv 1$, which follows from (1.6.5)). Then $\widetilde{\mu}(y_n) = 0$, whence $\widetilde{\mu}(0) = 0$, which is impossible. □

We now define the Gross–Sazonov topology. This topology is first defined on a separable Hilbert space H. On every finite-dimensional space F in H there arises the standard Gaussian measure γ_F as soon as we take an orthonormal basis $e_1, \ldots, e_n$ in F and use it to identify F with $\mathbb{R}^n$.

4.6.4. DEFINITION. A seminorm q on H is called *measurable* if the following condition is fulfilled: for every $\varepsilon > 0$, there exists $C > 0$ such that for every finite-dimensional subspace F in H one has the estimate $\gamma_F\bigl(x \in F\colon\ q(x) > C\bigr) < \varepsilon$.

The Gross–Sazonov topology on H is generated by the family of all measurable seminorms. On a general locally convex space E, the Gross–Sazonov topology is generated by the family of all seminorms of the form $q\circ T$, where $T\colon E \to l^2$ is a continuous linear operator and q is a measurable seminorm on l^2.

For the proof of the following theorem, see Bogachev, Smolyanov [**97**, Theorem 5.11.2].

4.6.5. THEOREM. *Let E be a locally convex space and let $\mathcal{P}$ be some set of probability measures on the dual space E^* equipped with the σ-algebra σ_E generated by the space E. For the uniform tightness of $\mathcal{P}$ on E^* with the topology $\sigma(E^*, E)$ it is sufficient that the set of the Fourier transforms of measures in $\mathcal{P}$ be equicontinuous in the origin in the Gross–Sazonov topology associated with the original topology of the space E.*

One can also use the weaker *Sazonov topology* $\mathcal{T}$ on E generated by the seminorms of the form $\|STx\|_{l^2}$, where $T\colon E \to l^2$ is a continuous linear operator and S is a Hilbert–Schmidt operator on l^2 (see § 2.5). The *Sazonov topology* $\mathcal{T}(X^*, X)$ on X^* is generated by the seminorms $\|SA^*x\|_{l^2}$, $A \in \mathcal{L}(l^2, X)$. If E itself is Hilbert, then the Sazonov topology on it is defined by the family of seminorms $\|Sx\|$, where S is a Hilbert–Schmidt operator on E.

If the topology of a locally convex space coincides with its Sazonov topology, then this space is called *nuclear*. For example, such are the classical spaces of functional analysis $\mathcal{S}(\mathbb{R}^d)$ and $\mathcal{S}'(\mathbb{R}^d)$. Nuclear Banach spaces are finite-dimensional.

The *strong topology* $\beta(X, X^*)$ on a locally convex space X is the topology of uniform convergence on sets in X^* bounded in the topology $\sigma(X^*, X)$. If a locally convex space X with its strong topology possesses the property that every continuous linear functional on its dual with the strong topology $\beta(X^*, X)$ (the topology of convergence on sets bounded in the topology $\sigma(X, X^*)$ on X) is represented by an element of X itself, then X is called *reflexive*. For example, such are the spaces $\mathbb{R}^\infty$, $\mathcal{S}(\mathbb{R}^d)$, $\mathcal{D}(\mathbb{R}^d)$ and topological duals $\mathcal{S}'(\mathbb{R}^d)$ and $\mathcal{D}'(\mathbb{R}^d)$.

The space $\mathcal{S}(\mathbb{R}^d)$ consists of infinitely differentiable functions with finite norms $p_{k,m}(f) = \sup_x (1+|x|^2)^k \|f^{(m)}(x)\|$, and $\mathcal{D}(\mathbb{R}^d)$ is the space of infinitely differentiable functions with compact support equipped with the topology of the inductive limit of increasing subspaces $\mathcal{D}_n$ of functions with support in the ball of radius n, and on $\mathcal{D}_n$ the topology is introduced by means of norms $p_m(f) = \sup_x \|f^{(m)}(x)\|$ (see Example 4.5.5, Bogachev, Smolyanov [**96**], and also Exercise 4.8.43).

4.6.6. THEOREM. *Let X be a locally convex space with its strong topology $\beta(X, X^*)$ and let $M \subset \mathcal{P}_r(X)$ be a family of measures such that their characteristic functionals are equicontinuous at the origin in the Sazonov topology $\mathcal{T}(X^*, X)$ on X^*. Then M has compact closure in the weak topology.*

For the proof, see Bogachev [**81**, § 7.13] and the proof of Theorem 4.1 in Vakhania, Tarieladze, Chobanyan [**629**, Chapter VI, § 4.2, p. 408].

4.6.7. EXAMPLE. Let M be a family of Radon probability measures on a Hilbert space X satisfying the following condition: for every $\varepsilon > 0$, there is a Hilbert–Schmidt operator S_ε such that if $\|S_\varepsilon y\| \leqslant 1$, then $\widetilde{\mu}(y) \geqslant 1-\varepsilon$ for all $\mu \in M$. Then M is uniformly tight and has compact closure in the weak topology.

4.6.8. COROLLARY. *Let X be a reflexive nuclear space and let M be a family of Radon probability measures on the space $\bigl(X^*, \beta(X^*, X)\bigr)$ with characteristic functionals equicontinuous at the origin. Then M has compact closure in the weak topology.*

This corollary applies to $X^* = \mathcal{S}'(\mathbb{R}^d)$ and $X^* = \mathcal{D}'(\mathbb{R}^d)$.

We draw the reader's attention to the fact that in the case of a sequence of measures the hypotheses of these theorems require the equicontinuity of their Fourier transforms, while the Lévy theorem does not contain this condition (although it contains the condition of continuity of the limit function, which in case of equicontinuity holds automatically). Such sequential generalizations of the Lévy theorem are also obtained for a number of important classes of infinite-dimensional spaces in the papers Feldman [**220**], Fernique [**224**], Meyer [**459**], Boulicaut [**115**], Banaszczyk [**40**], [**41**] (see the survey Heyer, Kawakami [**320**], where there is a discussion of analogs for groups). Let us present several typical results. First we observe that for an infinite-dimensional Hilbert space such an analog of the Lévy theorem is false: if $\{e_n\}$ is an infinite orthonormal sequence, then the Fourier transforms of the Dirac measures at the points e_n have the form $\exp i(y, e_n)$ and converge pointwise to 1, which is the Fourier transform of Dirac's measure at the origin, but there is no weak convergence of measures here.

4.6.9. PROPOSITION. (i) *Let $\{\mu_n\}$ be a sequence of probability measures on the cylindrical σ-algebra $\sigma(X)$ of a locally convex space X whose Fourier transforms converge pointwise to the Fourier transform of a probability measure μ on $\sigma(X)$. Then the measures μ_n converge weakly to μ provided that the space X is equipped with the weak topology $\sigma(X, X^*)$.*

(ii) *If in this situation X is a normed space, then the set of measures $\mu_n \circ l^{-1}$, where $\|l\| \leqslant 1$, is uniformly tight on the real line. Hence the Fourier transforms of the measures μ_n are equicontinuous.*

PROOF. (i) It is known that every function on X continuous in the weak topology has the form $f(l_1, l_2, \ldots)$, where $l_i \in X^*$ and $f \in C(\mathbb{R}^\infty)$ (see Bogachev,

Smolyanov [**97**, Theorem 5.12.69]). Hence our assertion reduces to the case of the space $\mathbb{R}^\infty$, in which it is known from Corollary 2.4.8.

(ii) Let $\varepsilon > 0$. For every $k \in \mathbb{N}$ the set

$$M_k(\varepsilon) := \bigl\{l \in X^*\colon\ \mu_n\bigl(l^{-1}([-k,k])\bigr) \geqslant 1-\varepsilon\ \forall n\bigr\}$$

is norm closed in X^*, since if a sequence of functionals $l_j \in M_k(\varepsilon)$ converges to a functional l at least pointwise, then the measures $\mu_n\circ l_j^{-1}$ converge weakly to the measure $\mu\circ l_j^{-1}$ for every fixed n, which gives the estimate $\mu_n\bigl(l^{-1}([-k,k])\bigr) \geqslant 1-\varepsilon$ implying that $l \in M_k(\varepsilon)$. By Baire's theorem there exists $k = k(\varepsilon)$ such that $M_k(\varepsilon)$ contains some ball W_1. Then $-W_1 \subset M_k(\varepsilon)$. Hence $W = (W_1 - W_1)/2 \in M_k(2\varepsilon)$, since for $l = (l_1 - l_2)/2$, where $l_1, l_2 \in M_k(\varepsilon)$, we have

$$l_1^{-1}([-k,k]) \cap l_2^{-1}([-k,k]) \subset l^{-1}([-k,k]),$$

which gives $\mu_n\bigl(l^{-1}([-k,k])\bigr) \geqslant 1-2\varepsilon$ for each n. Thus, $M_k(2\varepsilon)$ contains a ball of some radius r_ε centered at the origin. Therefore, if $\|l\| \leqslant 1$, then for each n we have $\mu_n\bigl(l^{-1}([-k/r_\varepsilon, k/r_\varepsilon])\bigr) \geqslant 1-2\varepsilon$. □

4.6.10. COROLLARY. *If the Fourier transforms of Radon probability measures μ_n on a Hilbert space H converge pointwise to the Fourier transform of a Radon probability measure μ, then, for every Hilbert–Schmidt operator T from H to a Hilbert space E, the measures $\mu_n\circ T^{-1}$ on the space E converge weakly to the measure $\mu\circ T^{-1}$.*

PROOF. The proposition above and Example 4.6.7 yield the uniform tightness of the measures $\mu_n\circ T^{-1}$; their Fourier transforms converge pointwise. □

4.6.11. THEOREM. *Suppose that the sequence of Fourier transforms of Radon probability measures μ_n on a nuclear locally convex space converges pointwise to the Fourier transform of a Radon probability measure μ. Then the measures μ_n converge weakly to the measure μ.*

PROOF. We first consider the special case of the space Σ of rapidly decreasing sequences, i.e., sequences with finite norms p_m defined by

$$p_m(x)^2 = \sum_{n=1}^\infty n^{2m} x_n^2.$$

This space is the intersection of a decreasing sequence of Hilbert spaces E_m generated by the norms p_m. The natural embeddings $E_{m+1} \to E_m$ are Hilbert–Schmidt operators. It is clear from the previous corollary that for Σ the assertion of the theorem is true. Indeed, if the condition is fulfilled for the measures μ_n on Σ, then these measures on E_m with $m > 1$ converge weakly to the measure μ, moreover, the sequence $\{\mu_n\}$ is uniformly tight on E_m. This implies their uniform tightness also on Σ due to the fact that the intersection of a sequence of compact sets from the spaces E_m is compact in Σ.

It follows from the reasoning above that the theorem is true for every finite power of Σ, hence also for the countable power. Therefore, it remains valid for an arbitrary power of Σ, since every continuous function on the power of Σ is a function of a finite or countable collection of variables (see Arkhangelskii, Ponomarev [**23**, Problem 386, p. 89, 116] or Engelking [**203**, 2.7.12(c)]). Finally, the general case follows from the known fact that any nuclear space is isomorphic to a linear subspace in some power of Σ (see Jarchow [**340**, Section 21.7, p. 500]). □

4.7. Prohorov spaces

Here we discuss some classes of topological spaces for which there hold certain implications (or their analogs) of the Prohorov theorem on equivalence of the relative weak compactness and uniform tightness of families of probability measures on complete separable metric spaces. In this section we consider only completely regular spaces.

4.7.1. DEFINITION. (i) A completely regular topological space X is called a Prohorov space if every set in the space of measures $\mathcal{M}_t^+(X)$ that is compact in the weak topology is uniformly tight.

(ii) A completely regular topological space X is called sequentially Prohorov if every sequence of nonnegative tight Baire measures weakly convergent to a tight measure is uniformly tight.

(iii) If in (i) or (ii) signed measures are considered, then we say that X is a *strongly Prohorov* or *strongly sequentially Prohorov* space, respectively.

In this definition one could speak of Radon measures, i.e., replace $\mathcal{M}_t^+(X)$ by $\mathcal{M}_r^+(X)$, since every tight Baire measure on X has a unique Radon extension.

The theorems of Prohorov and Le Cam proved in Chapter 2 can be formulated as follows.

4.7.2. THEOREM. *Every complete separable metric space is a Prohorov space. An arbitrary metric space is sequentially Prohorov.*

It is clear that any Prohorov space is sequentially Prohorov. We shall see below in Theorem 4.8.6 that the space $\mathbb{Q}$ of rational numbers is sequentially Prohorov, but is not Prohorov. Let us observe that the sequential Prohorov property is weaker than the requirement that weakly convergent sequences of tight measures be uniformly tight (because their limits need not be tight measures).

Theorem 3.1.9 says that complete metric spaces are strongly Prohorov.

The following useful observation is made in Hoffmann-Jørgensen [**325**].

4.7.3. LEMMA. *Suppose that we have continuous mappings f_n from a completely regular space X to Prohorov spaces X_n such that $\bigcap_{n=1}^\infty f_n^{-1}(K_n)$ is compact in X whenever the sets K_n are compact in X_n. Then X is a Prohorov space. An analogous assertion is true for the class of sequentially Prohorov spaces.*

PROOF. Let $\mathcal{P} \subset \mathcal{P}_r(X)$ be compact in the weak topology. Then its image in $\mathcal{P}_r(X_n)$ under the mapping $\widehat{f_n}\colon \mu \mapsto \mu\circ f_n^{-1}$ is compact. By our hypothesis, for every $\varepsilon > 0$, there exist compact sets $K_n \subset X_n$ for which $\mu\circ f_n^{-1}(X_n\backslash K_n) < \varepsilon 2^{-n}$ for all $\mu \in \mathcal{P}$. Also by hypothesis, the set $K = \bigcap_{n=1}^\infty f_n^{-1}(K_n)$ is compact in X. In addition, $\mu(X\backslash K) \leqslant \varepsilon$ for all $\mu \in \mathcal{P}$. The case of sequentially Prohorov spaces is analogous. □

4.7.4. THEOREM. *The class of Prohorov spaces is preserved by taking countable products and countable intersections and also by passing to closed subspaces and open subspaces, hence to G_δ-subsets. Analogous assertions are true for the class of sequentially Prohorov spaces.*

In addition, the space is Prohorov provided that every point has a neighborhood that is a Prohorov space (for example, this is true if the given space admits a locally finite cover by closed Prohorov subspaces).

PROOF. If X_n is a Prohorov space for all $n \in \mathbb{N}$, then the canonical projections f_n of their product X to X_n satisfy the hypothesis of the previous lemma, since the product of compact sets is compact by the Tychonoff theorem.

For the countable intersection of subspaces X_n of X with the Prohorov property the hypothesis of the lemma are satisfied by the identity inclusions of the intersection into X_n. For closed subsets our assertion is obvious (but it also suffices to take a single identity embedding in order to apply the lemma).

Let now U be an open set in X. Let $\mathcal{P} \subset \mathcal{P}_r(U)$ be compact and $\varepsilon > 0$. Then $\mathcal{P}$ is compact as part of $\mathcal{P}_r(X)$, hence by the Prohorov property of X there exists a compact set $K \subset X$ such that $\mu(X\backslash K) < \varepsilon$ for all $\mu \in \mathcal{P}$. Let us consider the family $\mathcal{V}$ of open sets $V \subset U$ with the property that the closure of the set V also belongs to U. Since X is completely regular, every point in U possesses a neighborhood with this property. Hence every compact set in U is contained in a set from $\mathcal{V}$. By Theorem 4.5.7 there exists a finite family of sets $V_1, \ldots, V_m \in \mathcal{V}$ such that for every measure $\mu \in \mathcal{P}$ there is a number $i \leqslant m$ for which $\mu(U\backslash V_i) < \varepsilon$, i.e., $\mu(\overline{V_i}) > 1 - \varepsilon$, where $\overline{V}$ denotes the closure of V. Therefore, $\mu(\overline{V_i} \cap K) > 1 - 2\varepsilon$. It remains to observe that the set $S = \bigcup_{i=1}^m \overline{V_i} \cap K$ is compact and contained in U, moreover, $\mu(S) > 1 - 2\varepsilon$ for all $\mu \in \mathcal{P}$. The case of sequentially Prohorov spaces is similar.

The assertion about G_δ-subsets follows from the assertion about open subsets and countable intersections.

Finally, suppose that in X every point x has a Prohorov neighborhood U_x. Then one can find a neighborhood V_x of x such that $\overline{V_x} \subset U_x$. The space $\overline{V_x}$ is also Prohorov. Suppose we are given a compact set $\mathcal{P} \subset \mathcal{P}_r(X)$ and $\varepsilon > 0$. Let us consider the family $\mathcal{V}$ of finite unions of neighborhoods V_x. Theorem 4.5.7 is applicable to it, which gives a finite collection of sets $V_1, \ldots, V_m \in \mathcal{V}$ such that for every measure $\mu \in \mathcal{P}$ there is a number $i \leqslant m$, for which $\mu(X\backslash V_i) < \varepsilon$, i.e., $\mu(\overline{V_i}) > 1 - \varepsilon$. The restrictions of the measures from $\mathcal{P}$ to the closed Prohorov sets $\overline{V_i}$ form a compact set of measures, hence there exist compact sets $K_i \subset \overline{V_i}$ such that $\mu(\overline{V_i}\backslash K_i) < \varepsilon/m$ for all $\mu \in \mathcal{P}$. Therefore, for the compact set $K = K_1 \cup \cdots \cup K_m$ we obtain $\mu(K) > 1 - 2\varepsilon$ for all $\mu \in \mathcal{P}$. □

4.7.5. COROLLARY. *All Čech complete spaces are Prohorov as G_δ-subsets of compact spaces. Hence all locally compact spaces are Prohorov.*

It is seen from Example 4.5.4 that the union of two Prohorov subspaces one of which is a point and the second one is countable (the set of natural numbers) need not be Prohorov. The same example shows that a countable union of closed Prohorov subspaces is not always Prohorov.

Let us consider some results and examples that enable one to construct broader classes of Prohorov and sequentially Prohorov spaces by means of the operations mentioned in Theorem 4.7.4.

4.7.6. PROPOSITION. *Suppose that a completely regular space X possesses a countable collection of closed subspaces X_n with the following property: a function on X is continuous precisely when its restriction to every X_n is continuous.*

(i) *If each X_n is Prohorov, then X is Prohorov as well.*

(ii) *If all X_n are either Polish or compact, then every weakly fundamental sequence in $\mathcal{M}_r(X)$ is uniformly tight. Hence X is strongly sequentially Prohorov.*

PROOF. (i) One can assume that $X_n \subset X_{n+1}$ by considering a new collection $X'_n = \bigcup_{i=1}^n X_i$. It is easily seen that it satisfies the hypothesis of the proposition.

Let $Y = \bigcup_{n=1}^{\infty} X_n$. It follows from our assumption that if a function continuous on Y is extended to X in an arbitrary way, then we obtain a continuous function. In particular, the indicator function of any subset $X\backslash Y$ is continuous. Hence $X\backslash Y$ is a discrete space functionally closed in X and its compact subsets are finite. Moreover, every subset $X\backslash Y$ is Baire in X. It is seen from this that, for every set $M \subset \mathcal{M}_r(X)$ compact in the weak topology, the restrictions of the measures from M to Y and to $X\backslash Y$ form sets compact in $\mathcal{M}_r(Y)$ and $\mathcal{M}_r(X\backslash Y)$, respectively. Hence it suffices to consider the case where $X = Y$, which we now assume.

Let $M \subset \mathcal{M}_r^+(X)$ be compact in the weak topology. We show that for every $\varepsilon > 0$ there is a number $n = n(\varepsilon)$ such that $\mu(X\backslash X_n) \leqslant \varepsilon$ for all $\mu \in M$. Indeed, otherwise for every n there exists a measure $\mu_{i_n} \in M$ with $\mu_{i_n}(X\backslash X_n) > \varepsilon$. We can assume that there are two increasing sequences of indices i_n and j_n with

$$\mu_{i_n}(X_{j_{n+1}}\backslash X_{j_n}) > \varepsilon, \quad \mu_{i_n}(X\backslash X_{j_{n+1}}) < \varepsilon/2.$$

The sequence $\{\mu_{i_n}\}$ possesses a limit point $\mu \in M$. Let us pick a number m for which $\mu(X\backslash X_m) < \varepsilon/2$. For every n there exists a compact set $K_n \subset X_{j_{n+1}}\backslash X_{j_n}$ such that $\mu_{i_n}(K_n) \geqslant \varepsilon$. We can assume that $j_1 > m$. There is a continuous function $f_n\colon X \to [0,1]$ such that $f_n|_{K_n} = 1$ and $f_n = 0$ on X_{j_n}. Let $f(x) = \sup_n f_n(x)$. Then $0 \leqslant f \leqslant 1$ and the function f is continuous, since its restriction to each X_k coincides with the maximum of finitely many functions f_n (hence is continuous). Therefore, $\displaystyle\int_X f\,d\mu < \varepsilon/2$, while $\displaystyle\int_X f\,d\mu_{i_n} \geqslant \varepsilon$. This contradiction shows that there exists $n = n(\varepsilon)$ with $\mu(X\backslash X_n) < \varepsilon$ for all $\mu \in M$. According to Corollary 4.5.8, the family M restricted to X_n has compact closure in the weak topology. By the Prohorov property for X_n this family of restrictions is uniformly tight.

(ii) Let $\{\mu_n\} \subset \mathcal{M}_r(X)$ be a weakly fundamental sequence. Then it converges weakly to some Baire measure μ. All measures μ_n are purely atomic on $X\backslash Y$. Let $A = \{a_n\}$ be the collection of all their atoms in $X\backslash Y$. Then $|\mu|\big(X\backslash(Y \cup A)\big) = 0$.

Indeed, otherwise there exists a Borel set $B \subset X\backslash(Y \cup A)$ such that $\mu(B)$ is either strictly positive or strictly negative. Then the function I_B is continuous on X, has zero integrals with respect to to all measures μ_n, but its integral with respect to μ is not zero, which leads to a contradiction. The same reasoning shows that the measures μ_n converge to μ on every set in A. Thus, as in assertion (i), it suffices to consider the case $X = Y$.

The reasoning completely analogous to the one used in the proof of Theorem 2.3.4 shows that for every $\varepsilon > 0$ there exists a number $n = n(\varepsilon)$ such that $|\mu_i|(X\backslash X_n) \leqslant \varepsilon$ for all i. Indeed, otherwise there exist increasing sequences of indices i_n and j_n for which

$$|\mu_{i_n}|(X_{j_{n+1}}\backslash X_{j_n}) > \varepsilon.$$

For every n there is a compact set $K_n \subset X_{j_{n+1}}\backslash X_{j_n}$ with $|\mu_{i_n}|(K_n) > \varepsilon$. There exists a continuous function ξ on X_{j_2} with values in $[1/2, 1]$ equal to 1 on K_1. This function can be extended to a continuous function on X_{j_3} with values in $[1, 1/2]$ and equal to $1/2$ on K_2. Consecutively extending ξ from X_{j_n} to $X_{j_{n+1}}$ such that the extension is continuous, takes values in $[1, 1/n]$ and equals $1/n$ on K_n, we obtain a function on all of X with values in $[0, 1]$. By the hypothesis of the proposition this function is continuous on the whole space. It is clear that the sets $U_n = \{1/n - \delta_n < \xi < 1/n + \delta_n\}$, where $\delta_n = (2n+1)^{-2}$, are open and disjoint. In addition, every point $x \in X$ possesses a neighborhood intersecting at most finitely many sets U_n. Hence for every choice of continuous functions φ_n with supports in U_n the series $\sum_{n=1}^{\infty}\varphi_n$ converges and defines a continuous function. For every n we take a continuous function f_n with values in $[-1,1]$ and support in U_n such that the integral of f_n with respect to the measure μ_n is larger than ε. Denote the integral of f_i with respect to the measure μ_n by a_n^i. Then $a_n=(a_n^1,a_n^2,\ldots)\in l^1$, since $\sum_{i=1}^{\infty}|f_i|\leqslant 1$. For every $\lambda=(\lambda_i)\in l^{\infty}$ the function $f^{\lambda}=\sum_{i=1}^{\infty}\lambda_i f_i$ is bounded and continuous. By assumption the sequence of the integrals of f^{λ} with respect to the measures μ_n converges. This means that the sequence $\{a_n\}$ is fundamental in the weak topology of l^1. By Corollary 1.3.7 the sequence $\{a_n\}$ converges in the norm of l^1, whence $\lim\limits_{n\to\infty}a_n^n=0$, a contradiction. In the case where all X_n are compact the proof is complete.

If all X_n are Polish spaces, it remains to verify the uniform tightness of the restrictions of μ_n to every space X_k. Suppose that these restrictions are not uniformly tight. The reasoning from the proof of the Prohorov theorem (Theorem 2.3.4) shows that for some $\varepsilon>0$ there is a subsequence of measures μ_{i_n} and a sequence of pairwise disjoint compact sets $K_n\subset X_k$ with the following properties: $|\mu_{i_n}|(K_n)>\varepsilon$ and the ε-neighborhoods of K_n (with respect to a complete metric generating the topology of X_k) are disjoint. Let us take a continuous function $\xi\colon X_k\to[0,1]$ equal to $1/n$ on K_n for each n. The same reasoning as above gives a contradiction. □

A completely regular space X is called *hemicompact* if it possesses a fundamental sequence of compact sets K_n (i.e., every compact set in X is contained in some K_n). If for the continuity of a function on X it suffices to have its continuity on all compact sets, then X is called a *$k_{\mathbb{R}}$-space.*

The latter property holds for any *k-space*, i.e., a space in which closed sets are the sets with closed intersections with all compact sets. By a theorem due to Arkhangelskii, every Čech complete space is a k-space (see Engelking [**203**, Theorem 3.9.5]). The class of hemicompact $k_{\mathbb{R}}$-spaces contains all locally compact σ-compact spaces.

4.7.7. EXAMPLE. In each of the following cases every weakly fundamental sequence of tight measures on X is uniformly tight:

(i) X is a hemicompact $k_{\mathbb{R}}$-space.

(ii) X is a locally convex space that is the inductive limit of an increasing sequence of separable Banach spaces E_n such that the embeddings $E_n\to E_{n+1}$ are compact operators. By definition, the space X is equipped with the strongest locally convex topology for which the embeddings $E_n\to X$ are continuous (see Example 4.5.5).

PROOF. Case (i) is clear from Proposition 4.7.6.

(ii) We observe that X is a k-space with a fundamental sequence of compact sets. For this we take an increasing sequence of closed balls U_n in the spaces X_n with $\bigcup_{n=1}^{\infty}U_n=X$ and denote by K_n the compact closure of U_n in X_{n+1}. If a set $A\subset X$ has closed intersections with all K_n, then it is easy to see that the set $A\cap E_n$ is closed in E_n. Let A have a limit point $a\notin A$. By induction we construct an increasing sequence of convex sets $V_n\subset E_n$ open in E_n such that $a\in V_n$ and $\overline{V_n}\cap A=\varnothing$. To this end, we observe that if a convex compact set K in a Banach space is disjoint with a closed set M, then K has a convex neighborhood of the closure which is disjoint with M. By definition, the set $V=\bigcup_{n=1}^{\infty}V_n$ is open in X. Since $a\in V$ and $A\cap V=\varnothing$, we arrive at a contradiction. □

4.7.8. EXAMPLE. Let X be a locally convex space that is the strict inductive limit of an increasing sequence of its closed subspaces X_n. Then X is a Prohorov space, provided that all spaces X_n are Prohorov. In particular, if each X_n is a separable Fréchet space, then every weakly fundamental sequence of nonnegative Baire measures on X is uniformly tight.

PROOF. According to Example 4.5.5, for every $\varepsilon > 0$, the measures from every weakly compact family M of nonnegative Radon measures on X are ε-concentrated on some subspace X_n. By Corollary 4.5.8, the restrictions of the measures from M to X_n form a set with compact closure. In order to obtain the last assertion of this example, we recall that the union of a sequence of separable Fréchet spaces is Souslin, hence Baire measures on this space are Radon. □

This example is contained in Hoffmann-Jørgensen [**325**, Corollary 7], where its justification is erroneous. Theorem 4.8.10 gives a modification of Proposition 4.7.6, which applies to the considered example too.

It is obvious that one can increase the number of such examples by taking countable products and passing to closed subsets. We observe that many classical spaces of functional analysis, such as $\mathcal{D}(\mathbb{R}^d)$, $\mathcal{D}'(\mathbb{R}^d)$, $\mathcal{S}(\mathbb{R}^d)$, and $\mathcal{S}'(\mathbb{R}^d)$, are Prohorov spaces, since they can be obtained by means of the indicated operations.

4.7.9. REMARK. The space $\mathcal{D}(\mathbb{R})$ is a Prohorov space, but is neither a k_R-space (Exercise 4.8.43), nor a hemicompact space (in addition, it is not σ-compact). The absence of a countable family of compact sets which is either fundamental or exhausting follows by Baire's theorem applied to the subspaces $\mathcal{D}_n(\mathbb{R})$ and the fact that every compact set in $\mathcal{D}(\mathbb{R})$ is contained in some of the subspaces $\mathcal{D}_n(\mathbb{R})$ (see Bogachev, Smolyanov [**96**, Chapter 8], [**97**, Chapter 2]).

The next result is obtained in Wójcicka [**661**]. In a sense it belongs to the subject of Chapter 5 about properties of spaces of measures, but it is placed here as an example of yet another operation that preserves the Prohorov property.

4.7.10. THEOREM. *Let X be a Prohorov space. Then the space $\mathcal{P}_r(X)$ with the weak topology is also Prohorov.*

PROOF. Let $S = \beta X$ be the Stone–Čech compactification of X. Then $\mathcal{P}_r(S)$ is compact in the weak topology, hence we have the following mapping considered below in Theorem 4.8.4 and Corollary 5.8.7:

$$T\colon \mathcal{P}_r\bigl(\mathcal{P}_r(S)\bigr) \to \mathcal{P}_r(S), \quad T(\Psi)(B) = \int_{\mathcal{P}_r(S)} m(B)\, \Psi(dm).$$

The spaces $\mathcal{P}_r(X)$ and $\mathcal{P}_r\bigl(\mathcal{P}_r(X)\bigr)$ are naturally embedded to $\mathcal{P}_r(S)$ and $\mathcal{P}_r\bigl(\mathcal{P}_r(S)\bigr)$, respectively. If $\Psi \in \mathcal{P}_r\bigl(\mathcal{P}_r(X)\bigr)$, then $T(\Psi) \in \mathcal{P}_r(X)$. Indeed, for every $\varepsilon > 0$ there is a compact set $Q \subset \mathcal{P}_r(X)$ with $\Psi(Q) > 1 - \varepsilon$. By our hypothesis there exists a compact set $K \subset X$ such that $q(K) > 1 - \varepsilon$ for all $q \in Q$. This gives $T(\Psi)(K) \geqslant (1-\varepsilon)^2$, i.e., $T(\Psi) \in \mathcal{P}_r(X)$. Let M be compact in $\mathcal{P}_r\bigl(\mathcal{P}_r(X)\bigr)$ and $\varepsilon > 0$. Then, as we have shown, the compact set $T(M)$ is contained in $\mathcal{P}_r(X)$, which by assumption gives compact sets $K_n \subset X$ with $K_n \subset K_{n+1}$ and

$$T(\Psi)(K_n) \geqslant 1 - \varepsilon^2 4^{-n} \quad \text{for all } \Psi \in M.$$

With the aid of Theorem 4.3.2 it is easy to verify that the sets

$$Q_n := \{q \in \mathcal{P}_r(S)\colon\ q(K_n) \geqslant 1 - \varepsilon 2^{-n}\}$$

are compact and $Q := \bigcap_{n=1}^{\infty} Q_n \subset \mathcal{P}_r(X)$. For all $\Psi \in M$ we have the bound $\Psi(Q_n) \geqslant 1 - \varepsilon 2^{-n}$, since $T(\Psi)(X\backslash K_n) \leqslant \varepsilon^2 4^{-n}$, whence

$$\Psi\bigl(q\colon\ q(X\backslash K_n) \geqslant \varepsilon 2^{-n}\bigr) \leqslant \varepsilon 2^{-n}.$$

We finally obtain $\Psi(Q) \geqslant 1-\varepsilon$, where the set Q is compact in the space $\mathcal{P}_r(X)$. □

No topological description of Prohorov spaces is known. The class of Prohorov spaces is not closed with respect to countable unions. In Example 4.5.4 we have a countable space X that is hemicompact, is Baire and is an F_σ-set in the Prohorov space βX, but is not Prohorov itself. An even more impressive (but also more difficult for justification) example is given by the set of rational numbers. It is presented in Theorem 4.8.6 in § 4.8(ii) along with some other interesting negative results and complementary remarks about the close class of Alexandroff spaces.

Yet another result from Fremlin, Garling, Haydon [**245**] strengthens the assertion from Example 4.7.7(i) with a similar proof.

4.7.11. THEOREM. *Let X be a hemicompact $k_{\mathbb{R}}$-space. Then every weakly compact subset of $\mathcal{M}_t(X)$ is uniformly tight, i.e., X is strongly Prohorov.*

PROOF. By our hypothesis, there exist compact sets $X_n \subset X_{n+1}$ such that every compact subset of X is contained in some of X_n, and the continuity of a function on every X_n implies its continuity on all of X. Suppose that a weakly compact set $M \subset \mathcal{M}_t(X)$ is not uniformly tight. As in the proof of assertion (ii) of Proposition 4.7.6, one can find measures $\mu_n \in M$ and increasing numbers j_n with

$$|\mu_n|(X_{j_{n+1}}\backslash X_{j_n}) > \varepsilon.$$

As in that proof, we take functions f_i for which $\sum_{i=1}^{\infty} |\lambda_i||f_i| \leqslant \|\lambda\|$ for all elements $\lambda = (\lambda_i) \in l^\infty$ and the integral of f_n with respect to μ_n is larger than ε. The mapping from M to l^1 associating to a measure μ the sequence of the integrals of f_i with respect to μ is continuous provided that both spaces M and l^1 are equipped with their weak topologies. The image of M under this mapping is weakly compact in l^1, which implies its compactness in norm. This contradicts the fact that the integral of f_n with respect to μ_n is larger than ε. □

This theorem yields the result from Fernique [**223**] that the strong dual to a Fréchet–Montel locally convex space is Prohorov.

4.7.12. COROLLARY. *Let a Fréchet space X be a Montel space: its bounded sets have compact closures. Then its strong dual X^*_β, i.e., the dual X^* with the topology of uniform convergence on bounded sets, which in this case coincides with the topology of uniform convergence on compact sets, is a Prohorov space.*

PROOF. Let $\{V_n\}$ be a base absolutely convex neighborhoods of zero in X. Then $X^* = \bigcup_{n=1}^{\infty} V_n^\circ$, where V_n° is the closed polar of V_n, i.e.,

$$V_n^\circ = \{f \in X^*\colon\ |f(x)| \leqslant 1\, \forall\, x \in V_n\}.$$

It is well-known that the sets V_n° are compact in the topology $\sigma(X^*, X)$ (see Bogachev, Smolyanov [**97**, Theorem 3.1.4]). In addition, by the Banach–Dieudonné theorem (see [**97**, Theorem 3.8.15]), in the topology of uniform convergence on compact sets (which in our case coincides with the strong topology by the assumption that X is Montel) the closed sets are precisely those sets whose intersections with equicontinuous sets are closed in the topology $\sigma(X^*, X)$; moreover, on equicontinuous sets both topologies agree. Equicontinuous sets in our situation are the subsets of the sets V_n°. Thus, on the sets V_n° the strong topology coincides with $\sigma(X^*, X)$, and these sets are compact. In addition, the cited theorem says that the continuity of a function in the strong topology is equivalent to the continuity of its restrictions to all sets V_n°. Furthermore, every compact set S in X^*_β is contained in some V_n°. Indeed, the polar S° of the set S in X is an absorbing closed absolutely convex set (its homothetic images cover the whole space X), whence it follows that $V_n \subset S^\circ$ for some n (this follows both from Baire's theorem and the known fact that all Fréchet spaces are barrelled). By the bipolar theorem this gives the desired inclusion $S \subset V_n^\circ$ (see [**97**, Theorem 3.1.1]). Therefore, the space X^*_β is a hemicompact k_R-space. Let us note by passing that the strong topology and the topology $\sigma(X^*, X)$ in our case have the same compact sets, hence the classes of Radon measures in both topologies are the same. □

In particular, for the countable power of the real line $\mathbb{R}^\infty$ the dual is the space of finite sequences $\mathbb{R}^\infty_0$, i.e., the countable union of finite-dimensional subspaces, moreover, in this case $\mathbb{R}^\infty_0$ is equipped with the strong topology that coincides with the topology of inductive limit, see Example 4.5.5. Thus, a nonmetrizable Prohorov space need not be Baire. If $\mathbb{R}^\infty_0$ is regarded as a subspace of the metrizable space $\mathbb{R}^\infty$ with its topology of coordinate-wise convergence (which is weaker than the strong topology), then we obtain an incomplete metric space that is not Prohorov (this follows from Theorem 4.8.9 below).

Note that infinite-dimensional Montel spaces are not normable. The spaces $\mathcal{S}(\mathbb{R}^d)$ and $C^\infty(\mathbb{R}^d)$ belong to this class.

4.8. Complements and exercises

4.8(i). Compactness in the space of signed measures

The next result from Varadarajan [**635**, Part 2, Theorem 3] is useful in the study of weak convergence of signed measures.

4.8.1. THEOREM. *Suppose that a net $\{\mu_\alpha\}$ of Baire measures converges weakly to a Baire measure μ. Then, for every functionally open set U, we have*

$$\liminf_\alpha |\mu_\alpha|(U) \geqslant |\mu|(U).$$

The net of measures $|\mu_\alpha|$ converges weakly to $|\mu|$ precisely when $|\mu_\alpha|(X) \to |\mu|(X)$.

PROOF. Let $\varepsilon > 0$. By the analog of Lemma 2.1.11 mentioned on p. 143 one can find a function $g \in C_b(X)$ such that $0 \leqslant g \leqslant 1$, $g = 0$ on $X\backslash U$ and

$$\int_X g\, d|\mu| > |\mu|(U) - \varepsilon.$$

It is readily seen that there exists a function $h \in C_b(X)$ such that one as the pointwise estimate $|h| \leqslant g$ and the inequality

$$\left|\int_X h\, d\mu\right| > \int_X g\, d|\mu| - \varepsilon.$$

It is clear that $|h| \leqslant 1$ and $h = 0$ on $X\backslash U$. In addition,

$$\left|\int_X h\,d\mu\right| > |\mu|(U) - 2\varepsilon.$$

Since

$$\int_X h\,d\mu_\alpha \to \int_X h\,d\mu,$$

we arrive at the estimates

$$\liminf_\alpha |\mu_\alpha|(U) \geqslant \lim_\alpha \left|\int_X h\,d\mu_\alpha\right| = \left|\int_X h\,d\mu\right| > |\mu|(U) - 2\varepsilon.$$

Letting $\varepsilon \to 0$, we obtain the first assertion of the theorem. If $|\mu_\alpha|(X) \to |\mu|(X) > 0$, then weak convergence of $|\mu_\alpha|$ to $|\mu|$ follows from the first assertion and Theorem 4.3.2. If $|\mu|(X) = 0$, then we even have convergence in variation. □

4.8.2. COROLLARY. *Suppose that a net $\{\mu_\alpha\}$ of Baire measures converges weakly to a Baire measure μ and*

$$\lim_\alpha |\mu_\alpha|(X) = |\mu|(X).$$

Let $\mu_\alpha = \mu_\alpha^+ - \mu_\alpha^-$ and $\mu = \mu^+ - \mu^-$. Then the nets $\{\mu_\alpha^+\}$ and $\{\mu_\alpha^-\}$ converge weakly to μ^+ and μ^-, respectively.

PROOF. We apply the proven theorem and the equality $\mu_\alpha^+ = (|\mu_\alpha| + \mu_\alpha)/2$, $\mu_\alpha^- = (|\mu_\alpha| - \mu_\alpha)/2$. □

4.8.3. THEOREM. *A set $M \subset \mathcal{M}_\sigma(X)$ is contained in a compact set in the weak topology precisely when*

$$\lim_{n\to\infty} \sup_{\mu\in M} \left|\int_X f_n\,d\mu\right| = 0$$

for every sequence of functions $f_n \in C_b(X)$ pointwise converging to zero.

PROOF. Suppose that M is compact and a sequence $\{f_n\} \subset C_b(X)$ converges pointwise to zero, but the absolute values of the integrals of the functions f_n with respect to measures from M do not tend to zero uniformly in $\mu \in M$. One can assume that $\|\mu\| \leqslant 1$ for all $\mu \in M$. Since the integrals of f_n with respect to each fixed measure tend to zero, by induction one can find increasing numbers n_i and measures $\mu_i \in M$ such that for some $\varepsilon > 0$

$$\left|\int_X f_{n_i}\,d\mu_i\right| \geqslant 3\varepsilon, \quad \int_X f_{n_j}\,d|\mu_i| \leqslant \varepsilon \ \ \forall j > i.$$

Let $g_i = \max(0, f_{n_i} - f_{n_{i+1}} - \varepsilon)$. Then

$$\int_X g_i\,d\mu_i \geqslant \varepsilon \quad \forall i,$$

because

$$\int_X \min(f_{n_i}, f_{n_{i+1}} + \varepsilon)\,d\mu_i \leqslant \int_X (f_{n_{i+1}} + \varepsilon)\,d|\mu_i| \leqslant 2\varepsilon.$$

Let $x \in X$. Let us find m with $f_{n_m}(x) < \varepsilon$ and then find a neighborhood U of x such that $f_{n_m}(y) < \varepsilon$ for all $y \in U$. Hence $g_i(y) = 0$ for all $i > m$ and $y \in U$, i.e., the sequence of functions g_i is locally finite. In addition,

$$\sum_{i=1}^\infty g_i(t) \leqslant f_{n_1}(t) \leqslant \|f_{n_1}\|_\infty.$$

For every element $\alpha = (\alpha_i) \in l^\infty$ we set $T\alpha := \sum_{i=1}^\infty \alpha_i g_i$. Obviously, $T\alpha \in C_b(X)$, moreover, the operator $T\colon l^\infty \to C_b(X)$ is linear and continuous. In case $\alpha_i \geqslant 0$ the functions $\sum_{i=n}^\infty \alpha_i g_i$ decrease pointwise to zero, whence it follows that for every measure $\nu \in \mathcal{M}_\sigma(X)$ their integrals with respect to ν tend to zero. This yields (by considering separately ν^+ and ν^-) that the series of the integrals of g_i with respect to ν converges absolutely. Thus, the mapping S on $\mathcal{M}_\sigma(X)$ taking ν to the sequence of integrals of the functions g_i with respect to ν takes values in l^1. This mapping S is adjoint for T when we equip $C_b(X)$ with the topology $\sigma(C_b, \mathcal{M}_\sigma)$ and equip l^∞ with the topology $\sigma(l^\infty, l^1)$. Therefore, the set $S(M)$ is weakly compact in l^1. Then it is also norm compact (see Exercise 1.7.15). This yields that the integrals of the functions g_i against measures from M converge to zero uniformly on M, which contradicts the fact that the integral of g_i with respect to μ_i is not smaller than ε in absolute value. The converse assertion is proved by the same reasoning as in one implication of Theorem 4.5.10 (where we did not use the nonnegativity of measures, as was noted there). □

The next result is obtained in Hoffmann-Jørgensen [**325**]. It is known from the theory of locally convex spaces that the closed absolutely convex hull of a compact set in a complete or quasicomplete locally convex space is compact. However, spaces of measures with the weak topology do not possesses such completeness. For example, in the unit ball in $\mathcal{M}(\mathbb{N})$ there is a fundamental net of probability measures that has no weak limit. Indeed, by the Hahn–Banach theorem, there is a functional on $C_b(\mathbb{N}) = l^\infty$ that is the limit at infinity for sequences having such a limit. This functional is not represented by a measure, but is a limit point in the weak-$*$ topology for the set of functionals represented by Dirac measures (a more general fact is considered in Exercise 4.8.48). Hence the next theorem does not follow trivially from general facts of the theory of locally convex spaces.

4.8.4. THEOREM. *Let X be a completely regular space. Then every weakly compact set M in $\mathcal{M}_\tau(X)$ is contained in a centrally symmetric convex weakly compact set and hence the closed convex hull of M is weakly compact.*

PROOF. According to Corollary 5.8.7 proved in Chapter 5, every Radon measure Ψ on the compact set M generates a τ-additive measure on X by the formula

$$T(\Psi)(B) = \int_M \mu(B)\,\Psi(d\mu).$$

For every function $f \in C_b(X)$ we have the equality

$$\int_X f(x)\,T(\Psi)(dx) = \int_M \int_X f(x)\,\mu(dx)\,\Psi(d\mu).$$

Hence the mapping $T\colon \mathcal{M}_r(M) \to \mathcal{M}_\tau(X)$ is continuous in the weak topology. The closed unit ball K in $\mathcal{M}_r(M)$ is compact in the weak topology. Hence $T(K)$ is centrally symmetric, convex and compact. Furthermore, $M \subset T(K)$, since we have $\mu = T(\delta_\mu)$ for all $\mu \in M$. □

As demonstrated by Example 4.5.9, the set of total variations of measures from a weakly compact set of signed measures on an arbitrary space is not always weakly compact. The next proposition from Fremlin, Garling, Haydon [**245**] gives a partial compensation.

For a set of measures M in some linear space $\mathcal{M}$ of measures on a common σ-algebra $\mathcal{A}$, the *solid hull* of $S(M)$ consists of all measures $\nu \in \mathcal{M}$ on $\mathcal{A}$ such that there exists a measure $\mu \in M$ for which $|\nu(A)| \leqslant |\mu|(A)$ for all $A \in \mathcal{A}$.

The solid hull is absolutely convex, but can be broader than the absolutely convex hull. For example, the solid hull of the set of probability measures is the unit ball in variation, which coincides with the absolutely convex hull, but the solid hull of a singleton equal to Lebesgue measure on $[0,1]$ consists of all measures whose densities are estimated in absolute value by 1, which is much larger than the interval in the space of measures joining Lebesgue measure and the same measure with minus (this interval is the absolutely convex hull of the point).

We recall that Corollary 3.1.10 says that in the case of a Polish space the solid hull of a set with a weakly compact closure has weakly compact closure too.

We shall say that a set $M \subset \mathcal{M}_\tau(X)$ is uniformly τ-smooth if, for every net of functions $f_\alpha \in C_b(X)$ decreasing to zero, the integrals of f_α with respect to measures in M tend to zero uniformly on M. Uniformly σ-smooth sets are defined similarly by replacing nets by usual sequences.

4.8.5. PROPOSITION. (i) *If a set $M \subset \mathcal{M}_\tau(X)$ is uniformly τ-smooth, then so is its solid hull in $\mathcal{M}_\tau(X)$. In addition, both sets are contained in compact sets in $\mathcal{M}_\tau(X)$ with the weak topology.* (ii) *If a set $M \subset \mathcal{M}_\sigma(X)$ is uniformly σ-smooth, then so is its solid hull in the space $\mathcal{M}_\sigma(X)$.*

PROOF. For justification of the first assertion it suffices to verify that the set $M^+ = \{\mu^+ \colon \mu \in M\}$ is uniformly τ-smooth. If this is false, then for some $\varepsilon > 0$ there exist a net of functions $f_\alpha \in C_b(X)$ decreasing to zero and measures μ_α^+ in M^+ such that the integral of f_α with respect to μ_α^+ is larger than ε. Then it is easy to show that there exist functions $g_\alpha \in C_b(X)$ for which $0 \leqslant g_\alpha \leqslant f_\alpha$ and the integral of g_α with respect to the measure μ_α is larger than ε. The net $\{g_\alpha\}$ tends to zero, but not necessarily monotonically, which is still not enough for obtaining a contradiction. But we find indices $\beta(\alpha) > \alpha$ such that the sum of the absolute value of the integral of $(f_{\beta(\alpha)} - g_\alpha)^+$ against the measure μ_α and ε will be smaller than the integral of g_α against μ_α. This is possible by the τ-additivity of μ_α and decreasing of the net of functions $(f_\beta - g_\alpha)^+$ to zero for each fixed α. Set

$$h_\alpha = \max(f_{\beta(\alpha)}, g_\alpha) = g_\alpha + (f_{\beta(\alpha)} - g_\alpha)^+.$$

It is clear that $h_\alpha \in C_b(X)$, $f_{\beta(\alpha)} \leqslant h_\alpha \leqslant f_\alpha$ and that the integral of h_α with respect to the measure μ_α is larger than ε by the foregoing estimates.

However, now with the aid of the constructed family of functions we can obtain the desired net monotonically decreasing to zero. This family will be used as a new index set directed by the usual pointwise comparison: for any functions h_α and h_γ with $\gamma \geqslant \alpha$, one can take β such that $\beta > \beta(\alpha)$ and $\beta > \beta(\gamma)$, which yields $h_\beta \leqslant h_\alpha$ and $h_\beta \leqslant h_\gamma$, since $f_\beta \leqslant f_{\beta(\alpha)}$ and $f_\beta \leqslant f_{\beta(\gamma)}$. As a result we obtain a net of functions decreasing to zero (because the net $\{h_\alpha\}$ tends to zero) for the integrals of which with respect to measures from M there is no uniform convergence. The second assertion is proved similarly. □

4.8(ii). More on Prohorov and Alexandroff spaces

Here we present a number of additional results on Prohorov spaces (mainly of a negative character) and also briefly discuss yet another interesting class of spaces close to Prohorov spaces.

The following surprising result of D. Preiss [**521**] became a fundamental achievement of topological measure theory.

4.8.6. THEOREM. *The space of rational numbers* $\mathbb{Q}$ *with its usual topology is not a Prohorov space.*

PROOF. 1) Let us show that there exists a nondecreasing sequence of nonempty compact sets X_k in $X = \mathbb{Q} \cap [0,1]$ with the union X and the property that the intersection $X_{k+1} \cap [x-\delta, x+\delta]$ is infinite for every k, $x \in X_k$ and $\delta > 0$. Let us write X as a sequence $\{q_k\}$. Set $X_1 = \{q_1\}$ and continue our construction inductively. If a compact set $X_k \subset X$ is already found, then, for every $m \in \mathbb{N}$, let $\mathcal{E}_m$ be a finite cover of this compact set X_k by intervals of length at most 2^{-m} intersecting X_k and let $I_{k,m} \subset X \backslash X_k$ be some finite subset that has nonempty intersection with each interval in $\mathcal{E}_m$. Set

$$X_{k+1} = X_k \cup \{q_{k+1}\} \cup \bigcup_{m \geqslant 1} I_{k,m}.$$

We verify that X_{k+1} is compact. Let $\mathcal{U}$ be a cover of X_{k+1} by open sets in $\mathbb{R}$. One can find a finite subcover $\mathcal{U}_0 \subset \mathcal{U}$ of the compact set X_k. There is also $m \in \mathbb{N}$ such that $[x - 2^{-m}, x + 2^{-m}] \subset \cup \mathcal{U}_0$ for every $x \in X_k$, hence $I_{k,l} \subset \cup \mathcal{U}_0$ for all $l \geqslant m$, moreover, $X_{k+1} \backslash \cup \mathcal{U}_0$ is finite, whence we obtain a finite subcover of all of X_{k+1}. If now $x \in X_k$ and $\delta > 0$, then for every $m \in \mathbb{N}$ there exists $x_0 \in X_{k+1} \backslash X_k$ such that $|x_0 - x| < 2^{-m}$, hence $[x - \delta, x + \delta] \cap X_{k+1}$ is infinite.

2) We now show that if we are given a sequence of numbers ε_k in $(0,1)$ and a countable closed set $F \subset [0,1]$, then there exists an element $z \in X \backslash F$ such that $\operatorname{dist}(z, X_k) < \varepsilon_k$ for all $k \in \mathbb{N}$. We can assume that $\varepsilon_k \to 0$. By induction we define sets $H_k \subset \mathbb{R}$ as follows. Let $H_1 = \mathbb{R}$. having a set H_k, let

$$H_{k+1} = H_k \cap \{x \colon \operatorname{dist}(x, X_k \cap H_k) < \varepsilon_k\}.$$

We observe that the sets H_k are open in $\mathbb{R}$ and $X_k \cap H_k \subset H_{k+1} \subset H_k$ for all k. Hence $E = \bigcap_{k \geqslant 1} H_k$ is a G_δ-set in $\mathbb{R}$ and $X_k \cap H_k \subset E$ for all k. The set $E \cap X$ contains q_1 and hence is not empty. Further, for every k we have $\operatorname{dist}(x, E \cap X_k) < \varepsilon_k$ for all $x \in H_{k+1}$, hence also for all $x \in E$, whence it follows that the set $E \cap X$ is everywhere dense in E. Moreover, if $x \in E \cap X$, then there exists $k \in \mathbb{N}$ such that $x \in X_k$; hence $x \in H_k$, so that in this case H_{k+1} is a neighborhood of x. Thus, every neighborhood of x contains infinitely many points from the set $H_{k+1} \cap X_{k+1} \subset E \cap X$. Therefore, $E \cap X$ has no isolated points, hence E also has no isolated points. This yields that E is uncountable (this follows from Baire's theorem, since E is a Polish space being a G_δ-set and points in E are nowhere dense). Then there exists a point $y \in E \backslash F$. Let $m \in \mathbb{N}$ be such that $\operatorname{dist}(y, F) \geqslant \varepsilon_m$. Since $y \in H_{m+1}$, there is a point $z \in H_m \cap X_m$ for which $|z - y| < \varepsilon_m$ and $z \notin F$. Let $k \in \mathbb{N}$. If $k \geqslant m$, then obviously $\operatorname{dist}(z, X_k) = 0 < \varepsilon_k$. If $k < m$, then $z \in H_{k+1}$, hence $\operatorname{dist}(z, X_k) \leqslant \operatorname{dist}(z, H_k \cap X_k) < \varepsilon_k$. Thus, the desired point z is found.

3) For $n, k \in \mathbb{N}$ set

$$G_{k,n} := \bigl\{x \in \mathbb{R} \backslash X_k \colon \operatorname{dist}(x, X_n) > 2^{-k}\bigr\}.$$

It is clear that the sets $G_{k,n}$ are open in $\mathbb{R}$. Let $\mathcal{K}$ be the set of all Radon probability measures μ on X for which $\mu(G_{k,n} \cap X) \leqslant 2^{-n}$ for all $n, k \in \mathbb{N}$.

4) Denote by $\mathcal{K}_1$ the set of Radon probability measures μ on the whole interval $[0,1]$ for which $\mu(G_{k,n} \cap [0,1]) \leqslant 2^{-n}$ for all $k, n \in \mathbb{N}$. The set $\mathcal{K}_1$ is closed in the

weak topology on $\mathcal{P}([0,1])$ (since the sets $G_{k,n}\cap[0,1]$ are open in the space $[0,1]$), hence it is weakly compact. We show that $\mu([0,1]\backslash X)=0$ for all measures $\mu\in\mathcal{K}_1$. Indeed, let $K\subset[0,1]\backslash X$ be compact and $n\in\mathbb{N}$. Then K and X_n are disjoint compact sets, hence there exists $k\in\mathbb{N}$ such that $|x-y|>2^{-k}$ for all $x\in X_n$ and $y\in K$. Then $K\subset G_{k,n}$, hence $\mu(K)<2^{-n}$. Since n was arbitrary, we obtain that $\mu(K)=0$, whence it follows that $\mu([0,1]\backslash X)=0$.

We now show that $\mathcal{K}$ is also compact in the weak topology on $\mathcal{P}(X)$. The identity mapping $j\colon X\to[0,1]$ induces the mapping $\widehat{j}\colon \mathcal{P}(X)\to\mathcal{P}([0,1])$, which is a homeomorphism between $\mathcal{P}(X)$ and the set $\bigl\{\mu\in\mathcal{P}([0,1])\colon \mu([0,1]\backslash X)=0\bigr\}$, see Proposition 4.3.15. It is clear that then $\widehat{j}$ is homeomorphically mapping $\mathcal{K}$ on $\mathcal{K}_1$, which proves compactness of $\mathcal{K}$.

5) Let us proceed to the proof of the fact that the obtained set of measures $\mathcal{K}$ on X is not uniformly tight. This is the most difficult part of justification of the whole construction. Let $K\subset X$ be compact. Let us consider the set W of functions $w\in[0,1]^X$ such that $w(x)=0$ if $x\in K$, $\sum_{x\in X}w(x)\leqslant 1$ and $\sum_{x\in G_{k,n}\cap X}w(x)\leqslant 2^{-n}$ for all $k,n\in\mathbb{N}$. It is clear that W is compact in $[0,1]^X$ with the Tychonoff product topology. In addition, if $D\subset W$ is a nonempty set directed by increasing (with the pointwise order), then its supremum in $[0,1]^X$ belongs to W. By Zorn's lemma the set W has some maximal element w (this means that in W there is no other element w_0 with $w(x)\leqslant w_0(x)$ for all $x\in[0,1]$).

We show that $\sum_{x\in X}w(x)=1$. Let $\sum_{x\in X}w(x)=\gamma<1$. For every $n\in\mathbb{N}$ there exists a finite set $L_n\subset X$ for which $\sum_{x\in L_n}w(x)>\gamma-2^{-n}$ and a number $m_n\in\mathbb{N}$ such that $L_n\subset X_{m_n}$. According to Step 2) there exists $z\in X\backslash K$ for which $\operatorname{dist}(z,X_n)<2^{-m_n}$ for all $n\in\mathbb{N}$. Let $r\in\mathbb{N}$ be such that $z\in X_r$ and $\gamma+2^{-r}\leqslant 1$. Set $w_0(z)=w(z)+2^{-r}$, $w_0(x)=w(x)$ if $x\in X\backslash\{z\}$. Then we have $w_0\in[0,1]^X$ and $\sum_{x\in X}w_0(x)\leqslant 1$. If $k,n\in\mathbb{N}$ and $z\notin G_{k,n}$, then

$$\sum_{x\in G_{k,n}\cap X} w_0(x)=\sum_{x\in G_{k,n}\cap X} w(x)\leqslant 2^{-n}.$$

If $z\in G_{k,n}$, then $n<r$ and $2^{-k}<\operatorname{dist}(z,X_n)<2^{-m_n}$, hence $m_n<k$ and $L_n\subset X_k$, whence we obtain

$$\sum_{x\in G_{k,n}\cap X} w(x)\leqslant \sum_{x\in X\backslash X_k} w(x)\leqslant \sum_{x\in X\backslash L_n} w(x)\leqslant 2^{-n-1}$$

and also

$$\sum_{x\in G_{k,n}\cap X} w_0(x)\leqslant 2^{-n-1}+2^{-r}\leqslant 2^{-n},$$

so that $w_0\in W$, hence w was not maximal.

Thus, $\sum_{x\in X}w(x)=1$, hence the measure μ generated by w is a probability measure on X. By the definition of w we have $\mu\in\mathcal{K}$ and $\mu(X\backslash K)=1$. Since K was arbitrary, the uniform tightness of $\mathcal{K}$ fails. Thus, $\mathbb{Q}\cap[0,1]$ is not a Prohorov space, hence $\mathbb{Q}$ is not a Prohorov space. □

We recall that by Theorem 4.7.2 the space $\mathbb{Q}$ of rational numbers is sequentially Prohorov. The first examples of separable metric spaces that are not Prohorov spaces were constructed in Choquet [**137**] and Davies [**159**].

As observed by Fernique [**223**], the space l^2 with its weak topology is not Prohorov.

4.8.7. EXAMPLE. The sequence of measures $\mu_n = n^{-3}\sum_{i=1}^{n^3}\delta_{ne_i}$, where $\{e_i\}$ is the standard orthonormal basis in l^2, converges weakly to Dirac's measure at the origin provided that l^2 is equipped with the weak topology, but obviously is not tight. For the proof of weak convergence it suffices to observe that for every set of the form $S = \{x\colon |(x,v)| < 1\}$, $v \in l^2$, one has the relation $\mu_n(S) \to 1$, which is seen from the estimates

$$\mu_n(l^2\backslash S) \leqslant \int_{l^2} |(x,v)|^2\,\mu_n(dx) \leqslant n^{-3}\sum_{i=1}^{n^3} n^2 v_i^2 \leqslant n^{-1}(v,v).$$

The space l^2 with the weak topology gives an example of a hemicompact σ-compact space that is not Prohorov. In the paper Fremlin, Garling, Haydon [**245**] this example was generalized as follows.

4.8.8. PROPOSITION. *If X is an infinite-dimensional Banach space, then $\bigl(X,\sigma(X,X^*)\bigr)$ and $\bigl(X^*,\sigma(X^*,X)\bigr)$ are not Prohorov spaces.*

PROOF. By the Dvoretzky–Rogers theorem (see, e.g., Day [**161**, Chapter IV, § 1, Lemma 2]) for every n one can find an n-dimensional subspace X_n in X and a linear operator $T_n\colon \mathbb{R}^n \to X_n$, where $\mathbb{R}^n$ is equipped with the standard norm, for which $\|T_n\| \leqslant 2$ and $\|Te_i\| = 1$ for the standard basis $\{e_i\}$ in $\mathbb{R}^n$. Set

$$\mu_n = n^{-1}\sum_{i=1}^{n}\delta_{y_i}, \quad y_i = n^{1/4}T_n e_i.$$

We show that $\mu_n \Rightarrow \delta_0$. As in Fernique's example, it suffices to verify that for every $f \in X^*$ we have $\mu_n\bigl(f^{-1}([-1,1])\bigr) \to 1$. The estimate

$$\sum_{i=1}^{n} |f(y_i)|^2 = n^{1/2}\sum_{i=1}^{n} |(e_i, T_n^* f)|^2 = n^{1/2}\|T_n^* f\|^2 \leqslant 4n^{1/2}\|f\|^2$$

yields that $|f(y_i)| > 1$ for at most $4n^{1/2}\|f\|^2$ indices i, hence we have the inequality

$$\mu_n\bigl(f^{-1}([-1,1])\bigr) \geqslant 1 - 4n^{-1/2}\|f\|^2,$$

as required. However, the sequence $\{\mu_n\}$ is not tight in the weak topology on X, since weakly compact sets are norm bounded in X, but the measure μ_n is concentrated on the sphere of radius $n^{1/4}$ (because we have $\|y_i\| = n^{1/4}$).

The case of X^* with the weak-$*$ topology is similar. It is worth noting that the dual space X^* with the stronger topology of uniform convergence on compact sets in X is Prohorov. Indeed, by the Banach–Dieudonné theorem (see Bogachev, Smolyanov [**97**, Theorem 3.8.15]) a set is closed in this topology precisely when its intersection with every closed ball is closed in the topology $\sigma(X^*,X)$, moreover, on balls both topologies coincide, which implies that the classes of compact sets in both topologies (hence the classes of Radon measures) are also the same. Therefore, in the new topology the space X^* is a k-space and remains hemicompact, so that Corollary 4.7.5 applies. □

Let us mention the following important result due to Preiss [**521**]. A metric space is called co-analytic if it is the complement of a Souslin set in a complete separable metric space. Preiss established that if such a space is not Polish, then it contains a G_δ-set homeomorphic to the set of rational numbers. This yields assertion (ii) in the theorem below.

4.8.9. THEOREM. (i) *A first category metric space cannot be Prohorov (unlike the space* $\mathbb{R}_0^\infty$ *with the inductive limit topology mentioned after Corollary* 4.7.12).

(ii) *Let* X *be a separable co-analytic metric space. Then* X *is a Prohorov space if and only if* X *is metrizable by a complete metric. An equivalent condition: the space* X *contains no countable* G_δ*-set dense in itself, i.e., having no isolated points.*

(iii) *Under the assumption of the continuum hypothesis, there exists a Prohorov separable metric space that cannot be metrized by a complete metric.*

PROOF. We justify only assertion (iii). Suppose that we have an uncountable set $Y \subset [0,1]$ such that it does not contain uncountable compact sets and the intersection of Y with each compact set in $\mathcal{R} = [0,1]\backslash\mathbb{Q}$ is at most countable. Then $X = [0,1]\backslash Y$ is a Prohorov space. Indeed, let $\mathcal{M} \subset \mathcal{P}_r(X)$ be compact and $\varepsilon > 0$. The set of extensions of measures from $\mathcal{M}$ to $\mathcal{R}$ is also compact. Since $\mathcal{R}$ is a Prohorov space (as a Polish space), there exists a compact set $K_1 \subset \mathcal{R}$ such that $\mu(K_1) > 1 - \varepsilon/2$ for all $\mu \in \mathcal{M}$. Similarly, since $K_1 \cap Y$ is finite or countable, the space $[0,1]\backslash(K_1 \cap Y)$ is also Prohorov, hence there exists a compact set $K_2 \subset [0,1]\backslash(K_1 \cap Y)$ such that $\mu(K_2) > 1 - \varepsilon/2$ for all $\mu \in \mathcal{M}$, where measures from $\mathcal{M}$ are also extended to $[0,1]\backslash(K_1 \cap Y)$. Then $K = K_1 \cap K_2$ is compact and obviously $K \subset X$ and $\mu(K) > 1-\varepsilon$ for all $\mu \in \mathcal{M}$. The space X is not Polish, since it cannot be a G_δ-set (otherwise its complement Y is a countable union of compact sets, which cannot be all countable).

We now construct a set Y with the stated properties. Under the continuum hypothesis the set Ω of all countable ordinals has the cardinality of the continuum, hence the collection of open sets in $[0,1]$ containing all rational numbers from $[0,1]$ can be enumerated by the elements of Ω in the form $\{U_\omega\}_{\omega\in\Omega}$. For every $\alpha \in \Omega$, by transfinite induction we choose points y_ω in the set $\bigcap_{\beta<\alpha} U_\beta \backslash \{y_\gamma\colon\ \gamma < \alpha\}$. Let $Y = \{y_\alpha\colon\ \alpha \in \Omega\}$. If K is compact in $\mathcal{R}$, then $[0,1]\backslash K$ is one of the sets U_ω. There is an at most countable set of points $y_\alpha \in Y$ with $\alpha < \omega$, since all these ordinal numbers are countable. If $\alpha \geqslant \omega$, then $y_\alpha \in U_\omega$, i.e., $y_\alpha \notin K$. Thus, $K \cap Y$ is at most countable. Finally, if we suppose that K is an uncountable compact set in Y, then the set $K \cap \mathcal{R}$ will be an uncountable G_δ-set, i.e., an uncountable Polish space, but then it contains an uncountable compact set (see Bogachev [**81**, Lemma 6.1.16]), which is impossible, as shown above. □

Since every countable space without isolated points is homeomorphic to $\mathbb{Q}$, assertion (ii) explains the role of $\mathbb{Q}$ in Theorem 4.8.6.

Under some additional set-theoretic assumption (the so-called axiom of constructibility of Gödel) there exists even a Souslin Prohorov subset of the interval that is not Polish (see Cox [**145**], Gardner [**257**]). It is an open question whether it is compatible with the ZFC that every universally measurable Prohorov space $X \subset [0,1]$ is topologically complete (i.e., Polish).

Hoffmann-Jørgensen [**325**] proved the following useful modification of Proposition 4.7.6, which can be applied in Examples 4.5.5 and 4.7.8.

4.8.10. THEOREM. *Suppose that a completely regular space* X *is the union of an increasing sequence of its closed subsets* X_n *and a family* $\mathcal{F} \subset C(X)$ *is given with the following properties:*

(i) *if* $f, g \in \mathcal{F}$, *then* $\max(f,g) \in \mathcal{F}$,

(ii) *for every* n *and every point* $x \notin X_n$, *there exists a function* $f \in \mathcal{F}$ *such that* $f \geqslant 0$, $f|_{X_n} = 0$, $f(x) \geqslant 1$,

(iii) *if a function* $\psi \geqslant 0$ *on* X *is such that for every* n *there is a function* $f_n \in \mathcal{F}$ *with* $f_n|_{X_n} = \psi|_{X_n}$, *then* ψ *is continuous.*

If $M \subset \mathcal{M}_r^+(X)$ *is compact in the weak topology, then for each* $\varepsilon > 0$ *there is* n *such that* $\mu(X\backslash X_n) < \varepsilon$ *for all* $\mu \in M$. *If all* X_n *are Prohorov, then* X *is also.*

PROOF. Otherwise for some $\varepsilon > 0$ there exist measures $\mu_n \in M$ such that $\mu_n(X\backslash X_n) > \varepsilon$. By assumption, for every $x \in X\backslash X_n$ there exists $f_{n,x} \in \mathcal{F}$ with $f_{n,x} \geqslant 0$, $f_{n,x}|_{X_n} = 0$, $f_{n,x}(x) \geqslant 1$. The sets $V_n(x) = \{y\colon f_{n,x}(y) > 1/2\}$ are open neighborhoods of x. By the Radon property of the given measures, for every n, there exists a finite collection of points $x(1,n), \ldots, x(k_n,n) \in X\backslash X_n$ for which

$$\mu_n\Bigl(V_n\bigl(x(1,n)\bigr) \cup \cdots \cup V\bigl(x(k_n,n)\bigr)\Bigr) > \varepsilon.$$

Set $h_n = \max\bigl(f_{n,x(1,n)}, \ldots, f_{n,x(k_n,n)}\bigr)$. Then we have $h_n \in \mathcal{F}$, $h_n \geqslant 0$, $h_m|_{X_n} = 0$ whenever $m \geqslant n$. In addition, $\mu_n\bigl(x\colon h_n(x) > 1/2\bigr) > \varepsilon$.

Let us consider the functions $f_n = \sup_{k\geqslant n} h_k$. If $x \in X_m$ and $m \geqslant n$, then $f_n(x) = \max\bigl(h_n(x), \ldots, h_m(x)\bigr)$. Hence our assumption yields the continuity of all functions f_n on all of X. Therefore, all sets $V_n = \{x\colon f_n(x) < 1/2\}$ are open, moreover, $X_n \subset V_n$, so they increase to X, hence every compact set is contained in some of them. Theorem 4.5.7 on account of weak compactness of M yields that there exists k such that $\mu(X\backslash V_k) < \varepsilon$ for all $\mu \in M$. Hence $\mu_k\bigl(x\colon f_k(x) \geqslant 1/2\bigr) < \varepsilon$, but then by the inequality $f_k \leqslant h_k$ we obtain $\mu_n\bigl(x\colon h_n(x) > 1/2\bigr) < \varepsilon$ contrary to the estimate obtained above. The last assertion of the theorem follows from the already proven facts, because the restrictions of the measures from M to X_n form a compact set by Corollary 4.5.8. □

Let us explain how one can use this theorem to derive Example 4.5.5, where a locally convex space X is the strict inductive limit of its increasing closed linear subspaces X_n. For $\mathcal{F}$ we take the set of all uniformly continuous functions $f \geqslant 0$ on X. It is obvious that condition (i) is fulfilled. To verify condition (ii) we apply the Hahn–Banach theorem, which gives a continuous linear functional on X equal to zero on X_n and to one at the point $x \notin X_n$. Let us verify condition (iii). Suppose that a function $\psi \geqslant 0$ on every X_n coincides with some function $f_n \in \mathcal{F}$. We verify the continuity of ψ on X. Let $x \in X$ and $\varepsilon > 0$. We can assume that $x = 0$ and $f(0) = 0$. By induction we can construct absolutely convex neighborhoods of zero $U_k \subset X_k$ for all $k \geqslant 1$ such that $U_{k+1} \cap X_k = U_k$ and $|f_k(u)| < \varepsilon + \varepsilon/2 + \cdots + \varepsilon/2^k$ for all $u \in U_k$. Indeed, if a neighborhood U_k is already found, then there exists an absolutely convex neighborhood of zero W in X_{k+1} such that

$$|f_{k+1}(y + w) - f_{k+1}(y)| < \varepsilon/2^{k+1} \quad \text{for all} \ \ y \in X_{k+1},\ w \in W.$$

Next, in the set $U_k + W$ we find an absolutely convex neighborhood of zero U_{k+1} such that $U_{k+1} \cap X_k = U_k$ (see Bogachev, Smolyanov [**97**, Lemma 1.3.12]). For any $u \in U_{k+1}$ we have $u = v + w$, where $v \in U_k$ and $w \in W$, whence

$$|f_{k+1}(u)| \leqslant |f_{k+1}(v + w) - f_{k+1}(v)| + |f_k(v)| \leqslant \varepsilon + \cdots + \varepsilon 2^{-k} + \varepsilon 2^{-k-1}.$$

The set $U = \bigcup_{k\geqslant 1} U_k$ is a neighborhood of the point x in the space X and we have $|\psi(x) - \psi(u)| < 2\varepsilon$ for all $u \in U$. We recall that merely the continuity of the restrictions to X_n is not enough for the continuity on X (see Exercise 4.8.43). Though, taking into account the given justification, the proof based on Theorem 4.8.10 does not differ much from the proof of Example 4.5.5.

In the papers Bouziad [**117**] and Choban [**135**] there are examples showing that the image of a Prohorov space X under a continuous open mapping need not be Prohorov (one can even find a countable space X and a compact mapping, i.e., with compact preimages of all points). The question about this was raised in Topsøe [**614**], where the following result was proved (see [**614**, Corollary 6.2]).

4.8.11. PROPOSITION. *Let $\pi\colon X \to Y$ be a surjection of completely regular spaces that is continuous and perfect, i.e., the preimages of points are compact and the images of closed sets are closed. Then X is Prohorov if and only if Y so is.*

PROOF. Let X be Prohorov and let $M \subset \mathcal{P}_r(Y)$ be compact in the weak topology. We have to verify the uniform tightness of M. For every measure μ in $\mathcal{P}_r(Y)$, there exists a measure $\nu \in \mathcal{P}_r(X)$ with $\mu = \nu\circ\pi^{-1}$. This follows from the surjectivity and perfectness of π, due to which the preimages of compact sets are compact, hence every compact set in Y turns out to be the image of a compact set from X, which ensures the existence of Radon preimages of measures (see Bogachev [**81**, Chapter 9]). We show that the set $S = \{\nu \in \mathcal{P}_r(X)\colon \nu\circ\pi^{-1} \in M\}$ is also weakly compact. It suffices to prove that this set is closed in $\mathcal{P}_r(\beta X)$. Suppose that a net of measures $\nu_\alpha \in S$ converges weakly to a measure $\nu \in \mathcal{P}_r(\beta X)$. By the compactness of M one can pick a subnet in $\{\nu_\alpha\circ\pi^{-1}\}$ converging weakly to some measure $\mu \in M$. It is known (see Engelking [**203**, Theorem 3.6.1, Theorem 3.7.15]) that the mapping π extends to a continuous mapping π_β from βX to βY and that $\pi_\beta(\beta X\backslash X) \subset \beta Y\backslash Y$. Since $Y = \pi(X)$ is everywhere dense in βY, the extension is also surjective. It is clear that the image of ν with respect to this extension equals μ. Since the measure μ is concentrated on a countable union of compact sets in Y and their preimages under π_β are contained in X and are compact, the measure ν is concentrated on X and is Radon on X. Thus, the set S is compact in the weak topology. By assumption this implies its uniformly tightness, whence we obtain the uniform tightness of M.

Conversely, let Y be Prohorov and let $M \subset \mathcal{P}_r(X)$ be compact in the weak topology. Then the image of M under the continuous mapping $\mu \mapsto \mu\circ\pi^{-1}$ is compact in the space $\mathcal{P}_r(Y)$. Therefore, by assumption, for every $\varepsilon > 0$ there exists a compact set $K \subset Y$ with $\mu\circ\pi^{-1}(K) > 1-\varepsilon$ for all $\mu \in M$. Since $\pi^{-1}(K)$ is compact, the set M is uniformly tight. □

It is interesting to compare the Prohorov and Skorohod properties (see § 2.6 and § 5.4). Example 5.4.18 (indicated in Bogachev, Kolesnikov [**87**]) shows that the space $\mathbb{R}_0^\infty$ of all finite sequences (with its natural topology of the inductive limit of an increasing sequence of finite-dimensional spaces) fails to have the Skorohod property, although it is Souslin and Prohorov. On the other hand, in § 5.4 we consider the class of almost metrizable spaces (spaces X for which there exists a bijective continuous proper mapping from a metric space onto X) and show that any almost metrizable space is sequentially Prohorov precisely when it has the strong Skorohod property for Radon measures.

Let $\mathcal{AL}$ be the class of Tychonoff spaces X such that every sequence of tight Baire measures μ_n without eluding load is uniformly tight; such spaces can be called *Alexandroff.*

Let $\mathcal{AL}_\mathcal{U}$ be the class of Tychonoff spaces X such that every sequence of Radon measures μ_n without eluding load on open sets is uniformly tight; such spaces can be called *$\mathcal{U}$-Alexandroff.*

According to the results obtained above, these classes are contained in the class of strongly sequentially Prohorov space, and the spaces of Radon measures on spaces from $\mathcal{AL}$ and $\mathcal{AL}_{\mathcal{U}}$ are weakly sequentially complete. For completely normal spaces both properties introduced by Alexandroff coincide. It is not difficult to show that complete metric spaces belong to $\mathcal{AL}$ and $\mathcal{AL}_{\mathcal{U}}$: actually, this is verified in the proof of the Prohorov theorem for complete metric spaces (see Theorem 2.3.4).

However, an incomplete separable metric space need not be Alexandroff (although it is always sequentially Prohorov). Say, one can take a Lebesgue nonmeasurable set X in $[0,1]$ of outer measure 1 and inner measure 0; on this set there is a sequence of measures with finite supports converging weakly to the non-Radon measure on E induced by Lebesgue measure; this sequence is not uniformly tight, but has no eluding load. One can easily verify the following properties of the classes $\mathcal{AL}$ and $\mathcal{AL}_{\mathcal{U}}$ (Exercise 4.8.44).

4.8.12. PROPOSITION. (i) *Closed subsets of perfectly normal spaces of class $\mathcal{AL}$ belong to $\mathcal{AL}$.* (ii) *The classes $\mathcal{AL}$ and $\mathcal{AL}_{\mathcal{U}}$ are stable under countable products.*

4.8.13. PROPOSITION. *Every k-space with a countable fundamental system of compact sets belongs to $\mathcal{AL}$ and $\mathcal{AL}_{\mathcal{U}}$.*

PROOF. Without loss of generality we can assume that the given space X is the union of a strictly increasing sequence of compact sets K_n and that every compact set in X is contained in some K_n. If X does not belong to $\mathcal{AL}_{\mathcal{U}}$, then one can construct a sequence of Radon measures μ_k and a sequence of disjoint compact sets Q_k with the following properties: $\mu_k(Q_k) \geqslant c > 0$ and every K_n intersects only finitely many sets Q_k. Hence every union of Q_k is closed. Since X is normal (Exercise 4.8.45), it remains to apply Remark 4.4.9. □

4.8.14. COROLLARY. *The inductive limit of an increasing sequence of locally convex spaces X_n with compact embeddings $X_n \to X_{n+1}$ belongs to $\mathcal{AL}$ and $\mathcal{AL}_{\mathcal{U}}$.*

For the proof it suffices to use that such inductive limits are k-spaces with countable families of compact sets (see Bogachev, Smolyanov [**97**, § 2.7]).

There are no explicit descriptions of the classes $\mathcal{AL}$ and $\mathcal{AL}_{\mathcal{U}}$ and it is not clear whether they differ.

4.8(iii). The central limit theorem

Several important classes of measures on infinite-dimensional spaces are introduced by means of independent random vectors or convolutions. In this and the next subsections we consider some examples, and here we are concerned with the central limit theorem (abbreviated as CLT).

Let X be a locally convex space and let $\{\xi_n\}$ be a sequence of X-valued independent centered random vectors with the same Radon distribution μ. Set

$$S_n = \frac{\xi_1 + \cdots + \xi_n}{\sqrt{n}}.$$

Note that the distribution of S_n coincides with the measure μ^{*n} defined by the equality

$$\mu^{*n}(A) = (\mu * \ldots * \mu)(n^{-1/2}A),$$

where the convolution is n-fold. The first problem which the central limit theorem deals with is the study of convergence of the sequence of random vectors S_n (in an appropriate sense).

If μ is a Radon probability measure on X such that $X^* \subset L^p(\mu)$, where $p > 0$, then μ is called a measure with *weak moment* of order p. If $q \in L^p(\mu)$ for every continuous seminorm q, then μ is called a measure with *strong moment* of order p. If μ is a Radon measure on X, $X^* \subset L^1(\mu)$ and there is a vector $m \in X$ such that

$$f(m) = \int_X f(x)\,\mu(dx) \quad \forall\, f \in X^*,$$

then m is called the *mean* or *barycenter* of the measure μ.

On any Fréchet space (in particular, on any Banach space) every measure with strong first moment has a mean and on a reflexive Banach space means exist for all measures with weak first moments (see Bogachev, Smolyanov [**97**, § 5.6]).

Every Radon Gaussian measure γ (see Definition 2.7.18) has a mean and is the shift of a centered Gaussian measure γ_0 by this mean, i.e., $\gamma(B) = \gamma_0(B - m)$.

4.8.15. DEFINITION. (i) A measure $\mu \in \mathcal{P}_r(X)$ with mean m on X is called *pre-Gaussian* if it has weak second moment and there is a Gaussian measure γ with mean m on X such that

$$\int_X f\,g\,d\mu = \int_X f\,g\,d\gamma \quad \forall\, f,\, g \in X^*.$$

(ii) A measure $\mu \in \mathcal{P}_r(X)$ with zero mean on X satisfies the central limit theorem (CLT) if the sequence $\{\mu^{*n}\}$ is uniformly tight. A measure $\mu \in \mathcal{P}_r(X)$ with mean m is said to satisfy the CLT if the shifted measure μ_{-m} with zero mean satisfies the CLT.

(iii) A space X is called a space with the CLT property if every measure μ in $\mathcal{P}_r(X)$ with zero mean and strong second moment satisfies the CLT; X is called a space with the strict CLT property if CLT is fulfilled for every measure $\mu \in \mathcal{P}_r(X)$ with zero mean and weak second moment.

Locally convex spaces with the strict CLT property were introduced in Bogachev [**77**]. Since for measures on the real line for the CLT it is necessary to have finite second moment (see p. 22 and the lemma below), in relation with the CLT we consider only measures with weak second moments.

4.8.16. LEMMA. *Let $\mu \in \mathcal{P}_r(X)$. If the sequence $\{\mu^{*n}\}$ is uniformly tight, then it converges weakly to some centered Radon Gaussian measure γ and μ is a pre-Gaussian measure.*

PROOF. If ξ_n are independent random variables with a common distribution and the distributions of the normed sums $S_n = (\xi_1 + \cdots + \xi_n)/\sqrt{n}$ are uniformly tight, then $\mathbb{E}\,\xi_n^2 < \infty$. Indeed, otherwise there are numbers $\varepsilon_n \to 0$ such that $\sup_x P(x \leqslant \sqrt{n} S_n \leqslant x + 1) \leqslant \varepsilon_n/\sqrt{n}$ (see Petrov [**510**, p. 46, Theorem 6]). Then for each number $r \in \mathbb{N}$ we have $P(-r \leqslant S_n \leqslant r) \leqslant 4r\varepsilon_n \to 0$. Hence there is no convergent subsequence of distributions of S_n. This gives the one-dimensional case (which can also be derived from Araujo, Giné [**19**, Theorem 4.7]). The general case reduces to it, because we obtain the weak second moment of the measure μ, and the uniform tightness of $\{\mu^{*n}\}$ yields the existence of a Radon probability measure γ that is a limit point for the set $\{\mu^{*n}\}$ in the weak topology. The measure γ is a Gaussian with zero mean, since all measures $\mu \circ f^{-1}$, where $f \in X^*$, are Gaussian with zero mean on the real line, which follows by the one-dimensional CLT. This also ensures the uniqueness of a limit point. Hence $\mu^{*n} \Rightarrow \gamma$. In addition, the covariance of the measure γ serves as the covariance of the measure μ. $\square$

On $\mathbb{R}^n$ any probability measure with weak second moment satisfies the CLT. Certainly, such a measure has also strong second moment. The situation is completely different in the infinite-dimensional case. For example, the space $C[0,1]$ does not have the CLT property; moreover, there exists a pre-Gaussian measure with compact support in $C[0,1]$ not satisfying the CLT. On the other hand, there is a measure with compact support in $C[0,1]$ that is not pre-Gaussian. Finally, there is a measure on $C[0,1]$ with weak second moment satisfying the CLT, but not having strong second moment (see Pauluaskas, Rachkauskas [**505**]). A Hilbert space has the CLT property. Since on a Hilbert space the covariance operator of a probability measure μ is nuclear precisely when μ has strong second moment, here the class of pre-Gaussian measures coincides with the class of measures with weak second moment satisfying the CLT (and also with the class of probability measures with strong second moment). The space $C[0,1]$ shows that these three classes of measures may differ in general Banach spaces. The coincidence of these classes characterizes Hilbert spaces: a Banach space is linearly homeomorphic to a Hilbert space if and only if the existence of the strong second moment of a probability measure is equivalent to the condition that it satisfies the CLT. Every probability measure with strong second moment on a Banach space X satisfies the CLT precisely when X is a space of type 2 (for the definitions of type and cotype, see Vakhania, Tarieladze, Chobanyan [**629**]). Thus, on non-Hilbert spaces of type 2 there are measures with weak second moments satisfying the CLT, but not having strong second moments. If every measure on X with weak second moment satisfying the CLT has strong second moment, then X is a space of cotype 2, and this property completely characterizes the spaces of cotype 2. The cotype 2 property is also equivalent to the condition that every pre-Gaussian measure on X satisfies the CLT. The proofs of these assertions and references can be found in [**505**, Chapter 3], Araujo, Giné [**19**], and Ledoux, Talagrand [**421**, Chapter 10].

The definition of the strict inductive limit of a sequence of locally convex spaces X_n is given in Example 4.5.5. If $X = \bigcap_n X_n$, where each X_n is a locally convex space, X_{n+1} is a linear subspace in X_n and the natural embedding $X_{n+1} \to X_n$ is continuous, then X is called the projective limit of a decreasing sequence of spaces X_n, provided that X is equipped with the weakest locally convex topology in which the embeddings $X \to X_n$ are continuous. A base of absolutely convex neighborhoods of zero in this topology consists of the intersections $X \cap V$, where V is an absolutely convex neighborhood of zero in X_n. If the topology of X_n is generated by a family of seminorms $\mathcal{P}_n$, then the topology of X is generated by the restrictions to X of the seminorms from the union of all $\mathcal{P}_n$. The space X is isomorphic to a closed linear subspace in the product $\prod_{n=1}^{\infty} X_n$ consisting of all elements of the form $(x, x, \ldots)$.

4.8.17. THEOREM. (i) *A Banach space has the strict* CLT *property precisely when it is finite-dimensional.*

(ii) *The strict* CLT *property is inherited by closed subspaces and preserved by strict inductive limits of increasing sequences of closed subspaces, countable products, arbitrary direct sums, and projective limits of decreasing sequences of spaces.*

PROOF. Justification of (i) can be found in [**505**]. The assertion in (ii) about closed subspaces is obvious.

The case of the strict inductive limit X of closed subspaces X_n follows by the fact that here every Radon measure μ with finite second moment (or any finite moment) is concentrated on some of X_n. Indeed, otherwise there exist increasing numbers n_k for which $\mu(X_{n_{k+1}}\backslash X_{n_k}) > 0$. Without loss of generality one can assume that $n_k = k$. By induction one can construct a functional $l \in X^*$ such that the integral of $|l|$ over the set $X_{k+1}\backslash X_k$ will be not smaller than 1. To this end, we take a point x_1 in the topological support of the measure μ on $X_2\backslash X_1$ and a functional $l_1 \in X_2^*$ with $l_1(x_1) = 1$. This is possible by the Hahn–Banach theorem. The integral J_1 of $|l_1|$ over $X_2\backslash X_1$ is not zero. Next l on X_2 is defined as $J_1^{-1}l_1$. Let l be defined on some X_k. Its extension to X_{k+1} is constructed precisely as above. It follows from the definition of the inductive topology that the obtained functional l is continuous on X. It is clear that it does not belong to $L^1(\mu)$.

The case of a finite or countable product is obvious, since in this case the uniform tightness of a sequence of measures on the product is equivalent to the uniform tightness of the projections to every fixed factor. The case of a projective limit follows from the case of a product by passing to a closed subspace. Finally, the case of direct sums reduces to that of a countable direct sum, since every compact set in a direct sum of locally convex spaces is contained in some finite sum. For a countable direct sum the assertion follows from the already proven facts, since such a sum is the limit of an increasing sequence of finite sums. □

The assertion about projective limits remains valid for general countable projective limits. For uncountable products the theorem is not valid. For example, the power of cardinality of the continuum of Lebesgue measure on $[-1/2, 1/2]$ extends to a Radon measure μ with compact support in the product of the continuum of real lines $\mathbb{R}^c$. However, the measure μ does not satisfy the CLT and even is not pre-Gaussian. Indeed, if there is a centered Radon Gaussian measure γ on $\mathbb{R}^c$ with the same covariance operator as μ, then the finite-dimensional CLT implies that γ is the extension of the product of the continuum of standard Gaussian measures on the real line. However, this product vanishes on every compact set in $\mathbb{R}^c$, since it vanishes on the product of the continuum of compact intervals. The latter is seen from the fact that a collection of intervals of cardinality of the continuum contains an infinite subcollection of intervals with length bounded by some common number N, but even the countable power of the standard Gaussian measure equals zero on the countable power of every interval $[-N, N]$.

4.8.18. COROLLARY. *Let X be a complete nuclear barrelled locally convex space. Then its dual X^* with the strong topology possesses the strict* CLT *property. For example, this is true if X is a nuclear Fréchet space. The following spaces have the strict* CLT *property: $C_0^\infty[a,b]$, $\mathcal{S}(\mathbb{R}^k)$, $\mathcal{S}'(\mathbb{R}^k)$, and $\mathbb{R}^\infty$.*

On stable and infinitely divisible measures on locally convex spaces, see Bogachev, Smolyanov [**97**, § 5.12(v)] and the literature cited therein.

4.8(iv). Shift-compactness and sums of independent random elements

In relation to limit theorems we mention here an interesting phenomenon of shift-compactness, which is the property that certain noncompact sets of measures on a linear space become compact after shifting each measure to some vector depending on this measure. For example, an arbitrary set of shifts of a single measure is always shift-compact, but need not be compact. The theorem proven below gives a simple sufficient condition for shift-compactness.

4.8.19. LEMMA. *Let μ and ν be two Radon probability measures on a locally convex space X and let μ be symmetric. Then for every set $B \in \mathcal{B}(X)$ one has*

$$2\mu * \nu(B) - 1 \leqslant \mu\Bigl(\frac{B-B}{2}\Bigr), \quad 2\mu * \nu(B) - 1 \leqslant \nu\Bigl(\frac{B+B}{2}\Bigr). \tag{4.8.1}$$

PROOF. It is obvious from the definition of convolution that there exists an element $x_1 \in X$ for which $\mu * \nu(B) \leqslant \mu(B - x_1)$. Since $(B - x_1) \cap (-B + x_1)$ is contained in $(B-B)/2$, on account of symmetry of μ we obtain the following chain of relations:

$$\begin{aligned}\mu\Bigl(\frac{B-B}{2}\Bigr) &\geqslant \mu((B - x_1) \cap (-B + x_1)) \\ &\geqslant \mu(B - x_1) + \mu(-B + x_1) - 1 = 2\mu(B - x_1) - 1 \geqslant 2\mu * \nu(B) - 1.\end{aligned}$$

For the proof of the second estimate we use the symmetry of μ in the representation

$$2\mu * \nu(B) = \int_X [\nu(B - x) + \nu(B + x)]\,\mu(dx),$$

which gives an element x_2 such that $2\mu * \nu(B) \leqslant \nu(B - x_2) + \nu(B + x_2)$. Now the inclusion $(B - x_2) \cap (B + x_2) \subset (B + B)/2$ gives the estimate

$$\nu\Bigl(\frac{B+B}{2}\Bigr) \geqslant \nu((B - x_2) \cap (B + x_2)) \geqslant \nu(B + x_2) + \nu(B - x_2) - 1,$$

but the right-hand side is not smaller than $2\mu * \nu(B) - 1$. □

4.8.20. THEOREM. *Let $\{\mu_\lambda\}$ and $\{\nu_\lambda\}$ be two families of Radon probability measures on a locally convex space X with indices from some set Λ such that the family $\{\mu_\lambda * \nu_\lambda\}$ is uniformly tight. Then, there exists a family $\{x_\lambda\}$ of points in X such that the measures $\mu_\lambda * \delta_{x_\lambda}$ are uniformly tight.*

If, in addition, the measures μ_λ or the measures ν_λ are symmetric, then both families $\{\mu_\lambda\}$ and $\{\nu_\lambda\}$ are uniformly tight.

PROOF. Let us find compact sets K_n with $\mu_\lambda * \nu_\lambda(K_n) \geqslant 1 - 4^{-n}$ for all λ. Let $A_n^\lambda = \{x\colon\ \mu_\lambda(K_n - x) > 1 - 2^{-n}\}$. Then

$$\begin{aligned}\mu_\lambda * \nu_\lambda(K_n) &\leqslant \int_{A_n^\lambda} \mu_\lambda(K_n - x)\,\nu_\lambda(dx)(1 - 2^{-n})\nu_\lambda(X\backslash A_n) \\ &\leqslant 1 - \nu_\lambda(A_n^\lambda) + (1 - 2^{-n})\nu_\lambda(X\backslash A_n^\lambda),\end{aligned}$$

whence we see that $\nu_\lambda(X\backslash A_n^\lambda) \leqslant 2^{-n}$. Hence $\nu_\lambda\bigl(\bigcap_n A_n^\lambda\bigr) > 0$, so there exists an element $x_\lambda \in \bigcap_n A_n^\lambda$. This gives the uniform tightness of the family of measures $\mu_\lambda(\,\cdot - x_\lambda) = \mu_\lambda * \delta_{x_\lambda}$. Let now the measures μ_λ be symmetric. The uniform tightness of both families follows from Lemma 4.8.1 taking into account the compactness of the sets $(K_n + K_n)/2$ and $(K_n - K_n)/2$. □

4.8.21. COROLLARY. *If X is a separable Fréchet space and two families of Borel probability measures $\{\mu_\lambda\}$ and $\{\nu_\lambda\}$ on X are such that the family of convolutions $\mu_\lambda * \nu_\lambda$ is contained in a compact set in the weak topology and the measures μ_λ are symmetric, then both families are uniformly tight and have compact closures in the weak topology.*

An important direction in the area of limit theorems is connected with the study of sums of independent random elements. Here is a typical result.

4.8.22. THEOREM. *Let $\{\xi_n\}$ be a sequence of independent random vectors with values in a separable Fréchet space. Then for the sums $S_n = \xi_1 + \cdots + \xi_n$ the following assertions are equivalent:*

(i) *the elements S_n converge almost surely;*

(ii) *the elements S_n converge in probability;*

(iii) *the distributions μ_{S_n} of the sums S_n converge weakly.*

PROOF. By Theorem 2.7.9 the case of a Fréchet space reduces to the case of a Banach space. The equivalence of (i) and (ii) is not connected with weak convergence and its proof can be found in Vakhania, Tarieladze, Chobanyan [**629**, Chapter V]. It is clear that (i) yields (iii). Let us prove that (iii) implies (ii). To this end, it suffices to show that the sequence $\{S_n\}$ is fundamental in probability. We have to show that for every $\varepsilon > 0$ there exists N such that $P(\|S_n - S_m\| \geqslant \varepsilon) \leqslant \varepsilon$ for all $n, m \geqslant N$. If this is false, then for some $\varepsilon > 0$ there exist increasing numbers n_k for which $P(\|S_{n_{k+1}} - S_{n_k}\| \geqslant \varepsilon) > \varepsilon$. The sequence of measures μ_{S_n} is uniformly tight. This yields the uniform tightness of the distributions of $S_n - S_m$, $n \geqslant m$, which will be denoted by $\mu_{n,m}$. Indeed, we obviously have the uniform tightness of the distributions μ_{-n} of the elements $-S_n$ that are obtained as the images of the measures μ_n under the mapping $x \mapsto -x$. Hence the set of convolutions $\mu_n * \mu_{-m} = \mu_{n,m}$ is also uniformly tight. Then the sequence of measures μ_{n_{k+1},n_k} contains a weakly convergent subsequence. One can assume that this sequence converges. Let ν be its limit and let μ be the limit of μ_{S_n}. Since $S_{n_{k+1}} = S_{n_{k+1}} - S_{n_k} + S_{n_k}$ and the elements ξ_i are independent, the measure $\mu_{S_{n_{k+1}}}$ is the convolution of the distribution of $S_{n_{k+1}} - S_{n_k}$ with $\mu_{S_{n_k}}$, which in the limit gives the equality $\mu = \nu * \mu$ (Exercise 4.8.49). Therefore, ν is Dirac's measure at the origin (Lemma 4.6.3). Then $P(\|S_{n_{k+1}} - S_{n_k}\| \geqslant \varepsilon) \to 0$ for each $\varepsilon > 0$, which contradicts our assumption. □

For $X = \mathbb{R}$ this theorem goes back to Paul Lévy, the case of a Banach space X was considered in Geffroy [**259**]. For random elements with symmetric distributions (for which $\mu(B) = \mu(-B)$) this assertion can be somewhat reinforced, as the following important theorem of Itô–Nisio–Buldygin asserts.

4.8.23. THEOREM. *If in the previous theorem the distributions μ_{ξ_n} of the elements ξ_n are symmetric, then assertions* (i)–(iii) *are also equivalent to the following assertions:*

(iv) *the measures μ_{S_n} are uniformly tight;*

(v) *there exist a Radon measure μ on X and a linear subspace $F \subset X^*$ separating points in X for which $\widetilde{\mu}(f) = \lim\limits_{n\to\infty} \widetilde{\mu_{\xi_n}}(f)\ \forall\, f \in F$.*

PROOF. The justification employs the fact that in the one-dimensional case the uniform tightness of the measures μ_{S_n} implies their weak convergence (see, for example, Neveu [**482**, Proposition IV.7.2]). Hence the Fourier transforms of the measures $\mu_{S_n} = \mu_{\xi_1} \cdots \mu_{\xi_n}$ cannot have different partial limits. This gives weak convergence under condition (iv). Let (v) be fulfilled. It is known (see Bogachev [**81**, Proposition 6.5.4]) that F contains a countable set of functionals f_n separating points. By means of the sequence $\{f_n\}$ we obtain a continuous injective linear embedding of X into $\mathbb{R}^\infty$. Since in $\mathbb{R}^\infty$ weak convergence of probability measures follows from weak convergence of their finite-dimensional projections, we see that in $\mathbb{R}^\infty$ the sums S_n converge almost surely and their limit S almost surely belongs to X, because its distribution is the measure μ on X. Considering the images under the projections $(f_1,\ldots,f_n)$ we conclude that the random elements $S-S_n$ and S_n are independent. Hence the convolution of their distributions is the distribution of S, which is μ. Corollary 4.8.21 implies the uniform tightness of $\{\mu_{S_n}\}$. $\square$

If the measures μ_{ξ_n} are not symmetric, then convergence of $\widetilde{\mu_{S_n}}$ to the Fourier transform of a Radon measure μ on X implies that there exist nonrandom vectors $a_n\in X$ for which the series $\sum_{n=1}^{\infty}(\xi_n-a_n)$ converges almost surely in X.

By the Glivenko–Cantelli theorem (Glivenko [**278**], Cantelli [**131**]) the empirical distribution functions $F_n(t):=n^{-1}\sum_{i=1}^{n}I_{(-\infty,t]}(\xi_i)$ constructed by samples of size n from independent random variables ξ_i with a common distribution function F converge almost surely to F uniformly on $\mathbb{R}$. An important role in statistics is played by Kolmogorov's result [**374**] on existence of an explicitly calculated limit probability for the inequality $\sup_t|F_t(t)-F(t)|<\lambda/\sqrt{n}$ as $n\to\infty$.

A vast literature is devoted to generalizations, see Dudley [**192**], Dudley, Giné, Zinn [**195**], Giné, Zinn [**276**], Shorack, Wellner [**582**], Talagrand [**608**], and van der Vaart, Wellner [**625**], [**626**]. A natural analog of empirical distribution functions in the case of independent random elements ξ_i with a common distribution P in a measurable space is the empirical measure

$$P_n(C)=n^{-1}\sum_{i=1}^{n}I_C(\xi_i).$$

The class $\mathcal{C}$ of measurable sets is called a Glivenko–Cantelli class if

$$\sup_{C\in\mathcal{C}}|P_n(C)-P(C)|\to 0$$

almost surely or in probability. Similarly one introduces functional Glivenko–Cantelli classes. These matters are thoroughly discussed in the cited literature, where one can find additional references.

Exercises

4.8.24. Let K be a compact set in a topological space X and let A be a subalgebra of functions in $C_b(X)$ containing 1 and separating points in K. Prove that, for every function $f\in C_b(X)$ and every $\varepsilon>0$, there exists a function $g\in A$ such that

$$\sup_{x\in K}|f(x)-g(x)|\leqslant\varepsilon \quad\text{and}\quad \sup_{x\in X}|g(x)|\leqslant\sup_{x\in X}|f(x)|.$$

HINT: let $\sup_{x\in X}|f(x)|=1$; by the Stone–Weierstrass theorem there exists a function $h\in A$ with $\sup_{x\in K}|f(x)-h(x)|\leqslant\varepsilon/2$; set $M=\sup_{x\in X}|h(x)|$; if $M>1$, then for the function $u(t)=\max\bigl(-1,\min(t,1)\bigr)$ with values in $[-1,1]$ we can find a polynomial ψ such that $\max_{t\in[-M,M]}|\psi(t)-u(t)|\leqslant\varepsilon/2$ and $\max_{t\in[-M,M]}|\psi(t)|\leqslant 1$; then we obtain $g=\psi(g)\in A$, $|g|\leqslant 1$, $|g(x)-f(x)|\leqslant\varepsilon$ for all $x\in K$.

4.8.25. Construct a net of continuous functions f_α on $[0,1]$ such that $0\leqslant f_\alpha\leqslant 1$, $\lim\limits_{\alpha}f_\alpha(x)=1$ for all x, but $\lim\limits_{\alpha}\int_0^1 f_\alpha(x)\,dx=0$.

HINT: the set of indices consisting of finite subsets α of $[0,1]$ is partially order by inclusion; for such a set α, consisting of n points, we find a continuous function f_α with $0\leqslant f_\alpha\leqslant 1$ that equals 1 on α and has the integral less than $1/n$.

4.8.26. Prove Dini's theorem: any sequence of continuous functions on a compact set pointwise decreasing to zero is uniformly decreasing.

4.8.27. Prove that a compact space is metrizable precisely when there is a countable family of continuous functions on it separating points.

HINT: having a sequence of continuous functions f_n on a compact space K separating points, consider the injective mapping $f(x) = \bigl(f_n(x)\bigr)_{n=1}^{\infty}$ to $\mathbb{R}^\infty$ and observe that f is a homeomorphism of the compact space K and the compact metric space $f(K)$.

4.8.28. Let $\{x_n\}$ be a sequence in a completely regular space X such that the sequence of Dirac measures δ_{x_n} is weakly fundamental. (i) Show that if X is normal or $\mathcal{P}_\sigma(X) = \mathcal{P}_t(X)$, then $\{x_n\}$ converges in X. (ii) Show that this is false for the Tychonoff planck from Example 4.5.9.

HINT: (i) $\{x_n\}$ has limit points in the Stone–Čech compactification βX and there are no different limit points (otherwise one can take a function $f \in C_b(\beta X)$ equal to 1 in a neighborhood of one limit point and equal to 0 in a neighborhood of another); hence, $x_n \to x \in \beta X$. If $x \notin X$, then $S = \{x_n\}$ is closed in X. If X is normal, we obtain a contradiction taking the function $f \in C_b(S)$ with $f(x_n) = (-1)^n$ (assuming that x_n are different) and extending it to X. If $\mathcal{P}_\sigma(X) = \mathcal{P}_t(X)$, then $\delta_{x_n} \Rightarrow \mu \in \mathcal{P}_r(X)$ and X contains a point $z \in \operatorname{supp}(\mu)$; it is readily seen that $z = x$. (ii) Here $\beta X = [0, \omega_1] \times [0, \omega_0]$, see Gillman, Jerison [**274**, p. 123], and $x_n = (\omega_1, n) \to (\omega_1, \omega_0)$, so that the sequence $\{\delta_{x_n}\}$ is fundamental (and converges to a Baire measure that is not tight on X). Note that in Bogachev [**81**, Exercise 8.10.67] the condition of normality of X is forgotten.

4.8.29. Suppose that Baire probability measures μ_n on a topological space X converge weakly to a measure μ and $f \geqslant 0$ is a continuous function. Show that

$$\int_X f\,d\mu \leqslant \liminf_{n\to\infty} \int_X f\,d\mu_n.$$

HINT: let $f_k = \min(f, k)$. Then $f_k \in C_b(X)$ and for all $k \in \mathbb{N}$ we have

$$\int_X f_k\,d\mu = \lim_{n\to\infty} \int_X f_k\,d\mu_n \leqslant \liminf_{n\to\infty} \int_X f\,d\mu_n.$$

4.8.30. Let μ be a Radon probability measure on a completely regular space X and let $\mathcal{E}$ be a class of Borel sets closed with respect to finite intersections. Suppose that for every open set U and every $\varepsilon > 0$ one can find sets $E_1, \ldots, E_k \in \mathcal{E}$ such that $\bigcup_{i=1}^k E_i \subset U$ and $\mu\bigl(U \backslash \bigcup_{i=1}^k E_i\bigr) < \varepsilon$. Prove that if a sequence of Radon probability measures μ_n satisfies the equality $\lim\limits_{n\to\infty} \mu_n(E) = \mu(E)$ for all $E \in \mathcal{E}$, then the measures μ_n converge weakly to μ. Prove an analogous assertion for Baire measures and Baire sets.

HINT: see Corollary 2.4.5.

4.8.31. (A.D. Alexandroff [**9**, § 17]) Suppose that a sequence of Baire measures $\mu_n \geqslant 0$ converges weakly to a measure μ. Let Z and Z_n, $n \in \mathbb{N}$, be functionally closed sets such that, for every n, there exists m such that $Z_{n+k} \subset Z_n$ for all $k \geqslant m$, and also $\mu(Z) = \lim\limits_{n\to\infty} \mu(Z_n)$. Prove that $\limsup\limits_{n\to\infty} \mu_n(Z_n) \leqslant \mu(Z)$.

HINT: for given $\varepsilon > 0$ find n_1 such that $\mu(Z_n) \leqslant \mu(Z) + \varepsilon$ for all $n \geqslant n_1$; next, find $n_2 > n_1$ such that $\mu_n(Z_{n_1}) \leqslant \mu(Z_{n_1}) + \varepsilon$ for all $n \geqslant n_2$; further, take n_3 such that $Z_n \subset Z_{n_1}$ for all $n \geqslant n_3$, which gives the inequality $\mu_n(Z_n) \leqslant \mu(Z_{n_1}) + \varepsilon$ for all $n \geqslant n_3$.

4.8.32. (i) Suppose that a sequence of Baire measures μ_n on a completely regular space X converges weakly to a tight Baire measure μ on X and is uniformly tight. Let $\Gamma \subset C_b(X)$ be a family of uniformly bounded and pointwise equicontinuous functions. Show that (compare with Ranga Rao's Theorem 2.2.8)

$$\lim_{n\to\infty} \sup_{f\in\Gamma} \Bigl| \int f\,d(\mu_n - \mu) \Bigr| = 0.$$

(ii) Show that in the case where (X, d) is a separable metric space and $\Gamma \subset C_b(X)$ is uniformly bounded and

$$\lim_{\varepsilon\to 0} \sup\bigl\{|f(x) - f(y)|\colon\ f \in \Gamma, d(x, y) \leqslant \varepsilon\bigr\} = 0,$$

the indicated equality is true for every weakly convergent sequence of Baire measures (not necessarily tight).

HINT: (i) one can assume that $|f| \leqslant 1$ for all $f \in \Gamma$ and $\|\mu_n\| \leqslant 1$. Assume that for some $\varepsilon > 0$ for some sequence $\{f_n\} \subset \Gamma$ we have $\Bigl|\int f_n\, d(\mu_n - \mu)\Bigr| > \varepsilon$. Find a compact set K such that $|\mu|(X\backslash K) + |\mu_n|(X\backslash K) < \varepsilon/4$ for all n. By the Ascoli–Arzelà theorem the sequence $\{f_n\}$ contains a subsequence uniformly converging on K to some function f. We can assume that this is the whole sequence $\{f_n\}$. There is a function $g \in C_b(X)$ with $g|_K = f|_K$ and $|g| \leqslant 1$. For sufficiently large n we obtain $\sup_{x\in K} |g(x) - f_n(x)| \leqslant \varepsilon/4$ and $\Bigl|\int g\, d(\mu_n - \mu)\Bigr| \leqslant \varepsilon/4$, which leads to a contradiction. From (i) we obtain (ii), since on the completion of X the sequence $\{\mu_n\}$ is uniformly tight and the functions from Γ can be extended by continuity to the completion and the family of their extensions satisfies the condition in (ii).

4.8.33. Let X be a locally convex space. Show that the following condition is sufficient for weak convergence of measures $\mu_n \in \mathcal{P}_r(X)$ to $\mu \in \mathcal{P}_r(X)$: for each bounded uniformly continuous function f, the integrals of f with respect to the measures μ_n converge to the integral of f with respect to μ (a function f is called uniformly continuous if, for every $\varepsilon > 0$, there exists a neighborhood of zero W such that $|f(x+w) - f(x)| < \varepsilon$ for all $x \in X$, $w \in W$).

HINT: consider functions that are Lipschitz with respect to seminorms from a family defining the topology and apply Theorem 4.3.10.

4.8.34. Let μ be a nonnegative Baire measure on a normal space X. Prove that for every closed set $C \subset X$ and every $\varepsilon > 0$ there exists a functionally closed set Z such that $C \subset Z$ and $\mu(Z) \leqslant \mu^*(C) + \varepsilon$.

HINT: there exists a functionally open set U such that $C \subset U$ and $\mu(U) \leqslant \mu^*(C) + \varepsilon$; since X is normal, there exists a functionally closed set Z with $C \subset Z \subset U$.

4.8.35. (Varadarajan [**635**]) Let μ be a τ-additive Borel measure on a paracompact space X (a Hausdorff space in each open cover $\{U_\alpha\}$ of which one can inscribe a locally finite open cover $\{W_\beta\}$, i.e., every point has a neighborhood intersecting only finitely many sets W_β). Prove that the topological support of μ is Lindelöf.

HINT: let S be the support of μ and let U_t, $t \in T$, be an open cover of S. Since S is closed, it is also paracompact. Hence in the given cover one can inscribe an open cover $\mathcal{V}$ of the form $\mathcal{V} = \bigcup_{n=1}^\infty \mathcal{V}_n$, where every the subfamily $\mathcal{V}_n$ consists of disjoint sets $V_{n,\alpha}$ (see Engelking [**203**, Theorem 5.1.12]). The definition of S yields that $|\mu|(V_{n,\alpha} \cap S) > 0$. Hence for each fixed n there is an at most countable family of nonempty sets $V_{n,\alpha_k} \cap S$, which gives a countable cover of S by the sets V_{n,α_k}, hence a countable subcover in $\{U_t\}$.

4.8.36. (Varadarajan [**635**]) Suppose that X is paracompact and a sequence of measures $\mu_n \in \mathcal{M}_\tau(X)$ converges weakly to a Baire measure μ. Prove that μ has a unique τ-additive Borel extension.

HINT: according to Exercise 4.8.35 the topological supports S_n of the measures μ_n are Lindelöf. Let Z be the closure of $\bigcup_{n=1}^\infty S_n$. Then Z is also Lindelöf. Indeed, let $\{U_t\}$ be an open cover of Z. As in Exercise 4.8.35, one can inscribe in it an open cover $\mathcal{V}$ consisting of a sequence of families $\mathcal{V}_k = \{V_{k,\alpha}\}$, where for each fixed k, the sets $V_{k,\alpha}$ are open and disjoint. For every k, there are at most countably many indices α_j with $Z \cap V_{k,\alpha_j} \neq \varnothing$, since this is true for every S_n in place of Z and the union of all S_n is everywhere dense in Z. We obtain a countable cover of Z by the sets V_{k,α_j}, which implies the existence of a countable subcover in $\{U_t\}$. By Exercise 4.8.34 we have $|\mu|_*(X\backslash Z) = 0$. Hence the measure μ is τ_0-additive. Indeed, if X is the union of an increasing net of functionally open sets G_α, then we can find a countable sequence $\{G_{\alpha_n}\}$ covering Z, which gives $|\mu|\bigl(X\backslash \bigcup_{n=1}^\infty G_{\alpha_n}\bigr) = 0$ according to what we have proved. Hence μ has a unique τ-additive Borel extension, see § 4.1.

4.8.37. Suppose that measures $\mu_n \in \mathcal{P}_\sigma(X)$ on a topological space X are given by densities ϱ_n with respect to a fixed measure $\nu \in \mathcal{P}_\sigma(X)$.

(i) Show that weak convergence of the functions ϱ_n to ϱ in $L^1(\nu)$ yields weak convergence of the measures μ_n to the measure $\mu = \varrho \cdot \nu$, but the converse is false.

(ii) Show that if $\{\varrho_n\}$ is uniformly integrable, then weak convergence of the measures μ_n implies weak convergence of $\{\varrho_n\}$ in $L^1(\nu)$.

(iii) Suppose that for some $p \in (1, +\infty)$ the sequence $\{\varrho_n\}$ is bounded in $L^p(\nu)$. Show that the functions ϱ_n converge in the weak topology of the space $L^p(\nu)$ to $\varrho \in L^p(\nu)$ precisely when the measures μ_n converge weakly to the measure $\varrho \cdot \nu$.

(iv) Give an example showing that in the case where ν is Lebesgue measure on the whole real line and ϱ is a probability density, weak convergence of ϱ_n to ϱ in $L^p(\nu)$ with $p > 1$ does not imply weak convergence of the measures μ_n to $\varrho \cdot \nu$, i.e., the boundedness of ν is important in (iii).

(v) Deduce from Exercises 1.7.22 that if $\mu_n \Rightarrow \mu$ and the integrals of $\varrho_n \ln \varrho_n$ against the measure ν converge to the integral of $\varrho \ln \varrho$, then $\|\mu_n - \mu\| \to 0$.

4.8.38. (Haydon [**315**]) The Stone–Čech compactification of $\mathbb{N}$ contains a set Z such that $\mathcal{P}_t(Z) \neq \mathcal{P}_\tau(Z)$, but every weakly compact set of measures in $\mathcal{P}_t(Z)$ is uniformly tight, i.e., Z is a Prohorov space.

4.8.39. (Kawabe [**353**], [**354**]) Let X be a Hausdorff space, let Y be completely regular, and let the space $\mathcal{P}_\tau(Y)$ be equipped with the weak topology.

(i) Prove that the mapping $\lambda\colon X \to \mathcal{P}_\tau(Y)$, $x \mapsto \lambda(x, \cdot\,)$, is continuous if and only if, for every open set U in $X \times Y$, the function $x \mapsto \lambda(x, U_x)$ on X is upper semicontinuous, where, as usual, $U_x = \{y \in Y\colon (x,y) \in U\}$.

(ii) Show that if the mapping λ in (i) is continuous, then, for every $B \in \mathcal{B}(X \times Y)$, the function $x \mapsto \lambda(x, B_x)$ is Borel on X. Hence for every Borel measure μ on X one can define the Borel measure

$$\mu \circ \lambda(B) := \int_X \lambda(x, B_x)\, \mu(dx), \quad B \in \mathcal{B}(X \times Y).$$

(iii) Show that if in (ii) the measure μ is τ-additive, then so is $\mu \circ \lambda$.

(iv) Suppose that X is a k-space (for example, locally compact or metrizable), Y is compact, $f \in C_b(X \times Y)$, $\lambda \in C\bigl(X, \mathcal{P}_\tau(Y)\bigr)$. Prove that the function $x \mapsto \displaystyle\int_Y f(x,y)\, \lambda(x, dy)$ is continuous on X.

(v) Suppose that X is a completely regular k-space (for example, locally compact or metrizable) and a net of mappings $\lambda_\alpha\colon X \to \mathcal{P}_\tau(Y)$ is equicontinuous on every compact set in X and, in addition, for every $x \in X$, the net of measures $\lambda_\alpha(x, \cdot\,)$ is uniformly tight and converges weakly to $\lambda(x, \cdot\,)$ for some continuous mapping $\lambda\colon X \to \mathcal{P}_\tau(Y)$. Prove that if a net of measures $\mu_\alpha \in \mathcal{P}_\tau(X)$ is uniformly tight and converges weakly to a measure $\mu \in \mathcal{P}_\tau(X)$, then the net of measures $\mu_\alpha \circ \lambda_\alpha$ converges weakly to the measure $\mu \circ \lambda$.

(vi) Let X and Y be the same as in (iv), let $P \subset \mathcal{P}_\tau(X)$ be a uniformly tight family, and let $Q \subset C\bigl(X, \mathcal{P}_\tau(Y)\bigr)$ be a family of mappings equicontinuous on every compact set in X such that for every $x \in X$ the family of measures $\lambda(x, \cdot\,) := Q(x)$ on Y is uniformly tight. Prove that for every net of measures $\mu_\alpha \circ \lambda_\alpha$, where $\mu_\alpha \in P$, $\lambda_\alpha \in Q$, one can find a measure $\mu \in \mathcal{P}_\tau(X)$, a mapping $\lambda \in C\bigl(X, \mathcal{P}_\tau(Y)\bigr)$ and a subnet in $\{\mu_{\alpha'} \circ \lambda_{\alpha'}\} \subset \{\mu_\alpha \circ \lambda_\alpha\}$ such that one has weak convergence $\mu_{\alpha'} \Rightarrow \mu$, $\lambda_{\alpha'}(x, \cdot\,) \Rightarrow \lambda(x, \cdot\,)$ for every $x \in X$ and $\mu_{\alpha'} \circ \lambda_{\alpha'} \Rightarrow \mu \circ \lambda$. In particular, the set $P \circ Q := \{\nu \circ \zeta\colon \nu \in P, \zeta \in Q\}$ has compact closure in the weak topology in $\mathcal{P}_\tau(X \times Y)$. Show also that if it is given in addition that Y is a Prohorov space, then the family of measures $P \circ Q$ is uniformly tight.

4.8.40. (Hoffmann-Jørgensen [**326**]) Prove assertion (ii) in Corollary 4.5.8.

HINT: take a net $\{\sigma_\alpha\}$ in $\mathcal{M}$ with projections converging to μ and ν that converges to σ in $\mathcal{M}_t(\beta X \times \beta Y)$; then σ has projections μ and ν, hence $\sigma \in \mathcal{M}_t^+(X \times Y)$; in [**326**], this assertion is deduced from a version of Theorem 4.5.7, see also the case of τ-additive measures in Kawabe [**352**].

4.8.41. (Ressel [**550**]) Suppose that X and Y are Hausdorff spaces, $\{\mu_t\}_{t\in T}$ is a net of Radon probability measures on $X\times Y$ such that their projections on X converge weakly to a Radon measure ν and the projections on Y converge weakly to Dirac's measure δ_a at a point $a \in Y$. Show that the net $\{\mu_t\}$ converges weakly to the Radon extension of the measure $\nu\otimes\delta_a$ on $\mathcal{B}(X\times Y)$. Prove an analogous assertion for τ-additive measures.

HINT: let $U \subset X\times Y$ be an open set the projection of which on Y contains a. For a given number $\varepsilon > 0$, one can find a compact set K in X such that $K\times a \subset U$ and $\nu\otimes\delta_a(K\times a) > \nu\otimes\delta_a(U) - \varepsilon$. Find open sets $V \subset X$ and $W \subset Y$ with $K\times a \subset V\times W \subset U$. By weak convergence of projections there exists t_1 such that $\mu_t(X\times W) > 1-\varepsilon$ whenever $t > t_1$, whence $\mu_t(V\times W) \geqslant \mu_t(V\times Y) - \mu_t\bigl(X\times(Y\backslash W)\bigr) > \mu_t(V\times Y) - \varepsilon$. There is $t_2 > t_1$ with $\mu_t(V\times Y) > \nu(V) - \varepsilon$ for all $t > t_2$. Then $\mu_t(U) \geqslant \mu_t(V\times W) > \nu(V) - 2\varepsilon > \nu\otimes\delta_a(U) - 3\varepsilon$.

4.8.42. Suppose that X is a Souslin space, Y is a Polish space, and random elements ξ and ξ_n, where $n \in \mathbb{N}$, on a probability space $(\Omega, \mathcal{A}, P)$ with values in X and Borel mappings $f, f_n\colon X \to Y$ satisfy the following condition: the distributions of the elements $f_n\circ\xi_n$ converge weakly to the distribution of $f\circ\xi$. Prove that there exist random elements $\widetilde{\xi}, \widetilde{\xi}_n$ in X such that $P\circ\xi^{-1} = P\circ\widetilde{\xi}^{-1}$, $P\circ\xi_n^{-1} = P\circ\widetilde{\xi}_n^{-1}$, and $f_n\circ\widetilde{\xi}_n \to f\circ\widetilde{\xi}$ a.e.

HINT: there exist random elements η and η_n with values in Y (possibly, with a changed probability space) such that $\eta_n \to \eta$ a.e. and $P\circ\eta^{-1} = P\circ(f\circ\xi)^{-1}$, $P\circ\eta_n^{-1} = P\circ(f_n\circ\xi_n)^{-1}$. By the measurable choice theorem (see [**81**, § 6.9]) one can find Borel mappings g and g_n from Y to X such that $f\bigl(g(y)\bigr) = y$ for $P\circ\eta^{-1}$-a.e. y, $f_n\bigl(g_n(y)\bigr) = y$ for $P\circ\eta_n^{-1}$-a.e. y. Set $\widetilde{\xi} = g\circ\eta$, $\widetilde{\xi}_n = g_n\circ\eta_n$. Then $f\circ\widetilde{\xi} = \eta$ and $f_n\circ\widetilde{\xi}_n = \eta_n$ a.e.

4.8.43. The space $\mathcal{D}(\mathbb{R})$ consists of all infinitely differentiable functions with compact support and is equipped with the locally convex topology τ generated by all norms

$$p_{\{a_k\}}(\varphi) = \textstyle\sum_{k=-\infty}^{\infty} a_k \max\bigl\{|\varphi^{(m)}(x)|\colon\ x \in [k, k+1], m \leqslant a_k\bigr\},$$

where for $\{a_k\}$ one takes arbitrary two-sided sequences of natural numbers. A sequence $\{\varphi_j\}$ converges to φ in this topology precisely when the functions φ_j vanish outside some common interval and all derivatives of φ_j converge uniformly to the respective derivatives of φ. This topology τ is the topology of the so-called locally convex inductive limit of the sequence of spaces $\mathcal{D}_n$ consisting of smooth functions with support in $[-n, n]$ and equipped with the sequence of norms $\max|\varphi^{(m)}(t)|$. The space of linear functions on $\mathcal{D}(\mathbb{R})$ continuous in the topology τ is denoted by $\mathcal{D}'(\mathbb{R})$ and called the space of distributions or generalized functions. Similarly one defines $\mathcal{D}(\mathbb{R}^d)$ and $\mathcal{D}'(\mathbb{R}^d)$.

(i) Prove that the topology τ is strictly weaker than the topology τ_1 on $\mathcal{D}(\mathbb{R})$ in which open sets are defined as sets whose intersection with $\mathcal{D}_n$ is open in $\mathcal{D}_n$ for each n. To this end, show that the quadratic function $F(\varphi) = \sum_{n=1}^{\infty} \varphi(n)\varphi^{(n)}(0)$ is discontinuous in the topology τ, but is continuous in τ_1.

(ii) Prove that the topology τ is strictly stronger than the topology τ_2 on $\mathcal{D}(\mathbb{R})$ generated by the norms $p_\psi(\varphi) = \sup|\psi(x)\varphi^{(m)}(x)|$, where one takes arbitrary nonnegative integers m and positive locally bounded functions ψ. For this verify that the function $F(\varphi) = \sum_{n=1}^{\infty} \varphi^{(n)}(n)$ is linear and continuous in the topology τ, but discontinuous in the topology τ_2.

(iii) Prove that $\bigl(\mathcal{D}(\mathbb{R}), \tau\bigr)$ is not a k_R-space (X is called a k_R-space if the continuity of a function on X follows from its continuity on all compact sets).

4.8.44. Prove Proposition 4.8.12.

4.8.45. Prove that every k-space X possessing a countable fundamental system of compact sets (which means that X is the union of increasing compact sets K_n with the property that every compact set X is contained in some K_n) is normal.

HINT: for any disjoint closed sets A and B, one can inductively find a continuous function f with $f|_A = 0$, $f|_B = 1$; such a function exists on the compact set K_1; if it is already constructed on K_n, then it extends to K_{n+1} such that $f|_{A\cap K_{n+1}}=0$, $f|_{B\cap K_{n+1}}=1$; for this we let $f = 0$ on $A \cap K_{n+1}$, $f = 1$ on $B \cap K_{n+1}$, and observe that the obtained function is continuous on the compact set $K_n \cup (A \cap K_{n+1}) \cup (B \cap K_{n+1})$, so that we can extend it to the set K_{n+1}.

4.8.46. Suppose that X is a topological space such that there exists a continuous injective mapping h from X to some metric space. Suppose that $A \subset X$ and every infinite sequence of points in A has a limit point in X. Show that the closure of A is metrizable and compact.

HINT: observe that $h(\overline{A}) = \overline{h(A)}$. Indeed, $h(\overline{A}) \subset \overline{h(A)}$ by the continuity of h. If $y \in \overline{h(A)}$, then $y = \lim_{n\to\infty} h(x_n)$, where $x_n \in A$. Hence either $y \in h(A)$ or one can assume that $\{x_n\}$ is infinite, then $y = h(x)$, where x is a limit point of $\{x_n\}$. It easily follows that the set $\overline{h(A)}$ in a metric space is compact. The same is true for every subset of A, whence it follows that the mapping $h^{-1}\colon \overline{h(A)} \to \overline{A}$ is continuous, since the preimages of closed sets from $\overline{A}$ are compact. Thus, h is a homeomorphism of $\overline{A}$ and $\overline{h(A)}$.

4.8.47. Let τ be an uncountable ordinal that is not a limit of any countable sequence (say, the smallest uncountable ordinal). Show that, for every continuous function f on the space $[0, \tau)$ with the order topology, there exists $\tau_0 < \tau$ such that f is constant on $[\tau_0, \tau)$.

HINT: for every natural number k, there exists $\alpha_k < \tau$ such that $|f(\alpha) - f(\beta)| < 1/k$ for all $\alpha, \beta > \alpha_k$. Otherwise, by induction one can find an increasing sequence of numbers $\alpha_{k_n} < \tau$ with $|f(\alpha_{k_{n+1}}) - f(\alpha_{k_n})| \geqslant 1/k$. This contradicts the continuity of f, since such a sequence converges to $\sup \alpha_{k_n}$. By the condition that τ is uncountable, there exists $\tau_0 < \tau$ such that $\alpha_k < \tau_0$ for all k. Clearly, τ_0 is the desired element.

4.8.48. Let X be a completely regular space. Show that every Radon measure on its Stone–Čech compactification βX is a limit in the weak topology of a net of Radon measures on X itself.

HINT: use that the set of finite linear combinations of Dirac measures is everywhere dense in $\mathcal{M}_r(\beta X)$ with the weak topology and also that X is dense in βX.

4.8.49. Suppose that two sequences of Radon probability measures μ_n and ν_n on a locally convex space converge weakly to Radon measures μ and ν, respectively. Prove that the convolutions $\mu_n * \nu_n$ converge weakly to $\mu * \nu$.

HINT: use weak convergence of the products $\mu_n \otimes \nu$ (Theorem 4.3.17).

4.8.50. (Gillman, Jerison [**274**, p. 97, 6P]) Let $\Lambda = (\beta\mathbb{R}\backslash\beta\mathbb{N}) \cup \mathbb{N}$, where β denotes the Stone–Čech compactification. Prove that the set $\mathbb{N}$ is closed in Λ, but is not functionally closed and that the points of $\mathbb{N}$ are functionally closed and possess pairwise disjoint functionally open neighborhoods.

HINT: the closedness of $\mathbb{N}$ is easily verified; to see that the points of $\mathbb{N}$ are functionally closed and functionally separated, consider the continuous extensions to $\beta\mathbb{R}$ of functions $\varphi_n \in C_b(\mathbb{R})$ such that $\varphi_n^{-1}(0) = n$ and $\varphi_n = 1$ outside $[n - 1/4, n + 1/4]$ (these extensions equal 1 outside $\mathbb{R}$); if f is a continuous function on Λ and $\mathbb{N} \subset f^{-1}(0)$, then there exists a point $p \in \beta\mathbb{R}\backslash\beta\mathbb{N}$ such that $f(p) = 0$; Indeed, there are points $x_n \in (n, n + 1/4)$ such that $|f(x_n)| < 1/n$; for p one can take an arbitrary point in the closure of $\{x_n\}$ in $\beta\mathbb{R}$, which does not contain points from $\beta\mathbb{N}$, since there is a function $\varphi \in C_b(\mathbb{R})$ such that $\varphi(n) = 0$ and $\varphi(x_n) = 1$ for all n.

4.8.51. Let X be a separable Banach space and $\mu \in \mathcal{P}(X)$. Prove that if X has a Schauder basis or the approximation property, then the measure μ is the weak limit of a sequence of its images under continuous finite-dimensional operators on X. Investigate whether this is true for an arbitrary separable Banach space X.

CHAPTER 5

Spaces of measures with the weak topology

In this chapter, naturally complementing the previous one, we discuss various topological properties of spaces of measures on topological spaces connected with other properties of the spaces on which measures are defined. In full accordance with the title of the book, almost always we deal with the weak topology, but for the reader's convenience this chapter includes a section concerned with setwise convergence of measures. Although this topic is a bit separate from the main theme of the book, it is still directly related to it. In relation to spaces of measures some other questions are also discussed, in particular, mappings between spaces of measures. The material of this chapter assumes some acquaintance with the basic notions of general topology, but does not assume knowledge of the results of the previous chapters, with the exception of the definition of the weak topology.

5.1. Properties of spaces of measures

In Chapter 3 we studied certain metrics on spaces of measures and we have seen that these metrics agree well with weak convergence on spaces of nonnegative or probability measures and that the original space is isometrically embedded into the space of probability measures. In this section we prove a number of results of a similar nature for completely regular spaces and spaces of probability measures equipped with the weak topology introduced and partly studied in Chapter 4. Here we discuss some basic topological properties of the space of measures on a topological space X; in particular, connections between properties of X and the corresponding properties of the space of measures. In applications, among such properties the following three are the most important: metrizability of the subspace of probability measures and its membership in the class of Souslin spaces; completeness and sequential completeness; conditions for compactness.

Since we are interested in the weak topology, it is reasonable to consider *completely regular* (or *Tychonoff*) spaces; that is, spaces in which for every point x and every neighborhood U of x, there exists a continuous function equal to 1 at x and vanishing outside U. Most of the results of this chapter deal with such spaces.

We recall that the weak topology can be considered on the space $\mathcal{M}_\sigma(X)$ of all Baire measures on a space X, on its linear subspace $\mathcal{M}_t(X)$ of tight Baire measures (see p. 142), on the space $\mathcal{M}(X)$ of all Borel measures, on its linear subspace $\mathcal{M}_\tau(X)$ of all τ-additive measures, on the smaller linear subspace in $\mathcal{M}_r(X)$ of all Radon measures, and also on their subsets consisting of nonnegative measures ($\mathcal{M}^+(X)$, etc.) or of probability measures ($\mathcal{P}(X)$, etc.) as well as on balls in the variation norm. Finally, it is sometimes even useful to consider the set of Dirac measures and the set of measures with countable supports. The most natural settings of problems and the most useful results for applications are usually connected with Baire and Radon measures.

It is worth recalling that in the case of a completely regular space X the sets $\mathcal{M}_t(X)$ and $\mathcal{M}_r(X)$ correspond to the same subsets of $C_b(X)^*$ (since every measure in $\mathcal{M}_t(X)$ has a unique Radon extension), although as sets of measures they do not coincide in the general case, since they are defined on different σ-algebras.

The first result in this section shows that some basic properties of the original space must follow from the properties of spaces of measures on it. For example, it is seen from this result that metrizable spaces of measures occur only in case the original space is metrizable, which returns us to Chapter 3.

5.1.1. LEMMA. *Let X be a completely regular space. Then it is homeomorphic to the set of all Dirac measures on X and this set is closed in $\mathcal{M}_\tau(X)$ and in $\mathcal{M}_t(X)$ and also in the corresponding subspaces of nonnegative and probability measures with the weak topology.*

PROOF. Set $j(x) = \delta_x$. The obtained mapping $j\colon X \to \mathcal{M}_\sigma(X)$ is a topological embedding. Indeed, according to Example 4.2.4 a net $\{x_\alpha\}$ converges to x precisely when the net $\{j(x_\alpha)\}$ converges to $j(x)$.

Suppose now that a τ-additive measure μ is a limit point of the set of Dirac measures in the weak topology. Then there exists a net $\{\delta_{x_\alpha}\}$ weakly convergent to μ, in particular, μ is a probability measure. Let us take an arbitrary point x in the support of μ (which exists by the τ-additivity). We show that the net $\{x_\alpha\}$ converges to x. If this is false, then outside some neighborhood U of the point x there is a subnet $\{x'_\alpha\}$ of the original net. There exists a bounded nonnegative continuous function f equal to 1 in some neighborhood V of the point x and equal to 0 outside U. Since $f(x'_\alpha) = 0$, the integral of f against μ is zero, hence $\mu(V) = 0$ contrary to that x belongs to the support. Thus, $x_\alpha \to x$, whence $\mu = \delta_x$. □

Exercise 5.8.21 suggests to construct an example of a completely regular space X for which the set of Dirac measures is not closed in the space $\mathcal{M}_\sigma^+(X)$.

5.1.2. THEOREM. (i) *Let X be a compact space. Then the spaces of probability measures $\mathcal{P}_\sigma(X) = \mathcal{P}_t(X)$ and $\mathcal{P}_\tau(X) = \mathcal{P}_r(X)$ are compact in the weak topology.*

(ii) *If X is completely regular and $\mathcal{P}_t(X)$ (or $\mathcal{P}_\tau(X)$) is compact in the weak topology, then X is compact as well.*

PROOF. Compactness of the space $\mathcal{P}_t(X)$ is an immediate corollary of the Banach–Alaoglu theorem on the weak-$*$ compactness of balls in the dual space and the Riesz theorem identifying the dual to $C(X)$ with $\mathcal{M}_r(X)$ (see Theorem 4.1.9). Here we also have $\mathcal{M}_r(X) = \mathcal{M}_\tau(X)$ by compactness of X, see § 4.1, in addition, $\mathcal{M}_r(X)$ and $\mathcal{M}_t(X)$ coincide as sets of functionals. The necessity of compactness of X in assertion (ii) follows from Lemma 5.1.1. □

Note that in (ii) the class $\mathcal{P}_t(X)$ cannot be replaced by $\mathcal{P}_\sigma(X)$. One can verify that the space from Exercise 5.8.21 serves as an example.

5.1.3. THEOREM. *Suppose that the space X is completely regular.*

(i) *The space $\mathcal{M}_\tau^+(X)$ with the weak topology is metrizable precisely when X is metrizable. The metrizability of $\mathcal{M}_\tau^+(X)$ by a complete metric is equivalent to the metrizability of X by a complete metric.*

Analogous assertions are valid for the spaces $\mathcal{P}_\tau(X)$, $\mathcal{P}_t(X)$, $\mathcal{M}_t^+(X)$ in place of $\mathcal{M}_\tau^+(X)$.

(ii) *If X is separable, then the spaces $\mathcal{M}_\sigma(X)$, $\mathcal{M}_\tau(X)$ and $\mathcal{M}_t(X)$ of signed measures with the weak topology are separable as well along with the corresponding subspaces of nonnegative and probability measures.*

PROOF. (i) Lemma 5.1.1 yields that the properties of spaces of measures mentioned in (i) imply the corresponding properties of X. The converse assertions follow from the results in Chapter 3 (see Theorems 3.1.4 and 3.2.2). (ii) If X contains an everywhere dense countable set of points x_j, then the countable set of finite linear combinations of the measures δ_{x_j} with rational coefficients is everywhere dense in $\mathcal{M}_\sigma(X)$ and its subset corresponding to nonnegative coefficients is everywhere dense in $\mathcal{M}_\sigma^+(X)$. Linear combinations with nonnegative coefficients whose sum is 1 give a countable everywhere dense set in $\mathcal{P}_\sigma(X)$. It is clear that this also yields the separability of $\mathcal{M}_\tau(X)$ and $\mathcal{M}_t(X)$ along with their subspaces $\mathcal{M}_\tau^+(X)$, $\mathcal{M}_t^+(X)$, $\mathcal{P}_\tau(X)$, and $\mathcal{P}_t(X)$. □

It should be noted that the separability of $\mathcal{P}_t(X)$ with the weak topology does not yield the separability of X, and the separability of the whole space $\mathcal{M}_t(X)$ with the weak topology does not yield the separability of $\mathcal{P}_t(X)$ even if X is compact (see § 5.8(i)).

The following result from Grömig [**300**] and Koumoullis [**386**] enables us to transfer some finer topological characteristics from the space of measures.

5.1.4. PROPOSITION. *Let X be completely regular. The space X^∞ is homeomorphic to a closed subset in $\mathcal{M}_\tau^+(X)$ and to a subset in $\mathcal{M}_t^+(X)$.*

PROOF. Let us consider the mapping

$$j\colon (x_n) \mapsto \sum_{n=1}^\infty 2^{-n}\delta_{x_n}$$

from X^∞ to one of the two indicated spaces of measures. This mapping is obviously continuous with respect to the weak topology on the space of measures, since the mapping $x \mapsto \delta_x$ is continuous on X. Moreover, this mapping is a homeomorphic embedding. Indeed, it is injective by the injectivity of the mapping $x \mapsto \delta_x$. The same mapping can also be defined for the Stone–Čech compactification of the space X, denoted by βX. In the latter case the continuity of the inverse mapping follows from the fact that the space $(\beta X)^\infty$ is compact, hence a continuous injective mapping is automatically a homeomorphic embedding. This yields that the inverse mapping is also continuous for the space X, since the natural embedding $X^\infty \to (\beta X)^\infty$ is a homeomorphic embedding by the definition of the product topology and the property that X is homeomorphically embedded into its compactification.

It remains to verify that the set $j(X^\infty)$ is closed in $\mathcal{P}_\tau(X)$. Here we use again the compactification βX, since in the compact case the image is compact. In order to use this, we have to show that the intersection of $j\big((\beta X)^\infty\big)$ with the image of $\mathcal{P}_\tau(X)$ under the natural embedding to $\mathcal{P}_\tau(\beta X)$ coincides with the image of $j(X^\infty)$ under the same embedding, which can be informally written as the equality $j(X^\infty) = j\big((\beta X)^\infty\big) \cap \mathcal{P}_\tau(X)$. Then convergence of a net of measures from the set $j(X^\infty)$ to a measure $\nu \in \mathcal{P}_\tau(X)$ will imply that $\nu \in j\big((\beta X)^\infty\big) \subset j(X^\infty)$. The desired equality follows from Lemma 4.1.13. If X is a metric space, then the mapping j is Lipschitz in the metric $d_{\rm KR}$ on $\mathcal{P}_t(X)$ and the metric on X^∞ indicated on p. 45. □

Thus, any topological property inherited by closed subsets, but not inherited by countable products does not extend from the space X to the spaces $\mathcal{M}_\tau^+(X)$ and $\mathcal{M}_t^+(X)$. Normality and the Lindelöf property are examples of this sort. For the same reason $\mathcal{M}_\tau^+(X)$ and $\mathcal{M}_t^+(X)$ need not be Radon spaces (spaces on which all Borel measures are Radon) for a Radon space X (even for a compact space).

5.1.5. REMARK. Suppose that a completely regular space X is homeomorphically embedded into a completely regular space Y. For every measure $\mu \in \mathcal{M}_\tau(X)$ let $\widehat{\mu}$ denote its extension to $\mathcal{B}(Y)$ defined by the formula $\widehat{\mu}(B) := \mu(B \cap X)$, $B \in \mathcal{B}(Y)$. Then $\widehat{\mu} \in \mathcal{M}_\tau(Y)$. The mapping $\mu \mapsto \widehat{\mu}$ on $\mathcal{M}_\tau^+(X)$ is a homeomorphic embedding, which is obvious from Corollary 4.3.3 and the fact that open sets in X are the intersections of X with open sets in Y. Moreover, according to Theorem 4.8.1, the same is true for the unit sphere in the variation norm on the space $\mathcal{M}_\tau(X)$. However, this mapping need not be a homeomorphic embedding of the whole space $\mathcal{M}_\tau(X)$. For example, as observed in Choquet [**137**], if we take $X = (0,1]$ and $Y = [0,1]$ with their standard topologies, then the sequence of measures $\delta_{1/(2n)} - \delta_{1/(2n+1)}$ converges weakly to zero on Y, but not on X, because there exists a bounded continuous function f on X such that $f\bigl(1/(2n)\bigr) = 1$ and $f\bigl(1/(2n+1)\bigr) = 0$ for all n. On the space $\mathcal{P}_\sigma(X)$ the mapping $\mu \mapsto \widehat{\mu}$ can fail to be even injective (see Wheeler [**655**, §14]). Though, if X is closed and Y is normal, then the embedding of $\mathcal{M}_\tau(X)$ into $\mathcal{M}_\tau(Y)$ is homeomorphic, since every bounded continuous function on X extends to a bounded continuous function on Y.

5.1.6. PROPOSITION. *Let E be a G_δ-set in a topological space X. Then $\mathcal{P}_r(E)$ is a G_δ-set in $\mathcal{P}_r(X)$ with the weak topology.*

PROOF. We can identify the space $\mathcal{P}_r(E)$ with the set P_E in $\mathcal{P}_r(X)$ consisting of measures vanishing outside $X\backslash E$, since the natural mapping of $\mathcal{P}_r(E)$ to P_E is a homeomorphism. Indeed, if a net of measures μ_α converges weakly to a measure μ in P_E, then, for every open set U in E, we have $\mu(U) \leqslant \liminf \mu_\alpha(U)$, since $U = E \cap W$, where W is open in X, whence we obtain $\mu_\alpha(U) = \mu_\alpha(W)$. If E is open, then $\mathcal{P}_r(X)\backslash P_E = \bigcup_{k=1}^\infty M_k$, where $M_k := \{\mu \in \mathcal{P}_r(X)\colon\, \mu(X\backslash E) \geqslant 1/k\}$. By the closedness of $X\backslash E$ the sets M_k are closed. Thus, $\mathcal{P}_r(E)$ is a G_δ-set. If $E = \bigcap_{k=1}^\infty E_k$, where each E_k is open, then $\mathcal{P}_r(E) = \bigcap_{k=1}^\infty \mathcal{P}_r(E_k)$. □

5.1.7. THEOREM. *If E is a Polish space, then so is the subspace $\mathcal{S}^1(E)$ in $\mathcal{M}(E) := \mathcal{M}_\sigma(E)$ consisting of all measures μ with $\|\mu\| = 1$.*

PROOF. We recall that E is homeomorphic to a G_δ-sets in the metrizable compact space $Q = [0,1]^\infty$. Hence one can assume that E itself is a G_δ-set in Q. Let $\mathcal{P}(E) = \mathcal{P}_\sigma(E)$, $\mathcal{M}(Q) = \mathcal{M}_\sigma(Q)$. The unit ball $T = \bigl\{\mu \in \mathcal{M}(Q)\colon\, \|\mu\| \leqslant 1\bigr\}$ in $\mathcal{M}(Q)$ is compact and metrizable in the weak topology. Our set $\mathcal{S}^1(E)$ in the metrizable compact space T is the union of the following three sets: $\mathcal{P}(E)$, $-\mathcal{P}(E)$ and $D := \mathcal{S}^1(E)\backslash\bigl\{\mathcal{P}(E) \cup \bigl(-\mathcal{P}(E)\bigr)\bigr\}$.

The first two sets, as we know, are Polish spaces and so are G_δ-sets (see §2.1). Let us verify that D is also a G_δ-set. Then the union of three G_δ-sets will be a set of the same type in a metrizable compact space, hence it will be a Polish space. We recall that, as noted in Remark 5.1.5, the weak topology on $\mathcal{S}^1(E)$ coincides with the induced weak topology in T. Since the space $\mathcal{P}(E)$ is Polish, then the space $Z := \mathcal{P}(E) \times \mathcal{P}(E) \times (0,1)$ is Polish as well. Let us consider the mapping $\psi\colon\, (\mu,\nu,\alpha) \mapsto \alpha\mu - (1-\alpha)\nu$ from Z to $\mathcal{M}(E)$. This mapping is continuous provided that the spaces of measures are equipped with the weak topology. The set $U_r = \{\mu \in \mathcal{M}(E)\colon \ \|\mu\| \leqslant r\}$ is closed in the weak topology. Let

$$H := \bigl\{(\mu, \nu, \alpha) \in Z\colon \ \|\alpha\mu - (1-\alpha)\nu\| = 1\bigr\}.$$

The set H is the intersection of the sequence of open sets $\psi^{-1}\bigl(\mathcal{M}(E)\backslash U_{1-1/n}\bigr)$, i.e., it is a G_δ-set and for this reason is a Polish space. Now it is important to observe that the mapping ψ homeomorphically maps H onto the set D. Indeed, if measures $\mu, \nu \in \mathcal{P}(E)$ are such that

$$\|\alpha\mu - (1-\alpha)\nu\| = 1,$$

then it is easy to see that they are mutually singular (Exercises 1.7.14). This shows that if the variation norm of $\alpha\mu - (1-\alpha)\nu = \alpha'\mu' - (1-\alpha')\nu'$ is 1 for some $\alpha, \alpha' \in (0,1)$ and $\mu, \mu', \nu, \nu' \in \mathcal{P}(E)$, then $\alpha = \alpha'$, $\mu = \mu'$ and $\nu = \nu'$. Thus, ψ is a one-to-one mapping of H onto the set D (the fact that $\psi(H) = D$ is obvious from the decomposition $\mu = \mu^+ - \mu^-$, where $\mu^+(E) + \mu^-(E) = 1$ and $\mu^+(E) > 0$, $\mu^-(E) > 0$). Finally, the mapping $\psi^{-1}\colon D \to H$ is continuous. Indeed, if a net of measures μ_τ from D converges weakly to a measure μ from D, then by Theorem 4.8.1 we obtain convergence $\mu_\tau^+ \to \mu^+$ and $\mu_\tau^- \to \mu^-$, which on account of what has been said above means convergence of the triples $\psi^{-1}(\mu_\tau)$ to the triple $\psi^{-1}(\mu)$. Thus, the set D is homeomorphic to the G_δ-set H in a Polish space, which completes the proof. □

It is worth noting that typically $S^1(E)$ is not closed in the weak topology.

5.1.8. THEOREM. *Let X be completely regular. If X is a Souslin (or Luzin) space, then so are the spaces $\mathcal{M}_\sigma(X)$, $\mathcal{M}_\sigma^+(X)$ and $\mathcal{P}_\sigma(X)$ with the weak topology (these spaces consist of Radon measures due to our assumption). Conversely, if some of the spaces $\mathcal{M}_t(X)$, $\mathcal{M}_t^+(X)$ or $\mathcal{P}_t(X)$ is Souslin (or Luzin), then so is X.*

PROOF. By assumption there exists a Polish space E and a continuous surjection $\varphi\colon E \to X$. The induced mapping $\widehat{\varphi}\colon \mathcal{M}_\sigma(E) \to \mathcal{M}_\sigma(X)$, taking μ to $\mu\circ\varphi^{-1}$, is continuous with respect to the weak topologies. According to Theorem 4.1.12, the mapping $\widehat{\varphi}$ is surjective. If φ is injective, then $\widehat{\varphi}$ is injective as well. Hence it remains to prove that the space $\mathcal{M}_\sigma(E)$ is Luzin. This follows at once from the previous theorem, since $\mathcal{M}_\sigma(E) = 0 \cup \bigl(\mathcal{S}^1(E)\times(0,\infty)\bigr)$. □

If X is not completely regular, then an analogous theorem is valid for the A-topology considered in § 5.8(iv).

5.1.9. COROLLARY. *If X is a completely regular Souslin space, then the Borel σ-algebra in $\mathcal{M}_\sigma(X)$ is generated by the functions $\mu \mapsto \mu(B)$, $B \in \mathcal{B}(X)$, moreover, by some countable family of such functions. It is also generated by some sequence of functionals of the form $\mu \mapsto \displaystyle\int_X f_n\, d\mu$, $f_n \in C_b(X)$.*

PROOF. The Borel σ-algebra of the Souslin space X is generated by any countable family of Borel sets separating points, and there are such countable collections indeed (see Bogachev [**81**, Theorem 6.8.9, Corollary 6.7.5]). Using such a collection, we can construct a countable family of Borel sets separating measures on X. □

We now show that each weakly fundamental sequence of Baire measures converges weakly to some Baire measure, i.e., the space of Baire measures is *weakly sequentially complete.*

5.1.10. THEOREM. *Suppose that a sequence of Baire measures μ_n on a topological space X is weakly fundamental. Then $\{\mu_n\}$ converges weakly to some Baire measure on X.*

PROOF. By the Banach–Steinhaus theorem the formula

$$L(\varphi) = \lim_{n\to\infty} \int \varphi\, d\mu_n, \quad \varphi \in C_b(X),$$

defines a continuous linear functional on $C_b(X)$. According to Theorem 4.1.9, this functional is given by a Baire measure under the following condition: $L(\varphi_j) \to 0$ for every sequence of functions $\varphi_j \in C_b(X)$ that decrease pointwise to zero. Suppose that this condition is not fulfilled, i.e., the sequence $L(\varphi_j)$ does not converge to zero. We can assume that $\varphi_1 \leqslant 1$ and then $0 \leqslant \varphi_n \leqslant 1$ for all n. Let us set $I = [0,1]^\infty$ and consider the mapping $F\colon X \to I$, $F(x) = \bigl(\varphi_j(x)\bigr)_{j=1}^\infty$. Let us equip the space $Y = F(X)$ with the topology induced from I (since I is metrizable, then Y is metrizable as well). It is clear that F is continuous as a mapping from X to Y, hence the sequence of measures $\nu_n := \mu_n\circ F^{-1}$ on Y is weakly fundamental (for all $\psi \in C_b(Y)$ we have $\psi\circ F \in C_b(X)$). The natural extensions of the measures ν_n to the space I will also be denoted by ν_n. It is clear that on the compact space I the measures ν_n converge weakly to some measure ν. In addition,

$$\int_I x_j\, \nu(dx) = \lim_{n\to\infty} \int_I x_j\, \nu_n(dx) = \lim_{n\to\infty} \int_X \varphi_j(x)\, \mu_n(dx) = L(\varphi_j).$$

In order to obtain a contradiction with our supposition that $L(\varphi_j) \not\to 0$, it suffices to show that the measure ν is concentrated on the set

$$I_0 := \bigl\{x = (x_j) \in I\colon \lim_{j\to\infty} x_j = 0\bigr\}.$$

This will be done once we verify that $|\nu|(K) = 0$ for every set compact $K \subset I\backslash I_0$. Let $\varepsilon > 0$. The set $U = I\backslash K$ is open. Since $Y \subset I_0 \subset U$, the measures ν_n on U form a weakly fundamental sequence too. We recall that U is a Polish space (as an open subset of a Polish space). By the Prohorov theorem the sequence $\{\nu_n\}$ is uniformly tight on U, i.e., there is a compact set $Q \subset U$ such that $|\nu_n|(U\backslash Q) < \varepsilon$ for all n. Then

$$|\nu|(K) \leqslant |\nu|(I\backslash Q) \leqslant \liminf_{n\to\infty} |\nu_n|(I\backslash Q) \leqslant \varepsilon$$

by virtue of weak convergence on the space I (see Theorem 4.8.1) and the equality $|\nu_n|(I\backslash Q) = |\nu_n|(U\backslash Q)$ for all n. □

This theorem does not extend to nets in place of sequences: we recall (see the example on p. 179) that in the space $\mathcal{P}(\mathbb{N})$ there is a weakly fundamental net that has no limit in $\mathcal{P}(\mathbb{N})$. Hence the space $\mathcal{M}(\mathbb{N})$ and its subset $\mathcal{P}(\mathbb{N})$ are not complete in the sense of the theory of locally convex spaces. In § 5.8(iii) one can find additional information about weak sequential completeness.

5.2. Mappings of spaces of measures

Any continuous mapping $f\colon X \to Y$ generates the mapping

$$\widehat{f}\colon \mathcal{M}_r(X) \to \mathcal{M}_r(Y), \quad \mu \mapsto \mu\circ f^{-1},$$

which is continuous in the weak topology. It is clear that we also obtain the mappings

$$\widehat{f}\colon \mathcal{M}_t(X) \to \mathcal{M}_t(Y),\ \widehat{f}\colon \mathcal{M}_\tau(X) \to \mathcal{M}_\tau(Y),\ \widehat{f}\colon \mathcal{M}_\sigma(X) \to \mathcal{M}_\sigma(Y)$$

and the mappings between the corresponding spaces of nonnegative or probability measures. Here we discuss a number of topological properties of the induced mapping $\widehat{f}$, the most important of which is openness (under certain conditions) for an open mapping f (i.e, taking open sets to open sets). Due to this property one can construct one-sided inverse mappings for mappings between spaces of measures, which is useful in applications.

It is readily seen that if the mapping f is injective, then so is the mapping $\widehat{f}\colon \mathcal{M}_r(X) \to \mathcal{M}_r(Y)$; the following more general fact holds.

5.2.1. LEMMA. *Let K_n, where $n \in \mathbb{N}$, be increasing compact sets in a Hausdorff space X and let $f\colon X \to Y$ be an injective mapping to a Hausdorff space Y such that f is continuous on each K_n. Then every Radon measure ν concentrated on the union of the compact sets $f(K_n)$ has a unique Radon preimage under f.*

PROOF. We observe that if μ_1 and μ_2 are two Radon preimages of ν, then they are concentrated on the union of K_n and their restrictions on every set K_n coincide. The latter is seen from the fact that if the Radon measures μ_1 and μ_2 on the union of K_n are different, then $\mu_1(S) \neq \mu_2(S)$ for some compact set S in some K_n, and then the compact set $f(S)$ has different measures with respect to their images by the injectivity of f. The existence of a Radon preimage can be easily derived from the fact that f homeomorphically maps K_n onto $f(K_n)$, hence the restriction of ν on $f(K_n)$ has a preimage. $\square$

Certainly, the same is true for the classes $\mathcal{M}_t$ as well, but not always for $\mathcal{M}_\tau$.

5.2.2. EXAMPLE. Let $S \subset [0,1]$ be a set with $\lambda^*(S) = 1$, $\lambda^*([0,1]\backslash S) = 1$, where λ is Lebesgue measure (see Example 1.1.2). Let f be the natural projection of the space $\bigl(S\times\{0\}\bigr) \cup \bigl(([0,1]\backslash S)\times\{1\}\bigr)$ to $[0,1]$. Then f is continuous and injective and Lebesgue measure on $[0,1]$ is the image under f of two different τ-additive probability measures μ_1 and μ_2 which are induced by Lebesgue measure on $S\times\{0\}$ and $([0,1]\backslash S)\times\{1\}$, respectively.

See also Remark 5.1.5 above.

Open mappings between spaces of measures have been studied in Ditor, Eifler [**172**], Eifler [**202**], Schief [**570**], [**571**], Banakh [**30**], Banakh, Radul [**39**], and Bogachev, Kolesnikov [**87**]. Below we present some results in this direction, which can be summarized as follows: for a continuous open surjection f of completely regular spaces X and Y, the mapping $\widehat{f}\colon \mathcal{P}_r(X) \to \mathcal{P}_r(Y)$ is also a continuous open surjection if $\widehat{f}\bigl(\mathcal{P}_r(G)\bigr) = \mathcal{P}_r\bigl(f(G)\bigr)$ for every open set G in X, moreover, in case of Souslin spaces the latter condition is fulfilled automatically. Actually, in place of continuity of f it suffices to have its Borel measurability (see Theorem 5.8.16 and Corollary 5.8.17), but we shall also be interested in the continuity of the induced mapping $\widehat{f}$ with the purpose of constructing its continuous inverse.

5.2.3. LEMMA. *Let μ be a Radon probability measure on a completely regular space X and let*

$$G_0 = \bigl\{\nu \in \mathcal{P}(X)\colon\ \nu(V_i) - \mu(V_i) > -\varepsilon_0,\ i = 1,\dots,n\bigr\},$$

where V_i are open sets in X and $\varepsilon_0 > 0$. Then, for every $\varepsilon > 0$, there exist a number $\delta > 0$ and open pairwise disjoint sets $W_1, W_2, \dots, W_m$ such that the following conditions are fulfilled:

(i) $\mu(V_i) - \sum_{\{j\colon\ W_j \subset V_i\}} \mu(W_j) < \varepsilon,\ i = 1, \dots, n,$
(ii) $G = \{\nu \in \mathcal{P}(X)\colon\ \nu(W_i) - \mu(W_i) > -\delta,\ i = 1, \dots, m\} \subset G_0.$

PROOF. We prove this assertion by induction on n. For $n = 1$ it is obvious. Suppose now that for the sets $V_1, \dots, V_{n-1}$ and the number $\varepsilon/2$ we have found pairwise disjoint open sets $\widetilde{W}_1, \dots, \widetilde{W}_{m_{n-1}}$ and a number δ_{n-1} such that (i) and (ii) are fulfilled with δ_{n-1} in place of δ. Let $0 < \varepsilon_1 < \dfrac{\min(\varepsilon, \varepsilon_0, \delta_{n-1})}{2m_{n-1}}$. By Lemma 4.3.7 we can construct open sets W_j^1, $j = 1, \dots, m_{n-1}$, such that

$$\overline{W_j^1} \subset \widetilde{W}_j, \quad \mu(\widetilde{W}_j \setminus W_j^1) < \varepsilon_1, \quad \mu(\partial W_j^1) = 0.$$

Set $W = V_n \setminus \bigcup_{j=1}^{m_{n-1}} \overline{W}_j^1$. Now for every j one can construct open sets W_j^2 such that

$$\overline{W_j^2} \subset W_j^1 \cap V_n, \quad \mu((W_j^1 \cap V_n) \setminus W_j^2) < \varepsilon_1, \quad \mu(\partial W_j^2) = 0.$$

Let $W_j^3 = W_j^1 \setminus \overline{W_j^2}$. Then W, W_j^2, W_j^3, $j = 1, \dots, m_{n-1}$, are desired sets. It is clear that they are pairwise disjoint. For $i = 1, 2, \dots, n-1$ we have

$$\begin{aligned}\mu(V_i) - \sum_{j\colon\ \widetilde{W}_j \subset V_i} (\mu(W_j^2) + \mu(W_j^3)) &= \mu(V_i) - \sum_{j\colon\ \widetilde{W}_j \subset V_i} \mu(W_j^1) \\ < \mu(V_i) - \sum_{j\colon\ \widetilde{W}_j \subset V_i} \mu(\widetilde{W}_j) + m_{n-1}\varepsilon_1 &< \frac{\varepsilon}{2} + m_{n-1}\varepsilon_1 < \varepsilon.\end{aligned}$$

We also have

$$\mu(V_n) - \mu(W) - \sum_j \mu(W_j^2) < m_{n-1}\varepsilon_1 < \varepsilon.$$

Thus, condition (i) is fulfilled.

Let $0 < \delta < \dfrac{\min(\varepsilon_0, \delta_{n-1})}{4(m_{n-1}+1)}$. For the proof of (ii) it remains to show that

$$\nu(\widetilde{W}_j) - \mu(\widetilde{W}_j) > -\delta_{n-1}, \quad j = 1, \dots, m_{n-1},$$

and also that $\nu(V_n) - \mu(V_n) > -\varepsilon_0$, if $\nu \in G$. Indeed,

$$\begin{aligned}\nu(\widetilde{W}_j) - \mu(\widetilde{W}_j) &> \nu(W_j^1) - \mu(W_j^1) - \varepsilon_1 \\ &\geqslant \nu(W_j^2) + \nu(W_j^3) - \mu(W_j^2) - \mu(W_j^3) - \varepsilon_1 > -2\delta - \varepsilon_1 > -\delta_{n-1},\end{aligned}$$

$$\begin{aligned}\nu(V_n) - \mu(V_n) &\geqslant \nu(W) + \sum_j \nu(W_j^1 \cap V_n) - \mu(W) - \sum_j \mu(W_j^1 \cap V_n) \\ &> \nu(W) + \sum_j \nu(W_j^2) - \mu(W) - \sum_j \mu(W_j^2) - m_{n-1}\varepsilon_1 \\ &> -(m_{n-1}+1)\delta - m_{n-1}\varepsilon_1 > -\varepsilon_0.\end{aligned}$$

The lemma is proven. □

5.2.4. THEOREM. *Let X and Y be completely regular spaces and let $f\colon X \to Y$ be an open surjective continuous mapping satisfying the condition of local conservativity: for every open set $V \subset X$ and every Radon probability measure ν on $f(V)$, there exists a Radon probability measure μ on V such that $\mu \circ f^{-1} = \nu$. Then the mapping $\widehat{f}\colon \mathcal{P}_r(X) \to \mathcal{P}_r(Y)$ is an open surjection. The same is true for the* *mapping* $\widehat{f}\colon \mathcal{P}(X) \to \mathcal{P}(Y)$ *if* X *and* Y *are perfectly normal and* f *satisfies the conditions above with replacement of "Radon measure" with "Borel measure".*

PROOF. The surjectivity of $\widehat{f}$ follows from the definition. Let $\mu_0 \in \mathcal{P}_r(X)$ and let

$$U = \bigl\{\mu\colon\ \mu(V_i) - \mu_0(V_i) > -\varepsilon,\ i = 1, \ldots, n\bigr\}$$

be a neighborhood of μ_0 in the weak topology, where the sets $V_i \subset X$ are open and $\varepsilon > 0$. According to Lemma 5.2.3, we can assume that the sets V_i are pairwise disjoint. Let $\eta_0 = \mu_0 \circ f^{-1}$. The sets $f(V_i)$ are open. By Lemma 5.2.3 there exist pairwise disjoint sets $W_1, \ldots, W_m \subset Y$ such that

$$\eta_0\bigl(f(V_i)\bigr) - \sum_{W_j \subset f(V_i)} \eta_0(W_j) < \frac{\varepsilon}{2}.$$

Let $0 < \varepsilon_1 < \varepsilon/(2m)$. Let us fix a neighborhood of the measure η_0 of the form

$$O = \bigl\{\eta \in \mathcal{P}_r(Y)\colon\ \eta(W_j) - \eta_0(W_j) > -\varepsilon_1,\ j = 1, \ldots, m\bigr\}.$$

We prove that every measure $\eta \in O$ has a preimage in U. Set

$$I_j = \{i\colon\ W_j \subset f(V_i)\}.$$

For every $i \in I_j$ we can find numbers $\alpha_{ij} \geqslant 0$ such that $\sum_{i \in I_j} \alpha_{ij} = 1$ and $\alpha_{ij}\eta_0(W_j) \geqslant \mu_0\bigl(f^{-1}(W_j) \cap V_i\bigr)$. By the hypotheses of the theorem for every $i \in I_j$ there exists a preimage μ_{ij} of the measure $\eta|_{W_j}$ with respect to the mapping $f|_{V_i}$, i.e., a nonnegative Radon measure μ_{ij} on V_i such that $\mu_{ij} \circ f^{-1} = \eta|_{W_j}$. Let $\widetilde{\mu}$ be a preimage of the measure $\eta|_{Y \setminus \bigcup_j W_j}$ with respect to f. Set

$$\mu = \widetilde{\mu} + \sum_j \sum_{i \in I_j} \alpha_{ij}\mu_{ij}.$$

It is clear that $\eta = \mu \circ f^{-1}$. We show that $\mu \in U$. Indeed,

$$\begin{aligned}
\mu(V_i) &\geqslant \sum_{j\colon i \in I_j} \alpha_{ij}\mu_{ij}(V_i) = \sum_{j\colon i \in I_j} \alpha_{ij}\eta(W_j) > \sum_{j\colon i \in I_j} \alpha_{ij}\Bigl(\eta_0(W_j) - \varepsilon_1\Bigr) \\
&\geqslant \sum_{j\colon i \in I_j} \alpha_{ij}\eta_0(W_j) - m\varepsilon_1 \geqslant \sum_{j\colon i \in I_j} \mu_0\Bigl(f^{-1}(W_j) \cap V_i\Bigr) - m\varepsilon_1 \\
&= \mu_0(V_i) - \mu_0\Bigl(V_i \setminus \bigcup_{j\colon i \in I_j} f^{-1}(W_j)\Bigr) - m\varepsilon_1 \\
&\geqslant \mu_0(V_i) - \eta_0\Bigl(f(V_i) \setminus \bigcup_{j\colon W_j \in f(V_i)} W_j\Bigr) - m\varepsilon_1 \\
&\geqslant \mu_0(V_i) - \frac{\varepsilon}{2} - m\varepsilon_1 > \mu_0(V_i) - \varepsilon.
\end{aligned}$$

The first assertion is proven. The proof of the second assertion is completely analogous. $\square$

Note that the conclusion of Theorem 5.2.4 is true for arbitrary spaces and Baire measures if f is a continuous surjection taking functionally open sets to functionally open sets and the aforementioned condition of local conservativity is fulfilled for Baire measures and functionally open sets.

The condition of local conservativity in Theorem 5.2.4 is fulfilled in many cases, for example, for Borel mappings of Souslin spaces (see § 2.1).

5.2.5. COROLLARY. *Let X and Y be completely regular Souslin spaces and let $f\colon X \to Y$ be an open surjective continuous mapping. Then the mapping $\widehat{f}\colon \mathcal{P}(X) \to \mathcal{P}(Y)$ is an open surjection in the weak topology.*

PROOF. It suffices to recall that $\mathcal{P}(X) = \mathcal{P}_r(X)$ and $\mathcal{P}(Y) = \mathcal{P}_r(Y)$ for Souslin spaces and that every continuous mapping of Souslin spaces is locally conservative in the sense of Theorem 5.2.4. □

5.2.6. COROLLARY. *Let $f\colon X \to Y$ be an open surjective continuous mapping of complete metric spaces. Then the mapping $\widehat{f}\colon \mathcal{P}_r(X) \to \mathcal{P}_r(Y)$ is an open continuous surjection.*

PROOF. Let $V \subset X$ be open and let ν be a Radon probability measure on $f(V)$. We have to show that there exists a Radon probability measure μ on the set V such that $\mu\circ f^{-1} = \nu$. Let $\varepsilon > 0$ and let $E \subset f(V)$ be a compact set such that $\nu(E) > 1-\varepsilon$. We recall that V is homeomorphic to a complete metric space (see Engelking [**203**, Theorem 4.3.23]). According to [**203**, 5.5.11], there exists a compact set $K \subset V$ such that $f(K) = E$. Now our assertion follows from Bogachev [**81**, Theorem 9.1.9]. □

5.2.7. REMARK. Theorem 5.2.4 implies the result of Ditor, Eifler [**172**], where the spaces X and Y are compact. Indeed, let $f\colon X \to Y$ be a continuous open mapping, let $U \subset X$ be open, and let ν be a Radon probability measure on the set $V = f(U)$. In order to prove that there exists a Radon probability measure μ on U such that $\mu\circ f^{-1} = \nu$, it suffices to show that for every $\varepsilon > 0$ there exists a compact set $S \subset U$ satisfying the inequality $\nu\bigl(f(S)\bigr) > 1-\varepsilon$ (see [**81**, Theorem 9.1.9]). Let us take a compact set $K \subset V$ with $\nu(K) > 1-\varepsilon$. For every point $k \in K$ we can find a point $u_k \in U \cap f^{-1}(k)$ and its neighborhood U_k such that $\overline{U_k} \subset U$. The open sets $f(U_k)$ cover K, hence one can extract a finite subcover $f(U_{k_1}), \ldots, f(U_{k_n})$. Then the set $S = \bigcup_{i=1}^n \overline{U_{k_i}}$ is compact and $K \subset f(S)$, whence we obtain the desired estimate $\nu\bigl(f(S)\bigr) > 1-\varepsilon$.

We recall that a continuous mapping is called *perfect* if it takes closed sets to closed sets and the preimages of points are compact. Then the preimages of compact sets are also compact (see Engleking [**203**, Theorem 3.7.2]). Perfect mappings between spaces generate perfect mappings between spaces of measures (Koumoullis [**386**]).

5.2.8. THEOREM. *Suppose that $f\colon X \to Y$ is a continuous surjection of completely regular spaces. Then, whenever $s = t$ or $s = \tau$, the induced mapping $\widehat{f}\colon \mathcal{M}_s^+(X) \to \mathcal{M}_s^+(Y)$, $\mu \mapsto \mu\circ f^{-1}$ is perfect if and only if f is perfect.*

As observed in [**386**], this assertion can be false for $s = \sigma$ and for spaces of signed measures.

This theorem and Frolík's result that a space X is Lindelöf and Čech complete precisely when it admits a perfect surjection onto a complete separable metric space were used in [**386**] to prove the following result for a completely regular space X.

5.2.9. COROLLARY. *The space $\mathcal{M}_s^+(X)$, where $s = t$ or $s = \tau$, is Lindelöf and Čech complete precisely when so is X. In addition, $\mathcal{M}_s^+(X)$ is paracompact and Čech complete precisely when so is X.*

5.3. Continuous inverse mappings

We now discuss conditions for the existence of continuous inverse mappings for the mapping $\widehat{f}$ of spaces of measures induced by a mapping $f: X \to Y$ in case of non-injective $\widehat{f}$, i.e., the existence of continuous selections $\nu \mapsto \mu_\nu \in \widehat{f}^{-1}(\nu)$. Below a *right inverse* for a mapping T is understood as a one-to-one mapping associating to every element from the image of T a point in its preimage.

We recall the following classical result, called Michael's selection theorem (see Michael [**463**] or Repovš, Semenov [**547**, p. 190]).

5.3.1. THEOREM. *Let M be a metrizable space, let P be a complete metrizable closed subset in a locally convex space E, and let $\Phi\colon M \to 2^P$ be a lower semicontinuous mapping with values in the set of nonempty convex closed subsets of P, i.e., for every open set $U \subset P$, the set $\Phi^{-1}(U) := \{x \in M\colon \Phi(x) \cap U \neq \varnothing\}$ is open. Then there is a continuous mapping $f\colon M \to P$ such that $f(x) \in \Phi(x)$ for all x.*

For our purposes it will be sufficient to deal with the case where E is a normed space; a short proof for this case can be found in Fedorchuk, Filippov [**217**, Chapter 6], Repovš, Semenov [**547**, A§1]. Namely, in our case P and M will be the sets of all Radon probability measures on Polish spaces X and Y; then the weak topologies on P and M generate the Kanorovich–Rubinshtein norms on $\mathcal{M}_r(X)$ and $\mathcal{M}_r(Y)$. Let us emphasize an important circumstance that the space E itself need not be complete. Note that Filippov [**236**] constructed an example showing that the closedness of P cannot be omitted even if P is a G_δ-set in a Hilbert space.

Let us mention a typical application of this theorem (see Bogachev, Smolyanov [**97**, Corollary 1.12.20]).

5.3.2. COROLLARY. *Let $T\colon P \to M$ be a continuous affine mapping from a complete metrizable closed convex set P in a locally convex space to a metrizable set M in a locally convex space such that T is open, i.e., takes open sets to open sets. Then T possesses a continuous right inverse.*

Thus, by Corollary 5.2.6 the following assertion holds.

5.3.3. THEOREM. *Let X and Y be nonempty Polish spaces and let a mapping $f\colon X \to Y$ be open, surjective and continuous. Then the induced mapping $\widehat{f}\colon \mathcal{P}(X) \to \mathcal{P}(Y)$ possesses a right inverse continuous in the weak topology, i.e., there exists a mapping $\Psi\colon \mathcal{P}(Y) \to \mathcal{P}(X)$ continuous in the weak topology such that $\widehat{f}(\Psi(\nu)) = \nu$ for all $\nu \in \mathcal{P}(Y)$. The same is true for arbitrary complete metric spaces X and Y if we consider $\mathcal{P}_r(X)$ and $\mathcal{P}_r(Y)$ in place of $\mathcal{P}(X)$ and $\mathcal{P}(Y)$, respectively.*

Let us recall that *universally measurable* sets are sets measurable with respect to all Borel measures on the given topological space.

5.3.4. COROLLARY. *For every universally measurable set Y in a Polish space Z there exists a universally measurable subset X of the space $\mathcal{R}$ of irrational numbers in $[0,1]$ and a continuous surjective mapping $f\colon X \to Y$ such that the mapping $\widehat{f}\colon \mathcal{P}(X) \to \mathcal{P}(Y)$ has a right inverse $g\colon \mathcal{P}(Y) \to \mathcal{P}(X)$ continuous in the weak topology. For an arbitrary set $Y \subset Z$ an analogous assertion, but without universal measurability of X, is true for the spaces $\mathcal{P}_r(X)$ and $\mathcal{P}_r(Y)$.*

PROOF. It is known that there exists an open surjection $F\colon \mathcal{R} \to Z$ (see Arkhangelskii, Ponomarev [**23**, Chapter VI, Problem 150]; we recall that $\mathcal{R}$ is homeomorphic to $\mathbb{N}^\infty$, see Engelking [**203**, Chapter 4, 4.3.G]). Each measure μ on Y extends to Z in the following way: $\mu(Z\backslash Y)=0$, which is possible by the universal measurability of Y. Let $\Psi\colon \mathcal{P}(Z)\to\mathcal{P}(\mathcal{R})$ be the mapping from Theorem 5.3.3. The set $X=F^{-1}(Y)$ is universally measurable in $\mathcal{R}$. Indeed, if $\nu\in\mathcal{P}(\mathcal{R})$, then by the universal measurability of Y there exist Borel sets $B_1, B_2\subset Z$ such that

$$B_1\subset Y\subset B_2 \quad\text{and}\quad \nu\circ F^{-1}(B_2\backslash B_1)=0.$$

Then the sets $A_1=F^{-1}(B_1)$ and $A_2=F^{-1}(B_2)$ are Borel, $A_1\subset X\subset A_2$ and $\nu(A_2\backslash A_1)=0$. If measures $\mu_n\in\mathcal{P}(Y)$ converge weakly to $\mu\in\mathcal{P}(Y)$ on Y, then the measures $\Psi(\mu_n)$ and $\Psi(\mu)$ are concentrated on X. Indeed, if K is a compact set in the complement of X, then $\mu_n\bigl(F(K)\bigr)=\mu\bigl(F(K)\bigr)=0$, whence

$$\Psi(\mu_n)\Bigl(F^{-1}\bigl(F(K)\bigr)\Bigr)=\Psi(\mu)\Bigl(F^{-1}\bigl(F(K)\bigr)\Bigr)=0$$

and hence $\Psi(\mu_n)(K)=\Psi(\mu)(K)=0$. In addition, the measures $\Psi(\mu_n)$ converge in the weak topology of the space $\mathcal{P}(X)$ to the measure $\Psi(\mu)$ (since $\Psi(\mu_n)\Rightarrow\Psi(\mu)$ on $\mathcal{R}$, it suffices to observe that every bounded uniformly continuous function on X extends to a bounded continuous function on $\mathbb{R}$). Therefore, for f one can take the restriction of F to X and for $g(\mu)$ one can take the measure $\Psi(\mu)$ on X. If we consider only Radon measures, then the reasoning above remains in force for every set Y (but X can fail to be universally measurable). □

5.3.5. Remark. (i) Lemma 2.6.3 along with what has been said before it gives the existence of a continuous surjection $F\colon C\to S$ for which the mapping $\widehat{F}$ has a linear right inverse continuous in the weak topology. This assertion is not a corollary of Theorem 5.3.3, since S can fail to be the image of C under an open mapping (for example, if $S=[0,1]$). In addition, it shows that for the existence of a continuous right inverse G for the mapping $\widehat{F}$ it is not necessary that F be open.

(ii) It should be noted that a mapping g that is a continuous right inverse to the linear mapping $\widehat{f}$ can be nonlinear. However, in the case where X and Y are metrizable compact sets and $f\colon X\to Y$ is an open surjection, according to Michael [**462**, Theorem 1.1], the mapping f possesses a regular averaging operator (see p. 76), hence the mapping $\widehat{f}\colon \mathcal{M}(X)\to\mathcal{M}(Y)$ has a linear continuous right inverse.

This remark leads to an analog of Corollary 5.3.4 with the Cantor set in place of the set of irrational numbers.

5.3.6. Proposition. *For every universally measurable set Y in a Polish space, there exists a universally measurable subset X of the Cantor set C and a continuous surjective mapping $f\colon X\mapsto Y$ such that the mapping $\widehat{f}\colon \mathcal{P}(X)\to\mathcal{P}(Y)$ possesses a right inverse $g\colon \mathcal{P}(Y)\to\mathcal{P}(X)$ continuous in the weak topology. For compact Y, the set X can be chosen compact. For arbitrary Y, an analogous assertion, but without universal measurability of X, is true for the spaces $\mathcal{P}_r(X)$ and $\mathcal{P}_r(Y)$.*

Proof. Let us embed Y homeomorphically into $S=[0,1]^\infty$, which is possible by the Urysohn theorem (see Engelking [**203**, Theorem 4.2.10]). Each measure μ on Y can be extended to S by setting $\mu(S\backslash Y)=0$, which is possible by the universal measurability of Y. Let $F\colon C\to S$ and $G\colon \mathcal{P}(S)\to\mathcal{P}(C)$ be the mappings mentioned in Remark 5.3.5. Set $X=F^{-1}(Y)$. If measures $\mu_n\in\mathcal{P}(Y)$ converge weakly to a measure $\mu \in \mathcal{P}(Y)$ on Y, then, as in Corollary 5.3.4, the measures $G(\mu_n)$, which are concentrated on X, converge to the measure $G(\mu)$ in the weak topology of the space $\mathcal{P}(X)$. Hence for f we can take the restriction of F to X and for $g(\mu)$ we can take the measure $G(\mu)$ on X. It is obvious that X is compact provided that Y is compact. The last assertion of the corollary is clear from the proof. □

Note that in Corollary 5.3.4 and in Proposition 5.3.6 the weak topology on the space $\mathcal{P}(Y)$ coincides with the topology induced from the space of probability measures on the Polish space containing Y.

5.4. Spaces with the Skorohod property

In § 2.6 we have already discussed representations of weakly convergent measures in the form of images of a single measure under pointwise convergent mappings and also a parametrization of the whole space of probability measures on a metric space by mappings with the indicated property. Here this theme continues for more general topological spaces. The notion of a space with the Skorohod property introduced in Definition 2.6.1 has the following topological version.

5.4.1. DEFINITION. We shall say that a Hausdorff topological space X possesses the strong Skorohod property for Radon measures (or, for brevity, the strong Skorohod property) if, to every Radon probability measure μ on X, one can associate a Borel mapping $\xi_\mu\colon [0,1]\to X$ such that μ is the image of Lebesgue measure under the mapping ξ_μ and if measures μ_n converge weakly to μ, then $\xi_{\mu_n}(t) \to \xi_\mu(t)$ almost everywhere.

If the indicated parametrization is possible for each uniformly tight family of Radon probability measures on X, then X will be called a space with the *UTS* ("uniformly tight Skorohod") property.

Below for brevity we omit the precision "for Radon measures".

One can also consider the "weak Skorohod property" admitting in place of the interval $[0,1]$ with Lebesgue measure an arbitrary probability space (see Bogachev, Kolesnikov [**87**]).

There is the following analog of Lemma 2.6.2 with the same proof.

5.4.2. LEMMA. *Let X be a space with the strong Skorohod property. Then*

(i) *every subset $Y \subset X$ possesses this property;*

(ii) *if F is a continuous mapping from X onto a Hausdorff space Y and there exists a mapping $\Psi\colon \mathcal{P}_r(Y) \to \mathcal{P}_r(X)$ continuous in the weak topology such that $\Psi(\nu)\circ F^{-1} = \nu$ for all $\nu \in \mathcal{P}_r(Y)$, then Y possesses the strong Skorohod property.*

The principal result of § 2.6 was Theorem 2.6.4, according to which all separable metric spaces possess the strong Skorohod property. This gives the *UTS* property for any metric space. Let us consider nonmetrizable spaces with the Skorohod property. We first show that the *UTS* property is preserved under bijective continuous proper mappings. In particular, the strong Skorohod property is preserved under bijective continuous proper mappings onto sequentially Prohorov spaces. We recall that a mapping $f\colon X \to Y$ between topological spaces is called *proper* if $f^{-1}(K)$ is compact for every compact set $K \subset Y$.

5.4.3. THEOREM. *Let X and Y be Hausdorff spaces such that there exists a bijective continuous proper mapping $F\colon X \to Y$ and X has the UTS property.*

Then Y possesses this property too. In particular, if in this situation the space Y is sequentially Prohorov, then it has the strong Skorohod property as well.

PROOF. Since F is proper and continuous, for every measure $\mu \in \mathcal{P}_r(Y)$ on Y, there exists a unique measure $\widehat{\mu} \in \mathcal{P}_r(X)$ with $\widehat{\mu}\circ F^{-1} = \mu$, moreover, $\widehat{\mu} = \mu\circ G^{-1}$, where $G = F^{-1}$ (see Bogachev [**81**, § 9.1]). Let us take a Skorohod parametrization $\nu \mapsto \xi_\nu$ of Radon probability measures on X by Borel mappings from $[0,1]$, which exists by assumption. Then $\mu \mapsto F\circ\xi_{\widehat{\mu}}$ is the desired parametrization on Y. Indeed, suppose that Radon probability measures μ_n converge weakly to a Radon measure μ on Y and the sequence $\{\mu_n\}$ is uniformly tight. Since F is a proper mapping, the sequence of measures $\widehat{\mu_n}$ is also uniformly tight. In addition, for every compact set $Q \subset X$ we have

$$\limsup_{n\to\infty} \widehat{\mu_n}(Q) = \limsup_{n\to\infty} \mu_n\bigl(F(Q)\bigr) \leqslant \mu\bigl(F(Q)\bigr) = \widehat{\mu}(Q).$$

Along with the uniform tightness this gives $\limsup\limits_{n\to\infty} \widehat{\mu_n}(Z) \leqslant \widehat{\mu}(Z)$ for every closed set $Z \subset X$, i.e., $\widehat{\mu_n} \Rightarrow \widehat{\mu}$. Hence $\xi_{\widehat{\mu_n}}(t) \to \xi_{\widehat{\mu}}(t)$ for almost every $t \in [0,1]$. For such points t we also have $\xi_{\mu_n}(t) \to \xi_\mu(t)$ by the continuity of F. □

5.4.4. REMARK. It is clear from the proof that if $F\colon X \to Y$ is bijective, continuous and proper and measures $\mu_n \in \mathcal{P}_r(Y)$ are uniformly tight and converge weakly to μ, then $\mu_n\circ G^{-1} \Rightarrow \mu\circ G^{-1}$ on X, where $G = F^{-1}$.

A completely regular (Tychonoff) space X will be called *almost metrizable* if there exists a bijective continuous proper mapping $f\colon M \to X$ from a metrizable space M. If M is discrete, then X will be called *almost discrete*.

It is not difficult to give examples of nonmetrizable almost metrizable spaces (such examples are given below). It is straightforward to verify that almost metrizable spaces and almost discrete space possess the following properties (for the definition of a k-space, see p. 174).

5.4.5. PROPOSITION. (i) *Each subspace of an almost metrizable space is almost metrizable.*

(ii) *A Tychonoff space is metrizable precisely when it is an almost metrizable k-space.*

(iii) *A Tychonoff space X is almost metrizable precisely when the strongest topology inducing the original topology on every compact subset of X is metrizable.*

(iv) *A Tychonoff space is almost discrete precisely when it contains no infinite compact sets.*

(v) *A countable product of almost metrizable spaces is almost metrizable.*

(vi) *The images of almost metrizable and almost discrete spaces under continuous bijective proper mappings belong to the respective classes. In addition, both classes are preserved by arbitrary topological sums.*

The next theorem characterizes almost metrizable spaces with the strong Skorohod property.

5.4.6. THEOREM. *Each almost metrizable space has the UTS property. In addition, an almost metrizable space has the strong Skorohod property precisely when it is sequentially Prohorov.*

PROOF. If the space X is almost metrizable and sequentially Prohorov, then it has the strong Skorohod property by Theorem 5.4.3. The same reasoning proves the first assertion. Suppose now that X is almost metrizable and has the strong Skorohod property. Let Radon probability measures μ_n converge weakly to a Radon measure μ on X. Let us take Borel mappings $\xi_{\mu_n}\colon [0,1] \to X$ that converge almost everywhere to a Borel mapping $\xi_\mu\colon [0,1] \to X$ and have the property that ξ_{μ_n} and ξ_μ transform Lebesgue measure λ on $[0,1]$ into the measures μ_n and μ, respectively. Let us also take a metric space M for which there is a proper bijective continuous mapping F onto X. As observed in the proof of Theorem 5.4.3, there exist unique Radon probability measures ν_n and ν whose images under F are μ_n and μ, respectively. Since the preimages under F of all compact sets in X are compact in M, it is easy to see that $G_n(t) := F^{-1}\xi_{\mu_n}(t) \to G(t) := F^{-1}\xi_\mu(t)$ in M for every point t such that $\xi_{\mu_n}(t) \to \xi_\mu(t)$. We observe that $\nu_n = \lambda\circ G_n^{-1}$ and $\nu = \lambda\circ G^{-1}$, since $(\lambda\circ G_n^{-1})\circ F^{-1} = \mu_n$, $(\lambda\circ G^{-1})\circ F^{-1} = \mu$ and the measures μ_n and μ have unique preimages under F. This shows that $\nu_n \Rightarrow \nu$. Taking into account that all considered measures are Radon and M is metrizable, we obtain that the sequence $\{\nu_n\}$ is uniformly tight (see Theorem 2.3.6), hence so is $\{\mu_n\}$. □

Since countable products of sequentially Prohorov spaces are sequentially Prohorov as well (see § 4.7), we arrive at the following assertion.

5.4.7. COROLLARY. *A countable product of almost metrizable spaces with the strong Skorohod property has the strong Skorohod property.*

For almost discrete spaces even stronger results hold. We recall that a topological space X is called *sequentially compact* if every sequence in X contains a convergent subsequence.

5.4.8. THEOREM. *For a Tychonoff space X the following conditions are equivalent:*

(i) *X is an almost discrete space;*

(ii) *every compact subset in X is sequentially compact and every uniformly tight weakly convergent sequence $\mu_n \Rightarrow \mu$ of Radon probability measures on X converges in variation (or, which is equivalent, one has convergence $\mu_n(x) \to \mu(x)$ for every point $x \in X$).*

PROOF. The implication (i) ⇒ (ii) follows from Remark 5.4.4, since every weakly convergent sequence of Radon measures on a discrete space converges in variation. In order to prove the converse implication suppose that the space X satisfies condition (ii). It suffices to show that every compact set K in X is finite. Suppose that this is false. By the sequential compactness of K one can find a nontrivial convergent sequence $x_n \to x_0$ in K. Then the sequence of the Dirac measures δ_{x_n} at the points x_n is uniformly tight and converges weakly to Dirac's measure δ_{x_0}. By our assumption $\delta_{x_n}(x_0) \to \delta_{x_0}(x_0) = 1$. Then $x_n = x_0$ for all n, excepting finitely many numbers. □

5.4.9. COROLLARY. *Let X be an almost discrete sequentially Prohorov space, let E be a completely regular space, and let a sequence of Radon probability measures μ_n on $X\times E$ converge weakly to a Radon probability measure μ. Then, for every $x \in X$, the restrictions of the measures μ_n to the set $x\times E$ converge weakly to the restriction of μ, i.e., $\mu_n|_{x\times E} \Rightarrow \mu|_{x\times E}$.*

PROOF. The projections η_n of the measures μ_n to X converge weakly to the projection η of the measure μ. By Theorem 5.4.8 we have $\eta_n(x) \to \eta(x)$, hence

$\mu_n(x\times E) \to \mu(x\times E)$ for every $x \in X$. Since every set $x\times E$ is closed in $X\times E$, weak convergence on $X \times E$ yields that $\limsup\limits_{n\to\infty} \mu_n(Z) \leqslant \mu(Z)$ for every closed subset of the space $x\times E$. Hence we have the desired weak convergence. □

An almost metrizable space need not be metrizable (below we give an example of a countable almost metrizable nonmetrizable space). A rather unexpected example is the Banach space l^1 with the weak topology. The space l^1 possesses the so-called *Schur property*, i.e., every weakly convergent sequence converges in norm.

5.4.10. THEOREM. *Let X be a Banach space with the Schur property and let τ be an intermediate topology between the norm and weak topologies on X. Then the space (X, τ) is almost metrizable and has the UTS property.*

PROOF. According to Theorem 5.4.6, it suffices to prove that the identity mapping $X \to (X, \tau)$ is proper. To this end, let us fix a compact set $K \subset (X, \tau)$. Then K is weakly compact and by the Eberlein–Shmulian theorem K is sequentially compact in the weak topology. Suppose that K is not norm compact. Then K contains a sequence $\{x_n\}$ without norm convergent subsequences. Since K is sequentially compact in the weak topology, this sequence contains a weakly convergent subsequence, which contradicts the Schur property. □

It is worth noting that, as shown in § 4.7, the space l^1 with the weak topology (as any infinite-dimensional Banach space with the weak topology) is not sequentially Prohorov.

We now construct an almost discrete space without the strong Skorohod property. Let Card$\,(A)$ be the cardinality of a set A.

5.4.11. EXAMPLE. Let X_n, where $n \in \mathbb{N}$, be pairwise disjoint finite sets in $\mathbb{N}$ with Card$\,(X_n) <$ Card$\,(X_{n+1})$ for every n. Let us fix a point $\infty \notin \bigcup_{n\geqslant 1} X_n$ and define a topology on the union $X = \{\infty\} \cup \bigcup_{n=1}^{\infty} X_n$ in the following way. All points excepting ∞ are isolated and a base of neighborhoods of the only non-isolated point ∞ consists of the sets $X \setminus F$, where $F \subset \bigcup_{n\geqslant 1} X_n$ is a subset for which there is $m \in \mathbb{N}$ such that Card$\,(F \cap X_n) \leqslant m$ for every n. We show that the space X is almost discrete and does not have the strong Skorohod property. It suffices to observe that X has no nontrivial convergent sequences. On the other hand, the sequence of measures μ_n, where every μ_n is concentrated on X_n and assumes the equal value $[\mathrm{Card}\,(X_n)]^{-1}$ on all points in X_n, converges weakly to Dirac's measure at ∞. The existence of a Skorohod parametrization of a subsequence in μ_n would give a nontrivial convergent sequence. For the same reason no subsequence in $\{\mu_n\}$ can be uniformly tight (otherwise such a subsequence could be Skorohod parametrized according to Remark 5.4.4).

We now show that nonmetrizable almost metrizable spaces with the strong Skorohod property exist. They will be constructed as subsets of extremally disconnected spaces. A topological space is called *extremally disconnected* if the closure of every open set in it is open (see Engelking [**203**, § 6.2, p. 368]). A standard example of an extremally disconnected non-discrete space is $\beta\mathbb{N}$, the Stone–Čech compactification of $\mathbb{N}$. Moreover, the Stone–Čech compactification βX of a Tychonoff space X is extremally disconnected precisely when the space X itself is extremally disconnected (see [**203**, Theorem 6.2.7, p. 368]. This class of spaces will also be encountered in § 5.6 (see p. 228).

5.4.12. THEOREM. *Any countable subspace X in an extremally disconnected Tychonoff space K is almost discrete and has the strong Skorohod property.*

PROOF. Without loss of generality we can assume that K is compact (by replacing it with the Stone–Čech compactification). Since an extremally disconnected space does not contain nontrivial convergent sequences, each countable subspace $X \subset K$ is almost discrete. By Theorem 5.4.6, in order to establish the strong Skorohod property of the space X, it suffices to verify that X is sequentially Prohorov. Suppose that a sequence of Radon probability measures μ_n on X converges weakly to a Radon measure μ. The measures μ_n, regarded as measures on K, converge weakly to the measure μ on K. Grothendieck's result from Proposition 5.6.15 below gives convergence of $\{\mu_n\}$ to μ on Borel sets, which yields uniform tightness. Actually, weak convergence of countable sequences of probability measures on X is equivalent to convergence on every point in $X = \{x_n\}$, hence convergence in variation. So in place of Theorem 5.4.6 we could refer to the fact that $\mathbb{N}$ has the strong Skorohod property. □

5.4.13. COROLLARY. *For any $p \in \beta N \backslash \mathbb{N}$, the space $X = \{p\} \cup \mathbb{N}$ with the induced topology is a nonmetrizable almost discrete space with the strong Skorohod property.*

A closer look at the proof of Theorem 5.4.12 shows that it is true for a broader class of spaces. A Tychonoff space X is called a *Grothendieck space* if the space $C_b(X)$ with its usual sup-norm is a Grothendieck Banach space, where a Banach space E is called a Grothendieck Banach space if the weak-$*$ convergence of countable sequences in E^* is equivalent to weak convergence (i.e., convergence in the topology $\sigma(E^*, E^{**})$). According to the aforementioned result of Grothendieck, each extremally disconnected Tychonoff space is a Grothendieck space. Since $C_b(X)$ and $C_b(\beta X)$ are isomorphic, one can pass here to βX.

5.4.14. THEOREM. *Each countable subspace X in a Grothendieck space K is almost discrete and has the strong Skorohod property.*

PROOF. Since K has no nontrivial convergent sequences, the space X is almost discrete. According to Theorem 5.4.6, in order to establish the strong Skorohod property for X, it suffices to verify that X is sequentially Prohorov. Suppose that a sequence of probability measures μ_n converges weakly to a measure μ on X. The measures μ_n can be regarded as elements of the dual space $C_b^*(K)$ to the Banach space $C_b(K)$, then convergence of the sequence $\{\mu_n\}$ corresponds to the weak-$*$ convergence in $C_b^*(K)$. Since $C_b(K)$ is a Grothendieck Banach space, the sequence $\{\mu_n\}$ converges in the weak topology in $C_b^*(K)$, i.e., in the topology $\sigma\big(C_b^*(K), C_b^{**}(K)\big)$. Let L be the closed linear subspace in $C_b^*(K)$ generated by Dirac measures δ_x with $x \in X$. It is readily verified that the space L is isometrically isomorphic to the Banach space l^1. It is clear that $\mu_n \in L$ for all $n \in \mathbb{N}$. By the known property of weak convergence in l^1 the sequence $\{\mu_n\}$ converges in norm (Exercise 1.7.15), which gives its uniform tightness. □

5.4.15. COROLLARY. *A subspace X in a Grothendieck space K is almost discrete and has the strong Skorohod property precisely when all compact sets in X are metrizable (or, which is equivalent, are finite). In particular, this is true if K is an extremally disconnected Tychonoff space.*

From Theorem 5.4.14 and Corollary 5.4.9 we obtain the following fact.

5.4.16. COROLLARY. *Let X be a countable subset of a Grothendieck space and let E be a Tychonoff space. Suppose that a sequence of Radon probability measures μ_n on $X\times E$ converges weakly to a Radon measure μ. Then, for every $x\in X$, the restrictions of the measures μ_n to the set $x\times E$ converge weakly to the restriction of μ, i.e., one has $\mu_n|_{x\times E}\Rightarrow \mu|_{x\times E}$.*

The space $\{p\}\cup\mathbb{N}$, where $p\in\beta\mathbb{N}\backslash\mathbb{N}$, appears to be the simplest example of a nonmetrizable space with the strong Skorohod property. The fact that it is not metrizable is clear: p belongs to the closure of $\mathbb{N}$, but there is no infinite convergent sequence with elements of $\mathbb{N}$ (if such a sequence $\{n_i\}$ converges, then the function f with $f(n_{2i})=0$, $f(n_{2i+1})=1$ has no continuous extension to $\beta\mathbb{N}$).

It is worth mentioning that although weak convergence of countable sequences of probability measures on the space X in Corollary 5.4.13 corresponds to the discrete metric on X, these two weak topologies on the space of probability measures are different (otherwise X would be metrizable in the topology from $\beta\mathbb{N}$).

Thus, in the class of nonmetrizable countable almost metrizable spaces with a unique non-isolated point there are spaces possessing the strong Skorohod property as well as spaces without this property (and in the latter case the space is not sequentially Prohorov).

On the other hand, all countable spaces with a unique non-isolated point have the UTS-property, which we now prove.

5.4.17. PROPOSITION. *Every countable space with a unique non-isolated point has the UTS-property.*

PROOF. Suppose that $\mathcal{M}$ is a uniformly tight family of probability measures on a countable space X with a unique non-isolated point x_∞. For every $n\in\mathbb{N}$, there exists a compact set $K_n\subset X$ such that $\mu(K_n)>1-2^{-n}$ $\forall\,\mu\in\mathcal{M}$. Without loss of generality we can assume that $x_\infty\in K_n\subset K_{n+1}$ for every n. The compact sets K_n are metrizable, being countable. Hence the topological sum $Y=\oplus_{n\geqslant 1}K_n$ is also metrizable. Let us consider the projection $Y\to\bigcup_{n=1}^\infty K_n\subset X$. Our assertion will be proved once we establish that the induced mapping $\mathcal{P}_r(Y)\to\mathcal{P}_r(X)$ between spaces of measures has a continuous section $\mathcal{M}\to\mathcal{P}_r(Y)$ (a right inverse). For this we decompose every measure $\mu\in\mathcal{M}$ into a series of the form $\sum_{n=1}^\infty \mu_n$, where μ_n is a measure on K_n such that $\mu_n(K_n)=2^{-n}$ and the correspondence $\mu\mapsto\mu_n$ is continuous in $\mu\in\mathcal{M}$. The construction will be defined by induction. Let

$$K_1\setminus\{x_\infty\}=\{x_i\colon\ 1\leqslant i<\mathrm{Card}\,(K_1)\}.$$

Since the space X has a unique non-isolated point x_∞, every compact set in X is either finite or has a unique non-isolated point x_∞. For every measure $\mu\in\mathcal{M}$ let $N(\mu)=\sup\{m\colon\ \sum_{i<m}\mu(x_i)<1/2\}$ and

$$\mu_1=\Bigl(\frac{1}{2}-\sum_{i<N(\mu)}\mu(x_i)\Bigr)\delta_{x_{N(\mu)}}+\sum_{i<N(\mu)}\mu(x_i)\delta_{x_i}.$$

It is readily verified that $\mu_1\leqslant\mu$, $\mu_1(K_1)=1/2$ and that the measure μ_1 defined in this way depends continuously on $\mu\in\mathcal{M}$ in the weak topology. For the proof of continuity let us consider a sequence of measures $\mu^m\in\mathcal{M}$ weakly convergent to $\mu\in\mathcal{M}$. Let us fix a continuous function f on X with $|f|\leqslant 1$ and $\varepsilon>0$. We can assume that K_1 is infinite, then the sequence $\{x_n\}$ converges to x_∞. Hence there is N such that $|f(x_n) - f(x_\infty)| < \varepsilon$ for all $n > N$. Then we have

$$\begin{aligned}
\Big|\int f\,d\mu_1 - \int f\,d\mu_1^m\Big| &\leqslant \Big|\int_{\{x_n\colon\, n\leqslant N\}} f\,[d\mu_1 - d\mu_1^m]\Big| \\
&+ \Big|\int_{\{x_n\colon\, n>N\}\cup\{x_\infty\}} [f - f(x_\infty)]\,[d\mu_1 - d\mu_1^m]\Big| \\
&+ \Big|\int_{\{x_n\colon\, n>N\}\cup\{x_\infty\}} f(x_\infty)\,[d\mu_1 - d\mu_1^m]\Big| \leqslant \sum_{n\leqslant N} |\mu_1(x_n) - \mu_1^m(x_n)| \\
&+ \varepsilon + \Big|\frac{1}{2} - \sum_{n\leqslant N}\mu_1(x_n) - \Big(\frac{1}{2} - \sum_{n\leqslant N}\mu_1^m(x_n)\Big)\Big| \\
&\leqslant \varepsilon + 2\sum_{n\leqslant N} |\mu_1(x_n) - \mu_1^m(x_n)|.
\end{aligned}$$

The right-hand side tends to zero as $m \to \infty$, since $\mu^m(x_n) \to \mu(x_n)$ for every index $n < \infty$. Applying this construction to the measure $\mu - \mu_1$, we find a measure $\mu_2 \leqslant \mu - \mu_1$ on K_2 with $\mu_2(K_2) = 1/4$. Continuing inductively, we obtain the desired decomposition $\mu = \sum_{n=1}^\infty \mu_n$. $\square$

5.4.18. EXAMPLE. Let $X = \mathbb{R}_0^\infty = \{(x_1, \ldots, x_n, 0, 0, \ldots)\}$ be the strict inductive limit of the spaces $\mathbb{R}^n$. Then X is a Prohorov Souslin space in which there is a weakly convergent sequence of probability measures without the Skorohod representation. Hence X does not have the UTS-property.

PROOF. Let J be a line segment with endpoints a and b that belongs to $\mathbb{R}^n$ for some n, but is not contained in $\mathbb{R}^{n-1}$. For a fixed number $\varepsilon > 0$ we define a mapping $F(J,\varepsilon)$ from J to $\mathbb{R}^{n+1}$ as follows: the point $(a+b)/2$ is mapped to the point $(a+b)/2 + \varepsilon e_{n+1}$, where $e_{n+1} = (0, \ldots, 0, 1) \in \mathbb{R}^{n+1}$, $F(J,\varepsilon)(a) = a$, $F(J,\varepsilon)(b) = b$, and on the line segments $[a, \frac{a+b}{2}]$ and $[\frac{a+b}{2}, b]$ we define $F(J,\varepsilon)$ by linearity. This transformation is determined by that the middle of the line segment is lifted by ε to the space of the next dimension. The image of the line segment J under this transformation will be denoted by $T(J,\varepsilon)$. Let us equip the interval $I = [0,1]$ with Lebesgue measure μ. We shall construct a sequence of polygonal chains $I_m \subset \mathbb{R}^{d_m} \subset X$, each of which consists of 2^{k_m} line segments, and a probability measure μ_m on I_m is defined as follows: on every line segment constituting I_m it coincides with the usual linear measure normalized such that the measure of this line segment equals 2^{-k_m}. These polygonal chains are constructed inductively; on every step the number of polygonal chains increases and vertices of some of them belong to larger and larger $\mathbb{R}^n$, and the number of the step is less by 1 than the maximal dimension achieved. In a sense the constructed polygonal chains approach the line segment I. At the first step we set $I_1 = T([0,1], 2^{-1})$. At the second step we construct a "two-dimensional" (i.e., contained in a plane, but not in a line) polygonal chain $T([0,1], 2^{-2})$ consisting of two line segments I_2^1 and I_2^2, and also two three-dimensional polygonal chains $T(I_2^1, 2^{-2}) \cup I_2^2$ and $I_2^1 \cup T(I_2^2, 2^{-2})$ obtained by bending the first line segment and keeping the second one fixed and by an analogous transformation of the second line segment keeping fixed the first one. At the next step we construct two-dimensional polygonal chains with $\varepsilon = 2^{-3}$ from which we obtain three-dimensional polygonal chains by consecutive bending its halves (with $\varepsilon = 2^{-3}$), and from the three-dimensional lines we obtain four-dimensional polygonal chains by consecutively applying the transformation F (with $\varepsilon = 2^{-3}$) to the line segments of three-dimensional polygonal chains. Step number n begins by constructing a flat polygonal chain $I_{n,1} := T([0,1], 2^{-n})$, from which we construct step-by-step three-dimensional polygonal chains and so on, as described above. In particular, if at the kth substep of this step we have constructed $(k+1)$-dimensional polygonal chains $I_{n,k,1},\ldots,I_{n,k,n_k}$ and $k+1 < n+1$, then on the $(k+1)$th substep from each polygonal chain $I_{n,k,j}$ we obtain $(k+2)$-dimensional polygonal chains $I_{n,k+1,i}$ in the following way: one of the line segments J' composing $I_{n,k,j}$ is transformed by $F(J', 2^{-n})$, while the rest line segments are kept fixed, moreover, this is done with every such J'. The measures on the constructed polygonal chains are defined as explained above. The obtained measures are numbered in their natural order, which gives a sequence $\{\mu_m\}$.

The sequence $\{\mu_m\}$ converges weakly to μ. Indeed, it suffices to prove that $\limsup\limits_{m\to\infty} \mu_m(U) \geqslant \mu(U)$ for every open set U. Let π_n be the projection to $\mathbb{R}^n$. It is readily verified that $\mu_m\circ\pi_n^{-1} \Rightarrow \mu$ as $m\to\infty$. Hence we obtain

$$\limsup_{m\to\infty} \mu_m\circ\pi_n^{-1}(U) \geqslant \mu(U),$$

but $|\mu_m(U) - \mu_m\circ\pi_n^{-1}(U)| \leqslant 2^{-n}$, whence we obtain the desired inequality ensuring weak convergence by Theorem 4.3.2.

Suppose now that there exists a Skorohod parametrization $\{\xi_{\mu_m}\}$ for $\{\mu_m\}$. For each n we consider the set $A_n := \{\omega\colon\ \xi_{\mu_m}(\omega) \in \mathbb{R}^n\ \forall\, m\in\mathbb{N}\}$. Since $\{\xi_{\mu_m}\}$ converges a.e. and every convergent sequence in the strict inductive limit $X = \bigcup_{n=1}^\infty X_n$ is contained in some of X_n (see [**97**, Proposition 2.6.3]), we have $P(\cup_n A_n) = 1$. Let us take N such that $P(A_N) > 0$. Restricting the random variables ξ_{μ_m} to A_N, we obtain a sequence of measures $\tau_m = P|_{A_N}\circ\xi_{\mu_m}^{-1}$ with the following properties: $\mu_m - \tau_m \geqslant 0$, the measures τ_m are concentrated on $\mathbb{R}^N$ and converge weakly in $\mathbb{R}^N$. But, as one can easily see, the limit τ of such measures can be only zero, which contradicts the fact that $P(A_N) > 0$. Indeed, the space $\mathbb{R}^N$ can be represented as the union $\mathbb{R}^N = \bigcup_{i=1}^{2^N} B_i \cup B$ of the sets

$$B_i = \Bigl\{x_1 \in \Bigl[\frac{i-1}{2^N}, \frac{i}{2^N}\Bigr]\Bigr\}, \quad B = \mathbb{R}^N \setminus \textstyle\bigcup_{i=1}^{2^N} B_i.$$

Since $\tau \leqslant \mu$, we have $\tau(\partial B_i) = 0$ and $\tau(B) = 0$. Hence $\tau(B_i) = \lim\limits_{m\to\infty} \tau_m(B_i)$. It remains to observe that there are infinitely many indices m_l for which $\tau_{m_l}(B_i) = 0$. This can be seen from the construction of our polygonal chains. For example, if $N = 2$, then on every step $n > 1$ there is a substep (with $k = 2$) at which we construct a polygonal chain not intersecting $(1/2, 1)$. For the corresponding random variable ξ_{m_l} this means that $P\bigl(A_2 \cap \{\xi_{m_l} \in (1/2,1)\}\bigr) = 0$. For $N > 2$ at the step with numbers larger than $N-1$ the situation is analogous. The fact that X is a Prohorov space is clear from Example 4.7.8. □

Certainly, in this example it is crucial that the union of $\mathbb{R}^n$ is equipped with the topology of the inductive limit (which in this case coincides with the strongest locally convex topology on $\mathbb{R}_0^\infty$, see Bogachev, Smolyanov [**97**, Example 2.4.11]). The situation changes if we equip $\mathbb{R}_0^\infty$ with the topology induced from $\mathbb{R}^\infty$. The same is true for every strict inductive limit of an increasing sequence of Banach spaces, which is seen from the following assertion (see Banakh, Bogachev, Kolesnikov [**32**]).

5.4.19. PROPOSITION. *Suppose that a bornological locally convex space X (where seminorms bounded on bounded sets are continuous) has a countable fundamental family of bounded sets (i.e., each bounded set is contained in a set of this family). Then X has the strong Skorohod property precisely when X is normable.*

As one can see from the presented results, the strong Skorohod property is rather rare outside the class of metric spaces, although there exist nonmetrizable spaces with this property. There exist even nonmetrizable compact spaces with this property. The proof of the following fact is given in [**32**].

5.4.20. PROPOSITION. *For every ordinal α, the interval of ordinals $[0, \alpha]$ with its order topology has the strong Skorohod property.*

5.5. Uniformly distributed sequences

An interesting notion connected with weak convergence of measures is that of a uniformly distributed sequence of points. Here we present some basic facts related to this notion in general topological spaces, addressing the reader for details to the books Kuipers, Niederreiter [**400**] and Hlawka [**322**], where there is an extensive bibliography. We only note that as early as the beginning of the XX century P. Bohl, W. Sierpiński and H. Weyl (see [**654**]) studied uniformly distributed sequences of numbers and in the 1950s a study of their analogs in topological spaces began (see Hlawka [**321**]).

5.5.1. DEFINITION. A sequence of points x_n in a topological space X is called uniformly distributed with respect to a Borel (or Baire) probability measure μ on X if the measures $(\delta_{x_1} + \cdots + \delta_{x_n})/n$ converge weakly to μ.

Thus, it is required that for all $f \in C_b(X)$

$$\lim_{n\to\infty} \frac{f(x_1) + \cdots + f(x_n)}{n} = \int_X f(x)\,\mu(dx).$$

In Exercise 5.8.36 one can find a number of equivalent characterizations (due to Weyl) of uniformly distributed sequences in $[0, 1]$. One of them is the condition that for every integer number $m \neq 0$ one has the equality

$$\lim_{N\to\infty} N^{-1} \sum_{n=1}^{N} \exp(2\pi i m x_n) = 0.$$

The most important example of a uniformly distributed sequence was given independently by Bohl, Sierpiński and Weyl.

5.5.2. EXAMPLE. (i) For every irrational number θ in the interval $(0, 1)$, the sequence $x_n := n\theta - [n\theta]$, where $[x]$ is the integer part of the number x, is uniformly distributed with respect to Lebesgue measure on the interval $[0, 1]$. This follows from the aforementioned criterion of Weyl, since

$$\Bigl|N^{-1} \sum_{n=1}^{N} \exp(2\pi i m n\theta)\Bigr| = \frac{|\exp(2\pi i m N\theta) - 1|}{N|\exp(2\pi m\theta) - 1|} \leqslant \frac{1}{N|\sin(\pi m\theta)|}$$

for all nonzero integer numbers m.

It is clear from the basic properties of weak convergence that for every uniformly distributed sequence $\{x_n\}$ in $[0,1]$ with Lebesgue measure the averages $n^{-1}\sum_{i=1}^{n} f(x_i)$ converge to the integral of f for all Riemann integrable functions f as well.

We observe that if $\{x_n\}$ is a uniformly distributed sequence for a Radon measure μ on a completely regular space X and $T\colon\ X \to Y$ is a Borel mapping to a space Y such that the set of points of discontinuity of T has μ-measure zero, then $\mu\circ T^{-1}$ is a Radon measure and the sequence $\{T(x_n)\}$ is uniformly distributed with respect to $\mu\circ T^{-1}$ (see Theorem 4.3.12). This simple observation enables one to construct uniformly distributed sequences on many spaces. The very fact of existence of such sequences can be simpler derived from the general theorem below, proved in Niederreiter [**484**] and based on the following combinatorial lemma.

5.5.3. LEMMA. *Let X be a nonempty set. For every probability measure ν with finite support $\{z_1,\ldots,z_k\}\subset X$, there exists a sequence $\{y_n\}$ with $y_n\in\{z_1,\ldots,z_k\}$ such that, for every set $M\subset X$ and every $N\in\mathbb{N}$, one has the inequality*

$$\Bigl|\frac{S_N(M,\{y_n\})}{N}-\nu(M)\Bigr|\leqslant\frac{C(\nu)}{N}, \tag{5.5.1}$$

where $S_N(M,\{y_n\}) := \sum_{n=1}^{N} I_M(y_n)$ and $C(\nu)=(k-1)k$.

PROOF. Suppose that we have a sequence $\{y_n\}$ such that

$$\Bigl|\frac{S_N(z_i,\{y_n\})}{N}-\nu(z_i)\Bigr|\leqslant\frac{k-1}{N}\quad \forall\, i\leqslant k, \forall\, N\geqslant 1. \tag{5.5.2}$$

Then for $C(\nu)$ one can take $(k-1)k$. Indeed, since $y_n\in\{z_1,\ldots,z_k\}$ and μ concentrated on $\{z_1,\ldots,z_k\}$, it suffices to verify (5.5.1) for sets $M\subset\{z_1,\ldots,z_k\}$, but then by (5.5.2) the left-hand side of (5.5.1) is estimated by $k(k-1)N^{-1}$. Now by induction on k we show that one can achieve (5.5.2). If $k=1$, then the sequence $y_n\equiv z_1$ is suitable. Suppose that our assertion is true for $k-1$. Let $\nu(z_i)=\lambda_i>0$, $i=1,\ldots,k$. Let us consider the probability measure ν' with support at the points $z_1,\ldots,z_{k-1}$ and $\nu'(z_i)=\lambda_i(1-\lambda_k)^{-1}$. By the inductive assumption, there exists a sequence $\{y_n'\}$ such that $y_n'\in\{z_1,\ldots,z_{k-1}\}$ and

$$\Bigl|\frac{S_N(z_i,\{y_n'\})}{N}-\nu'(z_i)\Bigr|\leqslant\frac{k-2}{N}\quad \forall\, i\leqslant k-1, \forall\, N\geqslant 1.$$

Now we define a sequence $\{y_n\}$ as follows: if $n=\bigl[m(1-\lambda_k)^{-1}\bigr]$ for some $m\in\mathbb{N}$, where $[p]$ is the integer part of p, then we set $y_n:=y_m'$; otherwise we set $y_n:=z_k$. Note that such a number m is unique. Let us verify (5.5.2). Let us consider the case $i\leqslant k-1$. Then $S_N(z_i,\{y_n\})$ equals the number of natural numbers m such that

$$\bigl[m(1-\lambda_k)^{-1}\bigr]\leqslant N \quad\text{and}\quad y_m'=z_i.$$

Hence $S_N(z_i,\{y_n\})=S_L(z_i,\{y_n'\})$, where $L=[(N+1)(1-\lambda_k)]-\varepsilon$ and $\varepsilon=1$ or $\varepsilon=0$ depending on whether the number $(N+1)(1-\lambda_k)$ is integer or not. Thus,

$$\begin{aligned}\Bigl|\frac{S_N(z_i,\{y_n\})}{N}-\nu(z_i)\Bigr| &= \Bigl|\frac{L}{N}\frac{S_L(z_i,\{y_n'\})}{L}-(1-\lambda_k)\nu'(z_i)\Bigr|\\ &\leqslant \frac{L}{N}\Bigl|\frac{S_L(z_i,\{y_n'\})}{L}-\nu'(z_i)\Bigr|+\nu'(z_i)\Bigl|\frac{L}{N}-(1-\lambda_k)\Bigr|\\ &\leqslant \frac{k-2}{N}+\frac{\nu'(z_i)}{N}\bigl|N(1-\lambda_k)-[(N+1)(1-\lambda_k)]+\varepsilon\bigr|.\end{aligned}$$

We observe that the second term in the right-hand side is estimated by N^{-1}, since the number $\bigl|N(1-\lambda_k)-[(N+1)(1-\lambda_k)]+\varepsilon\bigr|$ equals λ_k if $(N+1)(1-\lambda_k)$ is integer and otherwise this number does not exceed 1.

Finally, let us consider the point z_k. It is easy to see that one has the equality $S_N(z_k,\{y_n\}) = N - L$, where L is defined above. Hence

$$\Bigl|\frac{S_N(z_k,\{y_n\})}{N} - \nu(z_k)\Bigr| = \Bigl|\lambda_1 + \cdots + \lambda_{k-1} - \frac{L}{N}\Bigr| \leqslant \frac{1}{N},$$

which completes the proof. □

5.5.4. THEOREM. *Let μ be a Radon (or τ-additive) probability measure on a completely regular space X. There is a sequence uniformly distributed with respect to μ precisely when there is a sequence of probability measures with finite supports weakly convergent to μ.*

PROOF. If $\{x_n\}$ is a uniformly distributed sequence, then the arithmetic means $n^{-1}(\delta_{x_1}+\cdots+\delta_{x_n})$ have finite supports and converge weakly to μ. The converse assertion is not obvious. Suppose that probability measures μ_j with finite supports converge weakly to μ. By the lemma above, for every j, there exists a number $C_j := C(\mu_j)$ and a sequence $\{y_n^j\}$ such that for all $M\subset X$ and $N\in\mathbb{N}$ one has

$$\Bigl|\frac{S_N(M,\{y_n^j\})}{N} - \mu_j(M)\Bigr| \leqslant \frac{C_j}{N}.$$

For every j we take a natural number $r_j \geqslant j(C_1+\cdots+C_{j+1})$. Now we pick the desired sequence of points $\{x_n\}$ in the following way. Every natural number n can be uniquely written in the form $n = r_1+\cdots+r_{j-1}+s$, where $j\in\mathbb{N}$ and $0<s\leqslant r_j$, and we set $r_0 := 0$. Let $x_n := y_s^j$. The obtained sequence is the one we need. Indeed, suppose that a set M has boundary of μ-measure zero. Every natural number $N>r_1$ is written in the form $N = r_1+\cdots+r_k+r$, $0<r\leqslant r_{k+1}$. Then, as one can readily verify, we have

$$S_N(M,\{x_n\}) = \sum_{j=1}^{k} S_{r_j}(M,\{y_n^j\}) + S_r(M,\{y_n^{k+1}\}).$$

Therefore,

$$\frac{S_N(M,\{x_n\})}{N} - \mu(M) = \sum_{j=1}^{k}\frac{r_j}{N}\Bigl(\frac{S_{r_j}(M,\{y_n^j\})}{r_j} - \mu_j(M)\Bigr)$$
$$+\frac{r}{N}\Bigl(\frac{S_r(M,\{y_n^{k+1}\})}{r} - \mu_{k+1}(M)\Bigr) + \sum_{j=1}^{k}\frac{r_j}{N}\mu_j(M) + \frac{r\mu_{k+1}(M)}{N} - \mu(M),$$

which is estimated in absolute value by

$$\sum_{j=1}^{k}\frac{r_j}{N}\frac{C_j}{r_j} + \frac{r}{N}\frac{C_{k+1}}{r} + \Bigl|\frac{1}{N}\Bigl(\sum_{j=1}^{k} r_j\mu_j(M) + r\mu_{k+1}(M)\Bigr) - \mu(M)\Bigr|$$
$$\leqslant \frac{1}{r_k}\sum_{j=1}^{k+1} C_j + \Bigl|\frac{1}{N}\Bigl(\sum_{j=1}^{k} r_j\mu_j(M) + r\mu_{k+1}(M)\Bigr) - \mu(M)\Bigr|$$
$$\leqslant \frac{1}{k} + \Bigl|\frac{1}{N}\Bigl(\sum_{j=1}^{k} r_j\mu_j(M) + r\mu_{k+1}(M)\Bigr) - \mu(M)\Bigr|.$$

As $N \to \infty$, we also have $k \to \infty$. Hence the first term in the right-hand side of the obtained estimate tends to zero. The second term tends to zero too, since $\mu_j(M) \to \mu(M)$ by weak convergence and the equality $N = \sum_{j=1}^{k} r_j + r$. □

5.5.5. COROLLARY. *Let X be a completely regular space. The following conditions are equivalent:*

(i) *every Radon probability measure on X has a uniformly distributed sequence,*

(ii) *the sequential closure of the set of probability measures with finite supports is $\mathcal{M}_r(X)$.*

In particular, for every Borel probability measure on a completely regular Souslin space there exists a uniformly distributed sequence.

Note that in this corollary it is important to deal with the sequential closure (the set of limits of all convergent sequences), but not with the broader closure in the usual topological sense, which, as we know, always coincides with $\mathcal{M}_r(X)$. Not every Radon measure on an arbitrary compact space possesses a uniformly distributed sequence. Let us consider an example from Losert [**440**].

5.5.6. EXAMPLE. Let $X = \beta\mathbb{N}$ be the Stone–Čech compactification of the space of natural numbers $\mathbb{N}$. Then on X there exists a Radon probability measure possessing no uniformly distributed sequences.

PROOF. We show that if μ is an atomless Radon probability measure on $\beta\mathbb{N}$, then μ has no uniformly distributed sequences. The existence of atomless measures on $\beta\mathbb{N}$ can be easily seen from the fact that $\beta\mathbb{N}$ can be continuously mapped onto $[0,1]$ (for this we set $f(n) = r_n$, where $\{r_n\}$ is the set of all rational numbers in $[0,1]$ and then extend f to a continuous function on $\beta\mathbb{N}$ with values in $[0,1]$). Lebesgue measure has an atomless preimage under f (this is clear from the Riesz theorem for the functional on $C(\beta\mathbb{N})$ obtained by an extension of the functional that maps the function $\varphi\circ f$ to the integral of φ with respect to Lebesgue measure). If we had a sequence of discrete measures convergent weakly to μ, then by Proposition 5.6.15 below the measure μ would be concentrated on a countable set, which contradicts the fact that it is atomless. □

Losert [**441**] also proved the following interesting assertion.

5.5.7. PROPOSITION. *If X is a compact space such that there exists a continuous mapping of the space $\{0,1\}^{\aleph_1}$ onto X, where $\aleph_1$ is the smallest uncountable cardinality, then every Radon probability measure on X has a uniformly distributed sequence. In particular, this is true for $[0,1]^{\mathfrak{c}}$ under the continuum hypothesis.*

The last assertion means that every Radon measure on the cube $[0,1]^{\mathfrak{c}}$ is the limit of a sequence of linear combinations of Dirac measures (under the continuum hypothesis). Additional information about uniformly distributed sequences can be found in the papers cited above and also in Bogachev, Lukintsova [**92**], Losert [**442**], Mercourakis [**458**], Plebanek [**512**], Sun [**602**], and in Exercises 5.8.36–5.8.42.

5.6. Setwise convergence of measures

In this section we give an account of principal results about setwise convergence of measures. Their proofs can be found in Bogachev [**81**, Chapter 4], where additional references are given. Apart weak convergence, which is the main theme of this book, in applications one often encounters convergence in variation and setwise convergence of measures.

Let $(X, \mathcal{A})$ be a space with a σ-algebra and let $\mathcal{M}(X, \mathcal{A})$ be the space of all real countably additive measures on $\mathcal{A}$. It has already been noted that the space $\mathcal{M}(X, \mathcal{A})$ with the norm $\mu \mapsto \|\mu\|$ is a Banach space. The space $\mathcal{M}(X, \mathcal{A})$ can also be equipped with the norm

$$\mu \mapsto \sup_{A \in \mathcal{A}} |\mu(A)|,$$

which is equivalent to the variation norm (see § 1.1).

Let us proceed to consideration of setwise convergence of measures, i.e., convergence $\mu_\alpha(A) \to \mu(A)$ for each $A \in \mathcal{A}$. This is a weaker convergence than convergence in variation. For example, the sequence of measures μ_n on the interval $[0, 2\pi]$ given by densities $\sin nx$ with respect to Lebesgue measure converges to zero on every measurable set. This follows from the Riemann–Lebesgue theorem, according to which

$$\lim_{n \to \infty} \int_0^{2\pi} f(x) \sin nx \, dx = 0$$

for every integrable function f.

5.6.1. DEFINITION. Let M be a family of measures on a σ-algebra $\mathcal{A}$. This family is called uniformly countably additive if, for every sequence of pairwise disjoint sets A_i, the series $\sum_{i=1}^\infty \mu(A_i)$ converges uniformly in $\mu \in M$, i.e., for every $\varepsilon > 0$, there exists n_ε such that $\bigl|\sum_{i=n}^\infty \mu(A_i)\bigr| < \varepsilon$ for all $n \geqslant n_\varepsilon$ and all $\mu \in M$.

The next important result combines two remarkable facts of measure theory: the *Nikodym theorem* on convergence and the *Vitali–Lebesgue–Hahn–Saks theorem*.

5.6.2. THEOREM. *Let $\{\mu_n\}$ be a sequence of measures in the space $\mathcal{M}(X, \mathcal{A})$ such that a finite limit $\lim\limits_{n \to \infty} \mu_n(A)$ exists for every set $A \in \mathcal{A}$. Then*

(i) *the formula $\mu(A) = \lim\limits_{n \to \infty} \mu_n(A)$ defines a measure $\mu \in \mathcal{M}(X, \mathcal{A})$;*

(ii) *there exists a nonnegative measure $\nu \in \mathcal{M}(X, \mathcal{A})$ and a bounded nondecreasing nonnegative function α on $[0, +\infty)$ such that $\lim\limits_{t \to 0} \alpha(t) = 0$ and*

$$\sup_n |\mu_n(A)| \leqslant \alpha\bigl(\nu(A)\bigr) \quad \forall A \in \mathcal{A}. \tag{5.6.1}$$

In particular, $\sup_n \|\mu_n\| < \infty$ and $\{\mu_n\}$ is uniformly countably additive;

(iii) *if $\lambda \in \mathcal{M}(X, \mathcal{A})$ is a nonnegative measure such that $\mu_n \ll \lambda$ for all n, then*

$$\lim_{t \to 0} \sup\bigl\{\mu_n(A)\colon\ A \in \mathcal{A}, \lambda(A) \leqslant t, n \in \mathbb{N}\bigr\} = 0.$$

5.6.3. COROLLARY. *Let measures $\mu_n \in \mathcal{M}(X, \mathcal{A})$ be such that $\sup\limits_n |\mu_n(A)| < \infty$ for every set $A \in \mathcal{A}$. Then $\sup_n \|\mu_n\| < \infty$.*

Some conditions equivalent to the uniform countable additivity are collected in the next theorem.

5.6.4. THEOREM. *Let M be a family of bounded measures on a σ-algebra $\mathcal{A}$. The following conditions are equivalent:*

(i) *M is uniformly countably additive;*

(ii) *$\lim\limits_{i \to \infty} \sup\limits_{\mu \in M} |\mu(A_i)| = 0$ for every sequence of pairwise disjoint sets $A_i \in \mathcal{A}$;*

(iii) *for every decreasing sequence of sets A_i in $\mathcal{A}$ with $\bigcap_{i=1}^{\infty} A_i = \varnothing$ we have* $\lim\limits_{i\to\infty} \mu(A_i) = 0$ *uniformly with respect to $\mu \in M$;*

(iv) *if ν is a bounded nonnegative measure such that $\mu_n \ll \nu$ for all n, then*

$$\lim_{t\to 0} \sup\bigl\{\mu(A)\colon\ \mu \in M, A \in \mathcal{A}, \nu(A) \leqslant t\bigr\} = 0.$$

An interesting generalization of this theorem will be given below in Theorem 5.6.7.

Setwise convergence of measures considered in Theorem 5.6.2 can be defined by a topology: this convergence is precisely convergence in the topology $\sigma(\mathcal{M}, \mathcal{F})$, where $\mathcal{M} = \mathcal{M}(X, \mathcal{A})$ is the space of all bounded countably additive measures on $\mathcal{A}$ and $\mathcal{F}$ is the linear space of all simple $\mathcal{A}$-measurable functions. A basis of neighborhoods of a point μ_0 in this topology consists of sets of the form

$$W_{A_1,\ldots,A_n,\varepsilon}(\mu_0) = \bigl\{\mu \in \mathcal{M}(X, \mathcal{A})\colon |\mu(A_i) - \mu_0(A_i)| < \varepsilon, i = 1, \ldots, n\bigr\},$$

where $A_i \in \mathcal{A}$ and $\varepsilon > 0$ (see § 1.3 for the definition of this topology). If the σ-algebra $\mathcal{A}$ is infinite, then the topology $\sigma(\mathcal{M}, \mathcal{F})$ is not generated by a norm (Exercise 5.8.24). One more natural topology on $\mathcal{M}$ is generated by duality with the space $B(X, \mathcal{A})$ of bounded $\mathcal{A}$-measurable functions, i.e., this is the topology $\sigma\bigl(\mathcal{M}, B(X, \mathcal{A})\bigr)$. If the σ-algebra $\mathcal{A}$ is infinite, then this topology is strictly stronger than the topology $\sigma(\mathcal{M}, \mathcal{F})$. However, as it follows from Theorem 5.6.2, *for countable sequences* convergence in the topology $\sigma(\mathcal{M}, \mathcal{F})$ is equivalent to convergence in the topology $\sigma\bigl(\mathcal{M}, B(X, \mathcal{A})\bigr)$ (for the proof we also use the fact that every function in $B(X, \mathcal{A})$ can be uniformly approximated by simple functions).

Finally, since $\mathcal{M}$ is a Banach space, one can also equip it with the usual weak topology $\sigma(\mathcal{M}, \mathcal{M}^*)$ of a Banach space (see § 1.3), which in nontrivial cases is strictly stronger than the topology $\sigma(\mathcal{M}, \mathcal{F})$, but is strictly weaker than the topology generated by the variation norm (Exercise 5.8.25). It turns out that convergence of *countable sequences* in the topology $\sigma(\mathcal{M}, \mathcal{M}^*)$ is the same as in the topology of setwise convergence. In addition, in both topologies compact sets are the same.

5.6.5. THEOREM. *For every set $M \subset \mathcal{M}(X, \mathcal{A})$, the following conditions are equivalent.*

(i) *The closure of M in the topology $\sigma(\mathcal{M}, \mathcal{M}^*)$ is compact.*

(ii) *The set M is bounded in variation and there exists a nonnegative measure $\nu \in \mathcal{M}(X, \mathcal{A})$ (a probability measure if $M \neq \{0\}$) such that the family M is uniformly ν-continuous, i.e., for each $\varepsilon > 0$, there is $\delta > 0$ with the property that*

$$|\mu(A)| \leqslant \varepsilon \quad \textit{for all } \mu \in M \textit{ whenever } A \in \mathcal{A} \textit{ and } \nu(A) \leqslant \delta.$$

In this case the measures from M are absolutely continuous with respect to ν, the closure of the set $\{d\mu/d\nu\colon\ \mu \in M\}$ is compact in the weak topology of $L^1(\nu)$, and for ν one can take $\sum_{n=1}^{\infty} 2^{-n}|\mu_n|/(1 + \|\mu_n\|)$ for some finite or countable collection of measures $\{\mu_n\} \subset M$.

(iii) *The set M is bounded in variation and uniformly countably additive.*

(iv) *The set M has compact closure in the topology of convergence on sets in $\mathcal{A}$. This is also equivalent to compactness of its closure in the topology of convergence on every bounded $\mathcal{A}$-measurable function.*

(v) *Every sequence in M contains a subsequence converging on every set in $\mathcal{A}$.*

Yet another condition for compactness in the topology of setwise convergence is given in Exercise 5.8.26.

5.6.6. COROLLARY. *A sequence of measures $\mu_n \in \mathcal{M}(X,\mathcal{A})$ converges in the topology $\sigma(\mathcal{M},\mathcal{M}^*)$ precisely when it converges on every set in $\mathcal{A}$.*

We recall once again that on more general sets of measures all the three topologies considered in the above theorem are different.

In relation to the Vitali–Lebesgue–Hahn–Saks theorem and Theorem 5.6.4 the question arises of whether it would be enough to verify the required conditions not for all sets in $\mathcal{A}$, but only for sets in some algebra generating $\mathcal{A}$. For example, dealing with a cube in $\mathbb{R}^n$, for such an algebra it would be nice to take the algebra of elementary sets. Simple examples show that this can be possible not for all conditions equivalent in the case of a σ-algebra. More surprising is the following result, found by G.Ya. Areshkin [**21**] for nonnegative measures and extended by V.N. Aleksjuk to signed measures (see Areshkin, Aleksjuk, Klimkin [**22**] or Bogachev [**81**, Theorem 4.7.27]). Let $\mathfrak{R}$ be a ring of subsets of a space X and let $\mathfrak{S}$ be the generated σ-ring (the minimal ring containing $\mathfrak{R}$ and closed with respect to countable unions).

5.6.7. THEOREM. *Let $\{\mu_\alpha\}_{\alpha\in\Lambda}$ be a family of countably additive measures of bounded variation on $\mathfrak{S}$. Then the following conditions are equivalent.*

(i) *The measures μ_α are uniformly additive on $\mathfrak{R}$ in the following sense: for every sequence of pairwise disjoint sets R_n in $\mathfrak{R}$, we have $\lim\limits_{n\to\infty}\sum_{k=n}^\infty \mu_\alpha(R_k)=0$ uniformly with respect to $\alpha\in\Lambda$.*

(ii) *For every sequence $\{\mu_{\alpha_n}\}\subset\{\mu_\alpha\}$ and every sequence of pairwise disjoint sets $R_n\in\mathfrak{R}$ we have $\lim\limits_{n\to\infty}\mu_{\alpha_n}(R_n)=0$.*

(iii) *The family $\{\mu_\alpha\}$ is equicontinuous on $\mathfrak{R}$ in the following sense: for every sequence of sets $R_n\in\mathfrak{R}$ such that $R_{n+1}\subset R_n$ and $\bigcap_{n=1}^\infty R_n=\varnothing$, we have $\lim\limits_{n\to\infty}\mu_\alpha(R_n)=0$ uniformly in $\alpha\in\Lambda$.*

(iv) *conditions* (i)–(iii) (*or any of these conditions*) *are fulfilled on $\mathfrak{S}$.*

5.6.8. THEOREM. (i) *Let μ be a separable finite nonnegative measure. Then weakly compact subsets in $L^1(\mu)$* (*or, which is the same, uniformly integrable subsets of $L^1(\mu)$*) *are metrizable in the weak topology.*

(ii) *Let $\mathcal{A}$ be a countably generated σ-algebra. Then compact subsets of the space $\mathcal{M}$ of all bounded measures on $\mathcal{A}$ with the topology of convergence on sets in $\mathcal{A}$ are metrizable.*

PROOF. By Theorem 1.3.10 it suffices to consider a weakly compact set K. There is a countable family $\{\varphi_n\}\subset\mathcal{L}^\infty(\mu)$ with the following property: if f,g in $\mathcal{L}^1(\mu)$ are such that the integrals of $f\varphi_n$ and $g\varphi_n$ are equal for all n, then $f=g$ almost everywhere. The functions $f\mapsto\displaystyle\int f\varphi_n\,d\mu$ are continuous on K with the weak topology and separate points. Hence the compact set K is metrizable.

(ii) The same reasoning applies to functions $\mu\mapsto\mu(A_n)$ on $\mathcal{M}$, where the countable family $\{A_n\}$ generates $\mathcal{A}$. □

Let us mention an interesting theorem due to V.F. Gaposhkin [**255**], [**256**] on subsequences that converge "almost weakly in L^1" (later reproved in Brooks, Chacon [**120**] in terms of measures; the proof can be found in Bogachev [**81**, Theorem 4.7.23]).

5.6.9. THEOREM. *Suppose that* $(X, \mathcal{A}, \mu)$ *is a space with a finite nonnegative measure,* $\{f_n\} \subset \mathcal{L}^1(\mu)$ *and* $\sup_n \|f_n\|_{L^1(\mu)} < \infty$. *Then one can find a subsequence* $\{n_k\}$ *and a function* $f \in L^1(\mu)$ *such that* $\{f_{n_k}\}$ *converges to* f *almost weakly in* $L^1(\mu)$ *in the following sense: for every* $\varepsilon > 0$, *there exists a measurable set* X_ε *such that* $\mu(X\backslash X_\varepsilon) < \varepsilon$ *and the functions* $f_n|_{X_\varepsilon}$ *converge to* $f|_{X_\varepsilon}$ *in the weak topology of the space* $L^1(\mu|_{X_\varepsilon})$.

In the previous result we do not assume that the space X with the σ-algebra $\mathcal{A}$ is equipped with a topology. However, it turns out that the case where $\mathcal{A}$ is the Borel σ-algebra of a topological space has a specific feature: some special classes of sets appear such as open, closed, and compact, having no natural analogs for arbitrary σ-algebras. Back in 1916 G.M. Fichtenholz (see [**232**, §30], [**233**], and also [**80**] about his works) discovered a remarkable fact: if the integrals of functions f_n over every open set in the interval converge to zero, then the integrals over every Borel set converge to zero as well. Dieudonné [**170**] proved 35 years later that if a sequence of measures on a compact metric space converges on every open set, then it converges on every Borel set. Grothendieck [**303**] extended the Dieudonné theorem to locally compact spaces. The method of Fichtenholz can be modified for Radon measures, and from his result one can derive the result of Dieudonné. For this reason the assertion that any sequence of Radon measures convergent on open sets converges on all Borel sets could be naturally called the *Fichtenholz–Dieudonné–Grothendieck theorem.* Later this result was extended to more general cases. Let us give the generalization obtained in Pfanzagl [**511**] (for the proof, see also Bogachev [**81**, § 8.19(x)]), and then mention a number of other related results.

5.6.10. THEOREM. *Let* $\mathcal{U}_0$ *be a topology base in a Hausdorff space* X *closed with respect to countable unions. Suppose that a sequence of Radon measures* μ_n *converges on every set in* $\mathcal{U}_0$. *Then it converges on every Borel set.*

The measures μ_n are given by densities f_n with respect to some bounded Radon measure ν (for example, of the form $\sum_{n=1}^{\infty} c_n|\mu_n|$), whence it follows that the functions f_n are uniformly integrable and converge to some function $f \in L^1(\nu)$ in the weak topology of $L^1(\nu)$. In particular, convergence of μ_n holds on an even broader class than $\mathcal{B}(X)$. Moreover, the limit of $\{\mu_n\}$ is a Radon measure. Finally, the theorem also yields the fact, which is not obvious at all, that the measures μ_n are uniformly bounded. However, the latter can be obtain under a weaker condition.

5.6.11. COROLLARY. *Let* $\mathcal{U}_0$ *be a topology base of a Hausdorff space* X *closed with respect to countable unions and let* M *be a family of Radon measures on* X *such that*

$$\sup\{|\mu(U)|\colon\ \mu \in M\} < \infty \quad \textit{for each} \ \ U \in \mathcal{U}_0.$$

Then the family M *is bounded in variation.*

It is important that $\mathcal{U}_0$ be closed with respect to countable unions, but Schachermayer [**565**, p. 15] proved the following fact for the algebra Γ_P of continuity sets of a measure (see also Graves, Wheeler [**290**]).

5.6.12. THEOREM. *Let* X *be a completely regular space and let* $P \in \mathcal{P}_r(X)$ *be atomless. If* $\mathcal{M} \subset \mathcal{M}_r(X)$ *is such that* $\sup_{\mu\in\mathcal{M}} |\mu(E)| < \infty$ *for each set* $E \in \Gamma_P$, *then* $\sup_{\mu\in\mathcal{M}} \|\mu\| < \infty$.

5.6.13. THEOREM. *A bounded set M of Radon measures on a Hausdorff space X has compact closure in the topology of convergence on Borel sets precisely when* $\lim\limits_{n\to\infty}\sup\limits_{\mu\in M}|\mu(K_n)|=0$ *for every sequence of pairwise disjoint compact sets K_n.*

If X is regular, then this is also equivalent to the condition that for every sequence of pairwise disjoint open sets U_n one has $\lim\limits_{n\to\infty}\sup\limits_{\mu\in M}|\mu(U_n)|=0$.

5.6.14. THEOREM. (i) *If a set of Radon measures on a Hausdorff space is compact in the topology of convergence on Borel sets, then it is uniformly tight.*

(ii) *A set M of Radon measures on a Hausdorff space has compact closure in the space of all Radon measures on X in the topology of convergence on Borel sets precisely when M is bounded, uniformly tight and for every compact set K and every $\varepsilon>0$, there exists an open set $U\supset K$ such that $|\mu|(U\backslash K)<\varepsilon$ for all $\mu\in M$.*

(iii) *A sequence of Radon measures μ_n on a Hausdorff space X converges to a Radon measure μ on every Borel set precisely when it is uniformly tight and* $\lim\limits_{n\to\infty}\mu_n(K)=\mu(K)$ *for every compact set K.*

PROOF. (i) If the set M is compact in the indicated topology, then by Theorem 5.6.7 there exists a Radon probability measure μ_0 such that all measures in M are uniformly absolutely continuous with respect to μ_0.

The necessity of the conditions indicated in (ii) follows from (i) and the proof of Theorem 5.6.13 in [**81**, Theorem 8.10.58]. By virtue of the uniform tightness the sufficiency reduces to the case of a compact space, which also follows from the proof of the cited theorem.

(iii) The uniform tightness of a sequence of Radon measures μ_n converging to a Radon measure μ on Borel sets follows from assertion (iii) of Theorem 5.6.4, since the measures μ_n are absolutely continuous with respect to the Radon measure $\sum_{n=1}^{\infty}c_n|\mu_n|$, where $c_n=2^{-n}(1+\|\mu_n\|)^{-1}$. The sufficiency of the indicated conditions follows from Theorem 5.6.10 applied to the restrictions of the regarded measures to compact sets K_j chosen such that $|\mu|(X\backslash K_j)<2^{-j}$ for all $\mu\in M$. Moreover, for every compact set $K\subset K_j$ one has convergence on the set $K_j\backslash K$, and every set $U\subset K_j$ open in the relative topology has such a form. □

Note that Theorem 5.6.10 can fail for arbitrary Borel measures. Pfanzagl [**511**, Example 2] constructed an example of a sequence of Borel probability measures on a Hausdorff space converging on every open set, but not converging on some Borel set. However, the Radon property of the regarded measures can be relaxed at the expense of certain moderate restrictions on the space. Several authors studied the so-called "convergence classes", i.e., classes of sets such that convergence on them implies convergence on all Borel sets; see Adamski, Gänssler, Kaiser [**5**], Gänssler [**250**], [**251**], Landers, Rogge [**409**], Rogge [**554**], and Stein [**594**]. For example, by Theorem 5.6.10 the class of open sets is a convergence class for Radon measures. It was established (see [**5**], [**409**]) that

(i) the class $\mathcal{G}$ of all open sets is a convergence class for τ-additive measures on regular spaces;

(ii) the class $\mathcal{G}_0$ of all functionally open sets is a convergence class for Baire measures (and Baire sets) on Hausdorff spaces, for τ-additive measures on completely regular spaces and for regular Borel measures on normal spaces;

(iii) the class $\mathcal{G}_r$ of all regular open sets (i.e., sets that coincide with interiors of their closures) is a convergence class for τ-additive measures on regular spaces and for regular Borel measures on normal spaces. Additional information about setwise convergence of measures can be found in Bogachev [**81**], Dieudonné [**169**], Drewnowski [**180**], Dubrovskii [**183**], [**184**], de Lucia, Pap [**445**], Ferrando, Sánchez Ruiz [**230**], and Klimkin [**369**].

We know that setwise convergence implies weak convergence of measures, but the converse is in general false. However, there is a class of spaces for which the converse is also true. We recall that a topological space X is called *extremally disconnected* if the closure of every open set is open (see p. 214). This is equivalent to the property that the closures of disjoint open sets in X are disjoint (see Engelking [**203**, p. 368]). In this case X has a topology base consisting of sets that are simultaneously open and closed. Although among metric spaces only discrete spaces are extremally disconnected, this exotic class of spaces is interesting by that it includes the Stone–Čech compactifications of discrete spaces, say, the compact space $\beta\mathbb{N}$ is extremally disconnected. The next result is obtained in Grothendieck [**303**].

5.6.15. PROPOSITION. *Let X be an extremally disconnected compact space. Then every weakly convergent sequence of Radon measures converges on every Borel set. In particular, this is true if $X = \beta\mathbb{N}$ is the Stone–Čech compactification of* $\mathbb{N}$.

One more interesting Grothendieck's result (see Grothendieck [**304**, p. 229]) connects compactness in two different topologies: the Mackey topology and the weak-$*$ topology corresponding to the norm.

5.6.16. THEOREM. *Suppose that K is a compact space and a set M in the space $\mathcal{M} := \mathcal{M}_r(K) = C(K)^*$ has compact closure in the Mackey topology $\tau(\mathcal{M}, C(K))$ of uniform convergence on absolutely convex $\sigma(C(K), \mathcal{M})$-compact sets in $C(K)$. Then M has compact closure in the topology $\sigma(\mathcal{M}, \mathcal{M}^*)$ as well.*

PROOF. By the Eberlein–Shmulian theorem and Theorems 5.6.13 and 5.6.5, it suffices to show that $\lim\limits_{n\to\infty} \mu_n(U_n) = 0$ for every sequence of measures $\mu_n \in M$ and every sequence of disjoint open sets $U_n \subset K$. If this is false, then there exist functions $f_n \in C(K)$ such that $|f_n| \leqslant 1$, $f_n = 0$ outside U_n and the integral of f_n with respect to μ_n is not smaller than some $\varepsilon > 0$. The sequence $\{f_n\}$ converges to zero pointwise, hence in the weak topology in $C(K)$. Its closed convex hull is weakly compact by Theorem 1.3.12 (it suffices to have its version for the separable closed linear span of $\{f_n\}$). This contradicts the compactness of the closure of M in the topology of uniform convergence on absolutely convex weakly compact sets. □

In the next section our discussion will also be partly connected with setwise convergence.

5.7. Young measures and the *ws*-topology

In applications, yet another mode of convergence of measures is useful that combines weak convergence and setwise convergence. Here we briefly discuss this form of convergence along with an interesting related class of measures on products of two spaces, called Young measures.

Suppose we have a measurable space $(\Omega, \mathcal{A})$ and a topological space T. Let us consider the space $\mathcal{M}(\Omega\times T)$ of all bounded measures on the product $\Omega\times T$ equipped with one of the two σ-algebras $\mathcal{A}\otimes\mathcal{B}(T)$ or $\mathcal{A}\otimes\mathcal{B}a(T)$. The set of nonnegative measures in $\mathcal{M}(\Omega\times T)$ is denoted by $\mathcal{M}^+(\Omega\times T)$. We shall say that a net of measures

$\mu_\alpha \in \mathcal{M}(\Omega\times T)$ converges to a measure μ in the ws-topology if, for every bounded $\mathcal{A}$-measurable function ψ and every function $\varphi \in C_b(T)$, one has the equality

$$\lim_\alpha \int_{\Omega\times T} \psi(\omega)\varphi(t)\,\mu_\alpha(d\omega dt) = \int_{\Omega\times T} \psi(\omega)\varphi(t)\,\mu(d\omega dt). \tag{5.7.1}$$

Clearly, this convergence is generated by a topology: the space $\mathcal{M}(\Omega\times T)$ can be equipped with seminorms defined as the absolute values of integrals of such functions $\psi\varphi$. A basic neighborhood of an element μ in the space $\mathcal{M}(\Omega\times T)$ has the form

$$U_{\psi_1,\ldots,\psi_n;\varphi_1,\ldots,\varphi_n;\varepsilon}(\mu) := \Bigl\{\nu\colon \Bigl|\int \psi_j\varphi_j\,d(\nu-\mu)\Bigr| < \varepsilon, j=1,\ldots,n\Bigr\}, \tag{5.7.2}$$

where $\varepsilon > 0$, $\varphi_j \in C_b(T)$, ψ_j are bounded $\mathcal{A}$-measurable functions. Convergence of a uniformly bounded net (for example, of a net of probability measures) in the ws-topology is equivalent to equality (5.7.1) with functions ψ of the form $\psi = I_A$, where $A \in \mathcal{A}$. The same is true for nets of nonnegative measures on $\Omega\times T$. If $\mathcal{A}$ consists of only Ω and the empty set, then the ws-topology reduces to the weak topology on $\mathcal{M}(T)$, and if T is a singleton, then we obtain the topology of duality with the space of bounded $\mathcal{A}$-measurable functions.

5.7.1. THEOREM. *Suppose that the space T is completely regular and its compact subsets are metrizable. Suppose also that a net of measures $\mu_\alpha \in \mathcal{M}(\Omega\times T)$ converges to a measure $\mu \in \mathcal{M}(\Omega\times T)$ in the ws-topology and is uniformly bounded in variation. If the projections of the measures $|\mu_\alpha|$ and $|\mu|$ to the space T are uniformly tight and the projections of the measures $|\mu_\alpha|$ to the space Ω are uniformly countably additive, then, for every bounded $\mathcal{A}\otimes\mathcal{B}(T)$-measurable function f with the property that for every $\omega \in \Omega$ the function $t \mapsto f(\omega,t)$ is continuous, one has*

$$\lim_\alpha \int_{\Omega\times T} f\,d\mu_\alpha = \int_{\Omega\times T} f\,d\mu.$$

PROOF. Without loss of generality we can assume that $|f| \leqslant 1$, $\|\mu_\alpha\| \leqslant 1$, $\|\mu\| \leqslant 1$. Let us fix $\varepsilon > 0$. Let π_T and π_Ω denote the operators of projecting to T and Ω, respectively. By hypothesis, there exists a compact set $K \subset T$ such that

$$|\mu_\alpha|\circ\pi_T^{-1}(T\backslash K) + |\mu|\circ\pi_T^{-1}(T\backslash K) \leqslant \varepsilon \quad \text{for all } \alpha.$$

By the metrizability of K the space $C(K)$ is separable. For every $\omega \in \Omega$, let g_ω denote the continuous function $t \mapsto f(\omega,t)$ on K. It is clear that the mapping $g\colon \Omega \to C(K)$, $\omega \mapsto g_\omega$ is Borel measurable. Since the projections of our measures to Ω are uniformly countably additive, there exists a probability measure ν on $\mathcal{A}$ with respect to which they have uniformly integrable densities. By using the separability of the Banach space $C(K)$ and applying Luzin's theorem to the mapping g and the measure ν, we can find a finite partition of Ω into sets $A_1,\ldots,A_p,A_{p+1}\in\mathcal{A}$ and functions $f_1,\ldots,f_p \in C(K)$ such that $\|f_i\|_{C(K)} \leqslant 1$, $\|g_\omega - f_i\|_{C(K)} \leqslant \varepsilon$ whenever $\omega \in A_i$, $i \leqslant p$, and

$$|\mu_\alpha|\circ\pi_\Omega^{-1}(A_{p+1}) + |\mu|\circ\pi_\Omega^{-1}(A_{p+1}) \leqslant \varepsilon.$$

Since T is completely regular, every function f_i extends to T with the same maximum of absolute value. The extension will also be denoted by f_i. By assumption, there exists an index α_0 such that the absolute value of the difference of the integrals of $h(\omega,t) := \sum_{i=1}^p f_i(t)I_{A_i}(\omega)$ with respect to the measures μ_α and μ does not exceed ε for all $\alpha \geqslant \alpha_0$. We observe that $\sup_x |f(x)-h(x)| \leqslant 2$, $|f(x)-h(x)| \leqslant \varepsilon$

on $\bigcup_{i=1}^p A_i{\times}K$ and $|\mu_\alpha|\bigl(\Omega{\times}(T\backslash K)\bigr)+|\mu_\alpha|(A_{p+1}{\times}T)$ does not exceed 2ε. It remains to use the estimate

$$\int_{\Omega\times T}|f-h|\,d|\mu_\alpha|\leqslant\int_{\bigcup_{i=1}^p A_i\times K}|f-h|\,d|\mu_\alpha|+4\varepsilon\leqslant 5\varepsilon$$

and a similar estimate for μ. □

5.7.2. COROLLARY. *Suppose that a sequence of nonnegative measures μ_n on $\Omega\times T$ converges to a measure μ in the ws-topology and that T is a Polish space. Then the conclusion of Theorem 5.7.1 is true. More generally, the same is true if T is a Prohorov space in which all compact sets are metrizable and the projections of the considered measures to T are Radon.*

PROOF. By assumption $\mu_n=|\mu_n|$. The projections of the measures μ_n to Ω are uniformly countably additive by convergence on every set in $\mathcal{A}$. The projections to T converge weakly, hence are uniformly tight (in the case of Radon projections and a Prohorov space this follows by assumption). □

Under broad conditions, compact sets in the *ws*-topology are metrizable, but on the whole space this topology is not metrizable in nontrivial cases.

5.7.3. PROPOSITION. *Let T be a Polish space and $\mathcal{A}$ a countably generated σ-algebra. Then the ws-topology is metrizable on every compact set $M\subset\mathcal{M}^+(\Omega{\times}T)$.*

PROOF. We observe that the sets of measures M_Ω and M_T obtained from M by projecting to Ω and T are compact in the topology of convergence on sets in $\mathcal{A}$ and in the weak topology, respectively. Now we can use that the compact set M_Ω is metrizable in the topology of setwise convergence (Theorem 5.6.8), the compact set M_T is metrizable in the weak topology, and the compact set M is homeomorphic to its image under the natural mapping to the metrizable compact space $M_\Omega{\times}M_T$. □

The next result is obtained in Raynaud de Fitte [**542**].

5.7.4. THEOREM. *Let T be a metrizable Souslin space with a metric d. Then any of the following conditions is equivalent to convergence of a net of measures $\mu_\alpha\in\mathcal{M}^+(\Omega{\times}T)$ to a measure μ from $\mathcal{M}^+(\Omega{\times}T)$ in the ws-topology:*

(i) $\liminf\limits_\alpha\int_{\Omega\times T}f\,d\mu_\alpha\geqslant\int_{\Omega\times T}f\,d\mu$ *for every bounded $\mathcal{A}\otimes\mathcal{B}(T)$-measurable f with the property that for every $\omega\in\Omega$ the function $t\mapsto f(\omega,t)$ is lower semicontinuous;*

(ii) $\lim\limits_\alpha\int_{\Omega\times T}f\,d\mu_\alpha=\int_{\Omega\times T}f\,d\mu$ *for every bounded $\mathcal{A}\otimes\mathcal{B}(T)$-measurable function f with the property that for every $\omega\in\Omega$ the function $t\mapsto f(\omega,t)$ is continuous;*

(iii) *the equality in* (ii) *is true for all functions of the form $f(\omega,t)=I_A(\omega)\varphi(t)$, where $A\in\mathcal{A}$ and φ is a bounded Lipschitz function on T.*

It is not clear whether convergence in the *ws*-topology implies property (ii) for arbitrary completely regular spaces.

According to Castaing, Raynaud de Fitte, Valadier [**132**, Theorem 2.2.3], if in Theorem 5.7.1 the projection μ_T of the measure $\mu\in\mathcal{P}(\Omega\times T)$ is Radon and the projection μ_Ω is atomless, then μ is the limit in the *ws*-topology of a net of measures of the form $\mu_\Omega{\circ}F_\alpha^{-1}$, $F_\alpha(x)=\bigl(x,\varphi_\alpha(x)\bigr)$, for some measurable mappings $\varphi_\alpha\colon\,\Omega\to T$, i.e., Young measures (see below).

The ws-topology is also called the stable topology and convergence in it is called the stable convergence (see Rényi [**546**]). However, in many papers this terminology is connected with property (ii), equivalent in case of a Polish space T. About the ws-topology, see Balder [**29**], Castaing, Raynaud de Fitte, Valadier [**132**], Florescu, Godet-Thobie [**239**], Häusler, Luschgy [**313**], Jacod, Mémin [**333**], Lebedev [**414**], Letta [**423**], Raynaud de Fitte [**542**], and Schäl [**567**].

Let us proceed to Young measures.

Let $(\Omega, \mathcal{B})$ and $(S, \mathcal{A})$ be measurable spaces and let $\mu \geqslant 0$ be a bounded measure on $\mathcal{B}$. We denote by $\mathcal{Y}(\Omega, \mu, S)$ the set of all measures $\nu \geqslant 0$ on $\mathcal{B}\otimes\mathcal{A}$ such that the image of ν under the natural projection $\Omega\times S \to \Omega$ coincides with μ. Measures in $\mathcal{Y}(\Omega, \mu, S)$ are called *Young measures.* We observe that $\mu(\Omega) = \nu(\Omega\times S)$. A typical example of a Young measure is the measure $\nu := \mu\circ F^{-1}$, where $F\colon \Omega \to \Omega\times S$, $F(x) = \bigl(x, u(x)\bigr)$ and $u\colon \Omega \to S$ is a measurable mapping. Such a measure ν is called the *Young measure generated by the mapping* u. Young measures are useful in variational calculus; there is some connection between convergence of mappings and convergence of the generated Young measures. A simple instance of this connection is indicated in Proposition 5.7.7; additional information can be found in Castaing, Raynaud de Fitte, Valadier [**132**], Florescu, Godet-Thobie [**239**], Giaquinta, Modica, Souček [**265**], Pedregal [**507**], and Valadier [**627**], [**628**].

5.7.5. PROPOSITION. *Let μ be a Radon probability measure on a Hausdorff space Ω, let u_n be measurable mappings from Ω to a separable metric space S, and let ν_n be the corresponding Young measures. Suppose that a $\mathcal{B}\otimes\mathcal{B}(S)$-measurable function $\Psi\colon \Omega\times S\to\mathbb{R}$ has the property that the sequence of functions $x \mapsto \Psi\bigl(x, u_n(x)\bigr)$ is uniformly μ-integrable and, for every fixed x, the function $y \mapsto \Psi(x, y)$ is continuous. Suppose that the measures ν_n converge weakly to a measure ν. Then the function Ψ is ν-integrable and*

$$\int_{\Omega\times S} \Psi\, d\nu = \lim_{n\to\infty} \int_\Omega \Psi\bigl(x, u_n(x)\bigr)\, \mu(dx).$$

The proof can be found in Valadier [**627**, Theorem 17].

5.7.6. LEMMA. *Let μ be a Radon probability measure on a topological space Ω and let $U \subset L^1(\mu)$ be a norm bounded set. Then the corresponding set of Young measures $\{\nu_u\colon u\in U\}$ is uniformly tight on $\Omega\times\mathbb{R}$. If μ is concentrated on a countable union of metrizable compact sets, then, for every $\varepsilon > 0$, one can find a metrizable compact set $K_\varepsilon \subset \Omega\times\mathbb{R}$ such that $\nu_u\bigl((\Omega\times\mathbb{R})\backslash K_\varepsilon\bigr) < \varepsilon$ for all $u \in U$.*

PROOF. Let π_u be the projection of ν_u on $\mathbb{R}$. We observe that

$$\begin{aligned}\sup_{u\in U} \int_{\mathbb{R}} |t|\, \pi_u(dt) &= \sup_{u\in U} \int_{\Omega\times\mathbb{R}} |t|\, \nu_u(d\omega dt) = \sup_{u\in U} \int_\Omega |u|\, d\mu \\ &= \sup_{u\in U} \|u\|_{L^1(\mu)} < \infty.\end{aligned}$$

Hence the projections of the measures ν_u to $\mathbb{R}$ form a tight family. The common projection on Ω is the measure μ concentrated on a countable union of metrizable compact sets. It remains to observe that if for two nonnegative measures μ and λ there are metrizable compact sets K and S with the measures of complements less than ε, then, for any measure ν with projections μ and λ, the complement of the metrizable compact set $K\times S$ has ν-measure less than $\varepsilon(\|\mu\| + \|\lambda\|)$. □

We now see that for Young measures generated by mappings, their weak convergence is equivalent to convergence of mappings in measure (which gives subsequences convergent almost everywhere).

5.7.7. PROPOSITION. *Let (S,d) be a separable metric space, let Ω be a Hausdorff space with a Radon probability measure μ, let $\{u_n\}$ be a sequence of measurable mappings from Ω to S, and let u_∞ be a measurable mapping from Ω to S. The mappings u_n converge to u_∞ in measure precisely when the generated Young measures ν_n converge weakly to the Young measure ν_∞ generated by u_∞.*

PROOF. Convergence in measure implies convergence of the integrals for every bounded continuous function ψ on $\Omega\times S$, since the functions $\psi(x,u_n(x))$ converge in measure to $\psi(x,u(x))$ according to Exercise 2.7.47.

Conversely, suppose that we have weak convergence of Young measures. Let

$$\psi(x,y) = \min\bigl(1, d(u_\infty(x), y)\bigr).$$

Then the integral of ψ with respect to ν_∞ is zero and the integral with respect to ν_n equals

$$\int \min\bigl(1, d(u_n, u_\infty)\bigr)\, d\mu.$$

Therefore, in order to show that $u_n \to u_\infty$ in measure, it suffices to establish convergence of the integrals of ψ against ν_n to the integral against ν. If we replace u_∞ by a continuous mapping v, then this is obvious. In the general case we can assume that $S = \mathbb{R}^\infty$, because S is homeomorphic to a set in $\mathbb{R}^\infty$. Now it remains to apply the Luzin theorem, which for every $\varepsilon > 0$ gives a set $E \subset \Omega$ with $\mu(\Omega\backslash E) < \varepsilon$ and a continuous mapping $v\colon \Omega \to S$ such that $v = u_\infty$ on E. Then

$$\int_{\Omega\times S} \min\Bigl(1, d\bigl(v(x), y\bigr)\Bigr)\, \nu_n(dx) = \int_\Omega \min\bigl(1, d(v, u_n)\bigr)\, d\mu$$

differs from the integral of $\min\bigl(1, d(u_n, u_\infty)\bigr)$ against the measure μ by at most 2ε, and the same is true for u_∞ in place of u_n. □

The next result shows that Young measures not generated by mappings arise naturally as limits of Young measures generated by mappings (which can also be used to construct Young measures not generated by mappings).

5.7.8. PROPOSITION. *Let a Radon probability measure μ on a completely regular space Ω be concentrated on a countable union of metrizable compact sets. Suppose that a sequence $\{u_n\}$ converges to a function u weakly in $L^1(\mu)$, but does not converge in norm. Then the sequence of the corresponding Young measures ν_n on $\Omega\times\mathbb{R}^1$ has a subsequence that converges weakly to some Young measure ν not generated by a function.*

PROOF. The sequence $\{u_n\}$ is uniformly integrable. By our assumption, there exists $c > 0$ and a subsequence $\{u_{n_k}\}$ such that $\|u - u_{n_k}\|_{L^1(\mu)} \geqslant c$ for all k. By Lemma 5.7.6 and Theorem 4.5.3 one can find a further subsequence (also denoted by u_{n_k}) for which the corresponding Young measures converge weakly to some measure ν. It is clear that ν is a Young measure. Suppose that ν is generated by some measurable function v. According to Proposition 5.7.7, the sequence $\{u_{n_k}\}$ converges in measure, hence by the Lebesgue–Vitali theorem it converges in norm. Then its limit in norm equals u, which is a contradiction. □

5.8. Complements and exercises

5.8(i). Separability of spaces of measures

Here we mention some separability properties of spaces of measures with the weak topology. Such properties were studied in Koumoullis, Sapounakis [**390**], Pol [**514**], Talagrand [**607**], where one can find additional references. If X is a separable completely regular space, then $\mathcal{M}_t(X)$, $\mathcal{M}_t^+(X)$ and $\mathcal{P}_t(X)$ with the weak topology are separable as well (Theorem 5.1.3(ii)). The converse is false even for compact spaces: the next theorem was proved in Talagrand [**607**] under the continuum hypothesis and in Avilés, Plebanek, Rodríguez [**27**] in the ZFC.

5.8.1. THEOREM. *There exists a compact space K such that the space $\mathcal{M}_t(K)$ is separable in the weak topology, but $\mathcal{P}_t(K)$ and the unit ball in $\mathcal{M}_t(K)$ are not.*

We observe that for any completely regular space X the separability of $\mathcal{P}_t(X)$ in the weak topology is equivalent to the separability of the unit ball B_1 in $\mathcal{M}_t(K)$. Indeed, the former easily implies the latter. Conversely, let S be a countable set dense in B_1 in the weak topology and let $\mu \in \mathcal{P}_t(X)$. Then there is a net $\{\sigma_\alpha\} \subset S$ converging to μ. Clearly, $|\sigma_\alpha|(X) \to 1$, since otherwise $\mu(X) < 1$. By Theorem 4.8.1 we have $|\sigma_\alpha| \Rightarrow \mu$. Hence we can find a countable set dense in $\mathcal{P}_t(X)$.

Note that the separability of the unit ball in $\mathcal{M}_t(K)$ in the weak topology does not imply the metrizability of K: according to [**607**] (also under the CH), it can even happen that there are no separable measures with support K.

A set of measures $M \subset \mathcal{M}_\sigma(X)$ is called *countably separated* if there exists a sequence $\{f_n\} \subset C_b(X)$ such that, for every μ and ν in M, the equality $\mu = \nu$ is equivalent to the identity

$$\int_X f_n(x)\,\mu(dx) = \int_X f_n(x)\,\nu(dx) \quad \forall n \in \mathbb{N}.$$

If X is separable metrizable, then $\mathcal{M}_\sigma(X)$ is countably separated.

A subset $M \subset \mathcal{M}_\sigma(X)$ is called *countably determined* in $\mathcal{M}_\sigma(X)$ if there exists a sequence $\{f_n\} \subset C_b(X)$ such that, whenever $\mu \in M$ and $\nu \in \mathcal{M}_\sigma(X)$, the equality

$$\int_X f_n(x)\,\mu(dx) = \int_X f_n(x)\,\nu(dx) \quad \forall n \in \mathbb{N}$$

implies the inclusion $\nu \in M$. Similarly one defines the property to be countably determined in $\mathcal{M}_\sigma^+(X)$. It is readily seen that for any compact space X the set $\mathcal{M}_\sigma^+(X)$ is countably separated if and only if X is metrizable (see Exercise 5.8.22). The following simple lemma from Koumoullis [**386**] is useful in such considerations.

5.8.2. LEMMA. *Let H be a countable family of bounded Baire functions on a topological space X. Then there exists a countable set $K \subset C_b(X)$ with the following property: if for a pair of Baire measures μ and ν on X one has the equality*

$$\int_X \varphi(x)\,\mu(dx) = \int_X \varphi(x)\,\nu(dx) \quad \forall \varphi \in K,$$

then this equality is fulfilled for all $h \in H$ in place of φ.

PROOF. It suffices to consider the case where H consists of a single function h. The class $\mathcal{H}$ of bounded Baire functions h for which our assertion is true contains $C_b(X)$ and is a linear space that is closed with respect to taking pointwise limits of uniformly bounded sequences. By the monotone class theorem $\mathcal{H}$ coincides with the class of all bounded Baire functions (see Bogachev [**81**, Theorem 2.12.9]). □

It is clear from the lemma that in the definitions of countably separated and countably determined sets one can consider bounded Baire functions (or even sequences of Baire sets).

Since a compact space K is metrizable precisely when it possesses a countable family of continuous functions separating points in K (Exercise 3.2.5), it is clear that a compact (in the weak topology) set $M \subset \mathcal{M}_\sigma(X)$ is countably separated if and only if it metrizable. According to Koumoullis, Sapounakis [**390**, Proposition 2.3], a compact set $M \subset \mathcal{M}_\sigma(X)$ is countably determined if and only if it is a G_δ-subset of $\mathcal{M}_\sigma(X)$ (and similarly for sets in $\mathcal{M}_\sigma^+(X)$). These assertions can fail for noncompact sets (for example, usually $\mathcal{M}_\sigma(X)$ is not metrizable in the weak topology). The next result (the proof of which is given in Koumoullis, Sapounakis [**390**, Theorem 4.1]) describes the situation for the whole space of measures. We recall that a space Y is called *countably submetrizable* if there exists a sequence of continuous functions separating points in Y (equivalently, there exists a continuous injection $Y \to \mathbb{R}^\infty$).

5.8.3. THEOREM. *Let X be a Hausdorff space and let s be one of the symbols σ, τ or t. The following assertions are equivalent:*

(i) *$\mathcal{M}_s(X)$ is countably separated;*

(ii) *$\mathcal{M}_s^+(X)$ is countably separated;*

(iii) *$C_b(X)$ is separable with the topology $\sigma\bigl(C_b(X), \mathcal{M}_s(X)\bigr)$;*

(iv) *$\mathcal{M}_s(X)$ is countably submetrizable;*

(v) *every point in $\mathcal{M}_s(X)$ is a G_δ-set.*

In addition, for $s = t$ conditions (i)–(v) *are equivalent to the condition that the space X is countably submetrizable.*

5.8(ii). Measurability on spaces of measures

Here we present some results of Ressel [**549**] on measurability of mappings of the form $\mu \mapsto \mu(A)$. Let X be a Hausdorff space and let $\mathcal{K}(X)$ be the set of all its compact subsets. The space $\mathcal{K}(X)$ has a natural topology: the *Vietoris topology* (see Fedorchuk, Filippov [**217**, Chapter 4], Repovš, Semenov [**547**, p. 195]), it is generated by sets of two types: $\{K \in \mathcal{K}(X) \colon K \subset U\}$ and $\{K \in \mathcal{K}(X) \colon K \cap U \neq \varnothing\}$, where $U \subset X$ is open. A topology base is formed by sets of the form $W(U_1, \ldots, U_n)$ with open sets $U_i \subset X$ consisting of $K \in \mathcal{K}(X)$ such that $K \subset \bigcup_{i=1}^n U_i$ and $K \cap U_i \neq \varnothing$ for each i. If X is a Polish space, then $\mathcal{K}(X)$ with this topology is Polish as well. Let X, Y, Z be completely regular spaces (in the general case the formulations and proofs are completely analogous with the A-topology in place of the weak topology). We recall that the *universal measurability* on a measurable space $(E, \mathcal{E})$ means the measurability with respect to all measures on $\mathcal{E}$.

5.8.4. THEOREM. *Let the space X be Souslin and let the space $\mathcal{M}_r^+(X)$ be equipped with the weak topology.*

(i) *If the space Y is Polish and $f \colon Y \to X$ is a continuous mapping, then the function $(\mu, K) \mapsto \mu\bigl(f(K)\bigr)$ on $\mathcal{M}^+(X) \times \mathcal{K}(Y)$ is upper semicontinuous.*

(ii) *If $A \subset X$ is a Souslin set, then $\varphi_A\colon \mu \mapsto \mu(A)$ is an S-function on $\mathcal{M}_r^+(X)$, i.e., the set $\{\varphi_A > t\}$ is Souslin for each $t \in \mathbb{R}^1$. If A is a set in the σ-algebra generated by Souslin sets, then the function φ_A is measurable with respect to the σ-algebra generated by Souslin sets. Finally, if A is universally measurable, then so is φ_A.*

PROOF. (i) Theorem 4.3.17 on the continuity of the product gives the continuity of the mapping $(\mu, K) \mapsto \mu\otimes\delta_K$ on $\mathcal{M}_r^+(X)\times\mathcal{K}(Y)$. It is straightforward to verify the upper semicontinuity of the function $(x, K) \mapsto I_{f(K)}(x)$ on $X\times\mathcal{K}(Y)$ (it suffices to use the existence of disjoint neighborhoods of the point x and the compact set $f(K)$ for each $x \notin f(K)$). This yields the upper semicontinuity of the integral of $I_{f(L)}(x)$ with respect to the measure $\mu\otimes\delta_K$, regarded as a function on $\mathcal{M}_r^+(X)\times\mathcal{K}(Y)$.

(ii) There is a continuous surjection $f\colon E \to X$ of a Polish space E. We observe that $\mu(A) = \sup\{\mu(f(K))\colon K \in \mathcal{K}(E)\}$ for all $\mu \in \mathcal{M}_r^+(X)$. Hence $\{\varphi_A > t\}$ is the projection of the Borel set $\{(\mu, K) \in \mathcal{M}_r^+(X)\times\mathcal{K}(E)\colon \mu(f(K)) > t\}$ in a Polish space. Finally, if A is universally measurable and $\sigma \geqslant 0$ is a measure on a ball S in $\mathcal{M}_r^+(X)$ (with respect to the total variation norm), then we can consider the measure m_σ on $\mathcal{B}(X)$ defined by

$$m_\sigma(B) = \int_S \mu(B)\,\sigma(d\mu).$$

There are sets $B_1, B_2 \in \mathcal{B}(X)$ with $B_1 \subset A \subset B_2$ and $m_\sigma(B_1) = m_\sigma(B_2)$, i.e., $\sigma(\mu\colon \mu(B_2) > \mu(B_1)) = 0$. The measurability of $\{\varphi_A > t\}$ with respect to σ follows from the inclusions $\{\varphi_{B_1} > t\} \subset \{\varphi_A > t\} \subset \{\varphi_{B_2} > t\}$. □

5.8.5. THEOREM. (i) *Let X, Y, and Z be Souslin spaces and let $f\colon X\times Y \to Z$ be a universally measurable mapping (i.e., $f^{-1}(B)$ is measurable with respect to all Borel measures on $X\times Y$ for all $B \in \mathcal{B}(Z)$). Set $f_y(x) := f(x, y)$. Let us equip the space of measures with the weak topology. Then the mapping*

$$F\colon \mathcal{M}_r^+(X)\times Y \to \mathcal{M}_r^+(Z),\ (\mu, y) \mapsto \mu\circ f_y^{-1}$$

is universally measurable. In addition, if the mapping f is continuous or Borel measurable, then so is F. Finally, if f is measurable with respect to the σ-algebra generated by Souslin sets, then F also possesses this property.

(ii) *If, in addition, $Z = \mathbb{R}^1$ and the function f is bounded, then the function*

$$\Psi\colon \mathcal{M}_r^+(X)\times Y \to \mathbb{R}^1,\ (\mu, y) \mapsto \int_X f(x, y)\,\mu(dx),$$

is universally measurable. If f has one of the following properties: is $\mathcal{A}$-measurable, is an S-function, is Borel measurable, is upper semicontinuous, is continuous, then Ψ also possesses the respective property.

PROOF. (i) The operator $\psi\colon \sigma \mapsto \sigma\circ f^{-1}$ from $\mathcal{M}_r^+(X\times Y)$ to $\mathcal{M}_r^+(Z)$ is universally measurable by Corollary 5.1.9 and the previous theorem. The mapping F is the composition of ψ with the continuous mapping $(\mu, y) \mapsto \mu\otimes\delta_y$, $\mathcal{M}_r^+(X)\times Y \to \mathcal{M}_r^+(X\times Y)$. Assertion (ii) can be easily derived from (i) and the previous theorem. □

The following result on Borel measurability in spaces of measures, obtained in Hoffmann-Jørgensen [**325**], is useful for various averagings on spaces of measures. It also enables one to extend some of the results above to signed measures.

5.8.6. PROPOSITION. *If f is a Baire measurable bounded function on a topological space X, then on the space $\mathcal{M}_\sigma(X)$ with the weak topology the following functions are Borel measurable:*

$$F(\mu) = \int_X f\,d\mu, \quad F_+(\mu) = \int_X f\,d\mu^+, \quad F_-(\mu) = \int_X f\,d\mu^-.$$

If X is completely regular, then these functions are Borel measurable on $\mathcal{M}_\tau(X)$ and $\mathcal{M}_t(X)$ with the weak topology for every Borel bounded function f. Finally, if f is nonnegative and lower semicontinuous, then the functions F_+, F_- and F_++F_- are also lower semicontinuous on $\mathcal{M}_\tau(X)$ and $\mathcal{M}_t(X)$.

PROOF. It is easy to see that it suffices to prove our assertion for F_+. It reduces to the case of a simple function, hence to the case of the indicator function of a set.

Let $f = I_U$, where the set U is functionally open. Then

$$\mu^+(U) = \sup\Bigl\{\int_X \varphi\,d\mu\colon\ \varphi\in C_b(X),\, 0\leqslant\varphi\leqslant I_U\Bigr\}.$$

Indeed, for a given number $\varepsilon>0$ we find a functionally open set $W\subset U$ such that for the set X^+ from the Hahn decomposition we have $U\cap X^+\subset W$ and $|\mu|\bigl(W\backslash(U\cap X^+)\bigr)<\varepsilon$. Next, in $U\cap X^+$ we take a functionally closed set Z for which $|\mu|\bigl((U\cap X^+)\backslash Z\bigr)<\varepsilon$. There is a function $\varphi\in C_b(X)$ such that $0\leqslant\varphi\leqslant 1$, $\varphi|_Z = 1$, $\varphi|_{X\backslash W} = 0$. Then the integral of φ with respect to the measure μ differs from $\mu^+(U)$ by at most 3ε. The integral of φ defines a continuous functional on $\mathcal{M}_\sigma(X)$. Hence F_+ is lower semicontinuous. The class $\mathcal{E}$ of all sets $E\in\mathcal{B}a(X)$ for which the function $f = I_E$ generates a Borel function F_+ is σ-additive: it contains the whole space, is closed with respect to countable unions of disjoint sets and contains the difference $E_2\backslash E_1$ for any sets $E_1, E_2\in\mathcal{E}$ such that $E_1\subset E_2$. By the monotone class theorem (see [**81**, Theorem 1.9.3]) we obtain $\mathcal{E}=\mathcal{B}a(X)$, since the class of functionally open sets admits finite intersections and the σ-algebra generated by it is $\mathcal{B}a(X)$.

Let us consider $\mathcal{M}_\tau(X)$ for a completely regular space X. The previous reasoning remains in force if we now take all open sets U. Moreover, the indicated equality for $\mu^+(U)$ is true by the τ-additivity, since $\mu^+(U)$ equals $\sup\{\mu^+(V)\}$, where sup is taken over functionally open sets $V\subset U$. Finally, the assertion about semicontinuity is clear from the proof, because any lower semicontinuous nonnegative function f is uniformly approximated by finite linear combinations of the indicator functions of open sets with nonnegative coefficients (see the proof of Lemma 4.1.6). □

5.8.7. COROLLARY. *Let X be a completely regular space. Then for each τ-additive measure Ψ on $\mathcal{M}_\tau(X)$ with respect to which the function $q\mapsto\|q\|$ is integrable (where $\|q\|$ is the variation of the measure q), we obtain the following τ-additive measures on $\mathcal{B}(X)$:*

$$\sigma(B) := \int_{\mathcal{M}_\tau(X)} q(B)\,\Psi(dq), \quad \eta(B) := \int_{\mathcal{M}_\tau(X)} |q|(B)\,|\Psi|(dq).$$

Therefore, for every $B\in\mathcal{B}(X)$ and $\varepsilon>0$, there exists an open set $U\supset B$ such that $|\Psi|\bigl(q\colon\ |q|(U\backslash B)>\varepsilon\bigr)<\varepsilon$.

PROOF. By Proposition 5.8.6, for every Borel set B the functions $q\mapsto q(B)$ and $q\mapsto|q|(B)$ are Borel on $\mathcal{M}_\tau(X)$. By the integrability of $q\mapsto\|q\|$ both measures σ and η are well-defined. We show that $\eta\in\mathcal{M}_\tau(X)$. Suppose that a net of open sets $U_\lambda \subset X$ is increasing to an open set U. The net of functions $q \mapsto |q|(U_\lambda)$ is increasing to $q \mapsto |q|(U)$ by the τ-additivity of $|q|$, in addition, these functions are lower semicontinuous on $\mathcal{M}_\tau(M)$. Now we apply Lemma 4.1.6. The same reasoning applies to q^+ and q^- in place of $|q|$, which gives the τ-additivity of η. □

The next result from Hoffmann-Jørgensen [**325**] shows that any Radon measure on a compact set of measures is almost concentrated on a uniformly tight set. Certainly, for Prohorov spaces this is trivial.

5.8.8. THEOREM. *Suppose that X is a completely regular space such that $\mathcal{M}_\tau(X) = \mathcal{M}_r(X)$. Let Ψ be a Radon measure on $\mathcal{M}_\tau(X)$ with the weak topology. Then, for every Borel set of measures $M \subset \mathcal{M}_r(X)$ and every $\varepsilon > 0$, there exists a compact uniformly tight set $M_\varepsilon \subset M$ such that $|\Psi|(M\backslash M_\varepsilon) \leqslant \varepsilon$.*

PROOF. By our hypothesis, there exists a compact set $K \subset M$ such that $|\Psi|(M\backslash K) < \varepsilon/2$. By Corollary 5.8.7 the measure

$$\eta(B) = \int_K |\mu|(B)\, |\Psi|(d\mu), \ \ B \in \mathcal{B}(X)$$

is well-defined and τ-additive. By our assumption it is Radon. Hence there exist compact sets $C_n \subset X$ such that $\eta(X\backslash C_n) \leqslant \varepsilon 8^{-n}$. Set

$$K_n := \{\mu \in K\colon\ |\mu|(X\backslash C_n) \leqslant 2^{-n}\}.$$

The sets K_n are closed in the weak topology by Proposition 5.8.6, but in this particular case this fact can be easily verified directly by considering the integrals of continuous functions with values in $[-1, 1]$ vanishing on C_n with respect to measures in K_n. Then the set $M_\varepsilon := \bigcap_{n=1}^\infty K_n$ is compact in the weak topology and uniformly tight. Moreover, by the Chebyshev inequality we have

$$|\Psi|(K\backslash K_n) \leqslant 2^n \int_K |\mu|(X\backslash C_n)\, |\Psi|(d\mu) = 2^n \eta(X\backslash C_n) \leqslant \varepsilon 4^{-n}.$$

Thus, $|\Psi|(K\backslash M_\varepsilon) \leqslant \sum_{n=1}^\infty |\Psi|(K\backslash K_n) \leqslant \varepsilon/2$, as required. □

This theorem holds, for example, for completely regular Souslin spaces. Note that in this situation not every weakly compact set in $\mathcal{M}_t(X)$ is uniformly tight (by Theorem 4.8.6 a counterexample is given by the set of rational numbers).

5.8(iii). Weak sequential completeness

In Theorem 5.1.10 we established weak sequential completeness of the space of Baire measures on an arbitrary topological space. Let us make some remarks about weak sequential completeness of the space $\mathcal{M}_t(X)$ of tight measures. We start with a couple of trivial observations.

5.8.9. EXAMPLE. Let X be completely regular. The space of measures $\mathcal{M}_t(X)$ is weakly sequentially complete provided that either all Baire measures on X are tight or every weakly fundamental sequence in $\mathcal{M}_t(X)$ is uniformly tight.

PROOF. Use weak sequential completeness of $\mathcal{M}_\sigma(X)$ and Theorem 4.5.3. □

5.8.10. EXAMPLE. For every σ-compact completely regular space X, the space $\mathcal{M}_t(X)$ is weakly sequentially complete.

PROOF. This assertion follows from weak sequential completeness of $\mathcal{M}_\sigma(X)$, since all Baire measures on X are tight. □

From Proposition 4.7.6 (ii) and Example 5.8.9 we obtain one more interesting example (borrowed from Fremlin, Garling, Haydon [**245**]).

5.8.11. EXAMPLE. Let X be a completely regular space such that there exists a sequence of compact sets $K_n \subset X$ with the property that every function on X continuous on each K_n is continuous on all of X. Then the space $\mathcal{M}_t(X)$ is weakly sequentially complete.

The next result was obtained in Moran [**470**].

5.8.12. THEOREM. *Let X be a normal metacompact space (for each open cover of X there is a pointwise finite open cover whose elements are subsets of the original cover). Then the space $\mathcal{M}_\tau(X)$ is weakly sequentially complete. If, in addition, X is Čech complete (see p. 141), then $\mathcal{M}_t(X)$ is weakly sequentially complete.*

We recall that a locally convex space is called *complete* if every Cauchy net in it (the one which is Cauchy in every seminorm from the family defining the topology) has a limit. This property is much stronger than the sequential completeness, where convergence is required only for countable Cauchy sequences. For example, the infinite-dimensional Hilbert space with the weak topology is sequentially complete, but not complete, since every linear functional on it can be obtained as the limit of a net of continuous linear functionals (such a net is weakly fundamental, but has no limit in the set of continuous functionals). Similarly, in nontrivial cases the spaces of measures $\mathcal{M}_\sigma(X)$ are not complete in the weak topology (for the failure of completeness it suffices to have a discontinuous linear functional on the Banach space $C_b(X)$ with its sup-norm, which is always fulfilled for any infinite-dimensional $C_b(X)$). However, if we reinforce the topology on $\mathcal{M}_\sigma$, completeness can appear (say, this space is complete with the usual variation norm).

Let us now consider the Mackey topology $\tau(\mathcal{M}_\sigma)$ on the space $\mathcal{M}_\sigma(X)$ of Baire measures on a Hausdorff space X associated with the usual duality between $C_b(X)$ and $\mathcal{M}_\sigma(X)$. This duality defines the topology $\sigma(C_b, \mathcal{M}_\sigma)$ on $C_b(X)$, which in turn generates the class $\mathcal{V}$ of convex $\sigma(C_b, \mathcal{M}_\sigma)$-compact sets in $C_b(X)$. Now $\mathcal{M}_\sigma(X)$ can be equipped with the topology of uniform convergence on the sets from $\mathcal{V}$ defined by means of the seminorms

$$p_V(\mu) = \sup_{f\in V}\left|\int_X f\,d\mu\right|, \quad V \in \mathcal{V}.$$

This is the Mackey topology $\tau(\mathcal{M}_\sigma)$ we are interested in. This is the strongest locally convex topology on $\mathcal{M}_\sigma(X)$ for which the dual remains the space $C_b(X)$.

5.8.13. THEOREM. *The space $\mathcal{M}_\sigma(X)$ with the Mackey topology $\tau(\mathcal{M}_\sigma)$ is complete. Hence if X is a completely regular space such that $\mathcal{M}_\sigma(X) = \mathcal{M}_t(X)$, then $\mathcal{M}_t(X)$ with the Mackey topology is complete.*

PROOF. Let $\{\mu_\alpha\}$ be a Cauchy net in the Mackey topology. The functional

$$L(f) = \lim_\alpha \int_X f\,d\mu_\alpha$$

is defined on $C_b(X)$. We have to verify that it is represented by a measure from $\mathcal{M}_\sigma(X)$. For this we apply Theorem 4.1.9. Suppose that functions f_n from $C_b(X)$ are decreasing pointwise to zero. We have to show that $L(f_n) \to 0$. It suffices to prove that the integrals of the functions f_n with respect to the measures

μ_α tend to zero uniformly in α. The condition that our sequence of measures is fundamental in the Mackey topology yields that the integrals of the functions from convex $\sigma(C_b, \mathcal{M}_\sigma)$-compact sets in $C_b(X)$ against the measures μ_α are uniformly fundamental (Cauchy). Thus, it remains to prove that $\{f_n\}$ is contained in a convex $\sigma(C_b, \mathcal{M}_\sigma)$-compact set. Let us consider the operator

$$T\colon l^1 \to C_b(X), \quad Ta = \sum_{n=1}^{\infty} a_n f_n, \quad a = (a_n).$$

The closed unit ball U in l^1 with the weak-$*$ topology $\sigma(l^1, c_0)$ is compact by the Banach–Alaoglu theorem (here l^1 is identified with c_0^*). The operator T is continuous with respect to the topologies $\sigma(l^1, c_0)$ and $\sigma(C_b, \mathcal{M}_\sigma)$ due to the uniform boundedness of $\{f_n\}$. Hence $T(U)$ is convex and compact in the topology $\sigma(C_b, \mathcal{M}_\sigma)$ and we have $\{f_n\} \subset T(U)$. □

5.8(iv). The A-topology

There is another natural possibility to topologize the space of probability measures hinted by the theorem of A.D. Alexandroff and used if X is not completely regular or if just the class of Borel measures does not coincide with the class of Baire measures. Let $\mathcal{G}$ be the class of all open sets in the space X.

The A-topology on the space $\mathcal{P}(X)$ of all Borel probability measures (or on its subspaces $\mathcal{P}_r(X)$ and $\mathcal{P}_\tau(X)$) is defined by means of neighborhoods of the form

$$U(\mu, G, \varepsilon) = \{\nu\colon\ \mu(G) < \nu(G) + \varepsilon\}, \quad \mu \in \mathcal{P}(X),\ G \in \mathcal{G},\ \varepsilon > 0.$$

A net $\{\mu_\alpha\}$ converges in this topology to μ if and only if $\liminf_\alpha \mu_\alpha(G) \geqslant \mu(G)$ for every $G \in \mathcal{G}$. It is easy to see that the A-topology is Hausdorff and that in case of a completely regular space it coincides with the weak topology on $\mathcal{P}_\tau(X)$. In general the A-topology is stronger than the weak topology (which is trivial if there are no nonconstant continuous functions on X). Another advantage of the A-topology is that the weak topology is naturally connected with Baire measures and can fail to be Hausdorff on Borel measures. In order to define the A-topology on the space $\mathcal{M}^+(X)$ of all nonnegative Borel measures, in addition to the indicated neighborhoods we also take $U'(\mu, \varepsilon) = \{\nu\colon\ |\mu(X) - \nu(X)| < \varepsilon\}$. For this topology there are results analogous to those presented above for the weak topology (see, for example, Topsøe [**612**]). In particular, X is homeomorphic to the set of Dirac measures with the A-topology, which is closed in the spaces $\mathcal{P}_r(X)$ and $\mathcal{P}_\tau(X)$ with the A-topology. The following assertion is also valid.

5.8.14. THEOREM. *The space $\mathcal{M}_\tau^+(X)$ with the A-topology is regular, completely regular or second countable precisely when X has the respective property.*

Let us also mention the following result from Holický, Kalenda [**329**].

5.8.15. THEOREM. (i) *Let X be a set in a Hausdorff space Y equipped with the induced topology. Suppose that X is a set of one of the following types: G_δ, Borel, $\mathcal{F}$-Souslin or $\mathcal{B}$-Souslin (i.e., is obtained, respectively, from closed sets or Borel sets by the Souslin A-operation, see* [**81**]*). Then $\mathcal{M}^+(X)$ and $\mathcal{M}_r^+(X)$ are sets of the respective type in the spaces $\mathcal{M}^+(Y)$ and $\mathcal{M}_r^+(Y)$ with the A-topology.*

(ii) *If the space X is Čech complete (see p. 141), then the space $\mathcal{M}_r^+(X)$ with the A-topology is also.*

For completely regular spaces, the assertions for $\mathcal{M}_\tau^+(X)$ are valid, of course, for the weak topology. Let us mention some results from Schief [**571**], giving analogs of the results in § 5.2 for the A-topology.

5.8.16. THEOREM. *Let X and Y be Hausdorff spaces and let $f\colon X\to Y$ be a Borel surjection that is open, i.e., the images of open sets are open. Suppose that for every open set $G\subset X$ we have $\widehat{f}(\mathcal{M}^+(G))=\mathcal{M}^+(f(G))$. Then the mapping $\widehat{f}\colon \mathcal{M}^+(X)\to\mathcal{M}^+(Y)$ is open in the A-topology.*

5.8.17. COROLLARY. *Let X and Y be Souslin spaces and let $f\colon X\to Y$ be a Borel surjection that is open. Then the induced mapping $\widehat{f}\colon \mathcal{M}^+(X)\to\mathcal{M}^+(Y)$ and also the mapping $\widehat{f}\colon \mathcal{P}(X)\to\mathcal{P}(Y)$ are open in the A-topology.*

Exercises

5.8.18. Prove that the bilinear form on $C_b(X)\times\mathcal{M}_\sigma(X)$ associating to (f,μ) the integral of f against μ is continuous on bounded sets with respect to the sup-norm on $C_b(X)$ and the weak topology on $\mathcal{M}_\sigma(X)$ for any space X, but is not always continuous on the whole product (say, for $X=[0,1]$).

5.8.19. Let X be a completely regular space. Prove that a set $M\subset\mathcal{P}_\tau(X)$ has compact closure in the weak topology precisely when for every net of open sets U_α increasing to X one has $\sup\limits_\alpha\inf\limits_{\mu\in M}\mu(U_\alpha)=1$. This is also equivalent to the property that for every net of bounded continuous functions f_α on X decreasing to zero one has

$$\inf_\alpha\sup_{\mu\in M}\int_X f_\alpha\,d\mu=0.$$

HINT: the necessity is easily verified, sufficiency follows from the compactness of the unit ball in $C_b(X)^*$ in the weak-$*$ topology and Theorem 4.1.9.

5.8.20. Show that any uniformly tight set of Radon probability measures on a Hausdorff space X has compact closure in the A-topology.

HINT: for such a set M there are increasing compact sets K_n with $|\mu|(X\backslash K_n)\leqslant 1/n$ for all $\mu\in M$; every net in M contains a subnet $\{\mu_\alpha\}$ for which the restrictions of the measures μ_α to K_n converge weakly to a Radon measure ν_n on K_n with the property that the restriction of ν_{n+1} to K_n equals ν_n; this defines a Radon measure ν on X to which the net $\{\mu_\alpha\}$ converges in the A-topology.

5.8.21. Construct an example of a completely regular space X for which the set of Dirac measures is not closed in $\mathcal{M}_\sigma^+(X)$.

HINT: let ω_1 be the smallest uncountable ordinal number and let the interval of ordinals $X=[0,\omega_1)$ be equipped with the topology. For every continuous function f on X there exists $\tau<\omega_1$ such that f is constant on $[\tau,\omega_1)$ (Exercise 4.8.47). Let μ be the measure that equals 0 on countable sets and 1 on their complements. Then μ is defined on all Baire sets. The net of Dirac measures δ_α indexed by $\alpha<\omega_1$ converges weakly to μ, since if a continuous function f equals 1 on $[\tau,\omega_1)$, then its integral with respect to the measure μ and the measures δ_α with $\alpha\geqslant\tau$ is 1 (because the set $[0,\tau)$ is countable).

5.8.22. Let X be a compact space. Prove that the set $\mathcal{M}_\sigma^+(X)$ is countably separated precisely when the space $C_b(X)$ is norm separable, which, in turn, is equivalent to the metrizability of the compact space X.

5.8.23. Let X be a completely regular space in which there exists a sequence of closed subspaces X_n with the following properties: $\mathcal{M}_\sigma(X_n)=\mathcal{M}_t(X_n)$ and every function on X

that is continuous on each X_n is continuous on the whole space X. Suppose that each Baire subset of X_n is Baire in X. Then the space $\mathcal{M}_t(X)$ is weakly sequentially complete.

HINT: as in the proof of Proposition 4.7.6, the complement of the set $Y = \bigcup_{n=1}^{\infty} X_n$ is discrete and its subsets are Baire in X. We can replace the measures μ_n by their (unique) Radon extensions. All measures μ_n are purely atomic on $X\backslash Y$ and the collection of their atoms in $X\backslash Y$ is an at most countable discrete subset A in X. As in the cited proposition, the limiting Baire measure μ is tight on $X\backslash Y$ and $|\mu|\bigl(X\backslash(Y\cup A)\bigr) = 0$. The assumption yields that the restriction of μ (which is well-defined by the supposed character of embedding of X_m) is tight on each X_m. Hence μ is tight on Y, then also on X.

5.8.24. Prove that if a σ-algebra $\mathcal{A}$ is infinite, then the topology of convergence of measures on sets in $\mathcal{A}$ is not generated by a norm.

HINT: use that the dual to the space of measures with the topology of setwise convergence coincides with the linear space L of simple functions; the dual to a Banach space is Banach; if $\mathcal{A}$ is infinite, then L cannot be Banach with respect to a norm q, since in that case for all $A_n \in \mathcal{A}$ the function $\sum_{n=1}^{\infty} 2^{-n} q(I_{A_n})^{-1} I_{A_n}$ belongs to L, which is impossible, since there exist A_n such that this function assumes countably many values.

5.8.25. Let $\mathcal{A}$ be the Borel σ-algebra on $[0,1]$. Show that on the space $\mathcal{M}$ of all countably additive measures on $\mathcal{A}$ all the three topologies considered in § 5.6, i.e., the topology of convergence on sets in $\mathcal{A}$, the topology generated by duality with the space of all bounded $\mathcal{A}$-measurable functions and the topology $\sigma(\mathcal{M}, \mathcal{M}^*)$, are different, although the families of convergent countable sequences for them coincide.

HINT: the duals to the space $\mathcal{M}$ with the first two topologies are identified, respectively, with the space of all simple functions and the space of all bounded $\mathcal{A}$-measurable functions, but these two spaces are different for every infinite σ-algebra. If we take a non-Borel Souslin set A, then the functional $\mu \mapsto \mu(A)$ belongs to $\mathcal{M}^*$, but is not given by an $\mathcal{A}$-measurable function.

5.8.26. Let $\mathcal{A}$ be a σ-algebra of subsets of a space X. Prove that a set M in the space of all bounded measures on $\mathcal{A}$ has compact closure in the topology of convergence on sets from $\mathcal{A}$ precisely when for every uniformly bounded sequence of $\mathcal{A}$-measurable functions f_n pointwise convergent to 0 we have convergence of the integrals of f_n with respect to measures μ to zero uniformly in $\mu \in M$.

HINT: if M is compact, then one can apply Theorem 5.6.5(ii) and Egorov's theorem; if the foregoing condition is fulfilled, then condition (ii) in Theorem 5.6.4 is fulfilled too, hence we can apply Theorem 5.6.5(i).

5.8.27. (Y. Peres) Let us equip the spaces $\mathcal{P}([0,1])$ and $\mathcal{P}([0,1]^2)$ of all Borel probability measures on $[0,1]$ and $[0,1]^2$ with the topology τ_s of convergence on Borel sets. Show that the mapping $\mu \mapsto \mu\otimes\mu$ is sequentially continuous, but is not continuous at the point λ, where λ is Lebesgue measure (the question about this was raised by F. Götze).

HINT: the sequential continuity is obvious from Fubini's theorem and Lebesgue's theorem. In order to verify discontinuity at λ, take the set $A = \bigl\{(x,y)\in[0,1]^2\colon x-y\in\mathbb{Q}\bigr\}$. This set is Borel (as the preimage of a countable set under a continuous mapping) and we have $\lambda\otimes\lambda(A) = 0$. Observe that each neighborhood of λ in the topology τ_s contains a measure $\nu \in \mathcal{P}([0,1])$ such that $\nu\otimes\nu(A) = 1$. For this it suffices to show that for every finite partition $[0,1]$ into Borel parts B_i there exist points $x_i \in B_i$ such that $x_i - x_j \in \mathbb{Q}$ if $i \neq j$. Then for ν one can take $\nu := \sum_{i=1}^{n} \lambda(B_i)\delta_{x_i}$. The required points exist, since the set $B := \prod_{i=1}^{n} B_i$ in $\mathbb{R}^n$ has positive Lebesgue measure, hence $B - B$ contains a ball, so also a point with rational coordinates.

5.8.28. (Schief [**571**]) (i) There exist locally compact spaces X and Y and a continuous open surjection $f\colon X \to Y$ such that the mapping $\widehat{f}\colon \mathcal{P}(X) \to \mathcal{P}(Y)$, $\mu \mapsto \mu\circ f^{-1}$ is open, but not surjective. (ii) Under the continuum hypothesis there exists a Hausdorff space X

and a continuous open surjection $f\colon X \to \mathbb{R}^1$ such that the mapping $\widehat{f}\colon \mathcal{P}(X) \to \mathcal{P}(\mathbb{R}^1)$ is not surjective. In addition, $\widehat{f}$ can be surjective, but not open.

5.8.29. (Schief [**568**], [**569**]) Let X be a Hausdorff space. Show that the mapping $(\mu,\nu) \mapsto \mu - \nu$ is continuous in the A-topology on the set of pairs of nonnegative Borel measures (μ,ν) on X with $\mu - \nu \geqslant 0$. Prove that the mapping $(\mu,\nu) \mapsto \mu + \nu$ on the set of nonnegative Borel measures is open in the A-topology.

5.8.30. (Dellacherie [**163**, Chapter 4, Theorem 31]) Let X be a Polish space, let M be a Souslin subset of the space of probability measures $\mathcal{P}(X)$ with the weak topology, and let $A \subset X$ be a Souslin set such that $\mu(A) = 0$ for all $\mu \in M$. Prove that there exists a Borel set $B \subset X$ such that $A \subset B$ and $\mu(B) = 0$ for all $\mu \in M$.

5.8.31. Let X be a Polish space and let M be a compact subset of the space of probability measures $\mathcal{P}(X)$ with the weak topology.

(i) (Dellacherie [**163**]) Prove that the function $I(E) := \sup_{\mu\in M} \mu^*(E)$ is a Choquet capacity: $I(A) \leqslant I(B)$ if $A \subset B$, $\lim\limits_{n\to\infty} I(A_n) = I(A)$ if A_n are increasing to A, $\lim\limits_{n\to\infty} I(K_n) = I(K)$ if compact sets K_n are decreasing to K. Deduce from this that in case of compact X for every Souslin set $A \subset X$ and every $\varepsilon > 0$ there exists a compact set $K_\varepsilon \subset A$ with $I(K_\varepsilon) > I(A) - \varepsilon$. Observe that the latter can be false for noncompact X (consider l^2 and $M = \{\mu\}$, where $\mu(E) = 0$ if E is a first category set, $\mu(E) = 1$ else).

(ii) (Choquet [**137**]) Let S be a compact or σ-compact set in X such that $\mu(S) = 0$ for all $\mu \in M$. Prove that for every $\varepsilon > 0$, there exists an open set $U \supset S$ such that $\mu(U) < \varepsilon$ for all $\mu \in M$.

(iii) (Choquet [**137**]) Show that under the continuum hypothesis there exists a function $f\colon [0,1] \to [0,1]$ such that its graph S is measurable with respect to every Borel measure on $[0,1]^2$ and every atomless measure vanishes on S. Then the set M of all Borel probability measures on $[0,1]^2$ whose projections on the first factor coincide with Lebesgue measure is compact, but for M and S assertion (ii) is false.

(iv) (Choquet [**137**]) Show that on any uncountable power of $[0,1]$ there exists a sequence of Radon probability measures μ_n weakly convergent to Dirac's measure $\mu_0 = \delta_0$ and a G_δ-set S such that for S and $M = \{\mu_n\}_{n\geqslant 0}$ assertion (ii) is false.

HINT: (i) For every compact set K the function $\mu \mapsto \mu(K)$ is upper semicontinuous. This gives $I(K) = \lim\limits_{n\to\infty} I(K_n)$ for any sequence of compact sets K_n decreasing to K. If sets E_n are increasing to E, then the equality $I(E) = \lim\limits_{n\to\infty} I(E_n)$ is easily deduced by arguing from the contradiction and using the lower continuity of the outer measure. (ii) In case of compact S the assertion is easily deduced from (i) (or is proved directly by a similar reasoning), and if $S = \bigcup_{n=1}^\infty S_n$, where all S_n are compact, then one can take some sets U_n corresponding to S_n and $\varepsilon 2^{-n}$ and let $U = \bigcup_{n=1}^\infty U_n$.

5.8.32. (Kallenberg [**343**]) Suppose that $(X, \mathcal{A})$ is a measurable space, $\mathcal{P}(\mathcal{A})$ is the set of all probability measures on $\mathcal{A}$ equipped with the σ-algebra $\mathcal{F}$ generated by the functions $\mu \mapsto \mu(A)$, $A \in \mathcal{A}$. Given a sequence of $\mathcal{A}\otimes\mathcal{F}$-measurable functions f_n on $X\times\mathcal{P}(\mathcal{A})$, let Λ denote the set of all $\mu \in \mathcal{P}(\mathcal{A})$ for which the sequence of functions $x \mapsto f_n(x,\mu)$ converges in measure μ. Prove that $\Lambda \in \mathcal{F}$.

HINT: suppose first that $|f_n| \leqslant 1$; for fixed n, k, m, the set of all measures μ such that the integral of $|f_n(x,\mu) - f_k(x,\mu)|$ against μ is less than m^{-1} belongs to $\mathcal{F}$. The general case easily reduces to this one.

5.8.33. Let X and Y be Souslin spaces, f a bounded Borel function on $X\times Y$, and $y \mapsto \mu_y$ a Borel mapping to $\mathcal{M}_r(X)$ with the weak topology. Show that the function

$$y \mapsto \int_X f(x,y)\,\mu_y(dx) \quad \text{is Borel measurable on } Y.$$

HINT: use Theorem 5.8.5 and Proposition 5.8.6; alternatively, for a general topological space Y, use the monotone class theorem and the equality $\mathcal{B}(X\times Y)=\mathcal{B}(X)\otimes\mathcal{B}(Y)$.

5.8.34. (Lange [**412**]) Let X be a Polish space and let $\mu\in\mathcal{P}(X)$. (i) Prove that the following subsets of $\mathcal{P}(X)$ are Borel in $\mathcal{P}(X)$ with the weak topology: $\{\nu\in\mathcal{P}(X)\colon\ \nu\ll\mu\}$ and $\{\nu\in\mathcal{P}(X)\colon\ \nu\sim\mu\}$; use that $L^1(\mu)$ is a Polish space.

(ii) If X is locally compact, then the following subsets of $\mathcal{P}(X)$ are Borel as well: (a) measures with compact support, (b) measures with compact connected support, (c) measures with a given closed support, (d) measures with support contained in a given closed set, (e) measures with support containing a given closed set, (f) measures with support without inner points, (g) measures with support without isolated points, (h) measures with support containing at most k points.

For non locally compact spaces, this can be false.

(iii) If $X=\mathbb{R}^n$, then the set of probability measures with convex support and the set of probability measures having a finite moment of a fixed order p are Borel.

5.8.35. (Dubins, Freedman [**182**]) The set of Borel probability measures on $[0,1]$ with uncountable topological support is Souslin, but not Borel in $\mathcal{P}[0,1]$ with the weak topology. Hence the set of measures with countable topological support is not Souslin.

5.8.36. (Weyl [**654**]) Let $\{x_n\}\subset[0,1)$. The following conditions are equivalent:

(i) the sequence $\{x_n\}$ is uniformly distributed with respect to Lebesgue measure on $[0,1)$,

(ii) for each interval $[\alpha,\beta]$ in $[0,1]$, we have $\lim\limits_{N\to\infty} N^{-1}F(N,\alpha,\beta)=\beta-\alpha$, where $F(N,\alpha,\beta)$ is the total number of $n\leqslant N$ such that $\alpha\leqslant x_n<\beta$,

(iii) $\lim\limits_{N\to\infty}\sup\limits_{\alpha,\beta}|N^{-1}F(N,\alpha,\beta)-(\beta-\alpha)|=0$,

(iv) for all integer $m\neq 0$ we have $\lim\limits_{N\to\infty} N^{-1}\sum_{n=1}^N \exp(2\pi i m x_n)=0$.

HINT: the equivalence of (i)–(iii) is seen from the general properties of weak convergence; (iv) follows from (i); (iv) implies (i), since if a measure is a limit point in the weak topology for the sequence of measures $N^{-1}\sum_{n=1}^N\delta_{x_n}$, then it assigns the same integral to each finite linear combination of the functions $\exp(i2\pi m x)$ as Lebesgue measure does, hence it coincides with Lebesgue measure.

5.8.37. (Bogachev, Lukintsova [**92**]) A completely regular space X has property (ud) if every Radon measure on X possesses a uniformly distributed sequence; if such a sequence can be taken uniformly tight, then X has property (tud). Show that both properties are preserved by products with a space where compact sets are metrizable (e.g., a Souslin space). It is unknown whether properties (ud) and (tud) are equivalent.

5.8.38. (de Bruijn, Post [**124**]) Let f be a function on $[0,1]$ such that for every uniformly distributed sequence $\{x_n\}\subset[0,1]$ there exists a finite limit $\lim\limits_{N\to\infty} N^{-1}\sum_{n=1}^N f(x_n)$. Prove that the function f Riemann integrable in the proper sense.

5.8.39. (Losert [**441**]) Let X and Y be compact metric spaces.

(i) Let $\mu\in\mathcal{P}_r(X\times Y)$ and let $\pi_X\colon\ X\times Y\to X$ be the natural projection. Show that if a sequence $\{x_n\}\subset X$ is uniformly distributed with respect to $\mu\circ\pi_X^{-1}$, then Y contains a sequence $\{y_n\}$ such that the sequence (x_n,y_n) is uniformly distributed with respect to μ.

(ii) Show that (i) can be false for nonmetrizable compact spaces even if μ is the product of Radon measures on X and Y.

(iii) Let $\pi\colon X\to Y$ be a continuous surjection, let d be a metric on the space Y, let $\mu\in\mathcal{P}_r(X)$, and let $\nu=\mu\circ\pi^{-1}$. Then, for every sequence $\{y_n\}$ uniformly distributed with respect to ν, there is a sequence $\{x_n\}$ uniformly distributed with respect to μ such that $\lim\limits_{n\to\infty} d\bigl(\pi(x_n),y_n\bigr)=0$.

(iv) Let $\pi\colon X\to Y$ be a continuous surjection and $\mu\in\mathcal{P}_r(X)$. Let $\nu=\mu\circ\pi^{-1}$. Then the following conditions are equivalent: (a) for every sequence $\{y_n\}$ uniformly distributed with respect to ν there is a sequence $\{x_n\}$ uniformly distributed with respect to μ such that $y_n = \pi(x_n)$, (b) the set of all points x possessing neighborhoods the images of which under π are not open has μ-measure zero.

5.8.40. (Losert [**441**]) Show that under the continuum hypothesis there exists a Radon probability measure μ on $[0,1]^{\mathfrak{c}}$ possessing uniformly distributed sequences, but such a sequence cannot be found in the support of μ.

HINT: in $[0,1]^{\mathfrak{c}}$ there is a compact set homeomorphic to $\beta\mathbb{N}$, on $\beta\mathbb{N}$ there is a Radon measure μ without uniformly distributed sequences (Example 5.5.6), but this measure has uniformly distributed sequences in X according to Proposition 5.5.7.

5.8.41. (Losert [**441**]) Show that $\{0,1\}^{\mathfrak{c}}$ contains an everywhere dense set M such that M has no uniformly distributed sequences with respect to the measure μ that is the product of the measure assigning the value $1/2$ to the points 0 and 1.

5.8.42. (Hlawka [**321**]) Let X be a completely regular space such that there is a countable collection of functions f_j in $C_b(X)$ with the property that if for Radon probability measures μ_n and μ for all j we have the equality

$$\lim_{n\to\infty} \int_X f_j\, d\mu_n = \int_X f_j\, d\mu,$$

then the measures μ_n converge weakly to μ. Prove that almost every sequence in X^∞ with the countable power of μ is uniformly distributed with respect to μ.

HINT: by the law of large numbers, for every j the set of points (x_n) for which the sums $N^{-1}\sum_{n=1}^N f_j(x_n)$ tend to the integral of f_j with respect to the measure μ has μ^∞-measure 1.

5.8.43. Let μ_n, where $n\in\mathbb{N}$, be bounded real measures on a σ-algebra $\mathcal{A}$ and let $E_k\in\mathcal{A}$ be disjoint sets such that $\lim\limits_{n\to\infty}\mu_n(E_k) = 0$ for every k and $\inf_n|\mu_n(E_n)| > 0$. Prove that there exists a sequence of indices $\{n_j\}$ for which $\inf_{n\in\{n_j\}}\bigl|\mu_n\bigl(\bigcup_{j=1}^\infty E_{n_j}\bigr)\bigr| > 0$.

HINT: let $\inf_n|\mu_n(E_n)| = \delta$. It suffices to find a sequence $\{n_j\}$ with

$$\sum_{i=1}^{j-1}|\mu_{n_j}(E_{n_i})| < \delta/3, \qquad \sum_{n=n_j}^{\infty}|\mu_{n_{j-1}}(E_n)| < \delta/3,$$

which will give $\bigl|\mu_{n_j}\bigl(\bigcup_{i=1}^\infty E_{n_i}\bigr)\bigr| > \delta/3$. Letting $n_0 = 1$ construct n_j inductively. If $n_1,\ldots,n_j$ are already constructed, then find $n'_{j+1}\geqslant n_j+1$ for which $\sum_{i=1}^j|\mu_n(E_{n_i})| < \delta/3$ for all $n\geqslant n'_{j+1}$. Next find $n_{j+1} > n'_{j+1}$ with $\sum_{n=n_{j+1}}^\infty|\mu_{n_j}(E_n)| < \delta/3$.

5.8.44. (Landers, Rogge [**410**], [**411**]) Suppose that probability measures P_n on a measurable space $(X,\mathcal{A})$ converge in variation to a measure P, a uniformly bounded sequence of $\mathcal{A}$-measurable functions f_n converges in measure P to a $\mathcal{A}$-measurable function f, $\mathcal{A}_n$ are increasing sub-σ-algebras in $\mathcal{A}$ and $\mathcal{A}_0$ is the σ-algebra generated by them. Prove that the conditional expectations $\mathbb{E}_{P_k}^{\mathcal{A}_n} f_m$ with respect to the measures P_k and σ-algebras $\mathcal{A}_n$ converge in measure P to $\mathbb{E}_P^{\mathcal{A}_0} f$ as $k,m,n\to\infty$. See also Gänssler, Pfanzagl [**253**] and Steck [**592**].

5.8.45. Suppose that Radon measures μ_n on a completely regular space X converge weakly to a Radon measure μ. Prove that $\lim\limits_{n\to\infty}\mu_n(B) = \mu(B)$ for every Borel set B if and only if there is a Radon probability measure ν on X such that the measures μ_n have uniformly integrable Radon–Nikodym densities with respect to ν (e.g., uniformly bounded in some $L^p(\nu)$ with $p>1$). The same is true for Baire measures and Baire sets.

HINT: the latter condition is necessary, as explained in § 5.6. If it is fulfilled, then by Theorem 5.6.5 each subsequence in $\{\mu_n\}$ contains a further subsequence that converges setwise. Since the limit can be only μ (Radon measures and Baire measures are uniquely determined by the integrals of bounded continuous functions), the whole sequence converges setwise.

Comments

The theory of weak convergence of measures grew from analysis of distribution functions in probability theory and measure theory, applied and theoretical problems of mathematical statistics, and problems of statistical physics.

Among important works of the earlier period of development of this theory in the XX century, even preceding Kolmogorov's probability theory, one should mention the papers Helly [**316**], Radon [**535**], Bray [**119**], and a number of works by Paul Lévy, including his books [**428**] and [**429**], containing results about convergence of distribution functions. Close to them in the sense of ideas are the works of Gateaux [**258**] (note that in Gateaux's papers published after his death his name was spelled as Gâteaux, the French word meaning "cake", but according to Mazliak [**451**] the spelling in his official documents did not use the circumflex; his Comptes Rendus note of 1913 also did not bear the circumflex; however, Hadamard and Lévy call him Gâteaux in the preface to [**258**]) and Lévy [**427**] on averaging on functional spaces. Let us also mentioned the later book Glivenko [**280**] and the paper Fichtenholz, Kantorovitch [**234**]. Certainly, it would be fair to count the beginning from a yet earlier time; that of one of the most striking results on weak convergence of distributions, the Central Limit Theorem. However, it appears to be more appropriate to call "theory" the results obtained after the arrival of the Lebesgue integral and even after the creation of Kolmogorov's axiomatic probability theory.

The foundations of the modern theory of weak convergence of measures were laid by the outstanding works of A.D. Alexandroff (Aleksandrov) [**9**] and Yu.V. Prohorov [**524**], [**525**], from which a large number of the results in this book are borrowed. As pointed out by A.D. Alexandroff himself, a source of his abstract work in general measure theory was his research [**8**] on the geometry of convex bodies.

The subsequent development of the area was considerably influenced by works of Skorohod [**584**], [**585**], Le Cam [**417**], and Varadarajan [**635**]. Approximately during the same years, or a bit later, important papers of Donsker [**175**], [**176**], Doob [**178**], Erdős, Kac [**205**], Fortet, Mourier [**240**], Minlos [**465**], Sazonov [**564**], and Dudley [**194**] were published.

Beginning from the 1950s, in the theory of weak convergence of measures, apart from a purely probabilistic direction related to the study of asymptotic behavior of random variables, there has been intensive development of the direction laid by the above-mentioned works by A.D. Alexandroff and Yu.V. Prohorov and belonging rather to measure theory and functional analysis, but in many respects furnishing the foundations for the first (probabilistic) direction. The present book is concerned with this second direction.

The fundamentals of the theory of weak convergence of measures on metric spaces are well presented in many books, see the best known text Billingsley [**67**] and also Bergström [**55**], [**56**], Dalecky, Fomin [**153**], Dudley [**193**], Ethier, Kurtz [**208**], Gänssler [**252**], Gänssler, Stute [**254**], Gihman, Skorohod [**273**], Hennequin, Tortrat [**318**], Hoffmann-Jørgensen [**327**], Kruglov [**393**], Parthasarathy [**504**], [**503**], Pollard [**518**], Rotar [**556**], Shiryaev [**581**], Skorohod [**586**], Smolyanov, Shavgulidze [**591**], Stroock [**598**], and Stroock, Varadhan [**599**]. In Bogachev [**81**], Fremlin [**244**] and Vakhania, Tarieladze, Chobanyan [**629**] one can find detailed discussions of weak convergence of measures on topological spaces.

The idea of weak compactness, originating in the theory of functions and functional analysis, was employed in the problems of weak convergence already at an early stage. Radon [**535**] proved that every bounded sequence of signed measures on a compact set in $\mathbb{R}^n$ contains a weakly convergent subsequence (earlier in the one-dimensional case this had been shown by Helly [**316**] in terms of functions of bounded variation). The term "schwach konvergent" — weakly convergent — was used by Radon in [**536**]. The space of measures and weak convergence were applied by Radon in his study of operators adjoint to linear operators on spaces of continuous functions and in potential theory. Bogoliubov and Krylov [**101**] showed that a complete separable metric space is compact precisely when the space of probability measures on it is compact in the weak topology. In addition, for metric spaces with compact balls they established uniform tightness of weakly compact sets of probability measures. The space of probability measures with the weak topology was investigated also in Blau [**71**] (where the A-topology was considered).

Continuity sets of measures on $\mathbb{R}^n$ were introduced in Gunther [**305**, p. 13], Jessen, Wintner [**341**], and Cramér, Wold [**148**]. Romanovsky [**555**] studied locally uniform convergence of characteristic functionals on $\mathbb{R}^n$. Convergence of distribution functions on $\mathbb{R}^n$ was investigated also in Haviland [**314**]. The results in § 1.6 connected with the Lévy continuity theorem (Theorem 1.6.6) go back to the works of Lévy himself (see [**428**]) and Glivenko [**279**], Cramer, Wold [**148**], Romanovsky [**555**], and other researchers. An important role is played here by the Bochner theorem obtained in [**76**]. On the Fourier transform, see Bhattacharia, Ranga Rao [**64**], Bochner [**75**], Esseen [**207**], Grafakos [**288**], Kawata [**355**], Lukacs [**446**], [**447**], Sasvári [**562**], and Ushakov [**624**].

Mathematical statistics during the whole period of its development has remained an important source of problems connected with weak convergence, see, for example, Chentsov [**134**], Cramer [**147**], Dudley [**190**], [**191**], Gänssler [**252**], Giné, Nickl [**275**], Kosorok [**385**], Le Cam [**419**], and Morozova, Chentsov [**471**].

Weak convergence of measures is sometimes called narrow convergence (see, e.g., Bourbaki [**116**], "convergence étroite" in the original), but more often such terminology refers to convergence of duality with the space of continuous functions with compact support on a locally compact space. Note that the theorem of A.D. Alexandroff on weak convergence (Theorems 2.2.5 and 4.3.2) is often called the "portmanteau theorem". In English the word "portmanteau" (originally a French word, meaning a coat-hanger) has the archaic meaning of a large travelling bag and may also denote multipurpose or multifunction objects or concepts. I do not know who invented such a nonsensical name for Alexandroff's theorem.

Propositions 2.4.2, 4.3.6 and Theorems 2.4.4, 2.4.10 were obtained by Prohorov [**525**] for complete separable metric spaces, but their extension to more general cases did not meet any difficulties.

Historical remarks connected with Theorem 2.4.9 are given in Diaconis, Freedman [**167**]. It is necessary to add to the papers cited there the classical paper of Mehler [**455**], where the important formula now bearing his name was derived and, in addition, it was shown that certain multidimensional spherical integrals give in the limit integrals with respect to the Gaussian measure, from which Theorem 2.4.9 follows (in the form of convergence of integrals of cylindrical functions).

Additional results on the Skorohod parametrization of weakly convergent sequences of measures or of the entire space of probability measures can be found in Banakh, Bogachev, Kolesnikov [**31**], Banakh, Bogachev, Kolesnikov [**32**]–[**35**], Bogachev, Kolesnikov [**87**], Berti, Pratelli, Riga [**59**], Choban [**135**], Cuesta-Albertos, Matrán-Bea [**152**], Jakubowski [**337**], Letta, Pratelli [**424**], Lindvall [**433**], Schief [**572**], Szczotka [**603**], Tuero [**620**], and Wichura [**658**]. An interesting approach to parametrization of measures on $\mathbb{R}^n$ was used by Krylov [**397**], who obtained a parametrization with certain differentiability properties. This approach is connected with the Monge–Kantorovich problem. It is clear from § 2.6 and § 5.3 that for the proof of the strong Skorohod property for Polish spaces it suffices to find a compact metric space Z with the strong Skorohod property that can be mapped onto $[0,1]^\infty$ by an open continuous mapping. Such compact sets exist in $\mathbb{R}^3$ (see Anderson [**17**], Aleksandrov, Pasynkov [**10**, Chapter 9] or Fedorchuk, Chigogidze [**216**, Theorem 2.1.12]). Hence it suffices to verify the Skorohod property for the three-dimensional cube. Note that in Blackwell, Dubins [**70**] a very short sketch of the proof of Theorem 2.6.4 is given, but a detailed justification on this way with verification of all details is considerably longer (see Fernique [**226**] and Lebedev [**414**, Chapter 5]).

Investigations of weak compactness in spaces of measures and tightness conditions were considerably influenced by the already-mentioned Prohorov work [**525**], the ideas, methods, and concrete results of which are now presented in textbooks and which have for half a century been successfully applied by many researchers. Although in his paper the fundamental Prohorov theorem was proved for probability measures on complete separable metric spaces, the term "Prohorov theorem" is traditionally applied to numerous later generalizations of the whole theorem or only to its direct or inverse assertions. This is explained by the exceptional importance of the phenomenon discovered in the theorem, whose value in the theory and applications even in the case of very simple spaces is not overshadowed by deep and non-trivial extensions. The tightness of every Borel measure on a complete separable metric space had been known earlier (see Ulam [**622**]). A.D. Alexandroff [**9**] established the "absence of eluding load" for weakly convergent sequences of measures (see Proposition 4.4.6), which yields directly certain partial cases of the Prohorov theorem. The idea to apply weak convergence in l^1 to weak convergence of measures is also due to A.D. Alexandroff [**9**]. Dieudonné [**171**] established the uniform tightness of any weakly convergent sequence of Radon measures on a paracompact locally compact space and constructed an example showing that local compactness alone is not enough. Le Cam [**417**] proved that in the case of a locally compact σ-compact space X, a family of measures is relatively compact in $\mathcal{M}_t(X)$ with the weak topology precisely when it is uniformly tight. He also observed that this follows from the results of Dieudonné [**170**]. The fact that the uniform tightness of a family of measures implies the compactness of its closure in the case of general completely regular spaces was observed by several researchers (L. Le Cam,

P.-A. Meyer, L. Schwartz) soon after the appearance of Prohorov's work and under its influence. The proof of this fact is quite simple, unlike the less obvious converse assertion and the sequential compactness which hold for more narrow classes of spaces. Certainly, the consideration of signed measures brings additional difficulties. Example 4.5.4 is borrowed from Varadarajan [**635**]. Compactness conditions for capacities are considered in O'Brien, Watson [**494**]. Weak convergence and compactness are investigated in an important series of works by Topsøe (see [**610**]–[**616**], in particular, [**612**]). On weak compactness in spaces of measures see also Adamski, Gänssler, Kaiser [**5**], Fernique [**227**], [**228**], Gerard [**263**], [**264**], Haydon [**315**], and Pollard [**515**].

In the theory of Markov processes, for establishing uniform tightness of families of measures, the method of the so-called Lyapunov functions is used, see Nagaev [**476**], [**477**], Hasminskii [**359**], and also the recent books Bogachev, Krylov, Röckner, Shaposhnikov [**91**] and Hennion, Hervé [**319**].

Uniform convergence $\sup_{E\in\mathcal{E}} |P_n(E) - P(E)| \to 0$ on classes of sets $\mathcal{E}$ under weak convergence $P_n \Rightarrow P$ of probability measures (and a similar problem for integrals) is studied in Rao [**541**], Billingsley, Topsøe [**68**] and Bhattacharia, Ranga Rao [**64**, Chapter 1]. The weak topology on spaces of measures on compact spaces and properties of averaging operators are considered in Bade [**28**].

About weak convergence of probability measures on nonseparable metric spaces, see Dudley [**185**], [**187**], and van der Vaart, Wellner [**625**].

In relation to questions discussed in § 2.7(iv), see also Bogachev, Miftakhov [**93**], Grinblat [**297**], [**298**], [**299**], Ivanov [**332**], Sadi [**558**], and Tsukahara [**619**].

The Skorohod space was introduced in Skorohod [**583**], [**584**]. Complete metrics on this space were constructed by Prohorov [**525**] and by Kolmogorov [**376**]. The results in § 2.7(v) on weak convergence of distributions in the space of functions with the metric of convergence in measure can be useful also for the study of processes with trajectories in the Skorohod space D; certainly, here one should replace the natural convergence in D by convergence in measure. Similarly one defines Skorohod spaces of mappings with values in general completely regular spaces, see Jakubowski [**336**], [**338**], Bogachev [**79**], Kallianpur, Xiong [**346**], Kouritzin [**391**]. We refer to these papers and to Afanas'ev [**6**], Bass, Pyke [**43**], Fernández, Gorostiza [**222**], Gorostiza, Rebolledo [**284**], Grigelionis [**293**], Lebedev [**414**], Mandrekar [**450**], Mitoma [**466**], Pakshirajan [**502**], and Woodroofe [**663**] for additional information about Skorohod spaces.

On the invariance principle and related questions, see Afanas'ev [**6**], Billingsley [**67**], Borovkov, Mogulskii, Sakhanenko [**109**], Bulinskii, Shiryaev [**129**].

The Kantorovich–Rubinshtein metric goes back to Kantorovich [**348**]. Later a similar metric (in our notation $d_{\rm FM}$) was used in Fortet, Mourier [**240**] in the study of convergence of empirical distributions. A bit later, in relation to a certain extremal problem, the Kantorovich–Rubinshtein metric (equivalent to the Fortet–Mourier metric) was considered by these authors in [**350**], [**351**] in case of compact metric spaces (in a slightly different form); see also Kantorovich, Akilov [**349**, Chapter VIII, § 4]. The Kantorovich metric is often called the Wasserstein distance under the influence of the well-known paper by Dobrushin [**173**], who effectively used this metric and believed that it had been first introduced in the paper Vasershtein [**637**], where it was indeed applied in the form (3.2.5) with $p = 1$ (see p. 50 of the cited paper). Nevertheless, the historically incorrect name has become popular in the literature along with the distorted transcription of the name of the author of [**637**] (by the way, timely translated into English with the spelling of the author's name Vasershtein), so that with this terminology no author at all stands behind the name "Wasserstein". An extensive literature on this subject is given in Bogachev, Kolesnikov [**88**], Rachev [**529**], [**530**], see also Bernot, Caselles, Morel [**58**], Levin, Milyutin [**426**], Sudakov [**601**], Vershik [**638**], [**642**], Vershik, Kutateladze, Novikov [**645**], and Villani [**649**].

Various duality theorems related to the Kantorovich norm and similar functionals (see Proposition 3.2.6) are discussed in Beiglböck, Leonard, Schachermayer [**49**], [**50**], Beiglböck, Schachermayer [**51**], Edwards [**200**], Kellerer [**357**], Levin [**425**], Rachev, Shortt [**533**], and Ramachandran, Rüschendorf [**539**], [**540**]. Approximations by discrete measures are discussed in Graf, Luschgy [**286**] and other papers cited in § 3.5(iv).

Metrics on various spaces of measures (mostly on subsets of the class of probability measures) have been studied by many authors, see Belili, Heinich [**52**], Boissard, Le Gouic, Loubes [**103**], Dudley [**188**], [**189**], [**193**], Givens, Shortt [**277**], Hauray, Mischler [**312**], Kakosyan, Klebanov, Rachev [**342**], Liese, Vajda [**430**], Neininger, Sulzbach [**480**], Rachev, Klebanov, Stoyanov, Fabozzi [**531**], Rachev, Rüschendorf [**532**], Rachev, Stoyanov, Fabozzi [**534**], Senatov [**575**], Yurinskii [**666**], Yamukov [**664**], Zolotarev [**669**], [**670**], [**671**], [**673**], [**674**], and Zolotarev, Senatov [**675**]. The proof of Theorem 3.2.7 was given in Kantorovich, Rubinshtein [**351**]. Other proofs in more general cases were suggested by several authors, see Fernique [**225**], Szulga [**604**]. The Kantorovich norm on the space of signed measures was considered in Sadovnichii [**559**], Fedorchuk, Sadovnichii [**218**], Hanin [**309**] (note that in [**309**, Proposition 4] it is erroneously asserted that convergence in the Kantorovich norm is equivalent to weak convergence for uniformly bounded sequences of signed measures: see Example 3.2.3).

Isometries of spaces of probability measures with certain metrics such as the Prohorov metric or the Kantorovich metric have been studied in Bertrand, Kloeckner [**61**], [**62**], Gehér, Titkos [**260**], Kloeckner [**370**], and Molnár [**469**]. In particular, it is shown in [**260**] that all isometries of the space of probability measures on a separable Banach space X with respect to the Prohorov metric are generated by affine isometries of X.

The Kantorovich metric of order p was explicitly introduced in Zolotarev [**668**] (for one-dimensional probability distributions, it had been considered earlier, see Dall'Aglio [**155**]), see also Zolotarev [**669**], Rachev [**529**], Dall'Aglio [**156**], Rachev, Rüschendorf [**532**] and references therein. In Kusuoka, Nakayama [**403**], a metric analogous to the L^p-metric of Kantorovich–Rubinshtein type was introduced on the set of pairs (μ, ξ), where μ is a probability measure and ξ is a mapping.

Gromov metric triples and related Gromov–Prohorov type metrics briefly discussed in § 3.4 have become an area of intensive studies over the past two decades. The works cited at the end of § 3.4 contain additional references. A recent account is given in Shioya [**580**].

On Effros measurability in the space $\mathcal{F}(X)$ of closed subsets of a space X mentioned in Exercise 3.5.27, see Christensen [**139**].

In the papers Austin, Edgar, Ionescu Tulcea [**26**] and Baxter [**48**], almost everywhere convergence of random elements ξ_n is characterized as convergence of the integrals of functions $f(\xi_\tau)$ with bounded continuous functions f, where ξ_τ is defined by means of the directed set of stopping times τ.

The fundamental Theorem 5.1.10 on weak sequential completeness was established by A.D. Alexandroff [**9**] for Borel measures on perfectly normal spaces, but a similar proof applies to Baire measures on arbitrary spaces. The proof given in the text is due to Le Cam [**417**]. Theorem 5.1.3 was obtained in Varadarajan [**635**] (another proof was given in Granirer [**289**]). The property of the space of measures on a space X to be Souslin or Luzin in the weak topology under appropriate conditions on X was established in Varadarajan [**636**], Hoffmann-Jørgensen [**323**], Schwartz [**573**], and Oppel [**495**], [**496**]. The fact that the space of signed measures of unit variation norm on a Polish space is Polish in the weak topology was established in Oppel [**496**]. Theorem 5.8.13 is a modification of a result in Hoffmann-Jørgensen [**325**]. Additional results and references concerning properties of spaces of measures and connections with general topology can be found in Banakh [**30**], Banakh, Cauty [**37**], Banakh, Chigogidze, Fedorchuk [**36**], Banakh, Radul [**38**], [**39**], Brown [**122**], Brown, Cox [**123**], Constantinescu [**140**]–[**142**], Fedorchuk [**213**], [**215**], [**214**], Flachsmeyer, Terpe [**238**], Frankiewicz, Plebanek, Ryll-Nardzewski [**243**], Kirk [**363**], [**364**], [**365**], Koumoullis [**386**], Talagrand [**607**], and Valov [**631**].

A number of authors investigated locally convex topologies on the space $C_b(X)$ for which the dual spaces are spaces of measures; these investigations are also connected with the consideration of tight or weakly compact families of measures, see Conway [**143**], Hoffmann-Jørgensen [**324**], Koumoullis [**387**], Miller, Sentilles [**464**], Mosiman, Wheeler [**472**], Sentilles [**578**], Stegall [**593**], and the very informative survey Wheeler [**655**].

It is shown in Mohapl [**468**] that if X is a complete metric space, then the space $\mathcal{M}_r(X)$ of Radon measures coincides with the space of all bounded linear functionals l on the space of bounded Lipschitz functions on X such that the restriction of l to the unit ball in the sup-norm is continuous in the topology of uniform convergence on compact sets.

As has already been noted, Prohorov's work [**525**] had a decisive influence on the development of the theory of weak convergence, and the appearance of the concept of "Prohorov space" illustrates this. It is worth noting that in the literature one can find several different notions of "Prohorov space". Indeed, for generalizations of the Prohorov theorem one has at least the following possibilities: (1) to consider compact families of tight nonnegative Baire measures (as in Definition 4.7.1); (2) to consider compact families of not necessarily tight nonnegative Baire measures; (3) to consider weakly convergent sequences of tight nonnegative Baire measures with tight limits; (4) to consider countably compact families of type (1) or (2); (5) to consider in (1)–(4) completely bounded (i.e., precompact) families instead of compact; (6) to deal with signed measures in place of nonnegative ones. Certainly, there exist other reasonable possibilities. The situation with signed measures is less studied so far.

Prohorov spaces were also studied in Banakh, Bogachev, Kolesnikov [**31**], [**32**], Choban [**135**], [**136**], Cox [**145**], Koumoullis [**388**], [**389**], Mosiman, Wheeler [**472**], and Smolyanov [**589**]. Saint-Raymond [**560**] gave a somewhat simpler proof of the fact that $\mathbb{Q}$ is not a Prohorov space (Theorem 4.8.6). Yet another proof (still highly nontrivial) was suggested in Topsøe [**614**].

In addition to the already cited papers, weak convergence of measures and the weak topology are considered in Adamski [**4**], Baushev [**47**], Borovkov [**107**], Conway [**144**], De Giorgi, Letta [**162**], Donsker [**177**], Doss [**179**], Dudley [**186**], [**188**], Fernique [**223**], [**227**], [**228**], Kallianpur [**345**], Kimme [**361**], [**362**], Lamperti [**406**], [**407**], Léger, Soury [**422**], Mohapl [**467**], Nakanishi [**479**], Neuhaus [**481**], Pachl [**500**], Plebanek, Sobota [**513**], Pollard [**516**], [**517**], Prigarin [**522**], Sentilles [**577**], Stone [**596**], Varadarajan [**632**], [**633**], [**634**], Wells [**652**], Whitt [**656**], Wichura [**657**], and Wilson [**660**]. See Dupuis, Ellis [**198**] on connections of weak convergence of measures with the theory of large deviations.

Young measures were introduced by L.C. Young (see, e.g., [**665**]). They found applications in variational calculus, nonlinear equations, and measure theory, see Castaing, Raynaud de Fitte, Valadier [**132**], Florescu, Godet-Thobie [**239**], Giaquinta, Modica, Souček [**265**], Pedregal [**507**], and Valadier [**628**].

Deep and important results on convergence of measures on sets (in terms of convergence of integrals) were obtained already by Vitaly [**651**], Lebesgue [**415**], Radon [**535**], [**536**] (see also [**537**]), Fichtenholz [**232**], [**231**], [**233**], Hahn [**306**], [**307**], Nikodym [**485**], [**486**], [**487**], and Saks [**561**].

In some problems one has to consider spaces of locally finite measures on a locally compact space M with the topology of duality with $C_0(M)$. For example, the configuration space Γ_M is the set of measures of the form $\gamma = \sum_{n=1}^{\infty} k_n \delta_{x_n}$, where k_n are nonnegative integer numbers and $\{x_n\} \subset M$ has no limit points. On compactness conditions in Γ_M see Pugachev [**528**], Bogachev, Pugachev, Röckner [**94**].

Weak convergence of measures is directly connected with limit theorems in probability theory. There is a vast literature on this subject, see Arak, Zaitsev [**18**], Afanas'ev [**6**], Bingham, Goldie, Teugels [**69**], Borovkov, Mogulskii, Sakhanenko [**109**], Bulinskii [**127**], Bulinskii, Shashkin [**128**], Csörgő, Révész [**151**], Davidson [**158**], Greenwood, Shiryayev [**291**], Korolyuk, Borovskikh [**384**], Kruglov, Korolev [**394**], Nahapetian [**478**]. Convergence of distributions of various special random processes (Gaussian, Markov, martingales, point, etc.) is studied in Baushev [**45**], Borovskikh, Korolyuk [**113**], Brooks, Chacon [**121**], Chentsov [**133**], Daley, Vere-Jones [**154**], Ershov [**206**], Gihman, Skorohod [**273**], Grigelionis, Mikulyavichus [**294**]–[**296**], Jacod, Mémin, Métivier [**334**], Jacod, Shiryaev [**335**], Kallenberg [**344**], Liptser, Shiryaev [**436**], Meyer, Zheng [**460**], Nikunen [**489**], [**490**], Prigent [**523**], Rebolledo [**543**], [**544**], and Saulis, Statulyavichus [**563**].

There are many works on convergence of random elements and measures in linear spaces, see de Acosta [**2**], Araujo, Giné [**19**], Bojdecki, Gorostiza, Ramaswamy [**104**], Borovskikh [**112**], Buldygin [**125**], Buldygin, Solntsev [**126**], Fortet, Mourier [**241**], [**242**], Janson, Kaijser [**339**], Hoffmann-Jørgensen [**326**], Kruglov [**393**], Kuelbs [**398**], [**399**], Kwapień [**405**], Le Cam [**418**], Ledoux, Talagrand [**421**], Linde [**432**], Mourier [**473**], Mushtari [**474**], [**475**], Nguen Zui Tien, Tarieladze, Chobanyan [**483**], Perlman [**509**], Prohorov, Sazonov [**527**], Smolyanov, Fomin [**590**], and Vakhania, Tarieladze, Chobanyan [**629**]. The Fourier transform in infinite-dimensional spaces was introduced by Kolmogorov [**375**], later it was studied in Le Cam [**416**], Mourier [**473**], Prohorov [**525**], [**526**], Gelfand, Vilenkin [**261**], and Gross [**302**] and it became a standard tool for the investigation of weak convergence of measures; for a modern exposition, see Bogachev, Smolyanov [**97**]. Kolmogorov [**377**] made an important observation about estimates of measures of ellipsoids in terms of Fourier transforms, which was already contained in Prohorov [**525**] and on which the theorems of Minlos [**465**] and Sazonov [**564**] were based as well as their generalizations (see § 4.6).

Geometry of spaces of measures with metrics of Kantorovich type has become a very popular subject. Transport problems, transport distances and gradient flows connected with such metrics are studied in Ambrosio, Gigli [**11**], [**12**], Ambrosio, Gigli, Savaré [**13**]–[**16**], Bernard [**57**], Blower [**72**], Dolbeault, Nazaret, Savaré [**174**], Figalli, Gigli [**235**], Gentil [**262**], Gigli, Maas [**271**], Villani [**649**], and Wolansky [**662**]. Barycenters in spaces of measures are considered in Agueh, Carlier [**7**]. Various geometric objects such as curvature on spaces of measures and metric measure spaces are studied in Bertrand, Kloeckner [**61**], [**62**], Biehler, Pfaffelhuber [**65**], Erbar, Kuwada, Sturm [**204**], Gigli [**266**]–[**269**], Gigli, Kuwada, Ohta [**270**], Gigli, Ohta [**272**], Ketterer [**358**], Kloeckner [**372**], Kondo [**381**], Kuwada [**404**], Ledoux [**420**], Löhr [**438**], [**439**], Lott [**443**], Lott, Villani [**444**], Mémoli [**457**], Ozawa, Shioya [**498**], Shioya [**580**], Sturm [**600**], and Takatsu [**605**].

This topic as well as the Kantorovich problem of finding optimal plans mentioned in § 3.2 in case of the distance function has deep and fruitful connections with the general area of Monge and Kantorovich problems. It would be too ambitious to touch on this area in the present book, but at least the formulations of the Monge and Kantorovich problems are worth mentioning here. Let μ and ν be two Radon probability measures on completely regular spaces X and Y. Suppose we are given a nonnegative Borel function h on $X\times Y$, called a cost function. The Monge problem deals with finding the infimum $M_h(\mu,\nu)$ of

$$\int_X h\big(x,T(x)\big)\,\mu(dx)$$

over all Borel mappings $T\colon X\to Y$ such that $\nu=\mu\circ T^{-1}$. If there is a mapping T minimizing this integral, then it is called an optimal mapping. The Kantorovich problem deals with finding the infimum $K_h(\mu,\nu)$ of

$$\int_{X\times Y} h(x,y)\,\sigma(dx\,dy)$$

over all probability measures σ in the set $\Pi(\mu,\nu)$ of Radon probability measures on the product $X\times Y$ with projections μ and ν on the factors. In § 3.2 the function h was a metric. For every mapping T transforming μ into ν we obtain a measure in $\Pi(\mu,\nu)$ defined as the image of μ under the mapping $x\mapsto\big(x,T(x)\big)$. The integral of h with respect to this measure equals the integral of $\big(x,T(x)\big)$ with respect to μ. Hence $K_h(\mu,\nu)\leqslant M_h(\mu,\nu)$. However, even if h is bounded, this inequality can be strict (even if there are transformations of μ into ν). The Kantorovich problem is closer to the area of this book, since it has stronger connections with weak convergence. For example, if h is bounded and continuous, the set $\Pi(\mu,\nu)$ is compact in the weak topology, hence the infimum is attained. Conditions for the existence of Monge optimal mappings are much more restrictive. Surprisingly enough, for bounded continuous functions h and atomless measures μ and ν on Souslin spaces one always has the equality $K_h(\mu,\nu)=M_h(\mu,\nu)$ (see Pratelli [**520**], Lipchius [**435**], and Bogachev, Kalinin, Popova [**86**]). More about Monge and Kantorovich problems can be read in Ambrosio, Gigli [**12**], Bogachev, Kolesnikov [**88**], and Villani [**648**], [**649**], where there is also an extensive bibliography.

Bibliography

[1] Abraham, R., Delmas, J.-F., Hoscheit, P. *A note on the Gromov–Hausdorff–Prokhorov distance between (locally) compact metric measure spaces.*[1] Electron. J. Probab. 2013. V. 18, №14. 21 pp. DOI 10.1214/EJP.v18-2116. MR3035742 [*125*][2]

[2] de Acosta, A. D. *Existence and convergence of probability measures in Banach spaces.* Trans. Amer. Math. Soc. 1970. V. 152. P. 273–298. DOI 10.2307/1995651. MR0267614 [*251*]

[3] Adams, R. A., Fournier, J. J. F. Sobolev spaces. 2nd ed. Elsevier/Academic Press, Amsterdam, 2003; xiv+305 pp. MR2424078 [*127*]

[4] Adamski, W. *An abstract approach to weak topologies in spaces of measures.* Bull. Soc. Math. Grèce (N.S.). 1977. V. 18, №1. P. 28–68. MR528420 [*251*]

[5] Adamski, W., Gänssler, P., Kaiser, S. *On compactness and convergence in spaces of measures.* Math. Ann. 1976. B. 220. S. 193–210. DOI 10.1007/BF01431090. MR0399400 [*227, 248*]

[6] Afanas'ev, V. I. Random walks and branching processes. Lecture notes NEC, Math. Steklov Inst., Moscow, 2007; 187 pp. (in Russian). [*248, 251*]

[7] Agueh, M., Carlier, G. *Barycenters in the Wasserstein space.* SIAM J. Math. Anal. 2011. V. 43, №2. P. 904–924. DOI 10.1137/100805741. MR2801182 [*252*]

[8] Alexandroff (Aleksandrov), A. D. *On the theory of mixed volumes of convex bodies.* Mat. Sbornik. 1937. V. 2, №5. P. 947–972 (in Russian); English transl. in: Alexandrov A. D. Selected works. Part 1. Gordon and Breach, Amsterdam, 1996. [*245*]

[9] Alexandroff, A. D. *Additive set functions in abstract spaces.* Rec. Math. Mat. Sbornik N.S. 1940. V. 8(50). P. 307–348; ibid. 1941. V. 9(51). P. 563–628; ibid. 1943. V. 13(55). P. 169–238. MR0004078 [*148, 154, 155, 157, 158, 160, 194, 245, 247, 250*]

[10] Aleksandrov, P. S., Pasynkov, B. A. Introduction to dimension theory: an introduction to the theory of topological spaces and the general theory of dimension. Izdat. "Nauka", Moscow, 1973; 576 pp. (in Russian). MR0365524 [*247*]

[11] Ambrosio, L., Gigli, N. *Construction of the parallel transport in the Wasserstein space.* Methods Appl. Anal. 2008. V. 15, №1. P. 1–29. DOI 10.4310/MAA.2008.v15.n1.a3. MR2482206 [*252*]

[12] Ambrosio, L., Gigli, N. *A user's guide to optimal transport.* Lecture Notes in Math. 2013. V. 2062. P. 1–155. DOI 10.1007/978-3-642-32160-3_1. MR3050280 [*116, 252*]

[13] Ambrosio, L., Gigli, N., Savaré, G. Gradient flows in metric spaces and in the Wasserstein spaces of probability measures. 2nd ed. Lectures in Mathematics ETH Zürich, Birkhäuser, Basel, 2008; x+334 pp. MR2401600 [*116, 252*]

[14] Ambrosio, L., Gigli, N., Savaré, G. *Density of Lipschitz functions and equivalence of weak gradients in metric measure spaces.* Rev. Mat. Iberoam. 2013. V. 29, №3. P. 969–996. DOI 10.4171/RMI/746. MR3090143 [*252*]

[15] Ambrosio, L., Gigli, N., Savaré, G. *Metric measure spaces with Riemannian Ricci curvature bounded from below.* Duke Math. J. 2014. V. 163, №7. P. 1405–1490. DOI 10.1215/00127094-2681605. MR3205729 [*252*]

[16] Ambrosio, L., Gigli, N., Savaré, G. *Calculus and heat flow in metric measure spaces and applications to spaces with Ricci bounds from below.* Invent. Math. 2014. V. 195, №2. P. 289–391. DOI 10.1007/s00222-013-0456-1. MR3152751 [*252*]

[1] The paper titles are given in italics to distinguish them from the book titles.

[2] In square brackets we indicate all page numbers where the corresponding work is cited.

[17] Anderson, R. D. *Monotone interior dimension-raising mappings.* Duke Math. J. 1952. V. 19. P. 359–366. MR0048798 [*247*]

[18] Arak, T. V., Zaĭtsev, A. Yu. Uniform limit theorems for sums of independent random variables. Trudy Mat. Inst. Steklov. V. 174, Moscow, 1986; 217 pp. (in Russian); English transl.: Proc. Steklov Inst. Math. 1988. V. 174, no. 1; viii+222 pp. MR871856 [*251*]

[19] Araujo, A., Giné, E. The central limit theorem for real and Banach valued random variables. John Wiley, New York – Chichester – Brisbane, 1980; xiv+233 pp. MR576407 [*22, 188, 189, 251*]

[20] Arcones, M. A. *The class of Gaussian chaos of order two is closed by taking limits in distribution*, Advances in stochastic inequalities. AMS special session on stochastic inequalities and their applications, Georgia Inst. Techn., Atlanta, Georgia, USA, October 17–19, 1997, ed. Th. P. Hill et al., Contemp. Math. V. 234, pp. 13–19. Amer. Math. Soc., Providence, Rhode Island, 1999. MR1694760 [*130*]

[21] Areškin, G. Ja. *On the compactness of a family of completely additive set functions.* Leningrad. Gos. Pedagog. Inst. Učen. Zap. 1962. V. 238. P. 102–118 (in Russian). MR0165059 [*225*]

[22] Areškin, G. Ja., Aleksjuk, V. N., Klimkin, V. M. *Certain properties of vector measures.* Leningrad. Gos. Pedagog. Inst. Učen. Zap. 1971. V. 404. P. 298–321 (in Russian). MR0447518 [*225*]

[23] Arkhangel'skiĭ, A. V., Ponomarev, V. I. Fundamentals of general topology. Problems and exercises. Translated from the Russian. Reidel, Dordrecht, 1984; xvi+415 pp. MR785749 [*139, 141, 170, 209*]

[24] Ash, R. B. Probability and measure theory. 2nd ed. Harcourt/Academic Press, Burlington, Massachusetts, 2000; xii+516 pp. MR1810041 [*xi*]

[25] Athreya, S., Löhr, W., Winter, A. *The gap between Gromov-vague and Gromov–Hausdorff-vague topology.* Stoch. Processes Appl. 2016. V. 126. P. 2527–2553. DOI 10.1016/j.spa.2016.02.009. MR3522292 [*125*]

[26] Austin, D. G., Edgar, G. A., Ionescu Tulcea, A. *Pointwise convergence in terms of expectations.* Z. Wahrscheinlichkeitstheorie Verw. Geb. 1974. B. 30. S. 17–26. DOI 10.1007/BF00532860. MR0358945 [*249*]

[27] Avilés, A., Plebanek, G., Rodríguez, J. *A weak* separable $C(K)^*$ space whose unit ball is not weak* separable.* Trans. Amer. Math. Soc. 2014. V. 366, №9. P. 4733–4753. DOI 10.1090/S0002-9947-2014-05962-X. MR3217698 [*233*]

[28] Bade, W. G. The Banach space $C(S)$. Lecture Notes Series, №26. Matematisk Inst., Aarhus Universitet, Aarhus, 1971; ii+154 pp. MR0287293 [*248*]

[29] Balder, E. J. *On ws-convergence of product measures.* Math. Oper. Res. 2001. V. 26, №3. P. 494–518. DOI 10.1287/moor.26.3.494.10581. MR1849882 [*231*]

[30] Banakh, T. O. *Topology of spaces of probability measures. I. The functors P_τ and $\hat{P}$. II. Barycenters of Radon probability measures and the metrization of the functors P_τ and $\hat{P}$.* Matem. Studii (Lviv). 1995. №5. P. 65–87, 88–106 (in Russian). MR1691094, MR1691095 [*205, 250*]

[31] Banakh, T. O., Bogachev, V. I., Kolesnikov, A. V. *On topological spaces with the Prokhorov and Skorokhod properties.* Dokl. Ros. Akad. Nauk. 2001. V. 380, №6. P. 727–730 (in Russian); English transl.: Dokl. Math. 2001. V. 64, №2. P. 244–247. MR1874809 [*247, 250*]

[32] Banakh, T. O., Bogachev, V. I., Kolesnikov, A. V. *Topological spaces with the strong Skorokhod property.* Georgian Math. J. 2001. V. 8, №2. P. 201–220. DOI 10.1016/S0304-0208(04)80153-1. MR1851030 [*218, 247, 250*]

[33] Banakh, T. O., Bogachev, V. I., Kolesnikov, A. V. *Topological spaces with the strong Skorokhod property, II.* In: Functional Analysis and its Applications, V. Kadets and W. Zelazko eds., pp. 23–47. Elsevier, Amsterdam, 2004. DOI 10.1016/S0304-0208(04)80153-1. MR2098868 [*247*]

[34] Banakh, T. O., Bogachev, V. I., Kolesnikov, A. V. *Topological spaces with the Skorohod representation property.* Ukranian. Math. J. 2005. V. 57, №9. P. 1171–1186. DOI 10.1007/s11253-006-0002-z. MR2216039 [*247*]

[35] Banakh, T. O., Bogachev, V. I., Kolesnikov, A. V. *k^*-Metrizable spaces and their applications.* J. Math. Sci. (New York). 2008. V. 155. P. 475–522. DOI 10.1007/s10958-008-9231-z. MR2731965 [*247*]

[36] Banakh, T., Chigogidze, A., Fedorchuk, V. *On spaces of σ-additive probability measures.* Topology Appl. 2003. V. 133, №2. P. 139–155. DOI 10.1016/S0166-8641(03)00041-5. MR1997961 [*250*]

[37] Banakh, T. O., Cauty, R. *Topological classification of spaces of probability measures on co-analytic sets.* Matem. Zametki. 1994. V. 55, №1. P. 9–19 (in Russian); English transl.: Math. Notes. 1994. V. 55, №1–2. P. 8–13. DOI 10.1007/BF02110758. MR1275298 [*250*]

[38] Banakh, T. O., Radul, T. N. *Topology of spaces of probability measures.* Mat. Sbornik. 1997. V. 188, №7. P. 23–46 (in Russian); English transl.: Sbornik Math. 1997. V. 188, №7. P. 973–995. DOI 10.1070/SM1997v188n07ABEH000241. MR1474854 [*250*]

[39] Banakh, T. O., Radul, T. N. *Geometry of mappings of spaces of probability measures.* Matem. Studii (Lviv). 1999. V. 11, №1. P. 17–30 (in Russian). MR1686040 [*205, 250*]

[40] Banaszczyk, W. *The Lévy continuity theorem for nuclear groups.* Studia Math. 1999. V. 136, №2. P. 183–196. DOI 10.4064/sm-136-2-183-196. MR1716173 [*169*]

[41] Banaszczyk, W. *Theorems of Bochner and Lévy for nuclear groups.* In: Research and Exposition in Mathematics, V. 24, pp. 31–44. Heldermann Verlag, Berlin, 2000. MR1858141 [*169*]

[42] Bartoszyński, R. *A characterization of the weak convergence of measures.* Ann. Math. Stat. 1961. V. 32. P. 561–576. DOI 10.1214/aoms/1177705061. MR0125592 [*137*]

[43] Bass, R. F., Pyke, R. *The space $D(A)$ and weak convergence for set-indexed processes.* Ann. Probab. 1985. V. 13. P. 860–884. MR799425 [*248*]

[44] Bauer, H. Probability theory. De Gruyter, Berlin, 1996; xiv+523 pp. Translated from the fourth (1991) German edition. MR1385460 [*xi*]

[45] Baushev, A. N. *On the weak convergence of Gaussian measures.* Teor. Veroyatn. Primen. 1987. V. 32, №4. P. 734–742 (in Russian); English transl.: Theory Probab. Appl. V. 32, №4. P. 670–677. MR927254 [*251*]

[46] Baushev, A. N. *On the weak convergence of probability measures in Orlicz spaces.* Teor. Veroyatn. Primen. 1995. V. 40, №3. P. 495–507 (in Russian); English transl.: Theory Probab. Appl. 1995. V. 40, №3. P. 420–429. DOI 10.1137/1140047. MR1401981 [*83*]

[47] Baushev, A. N. *On the weak convergence of probability measures in a Banach space.* Zap. Nauchn. Sem. S.-Peterburg. Otdel. Mat. Inst. Steklov. 1999. V. 260. Veroyatn. i Statist. 3. P. 17–30 (in Russian); English transl.: J. Math. Sci. (New York). 2002. V. 109, №6. P. 2037–2046. DOI 10.1023/A:1014512929198. MR1759152 [*83, 251*]

[48] Baxter, J. R. *Pointwise in terms of weak convergence.* Proc. Amer. Math. Soc. 1974. V. 46, №3. P. 395–398. DOI 10.2307/2039935. MR0380968 [*249*]

[49] Beiglböck, M., Leonard, C., Schachermayer, W. *A general duality theorem for the Monge–Kantorovich transport problem.* Studia Math. 2012. V. 209, №2. P. 151–167. DOI 10.4064/sm209-2-4. MR2943840 [*249*]

[50] Beiglböck, M., Leonard, C., Schachermayer, W. *On the duality theory for the Monge–Kantorovich transport problem.* In: Optimal transportation, pp. 216–265. London Math. Soc. Lecture Note Ser., V. 413. Cambridge University Press, Cambridge, 2014. DOI 10.1017/CBO9781107297296.010. MR3328997 [*249*]

[51] Beiglböck, M., Schachermayer, W. *Duality for Borel measurable cost functions.* Trans. Amer. Math. Soc. 2011. V. 363, №8. P. 4203–4224. DOI 10.1090/S0002-9947-2011-05174-3. MR2792985 [*249*]

[52] Belili, N., Heinich, H. *Distances de Wasserstein et de Zolotarev.* C. R. Acad. Sci. Paris. Sér. I. 2000. T. 330. P. 811–814. DOI 10.1016/S0764-4442(00)00274-3. MR1769953 [*249*]

[53] Bentkus, V., Götze, F. *Optimal rates of convergence in the CLT for quadratic forms.* Ann. Probab. 1996. V. 24, №1. P. 466–490. DOI 10.1214/aop/1042644727. MR1387646 [*131*]

[54] Bergin, J. *On the continuity of correspondences on sets of measures with restricted marginals.* Econom. Theory. 1999. V. 13, №2. P. 471–481. DOI 10.1007/s001990050265. MR1678819 [*135*]

[55] Bergström, H. Limit theorems for convolutions. Almqvist & Wiksell, Stockholm – Göteborg – Uppsala; John Wiley & Sons, New York – London, 1963; 347 pp. MR0161363 [*246*]

[56] Bergström, H. Weak convergence of measures. Academic Press, New York – London, 1982; 245 pp. MR690579 [*246*]

[57] Bernard, P. *Young measures, superposition and transport.* Indiana Univ. Math. J. 2008. V. 57, №1. P. 247–275. DOI 10.1512/iumj.2008.57.3163. MR2400257 [*252*]

[58] Bernot, M., Caselles, V., Morel, J.-M. Optimal transportation networks. Models and theory. Lecture Notes in Math. V. 1955. Springer-Verlag, Berlin, 2009; x+200 pp. MR2449900 [*249*]

[59] Berti, P., Pratelli, L., Riga, P. *Skorokhod representation on a given probability space.* Probab. Theory Related Fields. 2007. V. 137. P. 277–288. [*247*]

[60] Berti, P., Pratelli, L., Riga, P. *A survey on Skorokhod representation theorem without separability.* Theory Stoch. Proc. 2015. V. 20, №2. P. 1–12. DOI 10.1007/s00440-006-0018-1. MR2278458 [*247*]

[61] Bertrand, J., Kloeckner, B. *A geometric study of Wasserstein spaces: Hadamard spaces.* J. Topol. Anal. 2012. V. 4, №4. P. 515–542. DOI 10.1142/S1793525312500227. MR3021775 [*249, 252*]

[62] Bertrand, J., Kloeckner, B. *A geometric study of Wasserstein spaces: isometric rigidity in negative curvature.* Int. Math. Res. Not. 2016. №5. P. 1368–1386. DOI 10.1093/imrn/rnv177. MR3509929 [*249, 252*]

[63] Besov, O. V., Il'in, V. P., Nikolskiĭ, S. M. Integral representations of functions and imbedding theorems. V. I, II. Translated from the Russian. Winston & Sons, Washington; Halsted Press, New York – Toronto – London, 1978, 1979; viii+345 pp., viii+311 pp. MR0519341, MR0521808 [*127*]

[64] Bhattacharya, R. N., Rao, R. Ranga Normal approximation and asymptotic expansions. Society for Industrial and Applied Mathematics, Philadelphia, 2010; xxii+316 pp. MR3396213 [*37, 97, 246, 248*]

[65] Biehler, H. F., Pfaffelhuber, P. *Compact metric measure spaces and Λ-coalescents coming down from infinity.* ALEA Lat. Am. J. Probab. Math. Stat. 2012. V. 9. P. 269–278. MR2923193 [*252*]

[66] Billingsley, P. Probability and measure. 3d ed. Wiley, New York, 1995; 593 pp. MR1324786 [*xi*]

[67] Billingsley, P. Convergence of probability measures. 2nd ed. Wiley, New York, 1999; x+277 pp. MR1700749 [*x, 85, 89, 94, 245*]

[68] Billingsley, P., Topsøe, F. *Uniformity in weak convergence.* Z. Wahrscheinlichkeitstheorie Verw. Geb. 1967. B. 7. S. 1–16. DOI 10.1007/BF00532093. MR0209428 [*97, 248*]

[69] Bingham, N. H., Goldie, C. M., Teugels, J. L. Regular variation. Cambridge University Press, Cambridge, 1989; xix+494 pp. MR1015093 [*251*]

[70] Blackwell, D., Dubins, L. E. *An extension of Skorohod's almost sure representation theorem.* Proc. Amer. Math. Soc. 1983. V. 89, №4. P. 691–692. DOI 10.2307/2044607. MR718998 [*75, 247*]

[71] Blau, J. H. *The space of measures on a given set.* Fund. Math. 1951. V. 38. P. 23–34. DOI 10.4064/fm-38-1-23-34. MR0047117 [*246*]

[72] Blower, G. Random matrices: high dimensional phenomena. Cambridge University Press, Cambridge, 2009; x+437 pp. MR2566878 [*252*]

[73] Blumberg, A. J., Gal, I., Mandell, M. A., Pancia, M. *Robust statistics, hypothesis testing, and confidence intervals for persistent homology on metric measure spaces.* Found. Comput. Math. 2014. V. 14, №4. P. 745–789. DOI 10.1007/s10208-014-9201-4. MR3230014 [*125*]

[74] Bobkov, S. G. *Proximity of probability distributions in terms of Fourier–Stieltjes transforms.* Uspehi Mat. Nauk. 2016. V. 71, №6. P. 37–98 (in Russian); English tranls.: Russian Math. Surveys. 2016. V. 71, №6. P. 1021–1079. DOI 10.4213/rm9749. MR3588939 [*131*]

[75] Bochner, S. Vorlesungen über Fouriersche Integrale. Leipzig, 1932. English transl.: Lectures on Fourier integrals: with an author's supplement on monotonic functions, Stieltjes integrals, and harmonic analysis. Princeton University Press, Princeton, 1959; 333 pp. MR0107124 [*246*]

[76] Bochner, S. *Monotone Funktionen, Stieltjessche Integrale und harmonische Analyse.* Math. Annal. 1933. B. 108. S. 378–410. DOI 10.1007/BF01452844. MR1512856 [*246*]

[77] Bogachev, V. I. *Locally convex spaces with the CLT property and supports of measures.* Vestnik Moskovsk. Univ. 1986. №6. P. 16–20 (in Russian); English transl.: Moscow Univ. Math. Bull. 1986. V. 41, №6. P. 19–23. MR872068 [*188*]

[78] Bogachev, V. I. Gaussian measures. Amer. Math. Soc., Rhode Island, Providence, 1998; xii+433 pp. MR1642391 [*89, 91*]

[79] Bogachev, V. I. *Measures on topological spaces.* J. Math. Sci. (New York). 1998. V. 91, №4. P. 3033–3156. DOI 10.1007/BF02432851. MR1654901 [*248*]

[80] Bogachev, V. I. *On G. M. Fichtenholz's works in the theory of integral.* Istor.–Matem. Issled. Ser. 2. 2005. №9(44). P. 252–264 (in Russian). [*226*]

[81] Bogachev, V. I. Measure theory. V. 1, 2. Springer, Berlin, 2007; xvii+500 pp., xiii+575 pp. MR2267655 [*ix, 1, 3, 6, 10, 15, 19, 30, 48, 50, 51, 66, 78, 86, 100, 141, 143, 144, 146, 153, 154, 156, 167, 169, 184, 186, 192, 197, 203, 208, 212, 222, 225, 234, 236, 246*]

[82] Bogachev, V. I. Differentiable measures and the Malliavin calculus. Amer. Math. Soc., Providence, Rhode Island, 2010; xvi+488 pp. MR2663405 [*89*]

[83] Bogachev, V. I. *Gaussian measures on infinite-dimensional spaces.* In: Real and Stochastic Analysis. Current Trends (M. M. Rao ed.), pp. 1–83. World Sci., Singapore, 2014. DOI 10.1142/9789814551281_0001. MR3220428 [*89*]

[84] Bogachev, V. I. *Distributions of polynomials on multidimensional and infinite-dimensional spaces with measures.* Uspehi Mat. Nauk. 2016. V. 71, №4. P. 107–154 (in Russian); English transl.: Russian Math. Surveys. 2016. V. 71, №4. P. 703–749. DOI 10.4213/rm9721. MR3588922 [*130*]

[85] Bogachev, V. I., Doledenok, A. N., Shaposhnikov, S. V. *Weighted Zolotarev metrics and the Kantorovich metric.* Dokl. Akad. Nauk. 2017. V. 473, №1. P. 12–16 (in Russian); English transl.: Dokl. Math. 2017. V. 95, №2. P. 113–117. DOI 10.1134/S1064562417020028. MR3701579 [*126*]

[86] Bogachev, V. I., Kalinin, A. N., Popova, S. N. *On the equality of values in the Monge and Kantorovich problems.* Zap. Nauchn. Sem. POMI. 2017. V. 457. P. 53–73 (in Russian); English transl.: J. Math. Sci. (New York). 2018. MR3723576 [*252*]

[87] Bogachev, V. I., Kolesnikov, A. V. *Open mappings of probability measures and the Skorohod representation theorem.* Teor. Veroyatn. Primen. 2001. V. 46, №1. P. 3–27 (in Russian); English transl.: Theory Probab. Appl. 2001. V. 46, №1. P. 20–38. DOI 10.1137/S0040585X97978701. MR1968703 [*75, 77, 186, 205, 211, 247*]

[88] Bogachev, V. I., Kolesnikov, A. V. *The Monge–Kantorovich problem: achievements, connections, and perspectives.* Uspehi Matem. Nauk. 2012. V. 67, №5. P. 3–110 (in Russian); English transl.: Russian Math. Surveys. 2012. V. 67, №5. P. 785–890. DOI 10.1070/RM2012v067n05ABEH004808. MR3058744 [*116, 249, 252*]

[89] Bogachev V. I., Kosov E. D., Popova S. N. *A characterization of Nikolskii–Besov classes via integration by parts.* Dokl. Akad. Nauk. 2017. V. 476, №3. P. 251–255 (in Russian): English transl.: Dokl. Math. 2017. V. 96, №2. P. 449–453. DOI 10.1134/S106456241705012X. MR3791375 [*127*]

[90] Bogachev, V. I., Kosov, E. D., Zelenov, G. I. *Fractional smoothness of distributions of polynomials and a fractional analog of the Hardy–Landau–Littlewood inequality.* Trans. Amer. Math. Soc. 2018. V. 370, №6. P. 4401–4432. DOI 10.1090/tran/7181. MR3811533 [*126, 130*]

[91] Bogachev, V. I., Krylov, N. V., Röckner, M., Shaposhnikov, S. V. Fokker–Planck–Kolmogorov equations. Amer. Math. Soc., Rhode Island, Providence, 2015; xii+480 pp. MR3443169 [*248*]

[92] Bogachev, V. I., Lukintsova, M. N. *On topological spaces possessing uniformly distributed sequences.* Dokl. Akad. Nauk. 2008. V. 418, №5. P. 587–591 (in Russian); English transl.: Dokl. Math. 2008. V. 77, №1. P. 102–106. DOI 10.1134/S1064562408010250. MR2477301 [*222, 243*]

[93] Bogachev, V. I., Miftakhov, A. F. *On weak convergence of finite-dimensional and infinite-dimensional distributions of random processes.* Theory Stoch. Process. 2016. №1. P. 1–11. MR3571407 [*248*]

[94] Bogachev, V. I., Pugachev, O. V., Röckner, M. *Surface measures and tightness of (r,p)-capacities on Poisson space.* J. Funct. Anal. 2002. V. 196. P. 61–86. DOI 10.1006/jfan.2002.3962. MR1941991 [*251*]

[95] Bogachev, V. I., Shaposhnikov, A. V. *Lower bounds for the Kantorovich distance.* Dokl. Akad. Nauk. 2015. V. 460, №6. P. 631–633 (in Russian); English transl.: Dokl. Math. 2015. V. 91, №1. P. 91–93. DOI 10.1134/S1064562415010299. MR3410641 [*126*]

[96] Bogachev, V. I., Smolyanov, O. G. Real and functional analysis: a university course. 2nd ed. Regular and Chaotic Dynamics, Moscow – Izhevsk, 2011; 728 pp. (in Russian; English transl.: Springer, to appear) [*14, 16, 17, 19, 45, 68, 77, 83, 139, 140, 164, 169, 175*]

[97] Bogachev, V. I., Smolyanov, O. G. Topological vector spaces and their applications. Springer, Cham, 2017; x+4560 pp. MR3616849 [*19, 82, 108, 162, 168, 170, 175, 176, 183, 185, 187, 188, 190, 209, 218, 251*]

[98] Bogachev, V. I., Wang, F.-Y., Shaposhnikov, A. V. *Estimates for Kantorovich norms on manifolds.* Dokl. Akad. Nauk. 2015. V. 463, №6. P. 633–638 (in Russian); English transl.: Dokl. Math. 2015. V. 92, №1. P. 494–499. DOI 10.1134/S1064562415040286. MR3443996 [*126*]

[99] Bogachev, V. I., Zelenov G. I. *On convergence in variation of weakly convergent multidimensional distributions.* Dokl. Akad. Nauk. 2015. V. 461, №1. P. 14–17 (in Russian); English transl.: Dokl. Math. 2015. V. 91, №2. P. 138–141. DOI 10.1134/S1064562415020039. MR3442783 [*130*]

[100] Bogachev, V. I., Zelenov, G. I., Kosov, E. D. *Membership of distributions of polynomials in the Nikolskii–Besov class.* Dokl. Akad. Nauk. 2016. V. 469, №6. P. 651–655 (in Russian); English transl.: Dokl. Math. 2016. V. 94, №2. P. 453–457. DOI 10.1134/S1064562416040293. MR3561348 [*126, 130*]

[101] Bogoliouboff, N. N., Kryloff, N. M. *La théorie générale de la mesure dans son application à l'étude de systèmes dynamiques de la mécanique non-linéaire.* Ann. Math. 1937. B. 38. S. 65–113. DOI 10.2307/1968511. MR1503326 [*246*]

[102] Bohman, H. *Approximate Fourier analysis of distribution functions.* Ark. Mat. 1961. V. 4, №2. P. 99–157. DOI 10.1007/BF02592003. MR0126668 [*131*]

[103] Boissard, E., Le Gouic, T., Loubes J.-M. *Distribution's template estimate with Wasserstein metrics.* Bernoulli. 2015. V. 21, №2. P. 740–759. DOI 10.3150/13-BEJ585. MR3338645 [*249*]

[104] Bojdecki, T., Gorostiza, L. G., Ramaswamy, S. *Convergence of $\mathcal{S}'$-valued processes and space-time random fields.* J. Funct. Anal. 1986. V. 66, №1. P. 21–41. DOI 10.1016/0022-1236(86)90078-9. MR829374 [*251*]

[105] Borel, E. Introduction géométrique à quelques théories physiques. Gauthier-Villars, Paris, 1914; vii+137 pp. [*66*]

[106] Borovkov, A. A. *The convergence of distributions of functionals on stochastic processes.* Uspehi Mat. Nauk. 1972. V. 27, №1. P. 3–41 (in Russian); English transl.: Russian Math. Surveys. 1972. V. 27, №1. P. 1–42. MR0400325 [*85*]

[107] Borovkov, A. A. *Convergence of measures and random processes.* Uspehi Mat. Nauk. 1976. V. 31, №2. P. 3–68 (in Russian); English transl.: Russian Math. Surveys. 1976. V. 31, №2. P. 1–69. MR0407921 [*85, 251*]

[108] Borovkov, A. A. Probability theory. Translated from the Russian. Gordon and Breach, Amsterdam, 1998; x+474 pp. (Russian ed.: Moscow, 1986). MR1711261 [*xi*]

[109] Borovkov, A. A., Mogul'skii, A. A., Sakhanenko, A. I. Limit theorems for random processes. Probability theory, 7, p. 5–194, Itogi Nauki i Tekhniki, Vsesoyuz. Inst. Nauchn. i Tekhn. Inform. (VINITI), Moscow, 1995 (in Russian); English transl.: Encyclopaedia of Math. Sci., V. 82 [Probability theory, VII], pp. 5–199. Springer-Verlag, Berlin, 1995. MR1492730 [*248, 251*]

[110] Borovkov, A. A., Sahanenko, A. I. *Remarks on convergence of random processes in nonseparable metric spaces and on the non-existence of a Borel measure for processes in $C(0,\infty)$.* Teor. Verojatnost. i Primenen. 1973. V. 18, №4. P. 812–815 (in Russian); English transl.: Theory Probab. Appl. 1974. V. 18, №4. P. 774–777. MR0328984 [*88*]

[111] Borovkov, K. A. *On the convergence of projections of uniform distributions on balls.* Teor. Verojatnost. i Primenen. 1990. V. 35, №3. P. 547–551 (in Russian); English transl.: Theory Probab. Appl. 1990. V. 35, №3. P. 546–550. MR1091211 [*98*]

[112] Borovskikh, Yu. V. U-statistics in Banach spaces. VSP, Utrecht, 1996; xii+420 pp. MR1419498 [*251*]

[113] Borovskikh, Yu. V., Korolyuk, V. S. Martingale approximation. VSP, Utrecht, 1997; xii+322 pp. MR1640099 [*251*]

[114] Borsuk, K. Theory of retracts. Panstw. Wydawn. Nauk., Warsaw, 1967; 251 pp. MR0216473 [*95*]

[115] Boulicaut, B. *Convergence cylindrique et convergence étroite d'une suite de probabilités de Radon.* Z. Wahrscheinlichkeitstheorie Verw. Geb. 1973. B. 28, S. 43–52. DOI 10.1007/BF00549293. MR0357725 [*169*]

[116] Bourbaki, N. Integration. II. Chapters 7–9. Springer, Berlin, 2004; viii+326 pp. Translated from the 1963 and 1969 French originals. MR2098271 [*246*]

[117] Bouziad, A. *Coincidence of the upper Kuratowski topology with the co-compact topology on compact sets, and the Prohorov property.* Topology Appl. 2002. V. 120. P. 283–299. DOI 10.1016/S0166-8641(01)00081-5. MR1897264 [*186*]

[118] Bovier, A. Gaussian processes on trees: from spin glasses to branching Brownian motion. Cambridge University Press, Cambridge, 2017; x+200 pp. MR3618123 [*43*]

[119] Bray, H. E. *Elementary properties of the Stieltjes integral.* Ann. Math. (2). 1918–1919. V. 20. P. 177–186. DOI 10.2307/1967867. MR1502551 [*245*]

[120] Brooks, J. K., Chacon, R. V. *Continuity and compactness of measures.* Adv. Math. 1980. V. 37, №1. P. 16–26. DOI 10.1016/0001-8708(80)90023-7. MR585896 [*225*]

[121] Brooks, J. K., Chacon, R. V. *Weak convergence of diffusions, their speed measures and time changes.* Adv. Math. 1982. V. 46, №2. P. 200–216. DOI 10.1016/0001-8708(82)90023-8. MR679908 [*251*]

[122] Brown, J. B. *Baire category in spaces of probability measures.* Fund. Math. 1977. V. 96, №3. P. 189–193. DOI 10.4064/fm-96-3-189-193. MR0444878 [*250*]

[123] Brown, J. B., Cox, G. V. *Baire category in spaces of probability measures. II.* Fund. Math. 1984. V. 121, №2. P. 143–148. DOI 10.4064/fm-121-2-143-148. MR765330 [*250*]

[124] de Bruijn, N. G., Post, K. A. *A remark on uniformly distributed sequences and Riemann integrability.* Indag. Math. 1968. V. 30. P. 149–150. MR0225946 [*243*]

[125] Buldygin, V. V. Convergence of random elements in topological spaces. Naukova Dumka, Kiev, 1980; 239 pp. (in Russian). MR734899 [*251*]

[126] Buldygin, V. V., Solntsev, S. A. Asymptotic behaviour of linearly transformed sums of random variables. Translated from the Russian. Kluwer, Dordrecht, 1997; xvi+500 pp. MR1471203 [*251*]

[127] Bulinskii, A. V. Limit theorems under conditions of weak dependence. Moscow State University, Moscow, 1989; 135 pp. (in Russian). [*251*]

[128] Bulinski, A., Shashkin, A. Limit theorems for associated random fields and related systems. World Sci., Hackensack, New Jersey, 2007; xii+436 pp. MR2375106 [*251*]

[129] Bulinskii, A. V., Shiryaev, A. N. Theory of random processes. Fizmatlit, Moscow, 2003; 400 pp. (in Russian) [*91, 248*]

[130] Burago, D., Burago, Y., Ivanov, S. A course in metric geometry. Amer. Math. Soc., Providence, Rhode Island, 2001; xiv+415 pp. MR1835418 [*123*]

[131] Cantelli, F.P. *Sulla determinazione empirica delle leggi di probabilita.* Giornale dell'Istituto Italiano degli Attuari. 1933. V. 4. P. 221–424. [*193*]

[132] Castaing, C., Raynaud de Fitte, P., Valadier, M. Young measures on topological spaces. With applications in control theory and probability theory. Kluwer, Dordrecht, 2004; xi+320 pp. MR2102261 [*230, 251*]

[133] Chentsov, N. N. *Weak convergence of stochastic processes whose trajectories have no discontinuities of the second kind and the "heuristic" approach to the Kolmogorov–Smirnov tests".* Teor. Veroyatn. Primen. 1956. V. 1. P. 154–161 (in Russian); English transl.: Theory Probab. Appl. 1956. V. 1, №1. P. 140–144. DOI 10.1137/1101013. MR0084220 [*251*]

[134] Chentsov (Čencov), N. N. Statistical decision rules and optimal inference. Translated from the Russian. Amer. Math. Soc., Providence, Rhode Island, 1982; viii+499 pp. (Russian ed.: Moscow, 1972). MR645898 [*246*]

[135] Choban, M. M. *Spaces, mappings and compact subsets.* Bul. Acad. Ştiinţe Rep. Moldova. 2001. №2(36). P. 3–52. MR1973595 [*186, 247, 250*]

[136] Choban, M. M. *Mappings and Prohorov spaces.* Topology Appl. 2006. V. 153, №13. P. 2320–2350. DOI 10.1016/j.topol.2005.07.003. MR2238733 [*250*]

[137] Choquet, G. *Sur les ensembles uniformément négligéables.* Séminaire Choquet, 9e année. 1970. №6. 15 pp. [*137, 182, 202, 242*]

[138] Chow, Y. S., Teicher, H. Probability theory. Independence, interchangeability, martingales. 3d ed. Springer-Verlag, New York, 1997; xxii+488 pp. MR1476912 [*xi*]

[139] Christensen, J. P. R. Topology and Borel structure. North-Holland, Amsterdam – London, Amer. Elsevier, New York, 1974; 133 pp. MR0348724 [*138, 249*]

[140] Constantinescu, C. Duality in measure theory. Lecture Notes in Math. V. 796. Springer, Berlin, 1980; 197 pp. MR574273 [*250*]

[141] Constantinescu, C. *Spaces of measures on topological spaces.* Hokkaido Math. J. 1981. V. 10. P. 89–156. MR662298 [*250*]

[142] Constantinescu, C. Spaces of measures. De Gruyter, Berlin – New York, 1984; 444 pp. MR748135 [*250*]

[143] Conway, J. *The strict topology and compactness in the space of measures.* Trans. Amer. Math. Soc. 1967. V. 126. P. 474–486. DOI 10.2307/1994310. MR0206685 [*250*]

[144] Conway, J. *A theorem on sequential convergence of measures and some applications.* Pacific J. Math. 1969. V. 28. P. 53–60. MR0238068 [*251*]

[145] Cox, G. V. *On Prohorov spaces.* Fund. Math. 1983. V. 116, №1. P. 67–72. DOI 10.4064/fm-116-1-67-72. MR713161 [*184, 250*]

[146] Cramér, H. Random variables and probability distributions. Cambridge University Press, Cambridge, 1937; viii+119 pp. (2nd ed.: 1962). MR0165599 [*xi*]

[147] Cramér, H. Mathematical methods of statistics. Princeton University Press, Princeton, 1946; xvi+575 pp. MR0016588 [*246*]

[148] Cramér, H., Wold, H. *Some theorems on distribution functions.* J. London Math. Soc. 1936. V. 11. P. 290–294. DOI 10.1112/jlms/s1-11.4.290. MR1574927 [*246*]

[149] Cremers, H., Kadelka, D. *On weak convergence of stochastic processes with Lusin path spaces.* Manuscripta Math. 1984. V. 45. P. 115–125. DOI 10.1007/BF01169769. MR724730 [*85*]

[150] Cremers, H., Kadelka, D. *On weak convergence of integral functionals of stochastic processes with applications to processes taking paths in* L_p^E. Stoch. Processes Appl. 1986. V. 21. P. 305–317. DOI 10.1016/0304-4149(86)90102-X. MR833957 [*85*]

[151] Csörgő, M., Révész, P. Strong approximations in probability and statistics. Academic Press, New York – London, 1981; 284 pp. MR666546 [*251*]

[152] Cuesta-Albertos, J. A., Matrán-Bea, C. *Stochastic convergence through Skorohod representation theorems and Wasserstein distances.* First International Conference on Stochastic Geometry, Convex Bodies and Empirical Measures (Palermo, 1993). Rend. Circ. Mat. Palermo (2). Suppl. No. 35, pp. 89–113, 1994. MR1297788 [*247*]

[153] Dalecky, Yu. L., Fomin, S. V. Measures and differential equations in infinite-dimensional space. Translated from the Russian. Kluwer, Dordrecht, 1991; xvi+337 pp. MR1140921 [*246*]

[154] Daley, D. J., Vere-Jones, D. An Introduction to the theory of point processes. V. I: Elementary theory and methods; V. II: General theory and structure. Springer, 2003, 2008; xxii+469 pp., xviii+573 pp. MR1950431, MR2371524 [*251*]

[155] Dall'Aglio, G. *Sugli estremi dei momenti delle funzioni di ripartizioni doppia.* Ann. Scuola Norm. Super. (3). 1956. V. 10. P. 33–74. MR0081577 [*110, 249*]

[156] Dall'Aglio, G. *Fréchet classes: the beginnings.* Advances in probability distributions with given marginals, pp. 1–12, Math. Appl., 67, Kluwer, Dordrecht, 1991. MR1215943 [*249*]

[157] D'Aristotile, A., Diaconis, P., Freedman, D. *On merging of probabilities.* Sankhyā. Ser. A. 1988. V. 50, №3. P. 363–380. MR1065549 [*135*]

[158] Davidson, J. Stochastic limit theory. An introduction for econometricians. The Clarendon Press, Oxford University Press, New York, 1994; xxii+539 pp. MR1430804 [*251*]

[159] Davies, R.O. *A non-Prokhorov space.* Bull. London Math. Soc. 1971. V. 3. P. 341–342; Addendum ibid. 1972. V. 4. P. 310. DOI 10.1112/blms/3.3.341. MR0293579 [*182*]

[160] Davydov, Y., Rotar, V. *On asymptotic proximity of distributions.* J. Theoret. Probab. 2009. V. 22, №1. P. 82–98. DOI 10.1007/s10959-008-0178-2. MR2472006 [*137*]

[161] Day, M. M. Normed linear spaces. 3d ed. Springer-Verlag, New York – Heidelberg, 1973; viii+211 pp. MR0344849 [*183*]

[162] De Giorgi, E., Letta, G. *Une notion générale de convergence faible pour des fonctions croissantes d'ensemble.* Ann. Scuola Norm. Sup. Pisa Cl. Sci. (4). 1977. V. 4, №1. P. 61–99. MR0466479 [*251*]

[163] Dellacherie, C. Un cours sur les ensembles analytiques. In: Analytic sets, pp. 184–316. Academic Press, New York, 1980. MR0608794 [*242*]

[164] Dembski, W. A. *Uniform probability.* J. Theoret. Probab. 1990. V. 3, №4. P. 611–626. DOI 10.1007/BF01046100. MR1067671 [*98*]

[165] Depperschmidt, A., Greven, A., Pfaffelhuber, P. *Marked metric measure spaces.* Electron. Commun. Probab. 2011. V. 16. P. 174–188. DOI 10.1214/ECP.v16-1615. MR2783338 [*125*]

[166] Dereich, S., Scheutzow, M., Schottstedt, R. *Constructive quantization: approximation by empirical measures.* Ann. Inst. H. Poincaré. Probab. Stat. 2013. V. 49. P. 1183–1203. DOI 10.1214/12-AIHP489. MR3127919 [*133*]

[167] Diaconis, P., Freedman, D. *A dozen de Finetti-stile results in search of a theory.* Ann. Inst. H. Poincaré. Probab. et Statist. 1987. V. 23, suppl. №2. P. 397–423. MR898502 [*98, 247*]

[168] Diestel, J. Geometry of Banach spaces – selected topics. Lecture Notes in Math. V. 485. Springer-Verlag, Berlin – New York, 1975; xi+282 pp. MR0461094 [*83*]

[169] Dieudonné, J. *Sur le théorème de Lebesgue–Nikodým. III.* Ann. Inst. Fourier (Grenoble) 1947–1948. V. 23. P. 25–53. MR0028924 [*228*]

[170] Dieudonné, J. *Sur la convergence des suites des mesures de Radon.* An. Acad. Brasil. Cienc. 1951. V. 23. P. 21–38; Addition: ibid., P. 277–282. MR0042496 [*226, 247*]

[171] Dieudonné, J. *Sur le produit de composition.* Compositio Math. 1954. V. 12. P. 17–34. MR0064323 [*247*]

[172] Ditor, S. Z., Eifler, L. Q. *Some open mapping theorems for measures.* Trans. Amer. Math. Soc. 1972. V. 164. P. 287–293. DOI 10.2307/1995975. MR0477729 [*205, 208*]

[173] Dobrushin, R. L. *Prescribing a system of random variables by conditional distributions.* Teor. Veroyat. Primen. 1970. V. 15, №3. P. 469–497 (in Russian); English transl.: Theory Probab. Appl. 1970. V. 15. P. 458–486. MR0298716 [*248*]

[174] Dolbeault, J., Nazaret, B., Savaré, G. *A new class of transport distances between measures.* Calc. Var. Partial Differ. Equ. 2009. V. 34, №2. P. 193–231. DOI 10.1007/s00526-008-0182-5. MR2448650 [*252*]

[175] Donsker, M. D. *An invariance principle for certain probability limit theorems.* Memoirs Amer. Math. Soc. 1951. V. 6. P. 1–12. MR0040613 [*93, 245*]

[176] Donsker, M. D. *Justification and extension of Doob's heuristic approach to the Kolmogorov–Smirnov theorems.* Ann. Math. Statist. 1952. V. 23. P. 277–281. MR0047288 [*245*]

[177] Donsker, M. D. *On the weak convergence of stochastic processes.* Math. Scand. 1961. V. 9. P. 43–54. DOI 10.7146/math.scand.a-10621. MR0133865 [*251*]

[178] Doob, J. L. *Heuristic approach to the Kolmogorov–Smirnov theorems.* Ann. Math. Statist. 1949. V. 20. P. 393–403. MR0030732 [*245*]

[179] Doss, S. *Sur la convergence stochastique dans les espaces uniformes.* Ann. Sci. École Norm. Super. 1954. V. 71. P. 87–100. MR0067395 [*251*]

[180] Drewnowski, L. *Equivalence of Brooks–Jewett, Vitali–Hahn–Saks, and Nikodym theorems.* Bull. Acad. Polon. Sci. Sér. Sci. Math., Astron. et Phys. 1972. V. 20. P. 725–731. MR0311869 [*228*]

[181] Drewnowski, L. *The metrizable linear extensions of metrizable sets in topological linear spaces.* Proc. Amer. Math. Soc. 1975. V. 51. P. 323–329. DOI 10.2307/2040316. MR0380336 [*134*]

[182] Dubins, L., Freedman, D. *Measurable sets of measures.* Pacific J. Math. 1964. V. 14. P. 1211–1222. MR0174687 [*243*]

[183] Dubrovskiĭ, V. M. *On some properties of completely additive set functions and their applications to a generalization of a theorem of H. Lebesgue.* Rec. Math. Mat. Sbornik. 1947. V. 20. P. 317–330 (in Russian). MR0021074 [*228*]

[184] Dubrovskiĭ, V. M. *On certain conditions of compactness.* Izv. Akad. Nauk SSSR. 1948. V. 12. P. 397–410 (in Russian). MR0026109 [*228*]

[185] Dudley, R. M. *Weak convergence of probabilities on nonseparable metric spaces and empirical measures on Euclidean spaces.* Illinois J. Math. 1966. V. 10. P. 109–126. MR0185641 [*248*]

[186] Dudley, R. M. *Convergence of Baire measures.* Studia Math. 1966. V. 27. P. 251–268; Addendum ibid. 1974. V. 51. P. 275. DOI 10.4064/sm-27-3-251-268. MR0200710 [*251*]

[187] Dudley, R. M. *Measures on nonseparable metric spaces.* Illinois J. Math. 1967. V. 11. P. 449–453. MR0235087 [*248*]

[188] Dudley, R. M. *Distances of probability measures and random variables.* Ann. Math. Statist. 1968. V. 39, №5. P. 1563–1572. DOI 10.1007/978-1-4419-5821-1_4. MR0230338 [*249, 251*]

[189] Dudley, R. M. Probability and metrics. Convergence of laws on metric spaces, with a view to statistical testing. Lecture Notes Series N 45, Matematisk Institut, Aarhus University, Aarhus, 1976; ii+126 pp. MR0488202 [*249*]

[190] Dudley, R. M. *A course on empirical processes.* Lecture Notes in Math. 1984. V. 1097. P. 2–141. DOI 10.1007/BFb0099432. MR876079 [*246*]

[191] Dudley, R. M. *An extended Wichura theorem, definition of Donsker class, and weighted empirical distributions.* Lecture Notes in Math. 1985. V. 1153. P. 141–178. DOI 10.1007/BFb0074949. MR821980 [*246*]

[192] Dudley, R. M. Uniform central limit theorems. 2nd ed. Cambridge University Press, New York, 1999; xii+472 pp. MR1720712 [*193*]

[193] Dudley, R. M. Real analysis and probability. 2nd ed. Cambridge University Press, Cambridge, 2003; viii+555 pp. MR1932358 [*xi, 135, 246, 249*]

[194] Dudley, R. M. Selected works of R. M. Dudley. Springer, New York, 2010; xxiv+481 pp. MR2768667 [*245*]

[195] Dudley, R. M., Giné, E., Zinn, J. *Uniform and universal Glivenko–Cantelli classes.* J. Theor. Probab. 1991. V. 4. P. 485–510. DOI 10.1007/BF01210321. MR1115159 [*193*]

[196] Dugundji, J. *An extension of Tietze's theorem.* Pacif J. Math. 1951. V. 1. P. 353–367. MR0044116 [*95*]

[197] Dunford, N., Schwartz, J. T. Linear operators, I. General theory. Interscience, New York, 1958; xiv+858 pp. MR0117523 [*16, 19, 112*]

[198] Dupuis, P., Ellis, R. S. A weak convergence approach to the theory of large deviations. John Wiley & Sons, New York, 1997; xviii+479 pp. MR1431744 [*251*]

[199] Edwards, D. A. *The structure of superspace.* In: Studies in topology (Proc. Conf., University North Carolina, Charlotte, 1974), pp. 121–133. Academic Press, New York, 1975. MR0401069 [*125*]

[200] Edwards, D. A. *On the Kantorovich–Rubinshtein theorem.* Expos. Math. 2011. V. 29. P. 387–398. DOI 10.1016/j.exmath.2011.06.005. MR2861765 [*249*]

[201] Edwards, R. E. Functional analysis. Theory and applications. Holt, Rinehart and Winston, New York – Toronto – London, 1965; xiii+781 pp. MR0221256 [*139*]

[202] Eifler, L. Q. *Open mapping theorems for probability measures on metric spaces.* Pacif. J. Math. 1976. V. 66. P. 89–97. MR0453960 [*205*]

[203] Engelking, P. General topology. Polish Sci. Publ., Warszawa, 1977; 626 pp. MR0500780 [*45, 77, 139, 140, 141, 160, 170, 174, 186, 195, 208, 210, 214, 228*]

[204] Erbar, M., Kuwada, K., Sturm, K.-T. *On the equivalence of the entropic curvature-dimension condition and Bochner's inequality on metric measure spaces.* Invent. Math. 2015. V. 201, №3. P. 993–1071. DOI 10.1007/s00222-014-0563-7. MR3385639 [*252*]

[205] Erdős, P., Kac, M. *On certain limit theorems in the theory of probability.* Bull. Amer. Math. Soc. 1946. V. 52. P. 292–302. DOI 10.1090/S0002-9904-1946-08560-2. MR0015705 [*245*]

[206] Ershov, M. P. *On weak compactness of continuous semi-martingales.* Teor. Veroyatn. Primen. 1979. V. 24, №1. P. 91–105 (in Russian); English transl.: Theory Probab. Appl. 1979. V. 24, №1. P. 91–106. MR522239 [*251*]

[207] Esseen, C.-G. *Fourier analysis of distribution functions. A mathematical study of the Laplace–Gaussian law.* Acta Math. 1945. V. 77. P. 1–125. DOI 10.1007/BF02392223. MR0014626 [*246*]

[208] Ethier, S. N., Kurtz, T. G. Markov processes. Characterization and convergence. John Wiley & Sons, New York, 1986; x+534 pp. MR838085 [*246*]

[209] Evans, C., Gariepy, R. F. Measure theory and fine properties of functions. CRC Press, Boca Raton – London, 1992; viii+268 pp. MR1158660 [*41*]

[210] Evans, S. N. Probability and real trees. Lecture Notes in Math. V. 1920. Springer, Berlin, 2008; xii+193 pp. MR2351587 [*125*]

[211] Evans, S. N., Molchanov, I. S. *The semigroup of metric measure spaces and its infinitely divisible probability measures.* Trans. Amer. Math. Soc. 2017. V. 369. P. 1797–1834. DOI 10.1090/tran/6714. MR3581220 [*125*]

[212] Evans, S. N., Winter, A. *Subtree prune and regraft: a reversible real tree-valued Markov process.* Ann. Probab. 2006. V. 34, №3. P. 918–961. DOI 10.1214/009117906000000034. MR2243874 [*125*]

[213] Fedorchuk, V. V. *Probability measures in topology.* Uspehi Mat. Nauk. 1991. V. 46, №1. P. 41–80 (in Russian); English transl.: Russian Math. Surveys. 1991. V. 46. P. 45–93. DOI 10.1070/RM1991v046n01ABEH002722. MR1109036 [*250*]

[214] Fedorchuk, V. V. *Functors of probability measures in topological categories.* J. Math. Sci. (New York). 1998. V. 91, №4. P. 3157–3204. DOI 10.1007/BF02432852. MR1654905 [*250*]

[215] Fedorchuk, V. V. *On the topological completeness of spaces of measures.* Izv. Ross. Akad. Nauk Ser. Mat. 1999. V. 63, №4. P. 207–223 (in Russian); English transl.: Izv. Math. 1999. V. 63, №4. P. 827–843. DOI 10.1070/im1999v063n04ABEH000253. MR1717684 [*250*]

[216] Fedorchuk, V. V., Chigogidze, A. Ch. Absolute retracts and infinite-dimensional manifolds. Nauka, Moscow, 1992; 232 pp. (in Russian). MR1202238 [*247*]

[217] Fedorchuk, V. V., Filippov, V. V. General topology. Basic constructions. Moskov. Gos. Univ., Mocsow, 1988; 252 pp. (in Russian). [*76, 139, 209, 234*]

[218] Fedorchuk, V. V., Sadovnichiĭ, Yu. V. *On some topological and categorical properties of measures with alternating signs.* Fundam. Prikl. Mat. 1999. V. 5, №2. P. 597–618 (in Russian). MR1803602 [*249*]

[219] Faĭnleĭb, A. S. *A generalization of Esseen's inequality and its application in probabilistic number theory.* Izvest. AN SSSR. 1968. V. 32, №4. P. 859–879 (in Russian); English transl.: Math. USSR-Izv. 1968. V. 2, №4. P. 821–844. MR0238782 [*131*]

[220] Feldman, J. *A short proof of the Levy continuity theorem in Hilbert space.* Israel J. Math. 1965. V. 3, №2. P. 99–103. DOI 10.1007/BF02760035. MR0192541 [*169*]

[221] Feller, W. An introduction to probability theory and its applications. V. I, II. 2nd ed. John Wiley & Sons, New York, 1957, 1971; xv+461 pp., xxiv+669 pp. MR0088081, MR0270403 [*xi, 40*]

[222] Fernández, B., Gorostiza, L. G. *A criterion of convergence of generalized processes and an application to a supercritical branching particle system.* Canad. J. Math. 1991. V. 43, №5. P. 985–997. DOI 10.4153/CJM-1991-055-2. MR1138577 [*248*]

[223] Fernique, X. *Processus linéaires, processus généralisés.* Ann. Inst. Fourier (Grenoble). 1967. V. 17. P. 1–92. MR0221576 [*176, 182, 251*]

[224] Fernique, X. *Généralisation du théorème de P. Lévy.* C. R. Acad. Sci. Paris, Sér. A. 1968. T. 266. P. 25–28. MR0223848 [*169*]

[225] Fernique, X. *Sur le théorème de Kantorovitch–Rubinstein dans les espaces polonais.* Lecture Notes in Math. 1981. V. 850. P. 6–10. MR622552 [*249*]

[226] Fernique, X. *Un modèle presque sûr pour la convergence en loi.* C. R. Acad. Sci. Paris, Sér. 1. 1988. T. 306. P. 335–338. MR934613 [*75, 247*]

[227] Fernique, X. *Convergence en loi de fonctions aléatoires continues ou cadlag, propriétés de compacité des lois.* Lecture Notes in Math. 1991. V. 1485. P. 178–195. DOI 10.1007/BFb0100856. MR1187780 [*248, 251*]

[228] Fernique, X. *Convergence en loi de variables aléatoires et de fonctions aléatoires, propriétés de compacité des lois. II.* Lecture Notes in Math. 1993. V. 1557, №12. P. 216–232. DOI 10.1007/BFb0087978. MR1308567 [*248, 251*]

[229] Fernique, X. Fonctions aléatoires gaussiennes, vecteurs aléatoires gaussiens. Université de Montréal, Centre de Recherches Mathématiques, Montréal, 1997; iv+217 pp. MR1472975 [*89*]

[230] Ferrando, J. C., Sánchez Ruiz, L. M. *A survey on recent advances on the Nikodým boundedness theorem and spaces of simple functions.* Rocky Mountain J. Math. 2004. V. 34, №1. P. 139–172. DOI 10.1216/rmjm/1181069896. MR2061122 [*228*]

[231] Fichtenholz, G. *Notes sur les limites des fonctions représentées par des intégrales définies.* Rend. Circ. Mat. Palermo. 1915. T. 40. P. 153–166. [*63, 251*]

[232] Fichtenholz, G. M. Theory of simple integrals dependent on a parameter. Petrograd, 1918; vii+333 pp. (in Russian). [*226, 251*]

[233] Fichtenholz, G. *Sur les suites convergentes des intégrales définies.* Bulletin International de l'Académie Polonaise des Sciences et des Lettres. Classe des Sciences Mathématiques et Naturelles. Sér. A: Sci. Math. Année 1923. P. 91–117. Cracovie, 1924. [*226, 251*]

[234] Fichtenholz, G., Kantorovitch, L. *Sur les opérations linéaires dans l'espace de fonctions bornées.* Studia Math. 1936. V. 5. P. 68–98. DOI 10.4064/sm-5-1-69-98. [*245*]

[235] Figalli, A., Gigli, N. *A new transportation distance between non-negative measures, with applications to gradients flows with Dirichlet boundary conditions.* J. Math. Pures Appl. (9). 2010. V. 94, №2. P. 107–130. DOI 10.1016/j.matpur.2009.11.005. MR2665414 [*252*]

[236] Filippov, V. V. *On a question of E. A. Michael.* Comm. Math. Univ. Carol. 2004. V. 45, №4. P. 735–738. MR2103087 [*209*]

[237] Fischer, H. History of the central limit theorem. From classical to modern probability theory. Springer, Berlin – New York, 2001; xvi+402 pp. MR2743162 [*22*]

[238] Flachsmeyer, J., Terpe, F. *Some applications of extension theory for topological spaces and measure theory.* Uspehi Mat. Nauk. 1977. V. 32, №5. P. 125–162 (in Russian); English transl.: Russian Math. Surveys. 1977. V. 32, №5. P. 133–171. MR0478106 [*250*]

[239] Florescu, L. C., Godet-Thobie, C. Young measures and compactness in measure spaces. De Gruyter, Berlin, 2012; xii+340 pp. MR2953092 [*231, 251*]

[240] Fortet, R., Mourier, E. *Convergence de la répartition empirique vers la répartition théorique.* Ann. Sci. École Norm. Sup. (3). 1953. V. 70. P. 267–285. MR0061325 [*245*]

[241] Fortet, R., Mourier, E. *Résultats complémentaires sur les éléments aléatoires prenant leurs valeurs dans un espace de Banach.* Bull. Sci. Math. (2). 1954. V. 78. P. 14–30. MR0063601 [*251*]

[242] Fortet, R., Mourier, E. *Les fonctions aléatoires comme éléments aléatoires dans les espaces de Banach.* Studia Math. 1955. V. 15. P. 62–79. DOI 10.4064/sm-15-1-62-79. MR0093052 [*251*]

[243] Frankiewicz, R., Plebanek, G., Ryll-Nardzewski, C. *Between Lindelöf property and countable tightness.* Proc. Amer. Math. Soc. 2001. V. 129, №1. P. 97–103. DOI 10.1090/S0002-9939-00-05489-7. MR1695139 [*250*]

[244] Fremlin, D. H. Measure theory. V. 1–5. University of Essex, Colchester, 2000–2003. MR2462519 [*246*]

[245] Fremlin, D. H., Garling, D. J. H., Haydon, R. G. *Bounded measures on topological spaces.* Proc. London Math. Soc. 1972. V. 25. P. 115–136. DOI 10.1112/plms/s3-25.1.115. MR0344405 [*137, 176, 179, 183, 238*]

[246] Fristedt, B., Gray, L. A modern approach to probability theory. Birkhäuser Boston, Boston, 1997; xx+756 pp. MR1422917 [*xi*]

[247] Fukaya, K. *Collapsing of Riemannian manifolds and eigenvalues of Laplace operator.* Invent. Math. 1987. V. 87, №3. P. 517–547. DOI 10.1007/BF01389241. MR874035 [*125*]

[248] Funano, K. *Estimates of Gromov's box distance.* Proc. Amer. Math. Soc. 2008. V. 136, №8. P. 2911–2920. DOI 10.1090/S0002-9939-08-09416-1. MR2399058 [*125*]

[249] Gadgil, S., Krishnapur, M. *Lipschitz correspondence between metric measure spaces and random distance matrices.* Int. Math. Res. Notes. IMRN. 2013. №24. P. 5623–5644. DOI 10.1093/imrn/rns208. MR3144175 [*125*]

[250] Gänssler, P. *A convergence theorem for measures in regular Hausdorff spaces.* Math. Scand. 1971. V. 29. P. 237–244. DOI 10.7146/math.scand.a-11049. MR0311862 [*227*]

[251] Gänssler, P. *Compactness and sequential compactness in spaces of measures.* Z. Wahrscheinlichkeitstheorie Verw. Geb. 1971. B. 17. S. 124–146. DOI 10.1007/BF00538864. MR0283562 [*227*]

[252] Gänssler, P. Empirical processes. Inst. Math. Statist., Hayward, California, 1983; ix+179 p. MR744668 [*246*]

[253] Gänssler, P., Pfanzagl, J. *Convergence of conditional expectations.* Ann. Math. Statist. 1971. V. 42, №1. P. 315–324. DOI 10.1214/aoms/1177693514. MR0282386 [*244*]

[254] Gänssler, P., Stute, W. Wahrscheinlichkeitstheorie. Springer-Verlag, Berlin – New York, 1977; xii+418 S. MR0501219 [*xi, 246*]

[255] Gaposhkin, V. F. *Lacunary series and independent functions.* Uspehi Mat. Nauk. 1966. V. 21, №6. P. 3–83 (in Russian); English transl.: Russian Math. Surveys. 1966. V. 21, №6. P. 1–82. MR0206556 [*225*]

[256] Gaposhkin, V. F. *Convergence and limit theorems for sequences of random variables.* Teor. Veroyatn. Primen. 1972. V. 17, №3. P. 401–423 (in Russian); English transl.: Theory Probab. Appl. 1972. V. 17. P. 379–400. MR0310948 [*225*]

[257] Gardner, R. J. *The regularity of Borel measures.* Lecture Notes in Math. 1982. V. 945. P. 42–100. MR675272 [*184*]

[258] Gâteaux (Gateaux), R. *Sur la notion d'intégrale dans le domaine fonctionnel et sur la théorie du potentiel.* Bull. Soc. Math. France. 1919. V. 47. P. 47–70. MR1504783 [*66, 245*]

[259] Geffroy, J. *Quelques extensions du théorème de P. Lévy sur la convergence presque sûre des séries à termes aléatoires indépendants.* C. R. Acad. Sci. Paris. 1959. T. 249. P. 1180–1182. MR0106506 [*192*]

[260] Gehér, Gy. P., Titkos, T. *Surjective Lévy–Prokhorov isometries.* ArXiv:1701.04267. [*249*]

[261] Gel'fand, I. M., Vilenkin, N. Ya. Generalized functions. V. 4. Applications of harmonic analysis. Translated from the Russian. Academic Press, New York – London, 1964; xiv+384 pp. MR0173945 [*251*]

[262] Gentil, I. *Dimensional contraction in Wasserstein distance for diffusion semigroups on a Riemannian manifold.* Potential Anal. 2015. V. 42, №4. P. 861–873. DOI 10.1007/s11118-015-9460-y. MR3339225 [*252*]

[263] Gérard, P. *Suites de Cauchy et compacité dans les espaces de mesures.* Bull. Soc. Roy. Sci. Liège. 1973. V. 42, №1–2. P. 41–49. MR0328564 [*248*]

[264] Gérard, P. *Un critère de compacité dans l'espace* $M_t^+(E)$. Bull. Soc. Roy. Sci. Liège. 1973. V. 42, №5–6. P. 179–182. MR0338750 [*248*]

[265] Giaquinta, M., Modica, G., Souček, J. Cartesian currents in the calculus of variations. V. I, II. Springer, Berlin – New York, 1998; xxiv+711 pp., xxiv+697 pp. MR1645086, MR1645082 [*44, 231, 251*]

[266] Gigli, N. *On the weak closure of convex sets of probability measures.* Rend. Mat. Appl. (7). 2009. V. 29, №2. P. 133–141. MR2604478 [*252*]

[267] Gigli, N. Introduction to optimal transport: theory and applications. Publ. Matemáticas do IMPA. Instituto Nacional de Matemática Pura e Aplicada (IMPA), Rio de Janeiro, 2011; 218 pp. MR2729922 [*252*]

[268] Gigli, N. *On the inverse implication of Brenier–McCann theorems and the structure of* $(\mathcal{P}_2(M), W_2)$. Methods Appl. Anal. 2011. V. 18, №2. P. 127–158. DOI 10.4310/MAA.2011.v18.n2.a1. MR2847481 [*252*]

[269] Gigli, N. *Second order analysis on* $(\mathcal{P}_2(M), W_2)$. Mem. Amer. Math. Soc. 2012. V. 216, no. 1018; xii+154 pp. DOI 10.1090/S0065-9266-2011-00619-2. MR2920736 [*252*]

[270] Gigli, N., Kuwada, K., Ohta, S.-I. *Heat flow on Alexandrov spaces.* Comm. Pure Appl. Math. 2013. V. 66, №3. P. 307–331. DOI 10.1002/cpa.21431. MR3008226 [*252*]

[271] Gigli, N., Maas, J. *Gromov–Hausdorff convergence of discrete transportation metrics.* SIAM J. Math. Anal. 2013. V. 45, №2. P. 879–899. DOI 10.1137/120886315. MR3045651 [*252*]

[272] Gigli, N., Ohta, S.-I. *First variation formula in Wasserstein spaces over compact Alexandrov spaces.* Canad. Math. Bull. 2012. V. 55, №4. P. 723–735. DOI 10.4153/CMB-2011-110-3. MR2994677 [*252*]

[273] Gihman, I.I., Skorohod, A.V. The theory of stochastic processes. V. 1–3. Springer, Berlin, 2004, 2007; viii+574 pp., xiii+441 pp., x+387 pp. MR0651015 [*85, 246, 251*]

[274] Gillman, L., Jerison, M. Rings of continuous functions. Van Nostrand, Princeton – New York, 1960; ix+300 pp. MR0116199 [*194, 198*]

[275] Giné, E., Nickl, R. Mathematical foundations of infinite-dimensional statistical models. Cambridge University Press, Cambridge, 2016; x+690 pp. MR3588285 [*246*]

[276] Giné, E., Zinn, J. *Some limit theorems for empirical processes.* Ann. Probab. 1984. V. 12. P. 929–989. MR757767 [*193*]

[277] Givens, C.R., Shortt, R.M. *A class of Wasserstein metrics for probability distributions.* Michigan Math. J. 1984. V. 31, №2. P. 231–240. DOI 10.1307/mmj/1029003026. MR752258 [*138, 249*]

[278] Glivenko, V. *Sulla determinazione empiric a della leggi di probabilita.* Giornale dell'Istituto Italiano degli Attuari. 1933. V. 4. P. 92–99. [*193*]

[279] Glivenko, V. *Sul teorema limite della teoria delle funzioni caratteristische.* Giornale dell'Istituto Italiano degli Attuari. 1936. V. 7. P. 160–167. [*246*]

[280] Glivenko, V. I. The Stieltjes integral. ONTI, Moscow – Leningrad, 1936; 216 pp. (in Russian) [*12, 38, 39, 245*]

[281] Gnedenko, B. V. The theory of probability theory. Translated from the fourth Russian edition. Chelsea, New York, 1967; 529 pp. MR0217824 [*xi, 40*]

[282] Gnedenko, B. V., Kolmogorov, A. N. Limit distributions for sums of independent random variables. GITTL, Moscow, 1949 (in Russian); English transl.: Addison-Wesley, Cambridge, Mass., 1954; ix+264 pp. MR0062975 [*22, 40*]

[283] Gohman, È. H. The Stieltjes integral and its applications. Fizmatgiz, Moscow, 1958; 192 pp. (in Russian). MR0099406 [*12*]

[284] Gorostiza, L. G., Rebolledo, R. *A random field approach to weak convergence of processes.* Statist. & Probab. Letters. 1993. V. 18. P. 49–55. DOI 10.1016/0167-7152(93)90098-4. MR1237623 [*248*]

[285] Goudon, T., Junca, S., Toscani, G. *Fourier-based distances and Berry–Esseen like inequalities for smooth densities.* Monatsh. Math. 2002. V. 135, №2. P. 115–136. DOI 10.1007/s006050200010. MR1894092 [*131*]

[286] Graf, S., Luschgy, H. Foundations of quantization for probability distributions. Lecture Notes in Math. V. 1730. Springer-Verlag, Berlin – New York, 2000; x+230 pp. MR1764176 [*132, 249*]

[287] Graf, S., Luschgy, H., Pagès, G. *Optimal quantizers for Radon random vectors in a Banach space.* J. Approx. Theory. 2007. V. 144, №1. P. 27–53. DOI 10.1016/j.jat.2006.04.006. MR2287375 [*133*]

[288] Grafakos, L. Classical and modern Fourier analysis. Pearson Education, Upper Saddle River, New Jersey, 2004; xii+931 pp. MR2449250 [*246*]

[289] Granirer, E. E. *On Baire measures on D-topological spaces.* Fund. Math. 1967. V. 60. P. 1–22. DOI 10.4064/fm-60-1-1-22. MR0208355 [*250*]

[290] Graves, W. H., Wheeler, R. F. *On the Grothendieck and Nikodym properties for algebras of Baire, Borel and universally measurable sets.* Rocky Mountain J. Math. 1983. V. 13, №2. P. 333–353. DOI 10.1216/RMJ-1983-13-2-333. MR702829 [*226*]

[291] Greenwood, P. E., Shiryayev, A. N. Contiguity and the statistical invariance principle. Gordon and Breach, New York, 1985; viii+236 pp. MR822226 [*251*]

[292] Greven, A., Pfaffelhuber, P., Winter, A. *Convergence in distribution of random metric measure spaces (Λ-coalescent measure trees).* Probab. Theory Related Fields. 2009. V. 145, №1-2. P. 285–322. DOI 10.1007/s00440-008-0169-3. MR2520129 [*125*]

[293] Grigelionis, B. *The relative compactness of sets of probability measures in* $D_{(0,\infty)}(\mathcal{X})$. Litovsk. Mat. Sb. 1973. V. 13, №4. P. 83–96 (in Russian). MR0353402 [*248*]

[294] Grigelionis, B., Mikulevicius, R. *On weak convergence of semimartingales.* Litovsk. Mat. Sb. 1981. V. 21, №3. P. 9–24 (in Russian); English transl.: Lithuanian Math. J. 1982. V. 21, №3. P. 213–224. MR637842 [*251*]

[295] Grigelionis, B., Mikulyavichyus, R. *On the weak convergence of random point processes.* Litovsk. Mat. Sb. 1981. V. 21, №4. P. 49–55 (in Russian); English transl.: Lithuanian Math. J. 1981. V. 21, №4. P. 297–301. MR641502 [*251*]

[296] Grigelionis, B., Mikulyavichus R. *Stably weak convergence of semimartingales and point processes.* Teor. Veroyatn. Primen. 1983. V. 28, №2. P. 320–332 (in Russian); English transl.: Theory Probab. Appl. 1983. V. 28, №2. P. 337–350. MR700212 [*251*]

[297] Grinblat, L. Š. *A limit theorem for measurable random processes and its applications.* Proc. Amer. Math. Soc. 1976. V. 61. P. 371–376. DOI 10.2307/2041344. MR0423450 [*248*]

[298] Grinblat, L. Š. *Convergence of probability measures on separable Banach spaces.* Proc. Amer. Math. Soc. 1977. V. 67. P. 321–323. DOI 10.2307/2041295. MR0494377 [*248*]

[299] Grinblat, L. Š. *Convergence of measurable random functions.* Proc. Amer. Math. Soc. 1979. V. 74. P. 322–325. DOI 10.2307/2043157. MR524310 [*248*]

[300] Grömig, W. *On a weakly closed subset of the space of τ-smooth measures.* Proc. Amer. Math. Soc. 1974. V. 43. P. 397–401. DOI 10.2307/2038903. MR0338758 [*201*]

[301] Gromov, M. Metric structures for Riemannian and non-Riemannian spaces. Translated from the French. Birkhäuser, Boston – Berlin, 1999; xix+585 pp. MR1699320 [*122*]

[302] Gross, L. *Harmonic analysis on Hilbert space.* Mem. Amer. Math. Soc. 1963. №46; ii+62 pp. MR0161095 [*251*]

[303] Grothendieck, A. *Sur les applications linéaires faiblement compactes d'espaces du type* $C(K)$. Canad. J. Math. 1953. V. 5. P. 129–173. MR0058866 [*226, 228*]

[304] Grothendieck, A. Topological vector spaces. Gordon and Breach, New York, 1973; 245 pp. MR0372565 [*228*]

[305] Gunther, N. M. Sur les intégrales de Stieltjes et leur application aux problèmes de la physique mathématique. Trudy Fiz.-Mat. V. A. Steklov Inst. Moscow, 1932; 494 pp. (2e éd.: Chelsey, New York, 1949). MR0031037 [*246*]

[306] Hahn, H. *Einige Anwendungen der Theorie der singulären Integrale.* Sitz. Akad. Wiss. Wien, Math.-naturwiss. Kl. IIa. 1918. B. 127, H. 9. S. 1763–1785. [*251*]

[307] Hahn, H. *Über Folgen linearer Operationen.* Monatsh. Math. Phys. 1922. B. 32. S. 3–88. DOI 10.1007/BF01696876. MR1549169 [*251*]

[308] Hammersley, J. M. *An extension of the Slutzky–Fréchet theorem.* Acta Math. 1952. V. 87. P. 243–257. DOI 10.1007/BF02392287. MR0050206 [*75*]

[309] Hanin, L. G. *An extension of the Kantorovich norm.* Monge–Ampère equations: applications to geometry and optimization, pp. 113–130, Contemp. Math., 226, Amer. Math. Soc., Providence, Rhode Island, 1999. DOI 10.1090/conm/226/03238. MR1660745 [*249*]

[310] Hardy, G. H., Landau, E., Littlewood, J. E. *Some inequalities satisfied by the integrals or derivatives of real or analytic functions.* Math. Z. 1935. B. 39, №1. S. 677–695. DOI 10.1007/BF01201386. MR1545530 [*126*]

[311] Hartman, S., Marczewski, E. *On the convergence in measure.* Acta Sci. Math. Szeged. 1950. V. 12. P. 125–131. MR0036821 [*99*]

[312] Hauray, M., Mischler, S. *On Kac's chaos and related problems.* J. Funct. Anal. 2014. V. 266, №10. P. 6055–6157. DOI 10.1016/j.jfa.2014.02.030. MR3188710 [*249*]

[313] Häusler, E., Luschgy, H. Stable convergence and stable limit theorems. Springer, Cham, 2015; x+228 pp. MR3362567 [*231*]

[314] Haviland, E.K. *On the theory of absolutely additive distribution functions.* Amer. J. Math. 1934. V. 56. P. 625–658. DOI 10.2307/2370960. MR1507048 [*246*]

[315] Haydon, R. *On compactness in spaces of measures and measure compact spaces.* Proc. London Math. Soc. 1974. V. 29, №1. P. 1–16. DOI 10.1112/plms/s3-29.1.1. MR0361745 [*196, 248*]

[316] Helly, E. *Über lineare Funktionaloperationen.* Sitz. Akad. Wiss. Wien, Math.-naturwiss. Kl. IIa. 1912. B. 12. S. 265–297. [*245, 246*]

[317] Hengartner, W., Theodorescu, R. Concentration functions. Academic Press, New York – London, 1973; xii+139 pp. MR0331448 [*99*]

[318] Hennequin, P.-L., Tortrat, A. Théorie des probabilités et quelques applications. Masson et Cie, Paris, 1965; viii+457 pp. MR0178481 [*xi, 22, 246*]

[319] Hennion, H., Hervé, L. Limit theorems for Markov chains and stochastic properties of dynamical systems by quasi-compactness. Springer, Berlin – New York, 2001; 145 pp. MR1862393 [*248*]

[320] Heyer, H., Kawakami, S. *Paul Lévy's continuity theorem: some history and recent progress.* Bull. Nara Univ. Educ. 2005. V. 54, №2. P. 11–21. MR2193086 [*169*]

[321] Hlawka, E. *Folgen auf kompakten Räumen.* Abh. Math. Sem. Univ. Hamburg. 1956. B. 20. S. 223–241. DOI 10.1002/mana.19580180122. MR0081368 [*219, 244*]

[322] Hlawka, E. Theorie der Gleichverteilung. Bibliogr. Inst., Mannheim, 1979; x+142 S. MR542905 [*219*]

[323] Hoffmann-Jørgensen, J. The theory of analytic spaces. Aarhus Various Publ. Series. №10. 1970; vi+314 pp. MR0409748 [*250*]

[324] Hoffmann-Jørgensen, J. *A generalization of the strict topology.* Math. Scand. 1972. V. 30, №2. P. 313–323. DOI 10.7146/math.scand.a-11087. MR0318857 [*250*]

[325] Hoffmann-Jørgensen, J. *Weak compactness and tightness of subsets of $M(X)$.* Math. Scand. 1972. V. 31, №1. P. 127–150. DOI 10.7146/math.scand.a-11420. MR0417369 [*163, 171, 175, 179, 184, 235, 237, 250*]

[326] Hoffmann-Jørgensen, J. Probability in Banach spaces. Lecture Notes in Math. 1976. V. 598. P. 1–186. MR0461610 [*196, 251*]

[327] Hoffmann-Jørgensen, J. Stochastic processes on Polish spaces. Aarhus Universitet, Matematisk Institut, Aarhus, 1991; ii+278 pp. MR1217966 [*246*]

[328] Hoffmann-Jørgensen, J. Probability with a view toward statistics. V. I, II. Chapman & Hall, New York, 1994; xi+589 pp., xiv+533 pp. MR1278485, MR1278486 [*xi*]

[329] Holický, P., Kalenda, O. *Descriptive properties of spaces of measures.* Bull. Polish Acad. Sci. Math. 1999. V. 47, №1. P. 37–51. MR1685676 [*239*]

[330] Holmes, R. *The universal separable metric space of Urysohn and isometric embeddings thereof in Banach spaces.* Fund. Math. 1991. V. 140, №3. P. 199–223. MR1173763 [*138*]

[331] Ibragimov, I. A., Linnik, Yu. V. Independent and stationary sequences of random variables. Wolters-Noordhoff Publ., Groningen, 1971; 443 pp. (Russian ed.: Moscow, 1965). MR0322926 [*22, 40*]

[332] Ivanov, A. V. Convergence of distributions of functionals of measurable random fields. Ukrainskii Matem. Zhurn. 1980. V. 32, №1. P. 27–34 (in Russian); English tranl.: Ukrainian Math. J. 1980. V. 32, №1. P. 19–25. MR561187 [*248*]

[333] Jacod, J., Mémin, J. *Sur un type de convergence intermédiare entre la convergence en loi et la convergence en probabilité.* Lecture Notes in Math. 1981. V. 850. P. 529–546; Rectification: Lecture Notes in Math. 1983. V. 986. P. 509–511. MR622586 [*231*]

[334] Jacod, J., Mémin, J., Métivier, M. *On tightness and stopping times.* Stochastic Process. Appl. 1983. V. 14, №2. P. 109–146. DOI 10.1016/0304-4149(83)90067-4. MR679668 [*251*]

[335] Jacod, J., Shiryaev, A. N. Limit theorems for stochastic processes. 2nd ed. Springer-Verlag, Berlin, 2003; xx+661 pp. MR1943877 [*251*]

[336] Jakubowski, A. *On the Skorokhod topology.* Ann. Inst. H. Poincaré. Probab. Statist. 1986. V. 22, №3. P. 263–285. MR871083 [*248*]

[337] Jakubowski, A. *The almost sure Skorokhod representation for subsequences in nonmetric spaces.* Theory Probab. Appl. 1997. V. 42, №1. P. 167–174. DOI 10.1137/S0040585X97976052. MR1453342 [*247*]

[338] Jakubowski, A. *A non-Skorohod topology on the Skorohod space.* Electron. J. Probab. 1997. V. 2, №4. P. 1–21. DOI 10.1214/EJP.v2-18. MR1475862 [*248*]

[339] Janson, S., Kaijser, S. *Higher moments of Banach space valued random variables.* Mem. Amer. Math. Soc. 2015. V. 238, №1127; vii+110 pp. DOI 10.1090/memo/1127. MR3402381 [*251*]

[340] Jarchow, H. Locally convex spaces. B. G. Teubner, Stuttgart, 1981; 548 pp. MR632257 [*170*]

[341] Jessen, B., Wintner, A. *Distribution functions and the Riemann zeta function.* Trans. Amer. Math. Soc. 1935. V. 38. P. 48–88. DOI 10.2307/1989728. MR1501802 [*246*]

[342] Kakosyan, A. V., Klebanov, L. B., Rachev, S. T. Quantitative criteria for the convergence of probability measures. Aiastan, Erevan, 1988; 249 pp. (in Russian). MR1024072 [*249*]

[343] Kallenberg, O. Foundations of modern probability. 2nd ed. Springer-Verlag, New York, 2002; xx+638 pp. MR1876169 [*xi, 242*]

[344] Kallenberg, O. Probabilistic symmetries and invariance principles. Springer, New York, 2005; xii+510 pp. MR2161313 [*251*]

[345] Kallianpur, G. *The topology of weak convergence of probability measures.* J. Math. Mech. 1961. V. 10, №6. P. 947–969. MR0132143 [*251*]

[346] Kallianpur, G., Xiong, J. Stochastic differential equations in infinite-dimensional spaces. Inst. Math. Statist., Hayward, California, 1995; vi+342 pp. MR1465436 [*248*]

[347] Kamke, E. Das Lebesguesche Integral. Eine Einführung in die neuere Theorie der reellen Funktionen. Teubner, Leipzig, 1925; 151 S. MR0081330 [*12*]

[348] Kantorovitch, L. V. *On the translocation of masses.* Dokl. Akad. Nauk SSSR. 1942. V. 37, №7–8. P. 227–229 (in Russian); English transl.: C. R. (Doklady) Acad. Sci. URSS. 1942. V. 37. P. 199–201. MR0009619 [*248*]

[349] Kantorovich, L. V., Akilov, G. P. Functional analysis. Translated from the Russian. 2nd ed. Pergamon Press, Oxford, 1982; xiv+589 pp. MR664597 [*248*]

[350] Kantorovich, L. V., Rubinshtein (Rubinštein), G. Sh. *On a functional space and certain extremum problems.* Dokl. Akad. Nauk SSSR. 1957. V. 115, №6. P. 1058–1061 (in Russian). MR0094707 [*248*]

[351] Kantorovich, L. V., Rubinstein (Rubinšteĭn), G. Sh. *On a space of completely additive functions.* Vestnik Leningrad. Univ. 1958. №7(2). P. 52–59 (in Russian). MR0102006 [*248, 249*]

[352] Kawabe, J. *A criterion for weak compactness of measures on product spaces with applications.* Yokohama Math. J. 1994. V. 42. MR1332005 P. 159–169. [*197*]

[353] Kawabe, J. *Convergence of compound probability measures on topological spaces.* Colloq. Math. 1994. V. 67, №2. P. 161–176. DOI 10.4064/cm-67-2-161-176. MR1305208 [*196*]

[354] Kawabe, J. *Weak convergence of compound probability measures on uniform spaces.* Tamkang J. Math. 1999. V. 30, №4. P. 271–288. MR1728475 [*196*]

[355] Kawata, T. Fourier analysis in probability theory. Academic Press, New York, 1972; xii+668 pp. MR0464353 [*246*]

[356] Kechris, A. S. Classical descriptive set theory. Springer, Berlin – New York, 1995; xviii+402 pp. MR1321597 [*138*]

[357] Kellerer, H. G. *Duality theorems for marginal problems.* Z. Wahrscheinlichkeitstheorie Verw. Geb. 1984. B. 67, №4. S. 399–432. DOI 10.1007/BF00532047. MR761565 [*249*]

[358] Ketterer, C. *Cones over metric measure spaces and the maximal diameter theorem.* J. Math. Pures Appl. (9). 2015. V. 103, №5. P. 1228–1275. DOI 10.1016/j.matpur.2014.10.011. MR3333056 [*252*]

[359] Khasminskii, R. Z. Stochastic stability of differential equations. 2nd ed. Springer, Heidelberg, 2012; xviii+339 pp. (Russian ed.: Moscow, 1969). MR2894052 [*248*]

[360] Khokhlov, V. I. *The uniform distribution on a sphere in* $\mathbf{R}^S$. *Properties of projections. I.* Teor. Veroyatn. Primen. 2005. V. 50, №3. P. 501–516 (in Russian); English tranls.: Theory Probab. Appl. 2006. V. 50, №3. P. 386–399. DOI 10.1137/S0040585X97981846. MR2223214 [*98*]

[361] Kimme, E. G. *On the convergence of sequences of stochastic processes.* Trans. Amer. Math. Soc. 1957. V. 84. P. 208–229. DOI 10.2307/1992898. MR0083843 [*251*]

[362] Kimme, E. G. *Some equivalence conditions for the convergence in distribution of sequences of stochastic processes.* Trans. Amer. Math. Soc. 1960. V. 95. P. 495–515. DOI 10.2307/1993570. MR0115207 [*251*]

[363] Kirk, R. B. *Measures in topological spaces and B-compactness.* Indag. Math. 1969. V. 31. P. 172–183. MR0246104 [*250*]

[364] Kirk, R. B. *Topologies on spaces of Baire measures.* Bull. Amer. Math. Soc. 1973. V. 79, №3. P. 542–545. DOI 10.1090/S0002-9904-1973-13193-3. MR0313772 [*250*]

[365] Kirk, R. B. *Complete topologies on spaces of Baire measures.* Trans. Amer. Math. Soc. 1973. V. 184. P. 1–29. DOI 10.2307/1996396. MR0325913 [*250*]

[366] Klartag, B. *A central limit theorem for convex sets.* Invent. Math. 2007. V. 168, №1. P. 91–131. DOI 10.1007/s00222-006-0028-8. MR2285748 [*98*]

[367] Klee, V. L., Jr. *Some topological properties of convex sets.* Trans. Amer. Math. Soc. 1955. V. 78. P. 30–45. DOI 10.2307/1992947. MR0069388 [*95*]

[368] Kliem, S., Löhr, W. *Existence of mark functions in marked metric measure spaces.* Electron. J. Probab. 2015. V. 20, №73. 24 pp. DOI 10.1214/EJP.v20-3969. MR3371432 [*125*]

[369] Klimkin, V. M. Introduction to the theory of set functions. Saratov. Gos. Univ., Kuibyshev, 1989; 210 pp. (in Russian). [*228*]

[370] Kloeckner, B. *A geometric study of Wasserstein spaces: Euclidean spaces.* Ann. Sc. Norm. Super. Pisa, Cl. Sci. (5). 2010. V. 9, №2. P. 297–323. MR2731158 [*249*]

[371] Kloeckner, B. *Approximation by finitely supported measures.* ESAIM: Control, Optim. Calc. Var. 2012. V. 18, №2. P. 343–359. DOI 10.1051/cocv/2010100. MR2954629 [*133*]

[372] Kloeckner, B. R. *A geometric study of Wasserstein spaces: ultrametrics.* Mathematika. 2015. V. 61, №1. P. 162–178. DOI 10.1112/S0025579314000059. MR3333967 [*252*]

[373] Kohn, R. V., Otto, F. *Upper bounds on coarsening rates.* Comm. Math. Phys. 2002. V. 229, №3. P. 375–395. DOI 10.1007/s00220-002-0693-4. MR1924360 [*126*]

[374] Kolmogorov, A. N. *Sulla determinazione empirica di una legge di distribuzione.* Giornale dell'Istituto Italiano degli Attuari. 1933. V. 4. P. 83–91 (English transl. in [**378**]). [*193*]

[375] Kolmogoroff, A. *La transformation de Laplace dans les espaces linéaires.* C. R. Acad. Sci. Paris. 1935. T. 200. P. 1717–1718 (English transl. in [**378**]). [*251*]

[376] Kolmogorov, A. N. *On Skorokhod convergence.* Teor. Veroyatn. Primen. 1956. V. 1, №2. P. 239–247 (in Russian); English transl.: Theory Probab. Appl. 1956. V. 1, №2. P. 215–222. MR0085638 [*248*]

[377] Kolmogorov, A. N. *A note on the papers of R. A. Minlos and V. V. Sazonov.* Teor. Veroyant. Primen. 1959. V. 4, №2. P. 237–239 (in Russian); English transl.: Theory Probab. Appl. 1959. V. 4. P. 221–223. [*251*]

[378] Kolmogorov, A. N. Selected works. V. I, II. Kluwer, Dordrecht, 1991, 1992. MR1175399, MR1153022, MR1228446 [*269*]

[379] Kolmogorov, A. N., Fomin, S. V. Introductory real analysis. V. 1. Metric and normed spaces. V. 2. Measure. The Lebesgue integral. Hilbert space. Transl. from the 2nd Russian ed. Corr. repr. Dover, New York, 1975; xii+403 pp. MR0377445 [*139, 164*]

[380] Kolmogoroff, A., Prochorow, Ju. W. *Zufällige Funktionen und Grenzverteilugssätze.* Bericht über die Tagung Wahrscheinlichkeitsrechnung und mathematische Statistik, Berlin. 1956. S. 113–126 (English transl. in [**378**]). MR0082221 [*66, 150*]

[381] Kondo, T. *Probability distribution of metric measure spaces.* Differential Geom. Appl. 2005. V. 22, №2. P. 121–130. DOI 10.1016/j.difgeo.2004.10.001. MR2122737 [*252*]

[382] Korolev, V. Yu., Shevtsova, I. G. *On the upper bound for the absolute constant in the Berry–Esseen inequality.* Teor. Veroyatn. Primen. 2009. V. 54, №4. P. 671–695 (in Russian); English transl.: Theory Probab. Appl. 2010. V. 54, №4. P. 638–658. DOI 10.1137/S0040585X97984449. MR2759643 [*37*]

[383] Korolev, V., Shevtsova, I. *An improvement of the Berry–Esseen inequality with applications to Poisson and mixed Poisson random sums.* Scand. Actuarial J. 2012. V. 2. P. 81–105. DOI 10.1080/03461238.2010.485370. MR2929524 [*37*]

[384] Koroljuk, V. S., Borovskikh, Yu. V. Theory of U-statistics. Translated from the Russian. Kluwer, Dordrecht, 1994; x+552 pp. MR1472486 [*251*]

[385] Kosorok, M. Introduction to empirical processes and semiparametric inference. Springer, New York, 2008; xiv+483 pp. MR2724368 [*246*]

[386] Koumoullis, G. *Some topological properties of spaces of measures.* Pacif. J. Math. 1981. V. 96, №2. P. 419–433. MR637981 [*201, 208, 233, 250*]

[387] Koumoullis, G. *Perfect, u-additive measures and strict topologies.* Illinois J. Math. 1982. V. 26, №3. P. 466–478. MR658457 [*250*]

[388] Koumoullis, G. *Cantor sets in Prohorov spaces.* Fund. Math. 1984. V. 124, №2. P. 155–161. DOI 10.4064/fm-124-2-155-161. MR774507 [*250*]

[389] Koumoullis, G. *Topological spaces containing compact perfect sets and Prohorov spaces.* Topol. Appl. 1985. V. 21, №1. P. 59–71. DOI 10.1016/0166-8641(85)90058-6. MR808724 [*250*]

[390] Koumoullis, G., Sapounakis, A. *Two countability properties of sets of measures.* Mich. Math. J. 1984. V. 31, №1. P. 31–47. DOI 10.1307/mmj/1029002959. MR736466 [*233, 234*]

[391] Kouritzin, M. A. *On tightness of probability measures on Skorokhod spaces.* Trans. Amer. Math. Soc. 2016. V. 368, №8. P. 5675–5700. DOI 10.1090/tran/6522. MR3458395 [*248*]

[392] Kreitmeier, W. *Optimal vector quantization in terms of Wasserstein distance.* J. Multivariate Anal. 2011. V. 102, №8. P. 1225–1239. DOI 10.1016/j.jmva.2011.04.005. MR2812740 [*133*]

[393] Kruglov, V. M. Supplementary chapters of probability theory. Vysš. Škola, Moscow, 1984; 264 pp. (in Russian). MR0756812 [*40, 246, 251*]

[394] Kruglov, V. M., Korolev, V. Yu. Limit theorems for random sums. Moskov. Gos. Univ., Moscow, 1990; 270 pp. (in Russian). MR1072999 [*40, 251*]

[395] Krugova, E. P. *On differentiability of convex measures.* Matem. Zametki. 1995. V. 57, №6. P. 51–61 (in Russian); English transl.: Math. Notes. 1995. V. 58, №6. P. 1294–1301. DOI 10.1007/BF02304888. MR1382094 [*130*]

[396] Krugova, E. P. *On translates of convex measures.* Mat. Sbornik. 1997. V. 188, №2. P. 57–66 (in Russian); English transl.: Sbornik Math. 1997. V. 188, №2. P. 227–236. DOI 10.1070/SM1997v188n02ABEH000201. MR1453259 [*129*]

[397] Krylov, N. V. *On SPDEs and superdiffusions.* Ann. Probab. 1997. V. 25. P. 1789–1809. DOI 10.1214/aop/1023481111. MR1487436 [*247*]

[398] Kuelbs, J. *Some results for probability measures on linear topological vector spaces with an application to Strassen's LogLog Law.* J. Funct. Anal. 1973. V. 14, №1. P. 28–43. MR0356157 [*251*]

[399] Kuelbs, J. *Fourier analysis on linear metric spaces.* Trans. Amer. Math. Soc. 1973. V. 181. P. 293–311. DOI 10.2307/1996634. MR0331455 [*251*]

[400] Kuipers, L., Niederreiter, H. Uniform distribution of sequences. Wiley, New York, 1974; xiv+390 pp. MR0419394 [*219*]

[401] Kuo, H. H. Gaussian measures in Banach spaces. Lecture Notes in Math. V. 463. Springer-Verlag, Berlin – New York, 1975; vi+224 pp. MR0461643 [*89*]

[402] Kuratowski, K. Topology. V. 1. Academic Press, New York – London, Polish Sci. Publ., Warsaw, 1966; xx+560 pp. MR0217751 [*139*]

[403] Kusuoka, S., Nakayama, T. *On a certain metric on the space of pairs of a random variable and a probability measure.* J. Math. Sci. Univ. Tokyo. 2001. V. 8, №2. P. 343–356. MR1837168 [*249*]

[404] Kuwada, K. *Space-time Wasserstein controls and Bakry–Ledoux type gradient estimates.* Calc. Var. Partial Differ. Equ. 2015. V. 54, №1. P. 127–161. DOI 10.1007/s00526-014-0781-2. MR3385156 [*252*]

[405] Kwapień, S. *Complément au théorème de Sazonov–Minlos.* C. R. Acad. Sci. Paris Sér. A-B. 1968. T. 267. P. A698–A700. MR0248893 [*251*]

[406] Lamperti, J. *A new class of probability limit theorems.* J. Math. Mech. 1962. V. 11. P. 749–772. MR0148120 [*251*]

[407] Lamperti, J. *On convergence of stochastic processes.* Trans. Amer. Math. Soc. 1962. V. 104. P. 430–435. DOI 10.2307/1993787. MR0143245 [*251*]

[408] Lamperti, J. W. Probability. A survey of the mathematical theory. 2nd ed. John Wiley & Sons, New York, 1996; xii+189 pp. MR1406796 [*xi*]

[409] Landers, D., Rogge, L. *Cauchy convergent sequences of regular measures with values in a topological group.* Z. Wahrscheinlichkeitstheorie Verw. Geb. 1972. B. 21. S. 188–196. DOI 10.1007/BF00538391. MR0310170 [*227*]

[410] Landers, D., Rogge, L. *Joint convergence of conditional expectations.* Manuscripta Math. 1972. V. 6. P. 141–145. DOI 10.1007/BF01369710. MR0293684 [*244*]

[411] Landers, D., Rogge, L. *Short proofs of two convergence theorems for conditional expectations.* Ann. Math. Statist. 1972. V. 43. P. 1372–1373. DOI 10.1214/aoms/1177692493. MR0315748 [*244*]

[412] Lange, K. *Borel sets of probability measures.* Pacific J. Math. 1973. V. 48. P. 141–161. MR0357723 [*243*]

[413] Larman, D. G., Rogers, C. A. *The normability of metrizable sets.* Bull. London Math. Soc. 1973. V. 5. P. 39–48. DOI 10.1112/blms/5.1.39. MR0320681 [*134*]

[414] Lebedev, V. A. Martingales, convergence of probability measures, and stochastic equations. Izdat. MAI, Moscow, 1996; 348 pp. (in Russian). MR1407742 [*231, 247, 248*]

[415] Lebesgue, H. *Sur les intégrales singulières.* Ann. Fac. Sci. Univ. Toulouse (3). 1909. V. 1. P. 25–117. MR1508308 [*251*]

[416] Le Cam, L. *Un instrument d'étude des fonctions aléatoires: la fonctionnelle caractéristique.* C. R. Acad. Sci. Paris. 1947. T. 224. P. 710–711. MR0019862 [*251*]

[417] Le Cam, L. *Convergence in distribution of stochastic processes.* Univ. Calif. Publ. Statist. 1957. V. 2. P. 207–236. MR0086117 [*57, 245, 247, 250*]

[418] Le Cam, L. *Remarques sur le théorème limite central dans les espaces localement convexes.* In: Les Probabilités sur les structures algébriques, pp. 233–249. Centre National des Recherches Scientifiques, Paris, 1970. MR0410832 [*251*]

[419] Le Cam, L. Asymptotic methods in statistical decision theory. Springer-Verlag, New York, 1986; xxvi+742 pp. MR856411 [*246*]

[420] Ledoux, M. The concentration of measure phenomenon. Amer. Math. Soc., Providence, Rhode Island, 2001; x+181 pp. MR1849347 [*98, 252*]

[421] Ledoux, M., Talagrand, M. Probability in Banach spaces. Isoperimetry and processes. Springer-Verlag, Berlin – New York, 1991; xii+480 pp. MR1102015 [*189, 251*]

[422] Léger, C., Soury, P. *Le convexe topologique des probabilités sur un espace topologique.* J. Math. Pures Appl. 1971. V. 50. P. 363–425. MR0464342 [*251*]

[423] Letta, G. *Convergence stable et applications.* Atti Sem. Mat. Fis. Modena (suppl.). 1998. V. 48. P. 191–211. MR1645717 [*231*]

[424] Letta, G., Pratelli, L. *Le théorème de Skorohod pour des lois de Radon sur un espace métrisable.* Rend. Accad. Naz. Sci. XL Mem. Mat. Appl. (5). 1997. T. 21. P. 157–162. MR1612812 [*247*]

[425] Levin, V. L. *The Monge–Kantorovich duality and its application in the theory of utility.* Economics Math. Methods. 2011. V. 47, №4. P. 143–165 (in Russian). [*249*]

[426] Levin, V.L., Miljutin, A.A. *The problem of mass transfer with a discontinuous cost function and a mass statement of the duality problem for convex extremal problems.* Uspekhi Matem. Nauk. 1979. V. 34, №3. P. 3–68 (in Russian); Engligh transl.: Russian Math. Surveys. 1979. V. 34, №3. P. 1–78. MR542237 [*249*]

[427] Lévy, P. Problèmes concrets d'analyse fonctionnelle. Paris, 1922 (2me éd.: Gauthier-Villars, Paris, 1951; xiv+484 pp.). MR0041346 [*245*]

[428] Lévy, P. Calcul des probabilités. Gautier-Villars, Paris, 1925; viii+352 p. [*245*]

[429] Lévy, P. Théorie de l'addition des variables aléatoires. Gautier-Villars, Paris, 1937; xvii+328 pp. (2e éd., 1954; 385 pp.). [*245*]

[430] Liese, F., Vajda, I. Convex statistical distances. B. G. Teubner, Leipzig, 1987; 224 pp. MR926905 [*249*]

[431] Lifshits, M. A. Gaussian random functions. Translated from the Russian. Kluwer, Dordrecht, 1995; xii+333 pp. MR1472736 [*89*]

[432] Linde, W. Probability in Banach spaces – stable and infinitely divisible distributions. Wiley, New York, 1986; 195 pp. MR874529 [*40, 251*]

[433] Lindvall, T. *Weak convergence of probability measures and random functions in the function space* $D[0,\infty)$. J. Appl. Probab. 1973. V. 10. P. 109–121. MR0362429 [*247*]

[434] Linnik, Yu. V., Ostrovskii, I. V. Decompositions of random variables and vectors. Translated from the Russian. Amer. Math. Soc., Providence, Rhode Island, 1977; ix+380 pp. MR0428382 [*40*]

[435] Lipchius, A. A. *A note on the equality in the problems of Monge and Kantorovich.* Teor. Veroyatn. Primen. 2005. V. 50, №4. P. 779–782 (in Russian); English transl.: Theory Probab. Appl. 2006. V. 50, №4. P. 689–693. DOI 10.1137/S0040585X97982074. MR2331990 [*252*]

[436] Liptser, R. Sh., Shiryayev, A. N. Theory of martingales. Translated from the Russian. Kluwer, Dordrecht, 1989; xiv+792 pp. MR1022664 [*251*]

[437] Loève, M. Probability theory. V. 1, 2. 4th ed. Springer, New York, 1977, 1978; xvii+425 pp., xvi+413 pp. MR0651017, MR0651018 [*xi, 22*]

[438] Löhr, W. *Equivalence of Gromov–Prohorov- and Gromov's $\square_\lambda$-metric on the space of metric measure spaces.* Electron. Commun. Probab. 2013. V. 18, №17. 10 pp. MR3037215 [*124, 252*]

[439] Löhr, W. *Equivalence of Gromov–Prohorov- and Gromov's box-metric on the space of metric measure spaces.* ArXiv:1111.5837v4. [*124, 252*]

[440] Losert, V. *Uniformly distributed sequences on compact, separable, non-metrizable groups.* Acta Sci. Math. (Szeged). 1978. V. 40, №1-2. P. 107–110. MR0476677 [*222*]

[441] Losert, V. *On the existence of uniformly distributed sequences in compact topological spaces. I.* Trans. Amer. Math. Soc. 1978. V. 246. P. 463–471. DOI 10.2307/1997987. MR515552 [*222, 243*]

[442] Losert, V. *On the existence of uniformly distributed sequences in compact topological spaces. II.* Monatsh. Math. 1979. B. 87, №3. S. 247–260. DOI 10.1007/BF01303079. MR536093 [*222*]

[443] Lott, J. *Some geometric calculations on Wasserstein spaces.* Comm. Math. Phys. 2008. V. 277, №2. P. 423–437. DOI 10.1007/s00220-007-0367-3. MR2358290 [*252*]

[444] Lott, J., Villani, C. *Ricci curvature for metric-measure spaces via optimal transport.* Ann. Math. (2). 2009. V. 169, №3. P. 903–991. DOI 10.4007/annals.2009.169.903. MR2480619 [*252*]

[445] de Lucia, P., Pap, E. *Convergence theorems for set functions.* In: Handbook of measure theory (Pap E., ed.). V. 1,2. P. 125–178. North-Holland, Amsterdam, 2002. DOI 10.1016/B978-044450263-6/50005-1. MR1954614 [*228*]

[446] Lukacs, E. Characteristic functions. 2nd ed. Hafner Publ., New York, 1970; x+350 pp. MR0346874 [*40, 246*]

[447] Lukacs, E. Developments in characteristic function theory. Griffin, London, 1983; viii+182 pp. MR810001 [*40, 246*]

[448] Lyapunov, A. M. Collected works. V. 1. Akad. Nauk USSR, Moscow, 1954; 452 pp. [*21*]

[449] Major, P. *On the invariance principle for sums of independent identically distributed random variables.* J. Multivar. Anal. 1978. V. 8. P. 487–517. DOI 10.1016/0047-259X(78)90029-5. MR520959 [*137*]

[450] Mandrekar, V. S. Weak convergence of stochastic processes: with applications to statistical limit theorems. De Gruyter, Berlin – Boston, 2016; vi+141 pp. MR3585321 [*248*]

[451] Mazliak, L. *The ghosts of the École Normale. Life, death and destiny of Réne Gateaux.* Statist. Sci. 2015. V. 30, №3. P. 391–412. DOI 10.1214/15-STS512. MR3383887 [*245*]

[452] McShane, E. J. *Linear functionals on certain Banach spaces.* Proc. Amer. Math. Soc. 1950. V. 1. P. 402–408. DOI 10.2307/2032394. MR0036448 [*83*]

[453] Meckes, E. S., Meckes, M. W. *On the equivalence of modes of convergence for log-concave measures.* Lecture Notes in Math. 2014. V. 2016. P. 385–394. DOI 10.1007/978-3-319-09477-9_24. MR3364698 [*40*]

[454] Medvedev, K. V. *Certain properties of triangular transformations of measures.* Theory Stoch. Process. 2008. V. 14, №1. P. 95–99. MR2479710 [*40*]

[455] Mehler, F. G. *Ueber die Entwicklung einer Function von beliebig vielen Variabeln nach Laplaceschen Functionen höherer Ordnung.* J. Reine Angew. Math. 1866. B. 66. S. 161–176. DOI 10.1515/crll.1866.66.161. MR1579340 [*66, 247*]

[456] Melleray, J., Petrov, F. V., Vershik, A. M. *Linearly rigid metric spaces and the embedding problem.* Fund. Math. 2008. V. 199, №2. P. 177–194. DOI 10.4064/fm199-2-6. MR2399498 [*125, 138*]

[457] Mémoli, F. *Gromov–Wasserstein distances and the metric approach to object matching.* Found. Comput. Math. 2011. V. 11, №4. P. 417–487. DOI 10.1007/s10208-011-9093-5. MR2811584 [*252*]

[458] Mercourakis, S. *Some remarks on countably determined measures and uniform distribution of sequences.* Monatsh. Math. 1996. B. 121, №1-2. S. 79–111. DOI 10.1007/BF01299640. MR1375642 [*222*]

[459] Meyer, P.-A. *Le théorème de continuité de P. Lévy sur les espaces nucléaires.* Séminaire N. Bourbaki. 1965–1966. Exp. 311 (1966). P. 509–522. MR1610982 [*169*]

[460] Meyer, P.-A., Zheng, W. A. *Tightness criteria for laws of semi-martingales.* Ann. Inst. H. Poincaré. 1984. V. 20, №4. P. 353–372. MR771895 [*251*]

[461] Michael, E. *A short proof of the Arens–Eells embedding theorem.* Proc. Amer. Math. Soc. 1964. V. 15. P. 415–416. DOI 10.2307/2034516. MR0162222 [*96*]

[462] Michael, E. *A linear mapping between function spaces.* Proc. Amer. Math. Soc., 1964. V. 15. P. 407–409. DOI 10.2307/2034514. MR0162128 [*210*]

[463] Michael, E. *A selection theorem.* Proc. Amer. Math. Soc. 1966. V. 17. P. 1404–1406. DOI 10.2307/2035751. MR0203702 [*209*]

[464] Miller, D., Sentilles, D. *Weak convergence of probability measures relative to incompatible topology and σ-field.* Z. Wahrscheinlichkeitstheorie Verw. Geb. 1978. B. 45. S. 239–256. DOI 10.1007/BF00535305. MR510028 [*250*]

[465] Minlos, R. A. *Generalized random processes and their extension to a measure.* Trudy Moskov. Mat. Obsc. 1959. V. 8. P. 497–518 (in Russian); English transl.: Math. Stat. Probab. 1959. V. 3. P. 291–314. MR0108851 [*245, 252*]

[466] Mitoma, I. *Tightness of probabilities on* $C([0,1];\mathcal{S}')$ *and* $D([0,1];\mathcal{S}')$. Ann. Probab. 1983. V. 11, №4. P. 989–999. MR714961 [*248*]

[467] Mohapl, J. *On weakly convergent nets in spaces of nonnegative measures.* Czech. Math. J. 1990. V. 40(115), №3. P. 408–421. MR1065020 [*251*]

[468] Mohapl, J. *The Radon measures as functionals on Lipschitz functions.* Czech. Math. J. 1991. V. 41, №3. P. 446–453. MR1117798 [*250*]

[469] Molnár, L. *Lévy isometries of the space of probability distribution functions.* J. Math. Anal. Appl. 2011. V. 380. P. 847–852. DOI 10.1016/j.jmaa.2011.02.014. MR2794437 [*249*]

[470] Moran, W. *Measures on metacompact spaces.* Proc. London Math. Soc. 1970. V. 20. P. 507–524. DOI 10.1112/plms/s3-20.3.507. MR0437706 [*238*]

[471] Morozova, E. A., Chentsov, N. N. *Natural geometry of families of probability laws.* Probability theory, 8 (Russian), pp. 133–265. Itogi Nauki i Tekhniki, Akad. Nauk SSSR, Vsesoyuz. Inst. Nauchn. i Tekhn. Inform., Moscow, 1991 (in Russian). MR1128374 [*246*]

[472] Mosiman, S. E., Wheeler, R. F. *The strict topology in a completely regular setting: relations to topological measure theory.* Canad. J. Math. 1972. V. 24, №5. P. 873–890. MR0328567 [*250*]

[473] Mourier, E. *Eléments aléatoires dans un espace de Banach.* Ann. Inst. H. Poincaré. 1953. T. 19. P. 161–244. MR0064339 [*251*]

[474] Mushtari, D. Kh. *Lévy type criteria for weak convergence of probabilities in Fréchet spaces.* Teor. Veroyatn. Primen. 1979. V. 24, №3. P. 580–585 (in Russian); English transl.: Theory Probab. Appl. 1980. V. 24, №3. P. 587–592. MR541370 [*251*]

[475] Mushtari, D. Kh. Probabilities and topologies on linear spaces. Kazan Mathematics Foundation, Kazan', 1996; xiv+233 pp. MR1658715 [*74, 251*]

[476] Nagaev, S. V. *Some limit theorems for stationary Markov chains.* Teor. Veroyatn. Primen. 1957. V. 2, №4. P. 389–416 (in Russian); English transl.: Theory Probab. Appl. 1957. V. 2, №4. P. 378–406. MR0094846 [*248*]

[477] Nagaev, S. V. *More exact statement of limit theorems for homogeneous Markov chains.* Teor. Veroyatn. Primen. 1961. V. 6, №1. P. 67–86 (in Russian); English transl.: Theory Probab. Appl. 1961. V. 6, №1. P. 62–81. MR0131291 [*248*]

[478] Nahapetian, B. Limit theorems and some applications in statistical physics. Teubner, Stuttgart, 1991; 244 pp. MR1125536 [*251*]

[479] Nakanishi, S. *Weak convergence of measures on the union of metric spaces. I.* Math. Jap. 1986. V. 31, №3. P. 429–447. MR854792 [*251*]

[480] Neininger, R., Sulzbach, H. *On a functional contraction method.* Ann. Probab. 2015. V. 43, №4. P. 1777–1822. DOI 10.1214/14-AOP919. MR3353815 [*249*]

[481] Neuhaus, G. *On weak convergence of stochastic processes with multidimensional time parameter.* Ann. Math. Statist. 1971. V. 42. P. 1285–1295. DOI 10.1214/aoms/1177693241. MR0293706 [*251*]

[482] Neveu, J. Bases mathématiques du calcul des probabilités. Masson et Cie, Paris, 1964; xiii+203 pp. English transl.: Mathematical foundations of the calculus of probability. Holden-Day, San Francisco, 1965; 231 pp. MR0198504 [*xi, 192*]

[483] Nguyen, Zui Tien, Tarieladze, V. I., Chobanjan, S. A. *On compactness of families of second-order measures in a Banach space.* Teor. Veroyatn. Primen. 1977. V. 22, №4. P. 823–828 (in

Russian); English transl.: Theory Probab. Appl. 1978. V. 22, №4. P. 805–810. MR0467870 [*251*]

[484] Niederreiter, H. *On the existence of uniformly distributed sequences in compact spaces.* Compositio Math. 1972. V. 25. P. 93–99. MR0316661 [*220*]

[485] Nikodym, O. *Sur les suites de fonctions parfaitement additives d'ensembles abstraits.* C. R. Acad. Sci. Paris. 1931. T. 192. P. 727. [*251*]

[486] Nikodym, O. *Sur les familles bornées de fonctions parfaitement additives d'ensemble abstrait.* Monatsh. Math. Phys. 1933. B. 40. S. 418–426. DOI 10.1007/BF01708879. MR1550216 [*251*]

[487] Nikodym, O. *Sur les suites convergentes de fonctions parfaitement additives d'ensemble abstrait.* Monatsh. Math. Phys. 1933. B. 40. S. 427–432. DOI 10.1007/BF01708880. MR1550217 [*251*]

[488] Nikol'skiĭ, S. M. Approximation of functions of several variables and imbedding theorems. Translated from the Russian. Springer, New York – Heidelberg, 1975; viii+418 pp. [*127*]

[489] Nikunen, M. *On the weak compactness of a family of measures corresponding to continuous strong Markov processes.* Teor. Veroyatn. Primen. 1980. V. 25, №1. P. 157–161 (in Russian); English transl.: Theory Probab. Appl. 1980. V. 25, №1. P. 155–159. MR560068 [*251*]

[490] Nikunen, M. *Limit theorems for certain continuous Markov processes.* Ann. Acad. Sci. Fenn. Ser. A I Math. Dissertationes. 1980. №28. P. 1–43. MR575532 [*251*]

[491] Nourdin, I., Nualart, D., Poly, G. *Absolute continuity and convergence of densities for random vectors on Wiener chaos.* Electron. J. Probab. 2013. V. 18, №22. P. 1–19. DOI 10.1214/EJP.v18-2181. MR3035750 [*130*]

[492] Nourdin, I., Peccati, G. Normal approximations with Malliavin calculus. From Stein's method to universality. Cambridge University Press, Cambridge, 2012; xiv+239 pp. MR2962301 [*130*]

[493] Nourdin, I., Poly, G. *Convergence in total variation on Wiener chaos.* Stochastic Process. Appl. 2013. V. 123, №2. P. 651–674. DOI 10.1016/j.spa.2012.10.004. MR3003367 [*130*]

[494] O'Brien, G. L., Watson, S. *Relative compactness for capacities, measures, upper semicontinuous functions and closed sets.* J. Theor. Probab. 1998. V. 11, №3. P. 577–588. DOI 10.1023/A:1022659912007. MR1633366 [*248*]

[495] Oppel, U. G. *Zur Charakterisierung Suslinscher und Lusinscher Räume.* Z. Wahrscheinlichkeitstheorie Verw. Geb. 1976. B. 34, №3. S. 183–192. DOI 10.1007/BF00532701. MR0399394 [*250*]

[496] Oppel, U. *Zur schwachen Topologie auf dem Vektorraum der Borel-Masse polnischer und Lusinscher Räume.* Math. Z. 1976. B. 147, №1. S. 97–99. DOI 10.1007/BF01214279. MR0407223 [*250*]

[497] Ozawa, R. *Distance between metric measure spaces and distance matrix distributions.* Tsukuba J. Math. 2015. V. 38, №2. P. 159–170. MR3336265 [*125*]

[498] Ozawa, R., Shioya, T. *Limit formulas for metric measure invariants and phase transition property.* Math. Z. 2015. B. 280, №3-4. S. 759–782. DOI 10.1007/s00209-015-1447-2. MR3369350 [*252*]

[499] Pachl, J. K. *Measures as functionals on uniformly continuous functions.* Pacif. J. Math. 1979. V. 82, №2. P. 515–521. MR551709 [*137*]

[500] Pachl, J. Uniform spaces and measures. Springer, New York; Fields Institute, Toronto, 2013; x+209 pp. MR2985566 [*251*]

[501] Padmanabhan, A.R. *Convergence in probability and allied results.* Math. Japon. 1970. V. 15. P. 111–117. MR0288798 [*78*]

[502] Pakshirajan, R.P. *A note on the weak convergence of probability measures in the* $D[0,1]$ *space.* Statist. Probab. Lett. 2008. V. 78, №6. P. 716–719. DOI 10.1016/j.spl.2007.09.034. MR2409536 [*248*]

[503] Parthasarathy, K. R. Probability measures on metric spaces. Academic Press, New York, 1967; xi+276 pp. MR0226684 [*246*]

[504] Parthasarathy, K. R. Introduction to probability and measure. Springer-Verlag, New York, Macmillan India, 1978; xii+312 pp.; 2nd rev. ed.: Hindustan Book Agency, New Delhi, 2005; 338 pp. MR2190360 [*246*]

[505] Paulauskas, V. I., Račkauskas, A. Yu. Approximation theory in the central limit theorem. Exact results in Banach spaces. Translated from the Russian. Kluwer, Dordrecht, 1989; xviii+156 pp. MR1015294 [*189*]

[506] Peccati, G., Taqqu, M. S. Wiener chaos: moments, cumulants and diagrams. A survey with computer implementation. Bocconi & Springer Series, 1. Springer, Milan; Bocconi University Press, Milan, 2011; xiv+274 pp. MR2791919 [*130*]

[507] Pedregal, P. Parametrized measures and variational principles. Birkhäuser, Basel – Berlin, 1997; xi+212 pp. MR1452107 [*231, 251*]

[508] Pełczyński, A. Linear extensions, linear averagings, and their applications to linear topological classification of spaces of continuous functions. Dissert. Math. (Rozprawy Mat.) 1968. V. 58. 92 pp. MR0227751 [*76*]

[509] Perlman, M. D. *Characterizing measurability, distribution and weak convergence of random variables in a Banach space by total subsets of linear functionals.* J. Multivar. Anal. 1972. V. 2, №3. P. 174–188. DOI 10.1016/0047-259X(72)90025-5. MR0307288 [*251*]

[510] Petrov, V. V. Sums of independent random variables. Translated from the Russian. Springer, New York, 1975; x+346 pp. MR0388499 [*22, 188*]

[511] Pfanzagl, J. *Convergent sequences of regular measures.* Manuscr. Math. 1971. V. 4, №1. P. 91–98. DOI 10.1007/BF01168906. MR0285683 [*226, 227*]

[512] Plebanek, G. *Approximating Radon measures on first-countable compact spaces.* Colloq. Math. 2000. V. 86, №1. P. 15–23. DOI 10.4064/cm-86-1-15-23. MR1799884 [*222*]

[513] Plebanek, G., Sobota, D. *Countable tightness in the spaces of regular probability measures.* Fund. Math. 2015. V. 229, №2. P. 159–170. DOI 10.4064/fm229-2-4. MR3315379 [*251*]

[514] Pol, R. *Note on the spaces $P(S)$ of regular probability measures whose topology is determined by countable subsets.* Pacif. J. Math. 1982. V. 100, №1. P. 185–201. MR661448 [*233*]

[515] Pollard, D. *Compact sets of tight measures.* Studia Math. 1976. V. 56, №1. P. 63–67. DOI 10.4064/sm-56-1-63-67. MR0415292 [*248*]

[516] Pollard, D. *Induced weak convergence and random measures.* Z. Wahrscheinlichkeitstheorie Verw. Geb. 1977. B. 37, №4. S. 321–328. DOI 10.1007/BF00533423. MR571672 [*251*]

[517] Pollard, D. *Weak convergence on nonseparable metric spaces.* J. Austral. Math. Soc. Ser. A. 1979. V. 28, №2. P. 197–204. MR550962 [*251*]

[518] Pollard, D. Convergence of stochastic processes. Springer, Berlin – New York, 1984; xiv+215 pp. MR762984 [*85, 246*]

[519] Polovinkin, E. S., Balashov, M. V. Elements of convex and strongly convex analysis. Fizmatlit, Moscow, 2004; 416 pp. (in Russian). [*41*]

[520] Pratelli, A. *On the equality between Monge's infimum and Kantorovich's minimum in optimal mass transportation.* Ann. Inst. H. Poincaré (B) Probab. Statist. 2007. B. 43, №1. P. 1–13. DOI 10.1016/j.anihpb.2005.12.001. MR2288266 [*252*]

[521] Preiss, D. *Metric spaces in which Prohorov's theorem is not valid.* Z. Wahrscheinlichkeitstheorie Verw. Geb. 1973. B. 27. S. 109–116. DOI 10.1007/BF00536621. MR0360979 [*181, 183*]

[522] Prigarin, S. M. *Weak convergence of probability measures in spaces of continuously differentiable functions.* Sibirsk. Mat. Zh. 1993. V. 34, №1. P. 140–144 (in Russian); English transl.: Sib. Math. J. 1993. V. 34, №1. P. 123–127. DOI 10.1007/BF00971248. MR1216843 [*251*]

[523] Prigent, J.-L. Weak convergence of financial markets. Springer-Verlag, Berlin, 2003; xiv+422 pp. MR2036683 [*251*]

[524] Prohorov, Yu. V. *Probability distributions in functional spaces.* Uspehi Mat. Nauk. 1953. V. 8, №3. P. 165–167 (in Russian). MR0057482 [*245*]

[525] Prohorov, Yu. V. *Convergence of random processes and limit theorems in probability theory.* Teor. Veroyatn. Primen. 1956. V. 1, №2. P. 177–238 (in Russian); English transl.: Theory Probab. Appl. 1956. V. 1. P. 157–214. MR0084896 [*59, 64, 73, 74, 245, 246, 248, 250, 252*]

[526] Prohorov, Yu. V. *The method of characteristic functionals.* Proc. 4th Berkeley Symp. Math. Statist. and Probab. V. 2. P. 403–419. University of California Press, Berkeley, 1960. MR0133846 [*251*]

[527] Prohorov, Yu. V., Sazonov, V. V. *Some results associated with Bochner's theorem.* Teor. Veroyatn. Primen. 1961. V. 6, №1. P. 87–93 (in Russian); English transl.: Theory Probab. Appl. 1961. V. 6. P. 82–87. MR0149239 [*74, 251*]

[528] Pugachev, O. V. *The space of simple configurations is Polish.* Matem. Zametki. 2002. V. 71, №4. P. 581–589 (in Russian); English transl.: Math. Notes. 2002. V. 71, №4. P. 530–537. DOI 10.1023/A:1014835916189. MR1913587 [*251*]

[529] Rachev, S. T. *The Monge–Kantorovich problem on mass transfer and its applications in stochastics.* Teor. Veroyatn. Primen. 1984. V. 29, №4. P. 625–653 (in Russian); English transl.: Theory Probab. Appl. 1984. V. 29, №4. P. 647–676. MR773434 [*249*]

[530] Rachev, S. T. Probability metrics and the stability of stochastic models. Wiley, Chichester, 1991; xiv+494 pp. MR1105086 [*249*]

[531] Rachev, S. T., Klebanov, L. B., Stoyanov, S. V., Fabozzi, F. J. The methods of distances in the theory of probability and statistics. Springer, New York, 2013; xvi+619 pp. MR3024835 [*249*]

[532] Rachev, S. T., Rüschendorf, L. Mass transportation problems. V. 1. Springer, New York, 1998; 508 pp. MR1619170 [*105, 116, 249*]

[533] Rachev, S. T., Shortt, R. M. *Duality theorems for Kantorovich–Rubinstein and Wasserstein functionals.* Dissertationes Math. (Rozprawy Mat.). 1990. V. 299. 35 pp. MR1074632 [*249*]

[534] Rachev, S. T., Stoyanov, S. V., Fabozzi, F. J. A probability metrics approach to financial risk measures. Wiley–Blackwell, Chichester, 2011; xvi+375 pp. [*249*]

[535] Radon, J. *Theorie und Anwendungen der absolut additiven Mengenfunktionen.* Sitz. Akad. Wiss. Wien, Math.-naturwiss. Kl. IIa. 1913. B. 122. S. 1295–1438. [*245, 246*]

[536] Radon, J. *Über lineare Funktionaltransformationen und Funktionalgleichungen.* Sitz. Akad. Wiss. Wien, Math.-naturwiss. Kl. IIa. 1919. B. 128, H. 7. S. 1–39. [*246, 251*]

[537] Radon, J. Gesammelte Abhandlungen, V. 1, 2. Birkhäuser, Basel – Boston, 1987. MR925206 [*251*]

[538] Ramachandran, B. Advanced theory of characteristic functions. Statist. Publ. Soc., Calcutta, 1967; vii+208 pp. MR0225356 [*40*]

[539] Ramachandran, D., Rüschendorf, L. *A general duality theorem for marginal problems.* Probab. Theory Related Fields. 1995. V. 101, №3. P. 311–319. DOI 10.1007/BF01200499. MR1324088 [*249*]

[540] Ramachandran, D., Rüschendorf, L. *Duality and perfect probability spaces.* Proc. Amer. Math. Soc. 1996. V. 124, №7. P. 2223–2228. DOI 10.1090/S0002-9939-96-03462-4. MR1342043 [*249*]

[541] Rao, R. R. *Relations between weak and uniform convergences of measures with applications.* Ann. Math. Statist. 1962. V. 33, №2. P. 659–680. DOI 10.1214/aoms/1177704588. MR0137809 [*55, 97, 248*]

[542] Raynaud de Fitte, P. *Compactness criteria for the stable topology.* Bull. Polish Acad. Sci. Math. 2003. V. 51, №4. P. 343–363. MR2025306 [*230, 231*]

[543] Rebolledo, R. *La méthode des martingales appliquée a l'étude de la convergence en loi de processus.* Bull. Soc. Math. France. 1979. Mem. 62. P. 1–125. [*251*]

[544] Rebolledo, R. *Topologie faible et méta-stabilité.* Lecture Notes in Math. 1987. V. 1247. P. 544–562. DOI 10.1007/BFb0077655. MR942004 [*251*]

[545] Reed, M., Simon, B. Methods of modern mathematical physics. I. Functional analysis. 2nd ed. Academic Press, New York, 1980; xv+400 pp. MR751959 [*139*]

[546] Rényi, A. *On stable sequences of events.* Sankhyā. Ser. A. 1963. V. 25. P. 293–302. MR0170385 [*231*]

[547] Repovš, D., Semenov, P. V. Continuous selections of multivalued mappings. Kluwer, Dordrecht, 1998; 356 pp. MR1659914 [*209, 234*]

[548] Reshetnyak, Yu. G. *General theorems on semicontinuity and convergence with a functional.* Sibirsk. Mat. Zh. 1967. V. 8, №5. P. 1051–1069 (in Russian); English transl.: Sib. Math. J. 1967. V. 8. P. 801–816. MR0220127 [*44*]

[549] Ressel, P. *Some continuity and measurability results on spaces of measures.* Math. Scand. 1977. V. 40, №1. P. 69–78. DOI 10.7146/math.scand.a-11676. MR0486384 [*234*]

[550] Ressel, P. *A topological version of Slutsky's theorem.* Proc. Amer. Math. Soc. 1982. V. 85, №2. P. 272–274. DOI 10.2307/2044295. MR652456 [*197*]

[551] Rio, E. *Distances minimales et distances idéales.* C. R. Acad. Sci. Paris. 1998. T. 326. P. 1127–1130. DOI 10.1016/S0764-4442(98)80074-8. MR1647215 [*125, 126*]

[552] Rio, E. *Upper bounds for minimal distances in the central limit theorem.* Ann. Inst. H. Poincaré. 2009. V. 45, №3. P. 802–817. DOI 10.1214/08-AIHP187. MR2548505 [*125, 126*]

[553] Rockafellar, R. T. Convex analysis. Princeton University Press, Princeton, New Jersey, 1970; xviii+451 pp. MR0274683 [*40*]

[554] Rogge, L. *The convergence determining class of regular open sets.* Proc. Amer. Math. Soc. 1973. V. 37, №2. P. 581–585. DOI 10.2307/2039489. MR0311872 [*227*]

[555] Romanovsky, V. *Sur un théorème limite du calcul des probabilités.* Mat. Sbornik. (Recuil Math. Soc. Math. Moscou). 1929. V. 36. P. 36–64. [*246*]

[556] Rotar, V. Probability theory. Translated from the Russian. World Sci., River Edge, New Jersey, 1997; xviii+414 pp. MR1641490 [*xi, 246*]

[557] Roussas, G. G. Contiguity of probability measures: some applications in statistics. Cambridge University Press, London – New York, 1972; xiii+248 pp. MR0359099 [*44*]

[558] Sadi, H. *Une condition nécessaire et suffisante pour la convergence en pseudo-loi des processus.* Lecture Notes in Math. 1988. V. 1321. P. 434–437. DOI 10.1007/BFb0084148. MR960538 [*248*]

[559] Sadovnichii, Yu. V. *On the Kantorovich norm for alternating measures.* Dokl. Ros. Akad. Nauk. 1999. V. 368, №4. P. 459–461 (in Russian); English transl.: Dokl. Math. 1999. V. 60, №2. P. 223–225. MR1748157 [*249*]

[560] Saint-Raymond, J. *Caractérisation d'espaces polonais. D'après des travaux récents de J. P. R. Christensen et D. Preiss.* Séminaire Choquet (Initiation à l'analyse). 11–12e années. 1971–1973. №5. 10 pp. MR0473133 [*250*]

[561] Saks, S. *On some functionals.* Trans. Amer. Math. Soc. 1933. V. 35, №2. P. 549–556; Addition: ibid., №4. P. 965–970. DOI 10.2307/1989603. MR1501728 [*251*]

[562] Sasvári, Z. Multivariate characteristic and correlation functions. De Gruyter, Berlin, 2013; x+366 pp. MR3059796 [*44, 246*]

[563] Saulis, L., Statulevicius, V. A. Limit theorems for large deviations. Translated from the Russian. Kluwer, Dordrecht, 1991; viii+232 pp. MR1171883 [*251*]

[564] Sazonov, V. V. *On characteristic functionals.* Teor. Veroyatn. Primen. 1958. V. 3, №2. P. 201–205 (in Russian); English transl.: Theory Probab. Appl. 1958. V. 3. P. 188–192. MR0098423 [*245, 252*]

[565] Schachermayer, W. *On some classical measure-theoretic theorems for non-sigma-complete Boolean algebras.* Dissertationes Math. (Rozprawy Mat.). 1982. V. 214. P. 1–33. MR673286 [*226*]

[566] Schaefer, H. H. Topological vector spaces. Springer-Verlag, Berlin – New York, 1971; xi+294 pp. MR0342978 [*162*]

[567] Schäl, M. *On dynamic programming: compactness of the space of policies.* Stoch. Processes Appl. 1975. V. 3. P. 345–364. MR0386706 [*231*]

[568] Schief, A. *The continuity of subtraction and the Hausdorff property in spaces of Borel measures.* Math. Scand. 1988. V. 63, №2. P. 215–219. DOI 10.7146/math.scand.a-12235. MR1018811 [*242*]

[569] Schief, A. *Topological properties of the addition map in spaces of Borel measures.* Math. Ann. 1988. B. 282, №1. S. 23–31. DOI 10.1007/BF01457010. MR960831 [*242*]

[570] Schief, A. *On continuous image averaging of Borel measures.* Topol. Appl. 1989. V. 31, №3. P. 309–315. DOI 10.1016/0166-8641(89)90027-8. MR997498 [*205*]

[571] Schief, A. *An open mapping theorem for measures.* Monatsh. Math. 1989. B. 108, №1. S. 59–70. DOI 10.1007/BF01300067. MR1018825 [*205, 240, 241*]

[572] Schief, A. *Almost surely convergent random variables with given laws.* Probab. Theory Related Fields. 1989. V. 81. P. 559–567. DOI 10.1007/BF00367303. MR995811 [*247*]

[573] Schwartz, L. Radon measures on arbitrary topological spaces and cylindrical measures. Oxford University Press, London, 1973; xii+393 pp. MR0426084 [*250*]

[574] Seis, C. *Maximal mixing by incompressible fluid flows.* Nonlinearity. 2013. V. 26, №12. P. 3279–3289. DOI 10.1088/0951-7715/26/12/3279. MR3141856 [*126*]

[575] Senatov, V. V. *On some properties of metrics on the set of distribution functions.* Mat. Sbornik. 1977. V. 102, №3. P. 425–434 (in Russian); English transl.: Math. USSR-Sbornik. 1977. V. 31, №3. P. 379–387. MR0436248 [*249*]

[576] Senatov, V. V. Normal approximation: new results, methods and problems. Translated from the Russian. VSP, Utrecht, 1998; viii+363 pp. MR1686374 [*22*]

[577] Sentilles, F. D. *Compactness and convergence in the space of measures.* Illinois J. Math. 1969. V. 13. P. 761–768. MR0247447 [*251*]

[578] Sentilles, F. D. *Bounded continuous functions on a completely regular space.* Trans. Amer. Math. Soc. 1972. V. 168. P. 311–336. DOI 10.2307/1996178. MR0295065 [*250*]

[579] Sevastyanov, B. A. *A class of limit distributions for quadratic forms of normal stochastic variables.* Teor. Veroyatn. Primen. 1961. V. 6. P. 368–372 (in Russian); English transl.: Theory Probab. Appl. 1961. V. 6. P. 337–340. [*130*]

[580] Shioya, T. Metric measure geometry. Gromov's theory of convergence and concentration of metrics and measures. Europ. Math. Soc., Zurich, 2016; 194 p. MR3445278 [*125, 249, 252*]

[581] Shiryaev, A. N. Probability. Translated from the Russian. Springer-Verlag, New York, 1996; xvi+623 pp. (3d Russian ed.: Moscow, 2004). MR1368405 [*xi, 246*]

[582] Shorack, G. R., Wellner, J. A. Empirical processes with applications to statistics. John Wiley, New York, 1986; xxxviii+938 pp. MR838963 [*193*]

[583] Skorohod, A. V. *On the limiting transition from a sequence of sums of independent random quantities to a homogeneous random process with independent increments.* Dokl. Akad. Nauk SSSR. 1955. V. 104, №3. P. 364–367 (in Russian). MR0077801 [*248*]

[584] Skorohod, A. V. *Limit theorems for stochastic processes.* Teor. Veroyatn. Primen. 1956. V. 1. P. 261–290 (in Russian); English transl.: Theory Probab. Appl. 1956. V. 1. P. 261–290. MR0084897 [*75, 88, 245, 248*]

[585] Skorokhod, A. V. Studies in the theory of random processes. Translated from the Russian. Addison-Wesley, Reading, Mass., 1965; viii+199 pp. MR0185620 [*75, 245*]

[586] Skorohod, A. V. Integration in Hilbert space. Translated from the Russian. Springer-Verlag, Berlin – New York, 1974; xii+177 pp. MR0466482 [*246*]

[587] Slutsky, E. *Über stochastische Asymptoten und Grenzwerte.* Metron. 1925. V. 5. P. 3–89. [*99*]

[588] Smith, M. A., Turett, B. *Rotundity in Lebesgue–Bochner function spaces.* Amer. Math. Soc. 1980. V. 237, №1. P. 105–118. DOI 10.2307/1998127. MR549157 [*83*]

[589] Smolyanov, O. G. Analysis on topological linear spaces and its applications. Moskov. Gos. Univ., Moscow, 1979; 86 pp. (in Russian). [*250*]

[590] Smolyanov, O. G., Fomin, S. V. *Measures on topological linear spaces.* Uspehi Mat. Nauk. 1976. V. 31, №4. P. 3–56 (in Russian); English transl.: Russian Math. Surveys. 1976. V. 31, №4. P. 1–53. MR0420764 [*251*]

[591] Smolyanov, O. G., Shavgulidze, E. T. Continual integrals. 2nd ed. URSS, Moscow, 2015; 336 pp. [*246*]

[592] Steck, G. P. *Limit theorems for conditional distributions.* Univ. California Publ. Statist. 1957. V. 2. P. 237–284. MR0091552 [*244*]

[593] Stegall, C. *The topology of certain spaces of measures.* Topology Appl. 1991. V. 41. P. 73–112. DOI 10.1016/0166-8641(91)90102-R. MR1129700 [*250*]

[594] Stein, J. D., Jr. *A uniform boundedness theorem for measures.* Michigan Math. J. 1972. V. 19, №2. P. 161–165. MR0299746 [*227*]

[595] Steutel, F. W., van Harn, K. Infinite divisibility of probability distributions on the real line. Marcel Dekker, New York, 2004; xii+546 pp. MR2011862 [*40*]

[596] Stone, C. *Weak convergence of stochastic processes defined on semi-infinite time intervals.* Proc. Amer. Math. Soc. 1963. V. 14. P. 694–696. DOI 10.2307/2034973. MR0153046 [*251*]

[597] Strassen, V. *The existence of probability measures with given marginals.* Ann. Math. Statist. 1965. V. 36. P. 423–439. DOI 10.1214/aoms/1177700153. MR0177430 [*105*]

[598] Stroock, D. W. Probability theory: an analytic view. Cambridge University Press, Cambridge, 1993; xv+512 pp. MR1267569 [*22, 246*]

[599] Stroock, D. W., Varadhan, S. R. S. Multidimensional diffusion processes. Springer, Berlin – New York, 1979; xii+338 pp. MR532498 [*246*]

[600] Sturm, K.-Th. *On the geometry of metric measure spaces, I, II.* Acta Math. 2006. V. 196. P. 65–131; 133–177. DOI 10.1007/s11511-006-0003-7. MR2237207 [*123, 252*]

[601] Sudakov, V. N. *Geometric problems of the theory of infinite–dimensional probability distributions.* Trudy Mat. Inst. Steklov. 1976. V. 141. P. 1–190 (in Russian); English transl.: Proc. Steklov Inst. Math. 1979. №2. P. 1–178. MR0431359 [*249*]

[602] Sun, Y. *Isomorphisms for convergence structures.* Adv. Math. 1995. V. 116, №2. P. 322–355. DOI 10.1006/aima.1995.1069. MR1363767 [*222*]

[603] Szczotka, W. *A note on Skorokhod representation.* Bull. Pol. Acad. Sci. Math. 1990. V. 38. P. 35–39. MR1194243 [*247*]

[604] Szulga, A. *On minimal metrics in the space of random variables.* Theory Probab. Appl. 1982. V. 27, №2. P. 424–430. MR657942 [*249*]

[605] Takatsu, A. *Wasserstein geometry of Gaussian measures.* Osaka J. Math. 2011. V. 48. P. 1005–1026. MR2871291 [*252*]

[606] Talagrand, M. *Les boules peuvent engendrer la tribu borélienne d'un espace métrizable non séparable?* Séminaire Choquet, 17e anée. 1977–1978. F. 5, №2, 2 pp. Paris, 1978. MR522993 [*96*]

[607] Talagrand, M. *Separabilité vague dans l'espace des mesures sur un compact.* Israel J. Math. 1980. V. 37, №1–2. P. 171–180. DOI 10.1007/BF02762878. MR599312 [*233, 250*]

[608] Talagrand, M. *The Glivenko–Cantelli problem.* Ann. Probab. 1987. V. 15. P. 837–870. MR893902 [*193*]

[609] Tikhonov, Yu. V., Shaposhnikov S. V., Sheipak, I. A. *On singularity of functions and quantization of probability measures.* Matem. Zametki. 2017. V. 102, №4. P. 628–631 (in Russian); English transl.: Math. Notes. 2017. V. 102, №3-4. P. 587–590. DOI 10.4213/mzm11618. MR3706880 [*138*]

[610] Topsøe, F. *Preservation of weak convergence under mappings.* Ann. Math. Statist. 1967. V. 38, №6. P. 1661–1665. DOI 10.1214/aoms/1177698600. MR0219097 [*248*]

[611] Topsøe, F. *A criterion for weak convergence of measures with an application to convergence of measures on* $D[0,1]$. Math. Scand. 1969. V. 25. P. 97–104. DOI 10.7146/math.scand.a-10944. MR0254910 [*248*]

[612] Topsøe, F. Topology and measure. Lecture Notes in Math. V. 133. Springer-Verlag, Berlin – New York, 1970; xiv+79 pp. MR0422560 [*239*]

[613] Topsøe, F. *Compactness in spaces of measures.* Studia Math. 1970. V. 36, №3. P. 195–212. DOI 10.4064/sm-36-3-195-212. MR0268347 [*248*]

[614] Topsøe, F. *Compactness and tightness in a space of measures with the topology of weak convergence.* Math. Scand. 1974. V. 34, №2. P. 187–210. DOI 10.7146/math.scand.a-11520. MR0388484 [*163, 186, 250*]

[615] Topsøe, F. *Some special results on convergent sequences of Radon measures.* Manuscripta Math. 1976. V. 19. P. 1–14. DOI 10.1007/BF01172334. MR0412374 [*248*]

[616] Topsøe, F. *Uniformity in weak convergence with respect to balls in Banach spaces.* Math. Scand. 1976. V. 38, №1. P. 148–158. DOI 10.7146/math.scand.a-11624. MR0407224 [*248*]

[617] Tortrat, A. Calcul des probabilités et introduction aux processus aléatoires. Masson, Paris, 1971; xiv+303 pp. MR0375403 [*xi*]

[618] Toruńczyk, H. *A short proof of Hausdorff's theorem on extending metrics.* Fund. Math. 1972. V. 77, №2. P. 191–193. DOI 10.4064/fm-77-2-191-193. MR0321026 [*95, 96, 133*]

[619] Tsukahara, H. *On the convergence of measurable processes and prediction processes.* Illinois J. Math. 2007. V. 51, №4. P. 1231–1242. MR2417423 [*248*]

[620] Tuero, A. *On the stochastic convergence of representations based on Wasserstein metrics.* Ann. Probab. 1993. V. 21, №1. P. 72–85. MR1207216 [*247*]

[621] Tyurin, I. S. *Improvement of the remainder in the Lyapunov theorem.* Teor. Veroyatn. Primen. 2011. V. 56, №4. P. 808–811 (in Russian); English transl.: Theory Probab. Appl. 2012. V. 56, №4. P. 693–696. DOI 10.1137/S0040585X9798572X. MR3137072 [*37*]

[622] Ulam, S. *On the distribution of a general measure in any complete metric space.* Bull. Amer. Math. Soc. 1938. V. 44. P. 786. [*247*]

[623] Urysohn, P. *Sur un espace métrique universel.* Bull. Sci. Math. 1927. V. 51. P. 1–38. [*123*]

[624] Ushakov, N. G. Selected topics in characteristic functions. VSP, Utrecht, 1999; x+355 pp. MR1745554 [*44, 246*]

[625] van der Vaart, A. W., Wellner, J. A. Weak convergence and empirical processes. With applications to statistics. Springer-Verlag, New York, 1996; xvi+508 pp. MR1385671 [*85, 193, 248*]

[626] van der Vaart, A. W., Wellner, J. A. *Preservation theorems for Glivenko–Cantelli and uniform Glivenko–Cantelli classes.* In: High Dimensional Probability II (Giné E., Mason D., Wellner J. A., eds.). Progress in Probab. V. 47. P. 115–133. Birkhäuser, 2000. MR1857319 [*193*]

[627] Valadier, M. *Young measures.* Lecture Notes in Math. 1990. V. 1446. P. 152–188. DOI 10.1007/BFb0084935. MR1079763 [*231*]

[628] Valadier, M. *A course on Young measures.* Workshop di Teoria della Misura e Analisi Reale (Grado, 1993). Rend. Istit. Mat. Univ. Trieste. 1994. T. 26, suppl., pp. 349–394. MR1408956 [*231*]

[629] Vakhania, N. N., Tarieladze, V. I., Chobanyan, S. A. Probability distributions in Banach spaces. Translated from the Russian. Kluwer, 1991; xxvi+482 pp. MR1435288 [*89, 169, 189, 192, 246, 251*]

[630] Vallander, S. S. *Calculation of the Wasserstein distance between probability distributions on the line.* Teor. Veroyatn. Primen. 1973. V. 18, №4. P. 824–827 (in Russian); English transl.: Theory Probab. Appl. 1974. V. 18, №4. P. 784–786. MR0328982 [*110*]

[631] Valov, V. *Probability measures and Milyutin maps between metric spaces.* J. Math. Anal. Appl. 2009. V. 350, №2. P. 723–730. DOI 10.1016/j.jmaa.2008.06.003. MR2474807 [*250*]

[632] Varadarajan, V. S. *Weak convergence of measures on separable metric spaces.* Sankyā. 1958. V. 19. P. 15–22. MR0094838 [*251*]

[633] Varadarajan, V. S. *On the convergence of probability distributions.* Sankyā. 1958. V. 19. P. 23–26. MR0094839 [*251*]

[634] Varadarajan, V. S. *Convergence of stochastic processes.* Bull. Amer. Math. Soc. 1961. V. 67. P. 276–280. DOI 10.1090/S0002-9904-1961-10584-3. MR0125625 [*251*]

[635] Varadarajan, V. S. *Measures on topological spaces.* Mat. Sbornik. 1961. V. 55. P. 35–100 (in Russian); English transl.: Amer. Math. Soc. Transl. (2). 1965. V. 48. P. 161–228. MR0148838 [*80, 97, 160, 177, 195, 245, 248, 250*]

[636] Varadarajan, V. S. *Groups of automorphisms of Borel spaces.* Trans. Amer. Math. Soc. 1963. V. 109. P. 191–220. DOI 10.2307/1993903. MR0159923 [*250*]

[637] Vasershtein, L. N. *Markov processes over denumerable products of spaces describing large system of automata.* Probl. Peredači Inform. 1969. V. 5, №3. P. 64–72 (in Russian); English transl.: Problems Inform. Transmission. 1969. V. 5, №3. P. 47–52. MR0314115 [*248*]

[638] Vershik, A. M. *Some remarks on the infinite-dimensional problems of linear programming.* Uspehi Matem. Nauk. 1970. V. 25, №5. P. 117–124 (in Russian); English transl.: Russian Math. Surveys. 1970. V. 25, №5. P. 117–124. MR0295754 [*249*]

[639] Vershik, A. M. *The universal Uryson space, Gromov metric triples and random metrics on the natural numbers.* Uspehi Matem. Nauk. 1998. V. 53, №5. P. 57–64 (in Russian); English transl.: Russian Math. Surveys. 1998. V. 53, №5. P. 921–928. DOI 10.1070/rm1998v053n05ABEH000069. MR1691182 [*125*]

[640] Vershik, A. M. *Classification of measurable functions of several variables and invariantly distributed random matrices.* Funkt. Anal. i Pril. 2002. V. 36, №2. P. 12–27 (in Russian); English transl.: Funct. Anal. Appl. 2002. V. 36, №2. P. 93–105. DOI 10.1023/A:1015662321953. MR1922015 [*122, 125*]

[641] Vershik, A. M. *Random metric spaces and universality.* Uspehi Matem. Nauk. 2004. V. 59, №2. P. 65–104 (in Russian); English transl.: Russian Math. Surveys. 2004. V. 59, №2. P. 259–295. DOI 10.1070/RM2004v059n02ABEH000718. MR2086637 [*125*]

[642] Vershik, A. M. *The Kantorovich metric: the initial history and little-known applications.* Zap. Nauchn. Sem. S.-Peterburg. Otdel. Mat. Inst. Steklov. 2004. T. 312. P. 69–85 (in Russian); English transl.: J. Math. Sci. (New York). 2006. V. 133, №4. P. 1410–1417. DOI 10.1007/s10958-006-0056-3. MR2117883 [*125, 249*]

[643] Vershik, A. M. *On classification of measurable functions of several variables.* Zap. Nauchn. Sem. S.-Peterburg. Otdel. Mat. Inst. Steklov. 2012. T. 403. P. 35–57 (in Russian); English transl.: J. Math. Sci. (New York). 2013. V. 190, №3. P. 427–437. MR3029579 [*125*]

[644] Vershik, A. M. *Two ways of defining compatible metrics on the simplex of measures.* Zap. Nauchn. Sem. S.-Peterburg. Otdel. Mat. Inst. Steklov. 2013. T. 411. P. 38–48 (in Russian); English transl.: J. Math. Sci. (New York). 2014. V. 196, №2. P. 138–143. MR3048267 [*138*]

[645] Vershik, A. M., Kutateladze, S. S., Novikov, S. P. *Leonid Vital'evich Kantorovich* (*on the 100th anniversary of his birth*). Uspehi Mat. Nauk. 2012. V. 67, №3. P. 185–191 (in Russian); English transl.: Russian Math. Suveys. 2012. V. 67, №3. P. 589–597. DOI 10.1070/RM2012v067n03ABEH004802. MR3024853 [*249*]

[646] Vershik, A., Petrov, F., Zatitskiy, P. *Geometry and dynamics of admissible metrics in measure spaces.* Central Europ. J. Math. 2013. V. 11, №3. P. 379–400. DOI 10.2478/s11533-012-0149-9. MR3016311 [*138*]

[647] Vershik, A. M., Zatitskii, P. B., Petrov, F. V. *Virtual continuity of measurable functions and its applications.* Uspehi Matem. Nauk. 2014. V. 69, №6. P. 81–114 (in Russian); English

transl.: Russian Math. Surveys. 2014. V. 69, №6. P. 1031–1063. DOI 10.4213/rm9628. MR3400556 [*125*]

[648] Villani, C. Topics in optimal transportation. Amer. Math. Soc., Rhode Island, 2003; 370 pp. MR1964483 [*105, 116, 252*]

[649] Villani, C. Optimal transport, old and new. Springer, New York, 2009; xxii+973 pp. MR2459454 [*116, 249, 252*]

[650] Visintin, A. *Strong convergence results related to strict convexity.* Comm. Partial Differ. Equ. 1984. V. 9, №5. P. 439–466. DOI 10.1080/03605308408820337. MR741216 [*44*]

[651] Vitali, G. *Sull'integrazione per serie.* Rend. Circ. Mat. Palermo. 1907. T. 23. P. 137–155. [*251*]

[652] Wells, B. B., Jr. *Weak compactness of measures.* Proc. Amer. Math. Soc. 1969. V. 20, №3. P. 124–134. DOI 10.2307/2035973. MR0238067 [*251*]

[653] Wentzell, A. D. A course in the theory of stochastic processes. Translated from the Russian. McGraw-Hill International Book, New York, 1981; x+304 pp. MR614594 [*91*]

[654] Weyl, H. *Über die Gleichverteilung von Zahlen mod. Eins.* Math. Ann. 1916. B. 77. S. 313–352. DOI 10.1007/BF01475864. MR1511862 [*219, 243*]

[655] Wheeler, R. F. *A survey of Baire measures and strict topologies.* Expos. Math. 1983. V. 1, №2. P. 97–190. MR710569 [*202, 250*]

[656] Whitt, W. *Weak convergence of probability measures on the function space* $C[0,\infty)$. Ann. Math. Statist. 1970. V. 41. P. 939–944. DOI 10.1214/aoms/1177696970. MR0261646 [*251*]

[657] Wichura, M. J. *Inequalities with applications to the weak convergence of random processes with multidimensional time parameter.* Ann. Math. Statist. 1969. V. 40, №2. P. 681–687. DOI 10.1214/aoms/1177697741. MR0246359 [*251*]

[658] Wichura, M. J. *On the construction of almost uniformly convergent random variables with given weakly convergent image laws.* Ann. Math. Statist. 1970. V. 41, №1. P. 284–291. DOI 10.1214/aoms/1177697207. MR0266275 [*247*]

[659] Wichura, M. J. *A note on the weak convergence of stochastic processes.* Ann. Math. Statist. 1971. V. 42, №5. P. 1769–1772. DOI 10.1214/aoms/1177693181. MR0378016 [*99*]

[660] Wilson, R. J. *Weak convergence of probability measures in spaces of smooth functions.* Stoch. Proc. Appl. 1986. V. 23, №2. P. 333–337. DOI 10.1016/0304-4149(86)90047-5. MR876056 [*251*]

[661] Wójcicka, M. *The space of probability measures on a Prohorov space is Prohorov.* Bull. Polish Acad. Sci., Math. 1987. V. 35, №11–12. P. 809–811. MR961721 [*175*]

[662] Wolansky, G. *Limit theorems for optimal mass transportation.* Calc. Var. Partial Differ. Equ. 2011. V. 42, №3-4. P. 487–516. DOI 10.1007/s00526-011-0395-x. MR2846264 [*252*]

[663] Woodroofe, M. *On the weak convergence of stochastic processes without discontinuities of the second kind.* Z. Wahrscheinlichkeitstheorie Verw. Geb. 1968. B. 11, №1. S. 18–25. DOI 10.1007/BF00538382. MR0243596 [*248*]

[664] Yamukov, G. I. *Estimates of generalized Dudley metrics in spaces of finite-dimensional distributions.* Teor. Veroyatn. Primen. 1977. V. 22, №3. P. 590–595 (in Russian); Theory Probab. Appl. 1978. V. 22, №3. P. 579–583. MR0458522 [*126, 249*]

[665] Young, L. C. Lectures on the calculus of variations and optimal control theory. W. B. Saunders, Philadelphia – London – Toronto, 1969; 338 pp. MR0259704 [*251*]

[666] Yurinskii, V. V. *A smoothing inequality for estimates of the Lévy–Prokhorov distance.* Teor. Veroyatn. Primen. 1975. V. 20, №1. P. 3–12 (in Russian); English transl.: Theory Probab. Appl. 1975. V. 20, №1. P. 1–10. MR0370697 [*249*]

[667] Zolotarev, V. M. *Estimates of the difference between distributions in the Lévy metric.* Trudy Mat. Inst. Steklov. 1971. V. 112. P. 224–231 (in Russian); English transl.: Proc. Steklov Inst. Math. 1971. V. 112. P. 232–240. MR0321156 [*131*]

[668] Zolotarev, V. M. *On the continuity of stochastic sequences generated by recurrent processes.* Teor. Veroyatn. Primen. 1975. V. 20, №4. P. 834–847 (Russian); English transl.: Theory Probab. Appl. 1976. V. 20, №4. P. 819–832. MR0400365 [*249*]

[669] Zolotarev, V. M. *Metric distances in spaces of random variables and their distributions.* Mat. Sbornik. 1976. V. 101, №3. P. 416–454 (in Russian); English transl.: Math. USSR-Sbornik. 1976. V. 30, №3. P. 373–401. MR0467869 [*249*]

[670] Zolotarev, V. M. *Ideal metrics in the problem of approximating distributions of sums of independent random variables.* Teor. Veroyatn. Primen. 1977. V. 22, №3. P. 449–465 (in

Russian); English transl.: Theory Probab. Appl. 1978. V. 22, №3. P. 433–449. MR0455066 [*125, 249*]

[671] Zolotarev, V. M. *Properties and relations of certain types of metrics.* Zap. Nauchn. Sem. Leningrad. Otdel. Mat. Inst. Steklov. (LOMI). 1979. V. 87. P. 18–35 (in Russian); English transl.: J. Soviet Math. 1981. V. 17, №6. P. 2218–2232. MR554598 [*125, 126, 249*]

[672] Zolotarev, V. M. One-dimensional stable distributions. Translated from the Russian. Amer. Math. Soc., Providence, Rhode Island, 1986; x+284 pp. MR854867 [*40*]

[673] Zolotarev, V. M. *Probabilistic metrics.* Teor. Veroyatn. Primen. 1983. V. 28, №2. P. 264–287 (in Russian); English transl.: Theory Probab. Appl. 1984. V. 28, №2. P. 278–302. MR700210 [*249*]

[674] Zolotarev, V. M. Modern theory of summation of random variables. Translated from the Russian. VSP, Utrecht, 1997; x+412 pp. MR1640024 [*40, 249*]

[675] Zolotarev, V. M., Senatov, V. V. *Two-sided estimates of Lévy's metric.* Teor. Veroyatn. Primen. 1975. V. 20, №2. V. 239–250 (in Russian); English transl.: Theory Probab. Appl. 1976. V. 20, №2. P. 234–245. MR0375420 [*249*]

[676] Zorich, V. A. *Multidimensional geometry, functions of very many variables, and probability.* Teor. Veroyatn. Primen. 2014. V. 59, №3. P. 436–451 (in Russian); English transl.: Theory Probab. Appl. 2015. V. 59, №3. P. 481–493. DOI 10.1137/S0040585X97T987181. MR3415979 [*98*]

Index

Notation

Selected Published Titles in This Series

234 **Vladimir I. Bogachev,** Weak Convergence of Measures, 2018

232 **Dmitry Khavinson and Erik Lundberg,** Linear Holomorphic Partial Differential Equations and Classical Potential Theory, 2018

231 **Eberhard Kaniuth and Anthony To-Ming Lau,** Fourier and Fourier-Stieltjes Algebras on Locally Compact Groups, 2018

230 **Stephen D. Smith,** Applying the Classification of Finite Simple Groups, 2018

229 **Alexander Molev,** Sugawara Operators for Classical Lie Algebras, 2018

228 **Zhenbo Qin,** Hilbert Schemes of Points and Infinite Dimensional Lie Algebras, 2018

227 **Roberto Frigerio,** Bounded Cohomology of Discrete Groups, 2017

226 **Marcelo Aguiar and Swapneel Mahajan,** Topics in Hyperplane Arrangements, 2017

225 **Mario Bonk and Daniel Meyer,** Expanding Thurston Maps, 2017

224 **Ruy Exel,** Partial Dynamical Systems, Fell Bundles and Applications, 2017

223 **Guillaume Aubrun and Stanisław J. Szarek,** Alice and Bob Meet Banach, 2017

222 **Alexandru Buium,** Foundations of Arithmetic Differential Geometry, 2017

221 **Dennis Gaitsgory and Nick Rozenblyum,** A Study in Derived Algebraic Geometry, 2017

220 **A. Shen, V. A. Uspensky, and N. Vereshchagin,** Kolmogorov Complexity and Algorithmic Randomness, 2017

219 **Richard Evan Schwartz,** The Projective Heat Map, 2017

218 **Tushar Das, David Simmons, and Mariusz Urbański,** Geometry and Dynamics in Gromov Hyperbolic Metric Spaces, 2017

217 **Benoit Fresse,** Homotopy of Operads and Grothendieck–Teichmüller Groups, 2017

216 **Frederick W. Gehring, Gaven J. Martin, and Bruce P. Palka,** An Introduction to the Theory of Higher-Dimensional Quasiconformal Mappings, 2017

215 **Robert Bieri and Ralph Strebel,** On Groups of PL-homeomorphisms of the Real Line, 2016

214 **Jared Speck,** Shock Formation in Small-Data Solutions to 3D Quasilinear Wave Equations, 2016

213 **Harold G. Diamond and Wen-Bin Zhang (Cheung Man Ping),** Beurling Generalized Numbers, 2016

212 **Pandelis Dodos and Vassilis Kanellopoulos,** Ramsey Theory for Product Spaces, 2016

211 **Charlotte Hardouin, Jacques Sauloy, and Michael F. Singer,** Galois Theories of Linear Difference Equations: An Introduction, 2016

210 **Jason P. Bell, Dragos Ghioca, and Thomas J. Tucker,** The Dynamical Mordell–Lang Conjecture, 2016

209 **Steve Y. Oudot,** Persistence Theory: From Quiver Representations to Data Analysis, 2015

208 **Peter S. Ozsváth, András I. Stipsicz, and Zoltán Szabó,** Grid Homology for Knots and Links, 2015

207 **Vladimir I. Bogachev, Nicolai V. Krylov, Michael Röckner, and Stanislav V. Shaposhnikov,** Fokker–Planck–Kolmogorov Equations, 2015

206 **Bennett Chow, Sun-Chin Chu, David Glickenstein, Christine Guenther, James Isenberg, Tom Ivey, Dan Knopf, Peng Lu, Feng Luo, and Lei Ni,** The Ricci Flow: Techniques and Applications: Part IV: Long-Time Solutions and Related Topics, 2015

205 **Pavel Etingof, Shlomo Gelaki, Dmitri Nikshych, and Victor Ostrik,** Tensor Categories, 2015